Lecture Notes in Earth Sciences

Lecture Notes in Earth Sciences

Edited by Somdev Bhattacharji, Gerald M. Friedman,
Horst J. Neugebauer and Adolf Seilacher

19

E. Groten   R. Strauß   (Eds.)

GPS-Techniques
Applied to Geodesy
and Surveying

Proceedings of the International GPS-Workshop
Darmstadt, April 10 to 13, 1988

Springer-Verlag
Berlin Heidelberg GmbH

Editors

Prof. Dr. Erwin Groten
Technische Hochschule, Institut für Physikalische Geodäsie
Petersenstr. 13, D-6100 Darmstadt, FRG

Dr. Ing. Robert Strauß
Hessisches Landesvermessungsamt
Schaperstraße 16, D-6200 Wiesbaden 1, FRG

Workshop sponsored by

International Association of Geodesy
Stiftung Volkswagenwerk
Hessischer Minister für Wissenschaft und Kunst
Technische Hochschule Darmstadt

ISBN 978-3-540-50267-8   ISBN 978-3-540-45962-0 (eBook)
DOI 10.1007/978-3-540-45962-0

© Springer-Verlag Berlin Heidelberg 1988
Originally published by Springer-Verlag Berlin Heidelberg New York in 1988

2132/3140-543210 – Printed on acid-free paper

Prof. Karl Rinner
dedicated on the occasion of his

76 birthday.

Prof. Rinner had planned to attend this meeting but was
finally unable to do so because of illness.

CONTENTS

Appendix: Field Measurements

List of Contributors

K. Aksnes
Norwegian defence research establishment
Mathematics section
Postboks 25
N-2007 Kjeller
Norway

P.H. Andersen
Norwegian defence research establishment
Mathematics section
Postboks 25
N-2007 Kjeller
Norway

V. Ashkenazi
The University of Nottingham
Department of Civil Engineering
University Park
Nottingham NG7 2 RD
U.K.

W. Augath
Niedersächsisches Landesverwaltungsamt
Abtlg. Landesvermessung
Warmbüchenkamp 2
Postfach 107
D-3000 Hannover 1
West-Germany

G. Baustert
Hochschule der Bundeswehr München
Astronomische u. Physikalische Geodäsie
Werner Heisenberg Weg 39
D-8014 Neubiberg
West-Germany

A. Beckmann
GEOsat GmbH
Löhberg 78
D-4330 Mülheim a. d. Ruhr 1
West-Germany

G. Beutler
Universität Bern
Astronomisches Institut
Sidlerstraße 5
CH-3012 Bern
Switzerland

G. Blewitt
Jet Propulsion Laboratory
California Institute of Technology
4800 Oak Grove Drive
Pasadena, CA 91109
U.S.A.

C. Boucher
Nivellement et Metrologie
Inst. Geographique National
2, Avenue pasteur
F-94160 Saint-Mande
France

B. Breuer
IfAG
Hennesenbergstr. 37
D-5303 Bornheim 1
West-Germany

J.D. Cain
Aero Service Division
Western Atlas International, Inc.
3600 Briarpark Drive
P.O. Box 1939
Houston, Texas 77042
U.S.A.

P. Cross
University of Newcastle upon Tyne
Dept. of Surveying
Newcastle upon Tyne NE1 7RU
U.K

H.-J. Euler
Technische Hochschule Darmstadt
Institut für Physikalische Geodäsie
Petersenstr. 13
D-6100 Darmstadt
West-Germany

J. Feltens
Technische Hochschule Darmstadt
Institut für Physikalische Geodäsie
Petersenstr. 13
D-6100 Darmstadt
West-Germany

R. Galas
Technische Universität München
Institut für Astr. und Physik. Geodäsie
Arcisstraße 21
D-8000 München 2
West-Germany

A. Geiger
Eidgenössische Techn. Hochschule Zürich
Institut für Geodäsie und Photogrammetrie
HIL-Gebäude
ETH Hönggerberg
CH-8093 Zürich
Switzerland

C. Goad
The Ohio State University
Department of Geodetic Science
1958 Neil Avenue
Columbus, Ohio 43210
U.S.A.

E.W. Grafarend
Universität Stuttgart
Geodätisches Institut
Keplerstraße 11
Postfach 560
D-7000 Stuttgart 1
West-Germany

E. Groten
Technische Hochschule Darmstadt
Institut für Physikalische Geodäsie
Petersenstr. 13
D-6100 Darmstadt
West-Germany

G. Groven
c/o GPS Servises A/S
Gravarsvegen
N-4300 Sandnes
Norway

W. Gurtner
Universität Bern
Astronomisches Institut
Sidlerstraße 5
CH-3012 Bern
Switzerland

Ph. Hartl
Universität Stuttgart
Institut für Navigation
Keplerstraße 11
D-7000 Stuttgart 1
West-Germany

G.W. Hein
Hochschule der Bundeswehr München
Astronomische u. Physikalische Geodäsie
Werner Heisenberg Weg 39
D-8014 Neubiberg
West-Germany

B. Hofmann-Wellenhof
Technische Universität
Institut für Angewandte Geodäsie
Rechbauerstraße 12
A-8010 Graz
Austria

G.J. Husti
Technische Hogeschool Delft
Afdeling der Geodesie
Thijsseweg 11
NL-2600 GA Delft
The Netherlands

R. Hyatt
Trimble Navigation
585 N. Mary
Sunnyvale Calif. 94086
U.S.A.

W. Jacoby
Johannes Gutenberg-Universität
Institut für Geowissenschaften
Saarstraße 21
D-6500 Mainz
West-Germany

J. Kakkuri
Finnish Geodetic Institute
Ilmalankatu 1A
SF-00240 Helsinki
Finland

B. Kolaczek
Polish Ac. of Sci. Space Research Centre
Department of Planetary Geodesy
ul. Bartycka 18
00-716 Warszawa
Poland

J. Kremers
Landesvermessungsamt NRW
Muffendorfer Straße 19 - 21
Postfach 560
D-5300 Bonn 2
West-Germany

H. Landau
Hochschule der Bundeswehr München
Astronomische u. Physikalische Geodäsie
Werner Heisenberg Weg 39
D-8014 Neubiberg
West-Germany

D. Lelgemann
TU Berlin - Sekr. H 12, FB 7
Inst. für Geodäsie und Photogrammetrie
Straße des 17. Juni 135
D-1000 Berlin 12
West-Germany

W. Lindlohr
Universität Stuttgart
Geodätisches Institut
Keplerstraße 11
Postfach 560
D-7000 Stuttgart 1
West-Germany

F.J. Lohmar
Amt für Militärisches Geowesen
FAF Geographic Office
Frauenberger Straße 250
D-5350 Euskirchen
West-Germany

A. Müller
Universität Bonn
Geodätisches Institut
Nußalle 17
D-5300 Bonn 1
West-Germany

G. Nard
SERCEL
B.P. 64
F-44471 Carquefou Cedex
France

P. Paquet
Royal Observatory of Belgium
Avenue Circulaire 3
1180 Brussels
Belgium

B. Remondi
Astech Telesis
1156-C Aster Ave
Sunnyvale CA 94086
U.S.A.

M. Rothacher
Universität Bern
Astronomisches Institut
Sidlerstraße 5
CH-3012 Bern
Switzerland

W. Schlüter
Institut für Angewandte Geodäsie
Richard-Strauss-Allee 11
D-6000 Frankfurt 70
West-Germany

O. Schuster
GEOsat GmbH
Löhberg 78
D-4330 Mülheim a. d. Ruhr 1
West-Germany

G. Seeber
Universität Hannover
Institut für Theoretische Geodäsie
Nienburger Straße 1
D-3000 Hannover 1
West-Germany

H. Seeger
Institut für Angewandte Geodäsie
Richard-Strauss-Allee 11
D-6000 Frankfurt 70
West-Germany

L.E. Sjöberg
The Royal Institute of Technology
Department of Geodesy
Drottning Kristinas väg 30
S-10044 Stockholm 70
Sweden

G. Stangl
Austrian Academy of Sciences
Institute of Space Research
Lustbuehelstraße 46
A-8042 Graz
Austria

A.H. Stiller
DFVLR-RF/TN3
Linder Höhe
Postfach 906058
D-5000 Köln 90
West-Germany

E. Stöcker-Meier
Universität Bonn
Inst. f. Theoretische Geodäsie
Nußallee 17
D-5300 Bonn 1
West-Germany

R. Strauß
Hessisches Landesvermessungsamt
Schaperstraße 16
Postfach 3249
D-6200 Wiesbaden 1
West-Germany

W. Torge
Universität Hannover
Institut für Theoretische Geodäsie
Nienburgerstraße 6
D-3000 Hannover
West-Germany

B.H.W. van Gelder
Technische Hogeschool Delft
Afdeling der Geodesie
Thijsseweg 11
NL-2600 GA Delft
The Netherlands

P. Vanicek
University of New Brunswick
Department of Surveying Engineering
P.O. Box 4400
Fredericton, N.B.
Canada E3B 5A3

P. Willis
Nivellement et Metrologie
Inst. Geographique National
2, Avenue pasteur
F-94160 Saint-Mande
France

G. Wübbena
Universität Hannover
Institut für Theoretische Geodäsie
Nienburger Straße 1
D-3000 Hannover 1
West-Germany

S.Y. Zhu
Technische Hochschule Darmstadt
Institut für Physikalische Geodäsie
Petersenstr. 13
D-6100 Darmstadt
West-Germany

J.B. Zielinski
Space Research Centre,Polish Acad.of.Sci
Department of Planetary Geodesy
ul. Bartycka 18
00-716 Warszawa
Poland

Opening Address

Opening address

by E. Groten

Local Organizing Committee

On behalf of the Local Organizing Committee, I welcome you all to the
first International Workshop on GPS-techniques in surveying and
geodesy held at this university. This workshop is designed to bring
together experts from various countries and also scientists who carry
out, analyze and interpret such measurements with those who work on
instrumental and theoretical problems. The workshop focuses hereby on
high-precision applications with emphasis on monitoring time-dependent
phenomena such as those relevant to geodynamics as well as men-made
constructions as those in civil engineering and similar fields. It is
astonishing to see how, in spite of all earlier satellite work over
the last two decades, GPS-methods became so fast a relevant new
technology, in its proper sense, in modern geodesy and surveying
besides VLBI and Satellite Laser Ranging (SLR). With the recent
development of new dual-frequency receivers the role of GPS-procedures
in monitoring large-scale phenomena over big distances will still expand; and the application of kinematical GPS-approaches is of utmost
interest in solving high-precision problems. It is indeed fascinating
to realize how GPS-methods have become in such a short time a
surprisingly efficient and effective, this means : fast, precise and
easy to apply, tool which is able to replace already now, after a few
years of existence and with an incomplete set of a few out of the 18
satellites (of the final stage), at least partially some expensive,
slow and cumbersome classical surveying methods.

On the other hand, it cannot be overemphasized that GPS-procedures are
still at their beginning and the full spectrum of their capabilities
still has to be explored. In Europe, for example, where excellent
classical surveying systems do exist the situation is quite different
from the situation in other countries such as Canada or the USA. Even
within Europe the application types of GPS-methods will vary; for
example, in Norway the situation is quite different from central
European countries.

It is often forgotten, that together with GPS we will have to
introduce new concepts and a new thinking in combination with other
modern satellite procedures. GPS itself can resolve only a small part
of the problems to be solved by modern geodesy but it will open the
way to a great variety of new applications and capabilities. Modern

global tectonics is just one of the new disciplines of high interest and great practical impact. I could continue in citing other similarly important new fields. GPS is, however, of special importance because it replaces old technologies and fills gaps where modern and efficient tools are most needed.

Consequently, also the optimal combination of GPS-methods with new auxiliary and also classical high-precision techniques is of great importance, mainly under the european conditions outlined above. Moreover, the real-time or almost-real-time use of GPS in combination with photogrammetry, inertial geodesy, gravity gradiometry or even classical surveying is of substantial interest.

It is indeed important to realize the new concepts in modern satellite and space methods and I, therefore, spoke above of a new "technology" which should be optimally developed as there is a worldwide need of such capabilities and tools.

In view of the few active NAVSTAR-satellites in sky in 1988 this is perhaps not the best year for GPS-applications but the right time for a review of the experience gained until now and using it as a base for the planning of the future.

This meeting is designed as a typical workshop with about 48 presentations and only a small informal opening session, a few social events in order to enable the participants to exchange ideas even in the evening and practical demonstrations and measurements. Thus, we face a heavy program.

I express my sincere thanks to the sponsoring organizations, who made this meeting of scientists from 18 countries possible. I am particularly thankful to the president of the host university, Prof. H. Boehme, that he found time to welcome you here. I also appreciate that IAG is represented here by its 1st Vice-president Prof. W. Torge. Furthermore, I am indebted to the Hessisches Landesvermessungsamt for the efficient technical support in view of the field demonstration.

Moreover, I welcome the representatives of the industry who help to make this a real application-oriented but truly scientific workshop with a lot of practical demonstrations.

Finally, I welcome especially the colleagues from developing countries such as Brazil, China, etc. I am also happy that many young scientists are here.

We have here speakers from 14 nations and an audience consisting of about 20 nations. As we are still at the beginning of a new era in modern geodesy it is so important to lead these new efforts into the right direction in order to make them as efficient as possible. We focus here on a particularly sensitive and crucial part of this approach. I wish a good, fruitful and enjoyable meeting to all of you!

Welcome Addresses

Ladies and Gentlemen,

it is a great pleasure for me to welcome you here on behalf of the Technische Hochschule Darmstadt. We are proud to be host to the International Workshop on Global Positioning System Techniques in Geodesy, and I express my warmest thanks to my colleague, Professor Groten, and his collaborators for preparing, organizing and leading this important meeting. Under the protection of the International Association for Geodesy you will discuss during the next few days quite new possibilities of measurement-methods by satellite and their applications which promise your science interesting and various aspects for the future. I hope that this workshop will be very successful and increase the scientific understanding.

You have come to Darmstadt from many nations all over the world, and with a few words I'll try to introduce you our town. Surrounded by the recreational areas of the Odenwald and the Bergstraße, the Taunus and Spessart, Darmstadt is favoured not only by nature, having a mild climate and fertile countryside, but is also particularly conveniently placed for travel, being in the centre of the Federal Republic, between the industrial regions of the Rhein-Main and Mannheim-Ludwigshafen, with easy access to the cities of Wiesbaden, Mainz and Frankfurt. The town, with 135.000 occupants, compact and friendly, promotes itself with the ambitious slogan, "the arts live in Darmstadt". Its air of culture as well as its predominantly white collar nature are a legacy from its past as a grand ducal residence. There is much which bears witness to this: the Art Nouveau ensemble at Mathildenhöhe; the Rosenhöhe artists' quarter, home of painters, sculptors and writers; the high-quality collections and museums; the educational system, highly developed at all levels, and the Land Theatre, whose prestige extends far beyond the city limits, and which is regarded as a springboard to the very top.

However, science is also well represented in Darmstadt. The Technische Hochschule Darmstadt counts more than 16.000 students and about 1.500 scientists. Besides the THD two Technical Colleges are well-respected centres of education. Darmstadt is home to the technical facilities of the Post Office. The accelerator of the Society for Heavy-Ion Research - situated in the woods outside the city - is the workplace for many scientists from near and far. Finally, on a truly international note, there is the Europeans Space Operations Centre - likewise in Darmstadt. At ESOC the take offs of the European carrier rockets and the functioning of geostationary satellites are monitored. Among other things, ESOC provides German television companies with satellite photographs of the weather.

Research and Development are however not a concern only of scientific bodies, but also of industry. Darmstadt is the site of some particularly future-orientated branches of industry. The chemical industry occupies first place with several multinational companies. Reputedly the specialized engineering firms can produce here free of competition. A great deal of innovative force is behind the young enterprises working in the fields of telecommunications and computer science. After 1945 the graphic trades (publishing, printing, paper processing) settled in Darmstadt. Naturally there is a great deal of very varied contact between the University and local business and industry, to their mutual benefit.

Ravaged in the war - in a single night of bombing in September 1944, the whole of the city centre was laid to waste - Darmstadt is once more a lively, cosmopolitan community with an air of innovation. Progressiveness is the factor uniting the city's "great sons", predecessors to which the city readily refers, even if they did not have it easy here in their own time: Justus Liebig, the founder of agricultural chemistry and reformer of scientific education; Georg Christoph Lichtenberg, the physicist and aphorist; and Georg Büchner, the doctor, poet and revolutionary. They all straddled the border between the exact sciences and the fine arts, and set standards to which the University also feels a commitment: the endeavour - to be taken up time and again - to reconcile engineering and the humanities, the union of technology and society.

I hope you will like your visit in Darmstadt and its Technische Hochschule.

Prof. Dr. H. Böhme

President of the Technische Hochschule Darmstadt

MORE THAN FIVE YEARS OF GPS-EXPERIMENTS - RETHINKING OF GEODESY

by

Wolfgang Torge

Dear Colleagues,

since GPS installation started 15 years ago, and with geodetic GPS-experiments performed over more than 5 years, we all are aware of the possibilities which this space based positioning system offers to geodesy, and we foresee - at least partially - the changes which classical geodetic dogmas valid for more than 100 years, are going to experience. The geodetic community has early recognized the challenge of this new technique, as documented by numerous research activities and GPS related scientific meetings, attracting geodesists from research and application orientated institutions, as we see here.

It is a great pleasure for me, to sketch in this introduction the impetus of GPS to geodesy, and to indicate some of the problems which you will discuss later in more detail. But before, I should like, as the first Vicepresident of the International Association of Geodesy, to deliver you the greetings and best wishes for a successful meeting, from the IAG President and IAG Bureau. As the Chairman of the German Geodetic Commission at the Bavarian Academy of Sciences, I add the welcome greetings of that representative body of geodesy in the Federal Republic of Germany. We are thankful to Professor Groten for his initiative to organize this workshop in our country, where intensive GPS research is under way since some years, and we are happy about this positive response, from inside our country and from abroad.

Now, let me try to indicate how the potential possibilities of GPS-techniques are going to change geodesy, although we certainly do not yet know the final result of the presently occuring collision between classical and modern concepts.

Since the first establishment of classical control networks for position and height, about 100 years ago, the pre-satellite era which lasted approximately until the 1960s , was characterized by

- employment of time-consuming terrestrial observation techniques, with days to weeks per first order trigonometric point determination, and 5 to 10 km first order levelling progress per day, direct line of sight between neighbouring stations being necessary,

- changing orientation provided by the local plumb line direction, corresponding to the use of numerous local astronomical systems,

- separation of position and height control systems, with geometric (ellipsoid) resp. physical (geoid) reference surfaces, thus in a clever manner minimizing the effect of the gravity field on the derived parameters, as a first order approximation allowing to neglect them,

- relative accuracies referring to the station distance of $\pm 10^{-5}$ for positions, only in local high-precision networks observed with electro-optical distance measurement equipment $\pm 10^{-6}$ has been reached, while $\pm 10^{-6}$ to $\pm 10^{-7}$ for heights could be obtained by first order levelling,

- global orientation through astronomic methods for position networks, and through oceanographic information for height networks, with discrepancies up to $\pm 10^{-4}$ in position resp. $\pm 10^{-7}$ in height with respect to a common global system,

- monitoring time variations of the orientation of a global reference system with a relative accuracy of $\pm 10^{-6}$, by an astronomic control system.

With these features, position and height control networks met, for more than 100 years, the needs of administration and development in countries being at the transition from rural to industrial societies, and even served most of the requirements in more industrialized regions. A number of severe drawbacks became obvious with the rapid changes which human society experienced after the 1950s.

These drawbacks of the classical geodetic control are:

- the extremely slow progress at the establishment of control networks, especially if a dense station distribution (e.g. 1 to 5 km) is demanded,

- the complex error accumulation at larger networks, leading to changes in network scale and orientation, and to unpredictable network distortions, and causing severe problems at subsequent surveys in highly developped and densely populated areas,

- the weak ties of the national position control systems to a common global reference, with discrepancies no longer acceptable for modern navigation,

- the inability to investigate recent crustal movements at global and regional scales, being the strongly required geodetic contribution to geodynamics research, and needed for monitoring movements in connection with seismic and volcanic events, as well as those produced by man-made environment changes.

The following pre-GPS satellite era from approximately 1960 to 1980 already brought partial improvements to this situation, first through optical and then - more effective - through Doppler satellite positioning methods. Main characteristics have been

- the more rapid point determination, taking few days per station at high accuracy demands, with no need for visibility between stations,

- the common orientation in a global reference system with $\pm 10^{-6}$ to 10^{-7},

- absolute point determination accuracies of 1 to 5 meters, and relative accuracies of some $\pm 10^{-6}$ for distances from 100 to 500 km.

Consequently, classical control networks could now be controlled at distances larger than 100 km, and transformed to a global reference system. With absolute point positioning accuracy, control point requirements for topographic mapping could be fulfilled now. But probably most important was that threedimensional satellite technology forced geodesy to a threedimensional way of thinking, and corresponding modelling of geodetic observations and parameters.

11

After that transition epoch from classical to space techniques, space geodesy era started about 1980, covering now local, regional, and global scales. For most users of geodetic products, GPS-results are of special interest, as for local and regional problems with distances between 1 and some 100 km, this technique now offers rapid solutions with accuracies sufficient for most purposes. The main characteristics of this space geodesy era may be described as follows:

- monitoring the motions of the global geodetic reference frame in space and of large-scale tectonic plate movements through advanced space techniques, with approximately $\pm 10^{-8}$ to 10^{-9} (Satellite Laser Ranging, Very Long Base-Line Interferometry) relative accuracy, but with still large investments in hardware and operational costs,

- three-dimensional positioning through GPS-methods, delivering within short time (few minutes to hours) relative accuracies of $\pm 10^{-6}$ to 10^{-7} for distances from 10 km to some 100 km, and with cm-accuracy at 1 to 10 km distances,

- employment of GPS-techniques in the kinematic mode, giving relative navigation accuracies of ± 1 m in position, and ± 0.1 m/s in velocity.

As one example of the high efficiency of GPS-methods, I mention the European north-south-GPS-traverse, which has been proposed for regional geoid control few years ago by a IAG-Special Study Group chaired by Professor Birardi. The central and northern part of this traverse, between Austria and northern Norway, has been observed in 1986/1987, under direction of our Institute, and in cooperation with Geodetic agencies and institutes of Austria, Denmark, Norway, Sweden, and the Federal Republic of Germany. Using two or more TI-4100 receivers, and simultaneously observing adjacent (appr. 50 km station distance) and overlapping connections, the evaluation was performed with the Hannover-software. Preliminary height results have been compared with the normal heights obtained from the Unified European Levelling Network, and the gravimetric geoid heights of our EGG1-solution. For the 3600 km long traverse part, the r.m.s. discrepancy was only ± 0.67 m, which reduced after a tilt (-0."13 corresponding to -0.6 m/1000 km) to ± 0.27 m, thus revealing a high accuracy of GPS heights and of the European geoid.

We have to state some consequences of this recent developments:

- highly efficient space techniques now have reached and, at regional and global scales, far exceeded the accuracies of classical geodetic methods,

- geodetic control systems established through space methods are a priori three-dimensional, with global geocentric orientation, and may be subdivided into a global reference network monitoring time variations of global character, and regional/global networks, eventually composed of a base network (GPS reference stations) and densification nets,

- with navigation potential offered by GPS,kinematic survey methods
  of operational or experimental stage may get a substantial support,
  with eventual drastic change of methods; this refers to inertial
  surveying (position updating and gravity vector separation), photogrammetry (orientation), and airborne gravimetry and gravity gradiometry (orientation, Eötvös-correction, separation of gravity and
  disturbing accelerations),

- geodetic contributions to geodynamics research become more efficient now, as high resolution data aquisition in space and time
  is possible, with the chance of eventual continuous monitoring the
  earth surface, at least in areas where large movements occur.

But, before handling the new tools with maximum efficiency, taking
the boundary conditions of existing survey systems into account, a
lot of problems still has to be solved. Let me mention some of them,
which are related to GPS:

- optimal GPS-network design, with optimization strategies for observation time per station, satellite constellation, and network configuration including the question of optimum station distance and
  overlapping connections, and use of one- and two-frequency-receivers,

- optimum combination with existing classical control networks, and
  available terrestrial survey methods, as electronic distance measurements, levelling, and inertial surveying, or eventual super-
  seding of them,

- software improvements for functional and stochastic models, in
  order to get reliable results at least at the "cm"-level, and
  realistic accuracy estimates, including the cycle-slip problem,
  closely connected to site selection,

- implication for the classical philosophy of "geometric" position
  and "physical" height control systems, including the question of
  site coincidence of the control points, and high-precision geoid
  determination,

- additional measures as improved orbit determination at regional
  geodynamics investigations including investigations about the
  long-time stability of the GPS-system,

- implication for kinematic survey methods, which eventually in
  further future may enable these methods also to monitor time
  variations of the earth surface and gravity field.

Many of these problems will be discussed here, and I hope that at
least for some problem areas, the outcome of this workshop will set
one step further to new concepts in operational geodesy. I wish you
fruitful discussions and full success for the workshop.

Welcome

Robert Strauß

President of the "Arbeitskreis Triangulation of the
"Arbeitsgemeinschaft der Vermessungsverwaltungen der
Länder der Bundesrepublik Deutschland"

Ladies and gentlemen,
dear colleagues,

It is a great pleasure to me to welcome you on behalf of all those
colleagues, who are responsible for the control network of the Federal
Republic of Germany. You certainly will know, that there is no central
administration, which is responsible for the surveys in our country.
Instead of that we have a working group which is called "triangulation"
and which was established in 1949 by the gouvernments of the federal
states. It is the task of this working group to achieve standard regu-
lations for keeping the field of trigonometric points in a similar
type in all the federal states. The last result of our cooperation is
a rough draft for applications of GPS in the control network of the
Federal Republic of Germany. You can read it in number 2 1988 of the
Zeitschrift für Vermessungswesen. For that reason I dropped my report,
which I prepared for this workshop. The essential goal of our draft is
to prevent the establishment of additional new reference systems by
private users of GPS. We suppose that this is possible by the establishment of a GPS-basenetwork and the determination of precise transformation parameters between the reference system of the GPS, the World
Geodetic System 1984 and the official reference system, represented by
our first order network. We think that a world-wide or at least an
european cooperation is necessary. That is why we are very interested
to discuss problems like orbit determination or threedimensional reference systems. The program of this workshop shows that these are not
only problems to be solved. We should take time by the forelock until
the 18. satellite starts developing theory and gathering practical
experience. There ist not a shadow of a doubt that then GPS-applications will spread out. The working group I am representing here then
no longer will be divided into two parts, some already using GPS and
others being eaten up with envy.

I wish three days of effective work, good results, a pleasant stay to
all participants and the expected success to Professor Groten.

First session: General Aspects

Chairman: Prof. Kakkuri, Helsinki

GEODETIC APPLICATIONS WITH GPS IN NORWAY
AS PART OF A GLOBAL COOPERATION

by

K. Aksnes, P. H. Andersen, S. Hauge

and

B. Engen

Abstract

The Norwegian Defence Research Establishment (NDRE) and
the Norwegian Mapping Authority (NMA) have undertaken a
joint project in satellite geodesy in cooperation with
several geodetic groups in Europe and the U.S. Simultaneous GPS tracking is now being routinely performed with
TI-4100 receivers located in Tromsø and at Onsala,
Wettzell and five North-American VLBI stations. In Norway,
NMA is in charge of data collection and is operating
GPS reference stations in Tromsø and at Onsala. An orbit
computation service is also being planned. NDRE is
responsible for development, testing, and special
applications of the GPS data analysis tools. At the
core of these applications is a computer program, GEOSAT,
for high-precision calculation of orbits and associated
geodetic parameters, based on a variety of satellite
tracking data. Results are presented from use of this
program on a TI-4100 data set acquired at three VLBI
stations in the U.S., 3-7 June 1986.

1. Introduction

Even during its current experimental phase, the Navstar
Global Positioning System (GPS) has already become an
important tool for ship navigation and marine charting
with the ice-going vessel R/V Lance of the Norwegian
Hydrographic Service. GPS is also rapidly replacing other
methods used by private surveying companies for precise
offshore positioning of oil rigs and of vessels engaged in
towing, pipe-laying or oil exploration on the Norwegian
continental shelf.

The usefulness of GPS was rather dramatically highlighted
a few years ago when sea level measurements indicated that
the oil rig Ekofisk in the North Sea was sinking at a rate
of several decimeters per year. By means of differential
GPS, it was found that this subsidence amounted to about
40 cm per year relative to a nearby oil rig which was
believed to be stable. It would have been much more
desirable to refer the Ekofisk motion to the Norwegian
mainland some 300 km away, but to achieve a relative
positioning accuracy of, say, 3 cm over this baseline
length would have required a 0.1 ppm performance. This can
only be done based on very precise Navstar satellite
ephemerides. This and other high precision positioning
needs of vital importance to the offshore oil industry
incited an early interest in GPS tracking and orbit
improvement in Norway.

But Norway also has a need for GPS to solve a much more
fundamental problem; namely the establishment of an
improved national first order geodetic network and its
representations in local, regional, and global datums,
respectively NGO-48, ED-50, and WGS-84. NGO-48 and ED-50
are based on essentially the same first order points in
Norway determined astronomically and by triangulation.
NGO-48 suffers from scaling errors up to 40 ppm, but ED-50
is more uniform. Maps and coordinates for offshore
navigation and positioning are based on ED-50. Under the
auspices of the RETRIG subcommission of IAG, improvements
have been made in the ED-50 system by means of tri-
lateration and satellite techniques. However, the time
now seems ripe to switch to an entirely new European Datum
based on VLBI, SLR, and GPS techniques (Landau and
Hein 1986, Boucher and Altamimi 1986).

In Norway the most stringent demands for accurate geodetic
reference systems come from oceanographers interested in
tracking currents, eddies and storm surges by means of
altimetry in satellites. This requires very precise
knowledge of the marine geoid, which is also of interest
to oil prospecting because of the relationships between
the shape of the geoid, gravity anomalies, and oil bearing
structures below the oceans.

2. National Cooperation and Goals

The main responsibility for establishing and maintaining
geodetic reference systems and for charting and map
production on Norwegian territory and in territorial
waters lies with the Norwegian Mapping Authority (NMA).
This includes both civilian and military applications.
NMA has recently acquired five TI-4100 receivers and has
partime access to two more. With this equipment NMA is now
engaged in an international measurement campaign to be
described later.

The Norwegian Defence Research Establishment (NDRE) is
responsible for developing and testing algorithms and
software for high precision analysis of GPS data.
Initially most data processing will be done at NDRE, but
routine data processing will later be taken over by NMA.
At the NDRE Mathematics Section, a computer program
(GEOSAT) for precise analysis and simulation of satellite
tracking data has been under development for several years
(Andersen 1986). A brief description of the GEOSAT program
along with some test results are given later.

NDRE is heading a joint Norwegian geodetic experiment
aimed at using laser, PRARE and altimetry data on ESA's
ERS-1 satellite for point positioning and orbit and geoid
determination. GPS measurements are an important pre-
launch part of this experiment. The joint GPS and ERS-1
goals include

- differential GPS positioning relative to VLBI sites of
  designated GPS and ERS-1 reference stations in Tromsø,
  Stavanger, Jan Mayen, and Svalbard

- orbit calculations for GPS and ERS-1 and precise
  positioning of secondary geodetic points by means of
  differential GPS and PRARE measurements relative to the
  reference stations

- calulation of precise geoid, datum parameters and
  transformations between local (NGO-48), regional
  (ED-50), and global (WGS-84) reference systems

3. International GPS Tracking

Since December 1987, two of NMA's TI-4100 receivers
together with six such receivers owned by other nations
have been in permanent operation at the following seven
VLBI sites plus Tromsø Satellite Station:

Westford, Massachusetts
Richmond, Florida
Austin, Texas
Mojave, California
Yellowknife, Canada
Onsala, Sweden
Wettzell, W. Germany
Tromsø, Norway

Additional VLBI sites on Hawaii and in Japan, Australia,
S. Africa, Spain and China will or may take part during
October 30 to November 19, 1988 in a campaign (Mader 1988)
referred to as the first GPS Global Orbit Tracking
Experiment (GOTEX-1). During this campaign there will be
locally organized regional campaigns in secondary
reference networks. The main goals of the campaign are :

- the evaluation of the primary Mark III VLBI network for
  the determination of GPS orbits

- the comparison of regionally determined orbits to these
  global orbits

- the first epoch measurements of new primary fiducial
  sites and the establishment of secondary reference
  networks

- a more accurate relationship between the WGS-84 and the
  VLBI systems.

In March 1988 representatives from the geodetic com-
munities in Norway, Greece, and Spain met in Norway
and issued a Memorandum of Understanding (MOU) concerning
a European Tracking Experiment (EUTREX) with GPS
receivers in Tromsø and at the Dionysos SLR station
and the Madrid VLBI station. During an initial experiment
in the second half of 1988, NMA will station three of its
TI-4100 receivers at these three sites. The MOU also
invites participation from other nations. At the time when
the MOU was issued we were unaware of the GOTEX-1 campaign
which, however, fits nicely in with our initial
experiment. NDRE's GEOSAT software will be used for data
analysis.

4. The GEOSAT software system

The GEOSAT system has been described elsewhere
(Andersen 1986) and we shall therefore mention only
a few of the main features. GEOSAT is implemented on a
Norwegian manufactured computer, ND-570, and it consists
of approximately 70000 FORTRAN statements.

The system can be applied in three different modes :
estimation, simulation (in which synthetic observations
are generated) or for error analysis. The GEOSAT system
includes what we believe are the best available mathe-
mathical models (generalizations of the MERIT standard)
formulated in a general relativistic PPN-framework.

The system is a multi-station and multi-satellite tool which can handle several types of modern tracking data, including laser and microwave range, doppler, phase and altimetry, in a simultaneous manner. Single, double and triple differences can be generated for most of these measurement types. In the future also satellite gradiometry and possibly surface gravimetry will be implemented.

The estimation scheme is a three-level partitioned Bayesian weighted least squares method in which the model parameters can be treated as either "solve-for" or "consider" global parameters (fixed values for the whole dataset), arc parameters (fixed values for a part of the dataset) or local parameters (fixed values for each observation set).

Among the parameters that can be treated as either solve-for or consider parameters are orbital elements, surface scaling parameters for radiation pressure modelling, polynomial coefficients and trigonometric amplitude and phase for modelling empirical accelerations, gravity parameters, station and satellite oscillator parameters, earth rotation parameters, phase biases and rates, tropospheric scaling, baseline vectors, absolute coordinates and tidal parameters.

4.1 The Dataset

The GPS dataset consisted of pseudorange and phase measurements on both L1 and L2 from three TI-4100 receivers located in Westford, Richmond and Fort Davis during five days in June 1986. The receivers were equipped with Hydrogen-Maser oscillators. The available dataset is presented in Table 4.1. Note that the D-passes (day) usually involve only one or two satellites during a short time period.

PASS	1001 6	8	9	11	12	13	1002 6	8	9	11	12	13	1003 6	8	9	11	12	13	TEST A	B	C
154N	*	*	*	*	*	*	*	*	*	*	*	*	*		*	*	*	*	A	B	
154D			*		*				*		*						*		A		C
155N	*	*	*	*	*	*	*	*	*	*	*	*	*	*	*	*	*	*	A		C
155D									*		*						*		A	B	
156N	*	*	*	*	*	*	*	*	*	*	*	*	*	*	*	*	*	*	A	B	
156D			*		*				*		*						*		A		C
157N	*	*	*	*	*	*	*	*	*	*	*	*	*	*	*	*	*	*	A		C
157D			*		*				*		*						*		A	B	
158N	*	*	*	*	*	*	*	*	*	*	*	*	*	*	*	*	*	*	A	B	
158D			*		*				*		*						*		A		C

Table 4.1   The dataset

Since GEOSAT is a general-purpose software system not restricted to any specific satellite system, the TI-measurements were preprocessed by a program called PREPARE. This program re-formats the data and calculates station oscillator polynomials and, optionally, also preliminary station coordinates using pseudorange. PREPARE also corrects the phase measurements for cycle-slips, and it can generate normal points at the frequency chosen by the program operator. Normal points were generated every 10 minutes in this investigation.

For the purpose of getting some realistic knowledge about the accuracy of the calculated orbits, we performed three separate tests denoted A, B and C (see Table 4.1). Note that test A contains all data and that the datasets of B and C are independent. Furthermore, note that test B contains three N-passes (night) and C only two N-passes during a period of five days. Only SV9 and SV12 are present in the D-passes.

4.2 Data Processing Strategy

There are several problems connected with the reduction of GPS measurements. Multipath is reported by many authors to be a significant problem (Evans 1986) for pseudorange measurements obtained with the standard TI-antenna. The RMS due to multipath on pseudorange is reported to be around 1.3 m or worse. We have used normal points every 10 minutes to try to smooth high-frequency multipath effects. So far, this procedure does not seem to give any major improvement, so we think that the dominant multipath effects must be of lower frequency. Papers presented at a GPS workshop at JPL in March 1988 seem to be consistent with this hypothesis.

Another very serious problem is that the TI-receivers sometimes lose lock on the satellite signal and this causes cycle-slips. Most of these cycle-slips are corrected in PREPARE to an accuracy which is always better than five cycles, and usually better than two cycles. The cycle-slips are detected by comparison of the pseudorange rate of change with the corresponding phase rates. Also single differences between stations are used in this process. The corrections are obtained by polynomial prediction and interpolation. The cycle corrections are usually checked on a graphical display, and in a very few cases it is necessary to do some manual editing. If necessary, it is possible to correct for earlier erroneous cycle-slip corrections in each iteration in GEOSAT. In the last few iterations, the model parameters are well determined and the conditions for sucessful cycle-slip correction are the very best.

In the first few iterations of GEOSAT, only pseudoranges are used to calculate orbital elements, radiation pressure parameters, ionospheric corrections, satellite oscillator polynomial coefficients (1. order polynomial) and station oscillator polynomial coefficients (2. order polynomial) using known reference stations. Then pseudorange and phase on both L1 and L2 are processed simultaneously to solve also for phase biases and improved ionospheric corrections. This procedure seems to reduce the RMS due to ionospheric errors by a factor of two.

The arc length was five days and the oscillator parameters for both stations and satellites were updated on every pass. Only raw measurements were applied with no differencing.

In this investigation we applied models for the earth gravity field (8 x 8, part of the WGS84 standard) and the MERIT models for ocean and solid earth tides. The station coordinates were supplied by NGS in the WGS72 system. We transformed the coordinates to WGS84 by adding a 4.5 m correction in the z-direction. Due to this procedure, the coordinates might be in error by several decimeters with respect to WGS84.

4.3   Results

We have determined the orbits for test sets A, B and C. The calculated orbits were stored on files and compared globally. Table 4.2 columns A-B, A-C and B-C shows the RMS of the corresponding orbit differences. Comparing A-C and B-C, we see that the numbers are quite similar. The A-B column shows small numbers and since test set A is expected to give the best results, we conclude that test C gives good orbits only for SV11 and SV13. Even though set B contains approximately 50 % of the data from set A, we believe that both A and B give orbits better than 10 meters (1 σ). We also see that the formal errors from the A, B and C tests are very realistic when compared with the "actual" orbit error. This leads us to conclude that the orbit errors for solution A might be in the range 2- 5 meters for all satellites, except possibly SV6 and SV8 for which there is only a small number of measurements.

SV	FORMAL ORBIT ERROR (M)			ACTUAL ORBIT ERROR (M)		
	A	B	C	A-B	A-C	B-C
6	4.99	5.86	85.94	11.58	99.55	98.18
8	9.77	13.03	37.97	5.96	29.65	29.50
9	2.67	3.46	4.67	3.06	12.04	13.98
11	2.72	3.30	12.99	5.44	3.51	7.97
12	2.61	4.16	3.71	4.95	25.77	27.78
13	3.35	4.10	8.11	4.43	4.49	8.01

Table 4.2   Orbital results

The a priori standard deviation of all the measurements were assumed to be 1 meter when used to calculate the formal errors.

Figures 4.1 - 4.3 show the orbit differences for test A and B in the radial, along-track, and cross-track directions. Note the short-period differences with a period of 12 hours and also the secular drift both in the along-track and in the radial directions. The "hairy" high-frequency oscillations starting at 100000 seconds are due to roundoff in the output of GEOSAT.

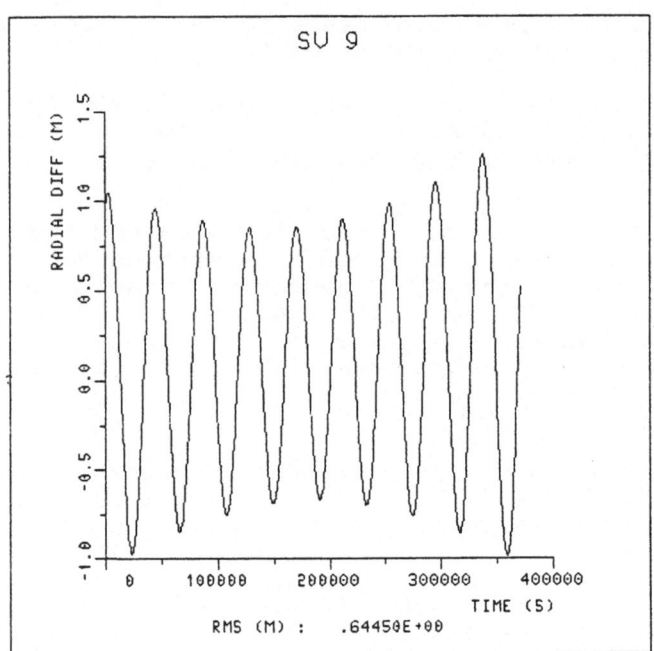

Figure 4.1   Radial orbit difference between test A and B for SV9

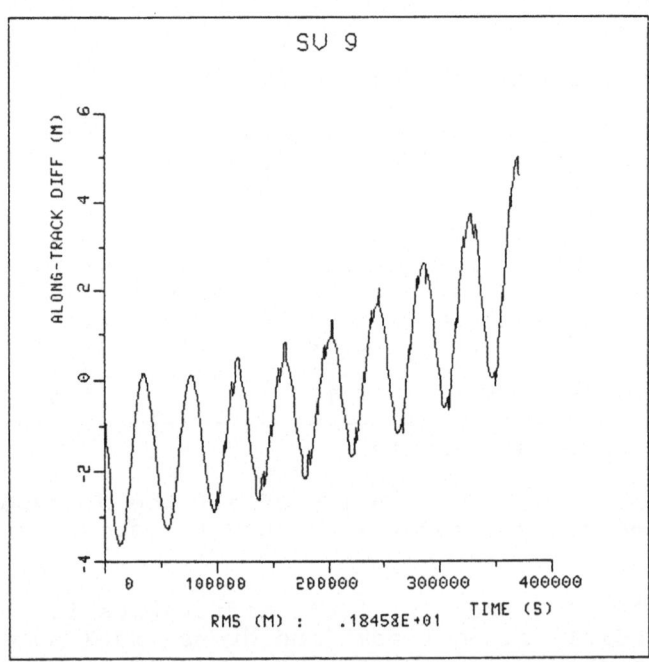

Figure 4.2   Along-track orbit difference between test A and B for SV9

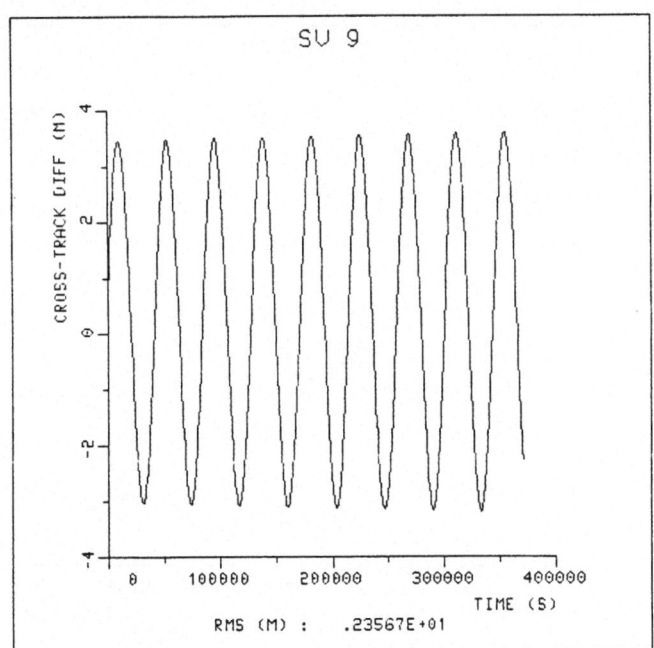

Figure 4.3   Cross-track orbit difference between test A and B
for SV9

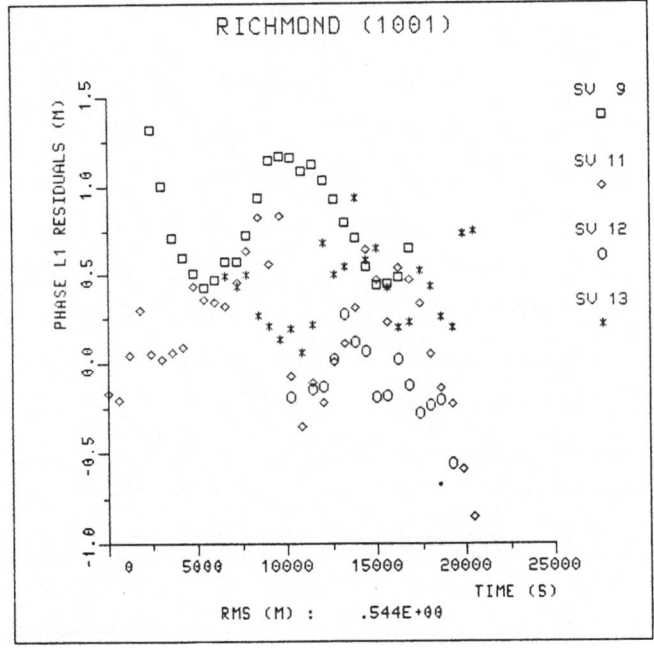

Figure 4.4   L1 phase residuals for Richmond

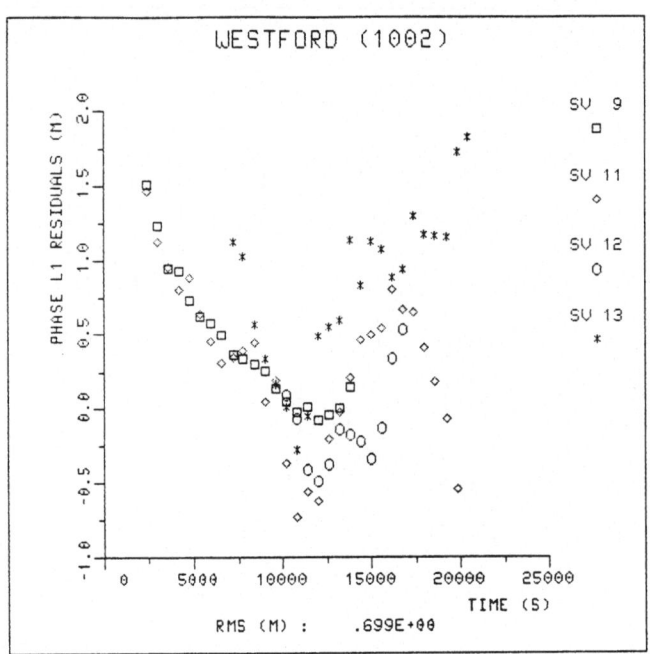

Figure 4.5   L1 phase residuals for Westford

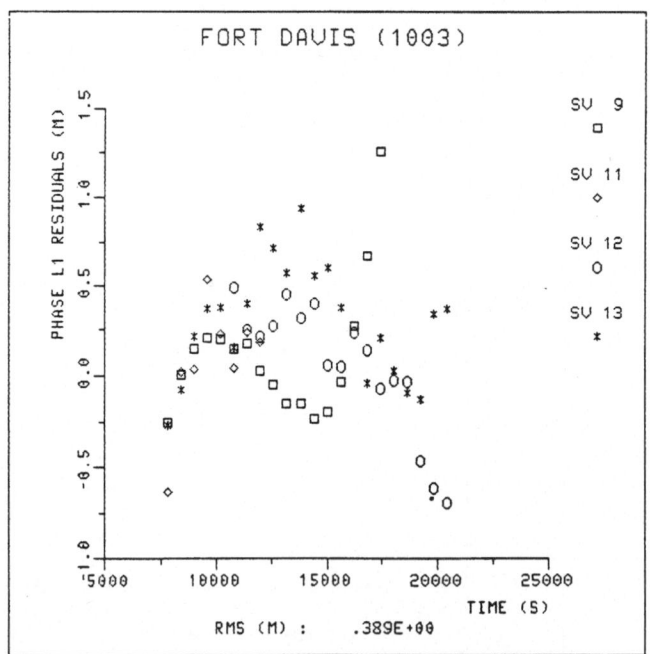

Figure 4.6   L1 phase residuals for Fort Davis

Typical undifferenced residuals are shown in Figures 4.4-4.6
for the L1 phase.

After the orbits were obtained, the coordinates of Richmond
were fixed (thereby defining the terrestrial system), and
the coordinates of the other two stations were solved for
together with all the other parameters. This is the way the
fiducial technique works. The results are shown in Table 4.3.
The B/C column shows the mean values of test B and C, and we
see that the mean values are very close to the values
calculated for test A. Furthermore, the formal standard
deviation compares quite favourably with the empirical
repetition accuracy in the range 16-24 cm. Note that this is
absolute positioning with only one reference station. Wendel
et al (1986) have calculated absolute station positions for
sites involved in the Spring 1985 GPS precision baseline
test. They obtained the following average errors : 1.07 m
east, 0.96 m north and 2.06 m vertical. Since our dataset is
very small, we want to emphasize that the results presented
here are preliminary, but they indicate that GEOSAT will be
able to produce very good results with GPS data.

STATION	TEST	SOLUTION (M)			ERROR (M)	
		X	Y	Z	FOR	EMP
WESTFORD	A	1492398.35	-4457293.19	4296818.17	0.18	
WESTFORD	B	1492398.28	-4457292.99	4296818.03	0.24	
WESTFORD	C	1492398.46	-4457293.38	4296818.24	0.26	
WESTFORD	B/C	1492398.37	-4457293.19	4296818.14		0.24
FT DAVIS	A	-1324206.15	-5332058.60	3232043.60	0.23	
FT DAVIS	B	-1324206.01	-5332058.50	3232043.59	0.30	
FT DAVIS	C	-1324206.23	-5332058.56	3232043.38	0.36	
FT DAVIS	B/C	-1324206.12	-5332058.53	3232043.49		0.16

Table 4.3   Absolute positioning results

4.4   Future Improvements

In order to take full advantage of the high accuracy
of the phase measurements, the time must be modeled
to an accuracy of about 30 ps. As an alternative, the
oscillator effects can be dramatically reduced using
differencing techniques. Clock modelling to the required
accuracy is very difficult even for Hydrogen-Masers. Thus
the next obvious step is to use doubly differenced phase
measurements in the last few iterations. In addition,
software must be developed for taking advantage of the
integer nature of the doubly differenced biases. JPL and
other institutions have demonstrated that improvement by a
factor 1.4 to 4 in relative positioning can be achieved
with properly fixed integer biases. A corresponding
improvement can be expected in the calculated orbits.

Some of the errors inherent in the results presented are due
to inconsistencies in the applied reference system. A new set

of coordinates with an internal consistency of about 10 cm has already been implemented (Murray and King 1988) together with a new gravity field GEM-T1 (Marsh et al 1988) and improved models for ocean and solid earth tides. The ROCK 4 radiation pressure model has been acquired but it is not implemented yet. It might also be necessary to include models for earth albedo and thermal radiation pressure in order to be able to model the orbits to a few decimeter level.

When data from a globally distributed network become available, it will be possible to solve for variations in the earth rotation and polar motion (ERP). ERP-values and nutation corrections determined from VLBI can be applied as an alternative.

5. Conclusions

GPS tracking and data analysis are well underway both on a national and international scale. It is hoped that the European Tracking Experiment 1988 (EUTREX-88) and the first GPS Global Orbit Tracking Experiment (GOTEX-1) in the second half of 1988 will evolve into permanent cooperative programs for inter-European and international GPS tracking, orbit determination, and reference system improvement. In Europe, the Bernese software (Gurtner et al 1985) is already well proven for this computational task and the GEOSAT software (Andersen 1986) is also very promising. Independent but coordinated GPS data analysis with these two software systems should be encouraged.

6. Acknowledgement

This work was supported in part by the Royal Norwegian Council for Scientific and Industrial Research SATKART program under contracts IT 6.61.19383 and IT 6.61.20761.

7. References

Aksnes K., Andersen P. H., Haugen E. (1988):
    A precise multipass method for satellite doppler positioning, submitted to Celestial Mechanics, Feb 1988.

Andersen P. H. (1986):
    GEOSAT - A computer program for precise reduction and simulation of satellite tracking data. In Proceedings of the Fourth International Geodetic Symposium on Satellite Geodesy, Austin, Texas, April 28 - May 2, 1986.

Boucher C. and Altamimi Z. (1986):
    The use of space techniques for the connection of geodetic datums. 62 th meeting of the Study Group on Geodynamics, Louvain-la-neuve 2-4 June, 1986.

Evans A. G. (1986):
Comparison of GPS pseudorange and bias doppler range
measurements to demonstrate signal multipath effects,
In Proceedings of the Fourth International Geodetic
Symposium on Satellite Geodesy, Austin, Texas, April 28 -
May 2, 1986.

Gurtner, W., Beutler G., Bauersima I., Schildknecht T.
Evaluation of GPS carrier difference observations:
The Bernese Second Generation Software Package.
In Proceedings of the First International Symposium
on Precise Positioning with the Global Positioning
System, Rockville, 1985.

Landau H. and Hein G. W. (1986):
Preliminary results of a feasibility study for a
European GPS tracking network. Fourth International
Geodetic Symposium on Satellite Positioning, Austin,
Texas, 28 April - 2 May, 1986.

Mader G. L. (1988):
GPS Global Orbit Tracking Experiment. Memorandum from
CSTG GPS Subcommission, issued at NOAA/NGS 24 Feb.

Marsh J. G. et al. (1988):
An improved model of the Earth's gravitational field :
GEM-T1, NASA Technical Memorandum 4019.

Murray M. H., King R. W. (1988):
SV3 coordinates of GPS receivers, MIT Internal Memorandum-
11 March 1988.

Wendel M., Hermann B. R., Swift E. (1986):
Absolute station position solutions for sites involved in
the Spring 1985 GPS Precision Baseline Test,
In Proceedings of the Fourth International Geodetic
Symposium on Satellite Geodesy, Austin, Texas, April 28 -
May 2, 1986.

GPS GEODESY WITH CENTIMETER ACCURACY

by

G. Blewitt, W. G. Melbourne, W. I. Bertiger, T. H. Dixon, P. M. Kroger,
S. M. Lichten, T. K. Meehan, R. E. Neilan, L. L. Skrumeda,
C. L. Thornton, S. C. Wu, and L. E. Young

Abstract

Centimeter-level accuracy is crucial for Global Positioning System (GPS) baseline measurements to be useful for many geophysical applications. This implies that baseline vector accuracy must be of the order of a few parts in 10^8 of baseline length for regional geodesy. The latest techniques developed at JPL for analyzing GPS data have indeed resulted in centimeter-level agreement with solutions determined by Very Long Baseline Interferometry (VLBI) in California, for baseline lengths of up to 1000 km.

The techniques we have found most promising for high accuracy geodesy are: (1) carrier phase ambiguity resolution, (2) multi-day orbit determination, (3) stochastic estimation of the zenith tropospheric delay, and (4) simultaneous use of carrier phase and pseudorange. The order of importance depends upon the scale of the network, and the approaches are often synergistic. For example, ambiguity resolution can depend upon the ability of the other techniques to improve precision.

The future of GPS looks bright if one considers that these results have been acheived despite a partial GPS constellation, no global tracking network, and pseudorange data plagued by multipath. A full GPS constellation and a global tracking network will not be realized until the 1990's, but steps are being made in the right direction. A new receiver/antenna prototype at JPL is showing promise of producing pseudorange observables accurate to 5 cm. Two of these receivers participated in the January 1988 CASA UNO experiment, which was managed by JPL in cooperation with about 30 other institutions. The purpose of this experiment was to accurately measure geodetically interesting baselines in South and Central America. Precise orbit determination for this experiment was enabled by tracking the GPS satellites from Australia, New Zealand, Hawaii, American Samoa, North America, and Europe. Results from this experiment should give valuable information on the potential of GPS.

1. Introduction

The Global Positioning System (GPS) data analysis team at the Jet Propulsion Laboratory (JPL) has recently estimated geodetic baselines which agree with Very Long Baseline Interferometry (VLBI) solutions at the level of 2 parts in 10^8 (2 cm per 1000 km of baseline length). Moreover, it has since been learned that there is a scaling difference between the GPS and VLBI solutions which arises from the use of inconsistent VLBI-inferred fiducial coordinates. It is expected that GPS and VLBI baselines will agree at the level of 1 part in 10^8 when this inconsistency is corrected. These results reflect the effectiveness of recently developed estimation techniques, which include new approaches to carrier phase ambiguity resolution, multi-day GPS orbit determination, stochastic troposphere estimation, and the use of simultaneous group and phase delay.

The goal of achieving centimeter-level geodesy has thus been realized for regional-sized GPS networks with good fiducial control. For networks of dimensions greater than 1000 km, centimeter-level GPS accuracy should be achievable anywhere on the globe considering forseeable upgrades in hardware, software, experimental design, and tracking networks.

2. State of the Art Techniques

2.1 Carrier Phase Ambiguity Resolution

The GPS carrier phase delay data are biased by an integer number of wavelengths. Resolving these biases is the well-known problem of "ambiguity resolution." Covariance studies show that if the biases are not resolved and are simply estimated as real-valued parameters, there is about a factor of 3–5 degradation in baseline precision [Melbourne, 1985].

Previous ambiguity resolution techniques have been limited by the ability to model the ionospheric delay. Our method uses the valuable information inherent in the simultaneous measurements of the pseudorange and carrier phase observables [Melbourne, 1985; Wubbena, 1985]. We start by using a time-average of a linear combination of the pseudorange data (P_1 and P_2) and carrier phase data (L_1 and L_2) to first determine the integer offsets, n_Δ, between the two carrier channels:

$$n_\Delta \equiv (n_1 - n_2)$$
$$\hat{n}_\Delta = \frac{L_1(\text{cm})}{19.029} - \frac{L_2(\text{cm})}{24.421} - \frac{P_1(\text{cm})}{153.35} - \frac{P_2(\text{cm})}{196.80} \tag{1}$$

where n_Δ can be inferred from the real-valued estimate, $\hat{n}_\Delta$. Using currently available TI-4100 receivers, it appears that half an hour is sufficient time to average down the pseudorange noise to a level at which these "widelane" biases are resolvable. With future high precision pseudorange receivers and antennas, widelane resolution should be possible in less than a minute.

We then proceed to form the ionosphere-free linear combination of the phase data:

$$L_C = L_1 + \alpha (L_1 - L_2)$$
$$L_C(\text{cm}) = \rho + 48.444\, n_\Delta + 10.695\, n_C + \nu \tag{2}$$

where $\alpha \simeq 1.5457$, ρ is the non-dispersive delay, and ν is the data noise. Each phase connected data arc is now biased by n_C, an integer number of 10.695 cm wavelengths.

In order to resolve these "narrowlane" biases, we estimate them simultaneously with ρ, which is modeled in terms of station coordinates, satellite ephemerides, clocks, and tropospheric delays. An algorithm based on the square-root information formalism (SRIF) is then applied which sequentially adjusts the biases to the nearest integer value in such a way that the global solution is updated at each step. In this way, ambiguity resolution

31

over longer baselines can be achieved by using the information contained in resolving the biases on the shorter baselines.

Based on our experience [Blewitt *et al.*, 1987] and that of others [Councelman, 1987; Dong and Bock, 1988], a well designed GPS network should contain several shorter length baselines (100–200 km) if ambiguity resolution is to be acheived over longer baselines. We have demonstrated ambiguity resolution on baselines up to 2000 km in length [Blewitt, 1988]. Network design to ensure ambiguity resolution should be an important consideration when planning GPS experiments.

2.2 Multi-Day Arc GPS Orbit Determination

The GIPSY software used at JPL for GPS data processing uses a pseudo-epoch state process noise filter to estimate all geodetic and orbital parameters simultaneously [Bierman, 1977; Thornton and Bierman, 1980]. Spacecraft motion is modeled in geocentric J2000 coordinates with a numerical integration of the variational equations and equations of motion. The GEM-L2 geopotential harmonic expansion (complete to degree and order 8 for GPS models) and effects from solar, lunar, and planet point masses are modeled. The ROCK4 solar radiation pressure model is used. General relativistic corrections for clock models and gravitational bending are included, and with this degree of model sophistication, satellite motions can be realistically and accurately estimated and propagated over arcs of two weeks or more.

The GPS filtering strategy for multi-day arc orbit determination has evolved at JPL in the past two years based on experience with the data taken during the 1985 and 1986 GPS experiments in North America [Lichten and Border, 1987]. Three solar pressure coefficients (including the Y-bias parameter) are estimated as constant parameters along with the orbital states.

Preliminary solutions indicate that with arcs of longer than two weeks results improve when the solar pressure coefficients are estimated stochastically. We believe that this may be evidence for unmodeled forces on the GPS satellites and this possibility is being intensively studied at this time.

2.3 Stochastic Troposphere Estimation

Water vapor radiometers (WVR's) can be used to calibrate GPS data for the wet tropospheric delay under certain meteorological conditions (e.g., when it is not raining) [Elgered *et al.*, 1987; Tralli *et al.*, 1988]. However, these conditions are not always available, and WVR's are bulky, expensive, and few in number. Consequently, most GPS data do not have the luxury of accurate tropospheric calibrations. Standard models using surface meteorological data are of limited value, and can be in error by more than ten centimeters.

At JPL, the tropospheric zenith delay can be estimated at every station. For stations with WVR calibrations, a tightly constrained constant calibration bias is estimated from the GPS data; typical corrections are 1-3 cm at zenith. We have found that estimating the tropospheric delays using a random walk stochastic model be very successful when WVR's are not available. The stochastic model is tuned so that the tropospheric delay can wander ∼ 4 cm over an 8-hour daily tracking session. From one day to the next, the troposphere zenith delay is reset to be independent from the previous day's value.

Daily repeatability for baselines and mapped orbit solutions are consistently improved by factors of up to 2 as compared to simply estimating a constant residual zenith delay [Lichten and Border, 1987; Tralli *et al.*, 1988]. We are currently investigating the application of stochastic troposphere estimation in Central America and the Caribbean.

2.4 Simultaneous Group and Phase Delay

Although the pseudorange group delay data type is typically two orders of magnitude noisier than phase delay, it provides an unambiguous differential range measurement. If

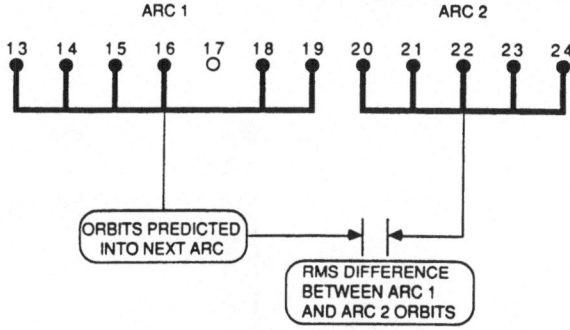

NOVEMBER 1985 MULTI-DAY ARCS

Figure 1. Definition of orbit repeatability for November 1985 multi-day arc test. Orbit solutions from arc 1 are mapped forward, and the RMS difference is taken at beginning of arc 2.

we assume that the pseudorange and carrier phase data are sampled simultaneously (to within $1\,\mu$sec), the pseudorange strengthens the carrier phase solution in two ways: (1) it allows for precise estimates of receiver synchronization offsets; (2) it constrains the carrier phase bias estimates.

Baseline accuracy is improved by about a factor of 2 with meter-level pseudorange from current TI-4100 receivers [Lichten and Border, 1987; Tralli and Dixon, 1988]. This factor will improve with the advent of high precision pseudorange receivers and antennas. Also, the phase biases are estimated more accurately, thus enabling ambiguity resolution with greater confidence. Covariance studies performed at JPL show that if carrier phase is combined with high precision pseudorange (< 10 cm for 6 minute points), it is nearly as effective as using carrier phase with the ambiguities resolved [Lichten and Border, 1987]. This may be important for longer baselines where ambiguity resolution is difficult.

3. GPS Accuracy

3.1 Daily Repeatability

One measure of system performance is repeatability of solutions. Strictly, only a lower bound on accuracy can be inferred from repeatability; however it is useful to compare baseline precision with that expected from formal errors. Moreover, GPS accuracy cannot be directly assessed on baselines which have not been previously estimated using VLBI, or satellite laser ranging.

We have analyzed data from 18 sites in the western U.S.A. which were occupied for up to 4 days in June 1986 [Blewitt *et al.*, 1987; Skrumeda *et al.*, 1987]. Station locations and satellite orbits were estimated independently for each day. Using combined carrier phase and pseudorange data, we stochastically estimated the residual zenith tropospheric delay and resolved 94% of the carrier phase ambiguities in the network. This resulted in subcentimeter baseline repeatability in the horizontal plane for all baselines up to 600 km in length. Repeatability is improved by a factor of 3 when compared to solutions without ambiguity resolution. In the vertical direction, repeatability is at the 3 cm level.

Multi-day orbit determination has proved to be important for improving the precision of longer baselines [Bertiger and Lichten, 1987]. Well-tracked GPS orbits show sub-meter repeatability when compared to orbits determined from separate, independent measurement

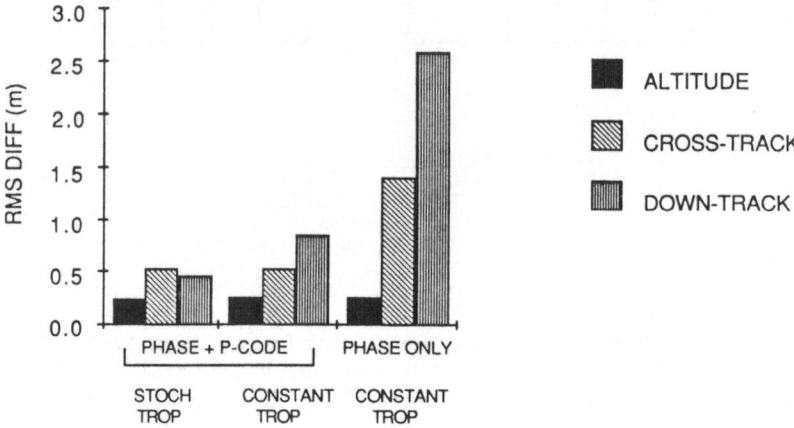

**GPS 8 ORBIT 1-DAY PREDICTION USING
INDEPENDENT 1-WEEK ARC SOLUTIONS
FROM NOVEMBER 1985**

Figure 2. Orbit repeatability of GPS 8 for three estimation strategies. The best strategy uses combined carrier phase with pseudorange, and stochastic troposphere estimation.

JUNE 1986 GPS EXPERIMENT: COLLOCATION WITH VLBI

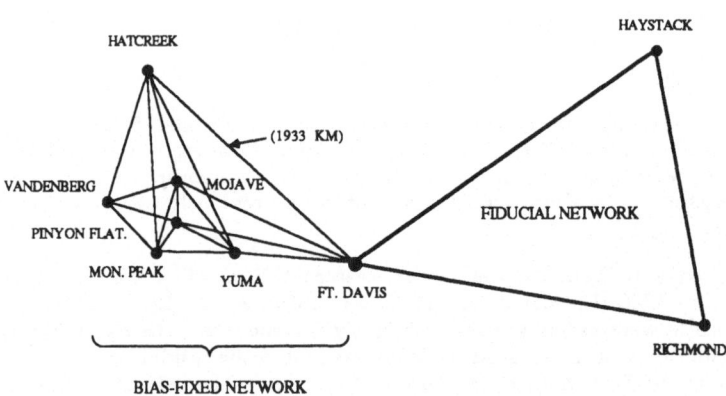

Figure 3. Network configuration of June 1986 GPS experiment for sites collocated with VLBI. The fiducial network was held fixed at the VLBI-inferred coordinates, and the other station locations were estimated using ambiguity resolution.

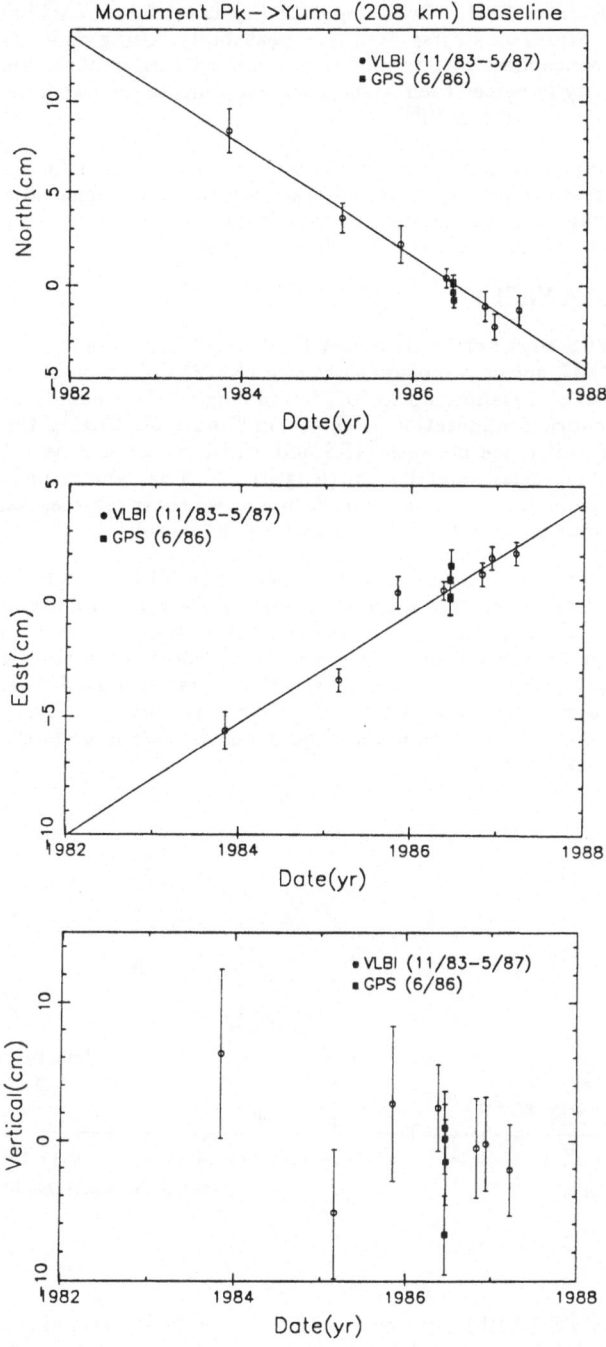

Figure 4. Example of daily GPS baseline solutions from the June 1986 test superimposed on a history of VLBI solutions for Monument Peak to Yuma (208 km).

sets [Lichten and Border, 1987]. Orbit repeatability is explained in Figure 1. Figure 2 shows the effect of various strategies on orbit repeatability. Using multi-day arcs for GPS orbits, stochastic troposphere estimation, and combined carrier phase and pseudorange, the daily repeatability is better than 3 cm in all components for baselines up to 2000 km in length (better than 2 parts in 10^8).

We are currently implementing carrier phase ambiguity resolution for solutions with simultaneously estimated multi-day arcs and station locations. Using all the inherent information in the system to accomplish this task is not trivial due to the large number of bias parameters (several hundred) which must be adjusted.

3.2 Comparison with VLBI

A comparison of GPS with both mobile and fixed VLBI baselines in the western U.S.A. [Skrumeda *et al.*, 1987] shows a root-mean square (RMS) difference of 2 cm for the horizontal components of 15 baselines up to 1086 km in length (Hatcreek, California to Yuma, Arizona). The network configuration is shown in Figure 3. Taking the mean over all baselines, the RMS difference between GPS and VLBI coordinates is 1.4 cm (east), 1.6 cm (north), 3.6 cm (vertical), and 1.2 cm (length). This agreement includes a factor of 2 improvement by ambiguity resolution. In Figure 4, we show representative examples of GPS baseline solutions along with the time-series of VLBI solutions.

The GPS length solutions are systematically longer than VLBI by an average of 0.8 cm. Although less well determined, the longer baselines in the range 1000–2000 km from Fort Davis, Texas, to stations in California, also show a systematic offset. Figure 5 shows (GPS-VLBI) baseline length solutions. It appears that there is a scaling difference at the level of about 1 to 2 parts in 10^8. We believe that these scaling differences arise from the use of coordinates at the fiducial sites (Ft. Davis, Richmond, Haystack) which were derived from an outdated VLBI analysis. This inconsistency is understood and will be corrected in future analyses.

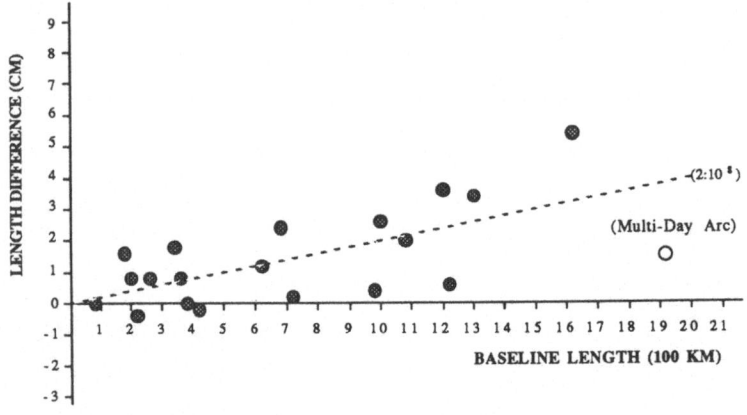

Figure 5. GPS-VLBI length comparison for single-day arc, bias-fixed solutions. Also shown is a multi-day arc solution for comparison. A scaling difference of 1–2 parts in 10^8 can be accounted for by the use of inconsistent fiducial coordinates.

For centimeter-level long baseline accuracy, multi-day GPS orbit determination is currently required. The comparison of GPS with VLBI is currently limited by the uncertainties in the VLBI-inferred fiducial baselines used to define the GPS reference frame. Nevertheless, baselines of 2000 km determined with these orbits agree with VLBI at the level of 0.3–5 cm in all vector components. The 1933 km GPS baseline between Hatcreek, California, to Fort Davis, Texas, agreed with VLBI to better than better than 2.6 cm for all components using data taken in November 1985 (see Figure 5). This solution was a result of a 6 day arc fit to the GPS orbits.

4.   Future Prospects

4.1   JPL Receiver/Antenna Development

A GPS receiver has been developed for NASA at JPL expressly for making high quality geodetic measurements [Meehan *et al.*, 1987a]. The ROGUE GPS receiver is based on a digital design that leads to very high accuracy phase and pseudorange measurements. The error contribution from the receiver itself is submillimeter for the carrier phase data, and is at the centimeter level for 2 minute P-code pseudorange.

In practice, however, pseudorange measurements are greatly degraded by multipath signals at the antenna [Meehan *et al.*, 1987b]. This leads to a situation where centimeter level pseudorange error contributions from system noise and other receiver errors are overshadowed by multipath error contributions of up to a meter. Earlier studies of multipath effects done at JPL and elsewhere determined that, for a given multipath environment, certain antenna/backplane configurations could significantly reduce the errors due to multipath.

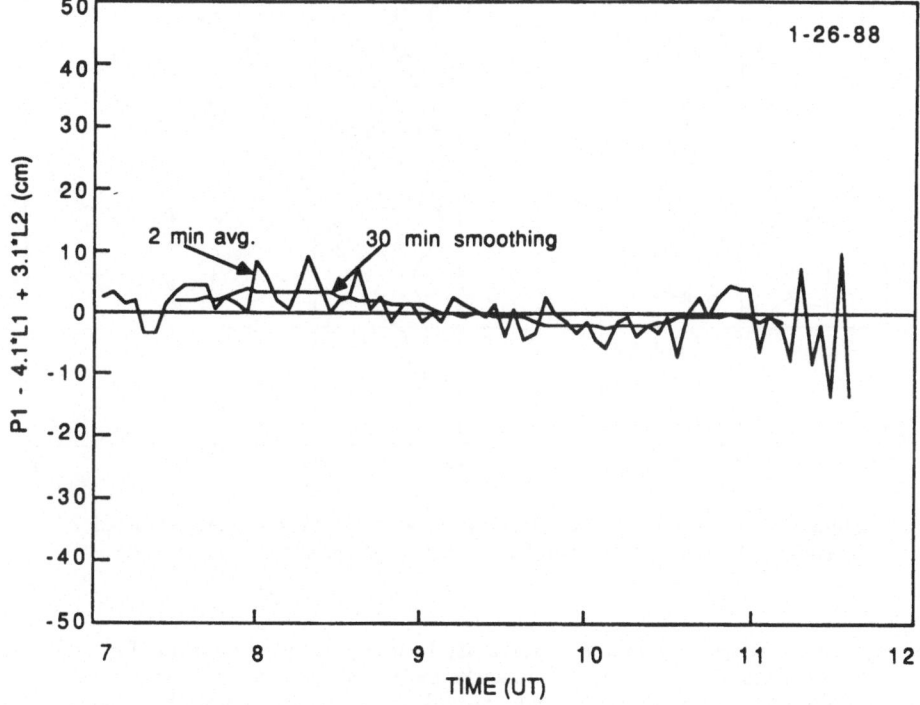

Figure 6.   A sample of single-band pseudorange noise from data taken at Owens Valley by the ROGUE receiver with a choke-ring backplane. Long-term multipath signatures are below the 5 cm level.

Field tests have shown that one such configuration, developed at JPL by D.J. Spitzmesser, can reduce long term multipath errors on pseudorange to about 5 cm. This configuration uses a drooped cross-dipole antenna to receive the L_1 and L_2 satellite signals and a conducting "choke-ring" backplane. Two prototype antenna/backplane combinations were used in conjunction with ROGUE receivers at the Mojave and Owens Valley Radio Observatory sites in California during the recent CASA UNO experiment. Figure 6 shows a sample of the pseudorange noise for data taken at Owens Valley. A linear combination of the data, which removes all signatures except for multipath and noise, is plotted with time. These preliminary results are encouraging, and more tests are in progress.

4.2   Global Tracking: The "CASA UNO" Experiment

The "CASA UNO" experiment provided three weeks of GPS data for an initial epoch measurement of a geodetic network in Central and South America managed by JPL, the CASA UNO experiment involved participants from about 30 institutions worldwide. In all, 24 sites in Central and South America were occupied by 16 receivers. An additional 13 receivers in the U.S.A. occupied 20 sites during the experiment. To provide effective fiducial control for such an extended network, a near-global tracking network of 12 receivers, most of which were collocated with VLBI sites (Figure 7). The locations of the non-collocated sites will be estimated simultaneously with the GPS orbit parameters. Covariance studies show that the consequences of not collocating all tracking sites with VLBI are insignificant due to the strong GPS data strength [Freymueller and Golombek, 1987]. In all, about 600 station-days of data are now being reduced.

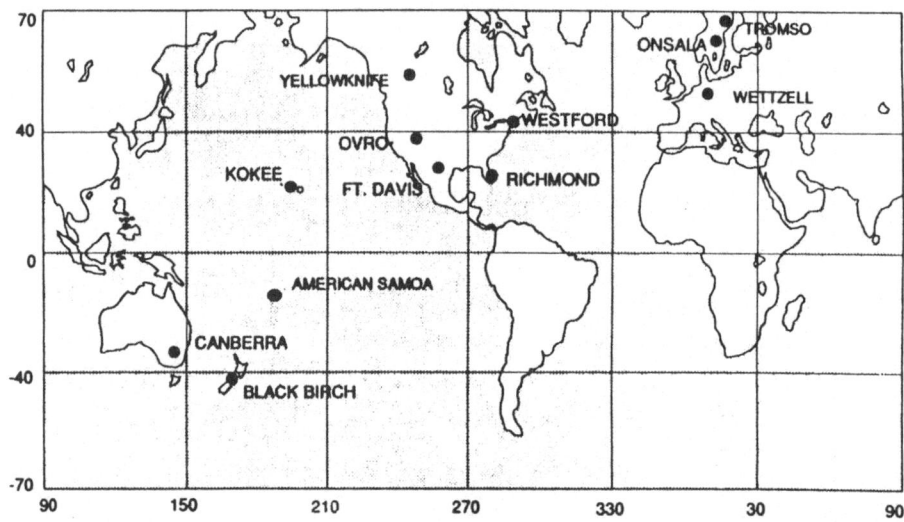

Figure 7. This near-global tracking network of 12 receivers acquired 3 weeks of GPS data during the CASA UNO experiment, January 1988.

The experiment provides a unique opportunity to investigate the benefits of global tracking using real data. So far, fiducial networks have been limited to sizes of the order of 3000 km. Results from CASA UNO will provide a valuable benchmark to calibrate covariance studies, so that we can reliably predict the performance of future global tracking systems.

Investigations are being undertaken to study the best estimation strategy applicable to global networks. Preliminary studies into the potential of GPS to determine earth orientation and the geocenter will be pursued with this data set.

5. Conclusions

GPS-based geodesy has reached the point where centimeter-level accuracy for regional baselines (< 1000 km) can be done routinely in areas of good fiducial control. Accuracies of 2 parts in 10^8 have been demonstrated for baselines up to 2000 km in length. This level of accuracy requires careful modeling of the GPS observables, resolution of carrier phase ambiguities, multi-day arc orbit determination, and stochastic estimation techniques.

Accuracies of 1 part in 10^8 of baseline length should become routine worldwide in the 1990's, as various aspects of GPS-based systems are improved. Important developments are precise pseudorange, receiver/antenna designs, global tracking networks, experiment design to enable ambiguity resolution, and the continual improvement of estimation strategies.

Acknowledgements

This work described in this paper was carried out at the Jet Propulsion Laboratory, California Institute of Technology, under contract with the National Aeronautics and Space Administration. The GPS data analysis was conducted at the Jet Propulsion Laboratory using the GIPSY analysis software. We thank the many people who have made key contributions to the GIPSY software and analysis effort, especially John M. Davidson, Ojars Sovers, James S. Border, Scott A. Stephens, and the late Gerald J. Bierman.

References

Bertiger, W.I. and S.M. Lichten (1987): Demonstration of 5 to 20 parts per billion repeatability for a continental baseline estimated with multi-day GPS orbits, EOS Trans. Am. Geo. U., Vol. 68, No. 44, p. 1238

Bierman, G.J. (1977): Factorization Methods for Discrete Sequential Estimation, Academic Press, Orlando, Florida

Blewitt, G. *et al.* (1987): Improved GPS network solutions using bias-optimizing techniques, EOS Trans. Am. Geo. U., Vol. 68, No. 44, p. 1236

Blewitt, G. (1988): Successful GPS carrier phase ambiguity resolution for baselines up to 2000 km in length, EOS Trans. Am. Geo. U., Vol. 69, No. 16, p. 325

Dong, D. and Y. Bock (1988): GPS network analysis: Ambiguity resolution, EOS Trans. Am. Geo. U., Vol. 69, No. 16, p. 325

Counselman III, C.C. (1987): Resolving carrier phase ambiguity in GPS orbit determination EOS Trans. Am. Geo. U., Vol. 68, No. 44, p. 1238

Tralli D.M. and T.H. Dixon (1988): A few parts in 10^8 geodetic baseline repeatability in the Gulf of California using the Global Positioning System, Geophysical Research Letters, Vol. 15, pp. 353–356.

Elgered G. *et al.* (1987): On the Weather Dependence of Baseline-Lengths Estimated by Very-Long-Baseline Interferometry, EOS Trans. Am. Geo. U., Vol. 68, No. 44, p. 1239

Freymueller J.T. and M.P. Golombek (1987): Effect of Fiducial Network Geometry on GPS Baseline Accuracy in South America, EOS Trans. Am. Geo. U., Vol. 68, No. 44, p. 1237

Lichten, S.M. and J.S. Border (1987): Strategies for high-precision global positioning system orbit determination, Journal of Geophysical Research, Vol. 92, No. B12, pp. 12751–12762

Meehan T.K. *et al.* (1987a): ROGUE: A new high accuracy, digital GPS receiver, Proceedings of the IUGG Conference, Vancouver, August 1987.

Meehan T.K. *et al.* (1987b): GPS multipath reduction using absorbing backplanes, EOS Trans. Am. Geo. U., Vol. 68, No. 44, p. 1238

Melbourne, W.M. (1985): The case for ranging in GPS based systems, Proceedings of the First Symposium on Precise Positioning with the Global Positioning System, Positioning with GPS–1985, Ed. C.C. Goad, Rockville, Maryland, pub. U.S. Department of Commerce, NOAA

Skrumeda, L.S. *et al.* (1987): Baseline results from the June 1986 Southern California and Caribbean GPS Experiments, EOS Trans. Am. Geo. U., Vol. 68, No. 44, p. 1236

Thornton, C.L., and G.J.Bierman (1980): UDU^T covariance factorization for Kalman filtering, Control and Dynamic Systems, Advances in Theory and Application, Ed. C.T. Leondes, Academic Press, Vol. 16, pp. 177–248

Tralli, D.M. *et al.* (1988), The effect of wet tropospheric path delays on estimation of geodetic baselines in the Gulf of California using the Global Positioning System, accepted for publication in Journal of Geophysical Research

Wubbena, G. (1985): Software developments for geodetic positioning with GPS using TI-4100 code and carrier measurements, Proceedings of the First Symposium on Precise Positioning with the Global Positioning System, Positioning with GPS–1985, Ed. C.C. Goad, Rockville, Maryland, pub. U.S. Department of Commerce, NOAA

RELATIVISTIC EFFECTS IN GPS

by

S. Y. Zhu', E. Groten

Abstract

Relativistic effects in GPS are twofold: first is the effect on orbit
and signal propagation, second is that on the clock. The first part
has an effect of up to 0.001 ppm in positioning. The second part
affects the clock frequency on the order of 10^{-10}, but only the
periodic fluctuation in it is of interest. This term is completely
canceled out by between-station differences, hence it is harmless for
relative positioning, but it directly affects the clock
synchronization and causes substantial error in single point
positioning. By adopting a Keplerian orbit, most of this fluctuation
can be corrected. The ammount of non-Keplerian part is estimated to be
less than 0.6 ns (18 cm).

1. Introduction

GPS is applied in geodetic positioning and in time transfer. Relativistic effects on both of them are discussed here, with emphasis on positioning.

There are two kinds of positioning by GPS, point positioning and relative positioning. The accuracy of the first is worse than 1 ppm, that of the latter is much higher. Recently, Bertiger and Lichten (1987) have demonstrated 5×10^{-9} repeatability for a continental baseline which was estimated using multi-day GPS orbits. Interstingly and fortunately, the magnitudes of relativistic effects are in agreement with the relevant accuracy; the relativistic effects on point positioning are pretty large (~1 ppm), those in relative positioning are much smaller (~10^{-9}). This correspondance between positioning accuracies and magnitudes of effects can be thought of as a major feature of relativistic effects in GPS positioning.

The earth's gravitation exerts direct relativistic effects on the orbital motions of satellites (relativistic perturbation) and on phase measurements (propagation correction); these effects are treated in section 2. The satellites carry high accuracy clocks on orbit; due to the gravitation of the earth and orbit motion of satellites the clocks will give time which is different from time given by the same kind of clocks at rest at the earth's surface. Its impact on time transfer and positioning is the topic of section 3.

Only the main results and relevant explanations are presented in this paper; all derivations are omitted. The results are based on general relativity. The treatment is carried out in the terrestrial (earth's center) system, see Zhu et al. (1987). The Schwarzschild metric of the earth is good enough for such considerations. Neglecting the effects of quadrupole of the earth's potential and the effects of the sun and other solar system bodies only causes errors in positioning of 10^{-5}ppm (or less) and errors in frequency of 10^{-14} (or less).

2. Relativistic dynamic effect and propagation correction

The relativistic perturbation reads (Zhu et al., 1987) :

$$\delta \ddot{r} = GE \; r \; [(4GE/r) - \dot{r}^2]/c^2 r^3 + 4GE(r \cdot \dot{r}) \dot{r}/c^2 r^3 , \qquad (1)$$

where r is the geocentric vector of the satellite. $\dot{r}$ is its velocity. δr is about 0.5×10^{-9} the Newtonian acceleration and amounts to 0.3×10^{-9} m/sec^2 for the Navstar satellite. Table 1 compares this effect with other Newtonian perturbations, the latters are taken from (Rizos and Stolz, 1985). From that table one sees that the relativistic perturbation is nearly as important as that of the earth's tidal potential or that of albedo pressure.

Table 1   Effect of Perturbing Forces on GPS Satellites

Source	Acceleration (m/sec^2)
Earth's non-sphericity :	
(a) C_{20}	5×10^{-5}
(b) other harmonics	3×10^{-7}
Point-mass effects of sun and moon	5×10^{-6}
Earth's tidal potential :	
(a) earth tides	1×10^{-9}
(b) ocean tides	1×10^{-9}
Solar radiation pressure	1×10^{-7}
Albedo pressure	1×10^{-9}
Relativistic effect	0.3×10^{-9}

Taking phase measurements as an example, the relativistic propagation correction is (cf. Holdridge, 1967).

$$\delta t = (2GE/c^3) \ln [(r + R + \rho)/(r + R - \rho)]. \qquad (2)$$

in which r and R are geocentric distances of satellite and station, respectively. ρ is the range distance. The corresponding correction for range measurement is $c\delta t$. For single phase measurement, the maximum value for $c\delta t$ is 19 mm.

This propagation correction depends on the geometry between station, satellite and geocenter. The combination of observations (differences) could hardly be of any help to reduce this error in a relative sense. Table 2 gives the order of magnitude of this error.

Table 2   Relativistic propagation error

Types of observation	max absolute error (mm)	max. relative error (ppm)
single observation	19	10^{-3}
differences between:		
- satellites	7*	10^{-3}
- stations	7*	10^{-3}
- epochs	**	10^{-3}

* with baseline length ~7000 km
** depends on the time interval of the two adjacent epochs

Both dynamic and propagation effects cause errors up to 0.001 ppm in positioning. These errors can not be eliminated (from relative point of view) by any kind of differences. To ensure ultimate 0.01 ppm accuracy, they should better be taken into account, although for most present applications they can be ignored.

3. Effects on clock frequency

If the transmitting satellite clock frequency is f_t when received at the station receiver it is shifted into f_r; after the usually Doppler shift has been accounted for, the relation between f_t and f_r becomes [cf. Table 1-1 of (Spilker, 1980); note that some printing errors in this reference should be corrected] :

$$f_r/f_t = 1 - (\phi_t + \underline{v}_t^2/2 - \phi_o + (\phi_o - \phi_r - \underline{v}_r^2/2))/c^2 \qquad (3)$$

where $\phi_t = GE/r$, $\phi_r = GE/R$, $\underline{v}_t = \dot{\underline{r}}$ is the velocity of satellite. $\underline{v}_r$ is the velocity of the station (due to the earth rotation). ϕ_o is the value of $(\phi_r + \underline{v}_r^2/2)$ on the geoid. It is introduced since the International Atomic Time (TAI) is defined on the geoid. The last part in eq. (3) is the frequency offset of the station clock with respect to TAI, which is constant at each station and could be easily corrected. $(\phi_t + \underline{v}_t^2/2)/c^2$ contains a constant part ($\sim 2.5 \times 10^{-10}$) and a periodic fluctuation, the latter is mostly due to the non-zero eccentricity. $(\phi_o/c^2) - \langle(\phi_t + v_t^2/2)/c^2\rangle_{const.}$ is about 4.465×10^{-10}, which can be calculated from the semi-major axis of the orbit and has already been removed by offsetting the GPS clock frequency prior to launch. A residual constant offset due to an off nominal semi-major axis of the actual orbit can be corrected as constant time and frequency offset. Only the periodic part is problematic; which could be as large as 46 ns for eccentricity $e = 0.002$; the corresponding range error is about 14 m. This effect must be taken into account.

Currently this periodic effect is calculated by considering the orbit as Keplerian cf. (Van Dierendonck et al., 1980; Jorgensen, 1986; Ashby, 1987). The result is

$$\Delta t_{sv} = -4.4428 \times 10^{-10} \ (sec/\sqrt{meter}) \ e\sqrt{A} \ (\sin E(t) - \sin E(t_{oc})) \qquad (4)$$

where A is the semi-major axis, $E(t)$ is the eccentric anomaly at time t. $E(t_{oc})$ is the value at a certain initial epoch t_{oc}. The $\sin E(t_{oc})$ term is a constant bias which can be corrected by a given time and frequency offset of the satellite clock with respect to GPS time. The only correction in the user's equipment is

$$\Delta t'_{sv} = -4.4428 \times 10^{-10} (sec/\sqrt{meter}) e\sqrt{A} \sin E(t) \doteq -2290 e \sin E (ns) \qquad (5)$$

The actual orbit is certainly not Keplerian. However, generally speaking perturbations of GPS orbits are small. But one should be aware that perturbations such as C_{20} in Table 1 may cause secular changes in Ω, ω and M, hence actual $E(t) - E(t_o)$ may significantly differ from its Keplerian value after a long time span. If we express the actual Δt_{sv} as

$$\Delta t_{sv} \ (actual) = \Delta t_{sv} \ (Keplerian) + \Delta t_{sv} \ (perturbation) \qquad (6)$$

then the amount of Δt_{sv} (perturbation) depends on the time span of ($t - t_{oc}$). t_{oc} is an epoch of time at which the time and frequency offsets of the satellite clock relative to GPS time are redetermined. If the offset is redetermined every two weeks, then the largest value of ($t -$

t_{oc}) is 604800 sec (Van Dierendonck et al., 1980). For e = 0.02 the perturbation Dt_{sv} is estimated as 1.2 ns (this equals 36 cm in range measurement), which is negligible in comparison to the current accuracy of point positioning.

For relative positioning Dt_{sv} is completely canceled out in the between-station differences and it is therefore harmless.

This periodic fluctuation not only causes a time offset Dt_{sv}, which affects the range measurements but also frequency offsets which affect the range rate measurements. From eq. (4) the average value of $\phi Df/f\phi$ is 2.1 x 10^{-12} and the maximum value of it is 6.7 x 10^{-12} which appears at perigee and apogee. Corresponding average and maximum range rate effects are 0.6 and 2 mm/sec, respectively, Again it is negligible for point positioning, and vanishes in relative positioning. Since the in-orbit stability of a satellite clock is better than 1 x 10^{-13}, this offset might be harmful for future high accuracy time and frequency transfer. To estimate and correct it, one must use the ephemeris message.

Table 3   Relativistic effect on clock

Secular drift :

- nominal drift   ~4.45 x 10^{-10} , calculated from nominal orbit
                       parameters, corrected prior to launch

- residual drift   <1 x 10^{-12} , due to an off nominal semi-major
                       axis of the orbit, is a constant time and
                       frequency offset, correctable

Periodic time offset :

- initial value   - k sinE (t_{oc}), k $\mp$ 46 ns for e = 0.02,
  (Keplerian)   constant time offset, correctable

- time variable   k sinE (t), can be corrected by using GPS
  (Keplerian)   Navigation Message (a_0, a_1, a_2 coefficients)

- Non-Keplerian   <1.2 ns (36 cm) for e=0.02 (t-t_{oc}) <604800 sec.

Periodic frequency   for e=0.02 $\phi Df/f\phi$ave=2.1x10^{-12}
offset   range rate:0.6mm/sec
                       $\phi Df/f\phi$max=6.7x10^{-12} range rate: 2 mm/sec

4. Scale problem in comparison with other techniques

When comparing GPS baseline solutions with those of other techniques one should be aware of the scale differences caused by the adoption of different relativistic models. For instance, LLR and VLBI data are usually treated in the solar system barycetric coordinate system, while SLR and GPS data are processed in the geocentric system. By definition the station coordinates and baselines in the barycetric system are different from those in the geocentric system. The difference may amount to $1.5*10^{-8}$. Some LLR and VLBI data analysing centers have already taken this into account, but other centers have not yet removed this definition discrepancy. In addition, sometimes the orbit of GPS satellite is determined by using the observations on a fiducial network which takes VLBI station coordinates as a priori, the scale error of these VLBI stations will affect the GPS baseline determination.

References

ASHBY, N., (1987): Relativistic Effects in the Global Positioning System, Presented at IAG Symp. GSA, 1987, Aug., Vancouver, Canada

BERTIGER, W. and S.M. LICHTEN (1987): ESO, Vol. 68, No. 44, p. 1238

JORGENSEN, P.S.(1986): Proceedings of the IEEE Position Location and Navigation Symposium, Las Vegas, Nevada, Nov. 4-7, pp. 177-183

RIZOS, C.K. and A. STOLZ (1985): Proc. First Int. Symp. on Precise Positioning with GPS, NOAA, Rockville, Md., May 1985, pp. 87-98

SPILKER, J.J. (1980): Global Positioning System, The Institute of Navigation, Washington, D.C., pp. 25-57

VAN DIERENDONCK, A.J., S.S. RUSSELL, E.R. KOPITZKE and M. BIRNBAUM (1980): ibid, pp. 55-73

ZHU, S.Y., E. GROTEN, R.S. PAN, H.J. YAN, Z.Y. CHENG, W.Y. ZHU, C. HUANG and M. YAO (1987): Motion of Satellite - the Choice of Reference Frames, Astrophys. Space Science Series, The Few Body Problem, D. Reidel

RELATIVISTIC MODELS OF PHASE AND DOPPLER OBSERVATIONS
OF ELECTROMAGNETIC SIGNALS

by

Elke Stöcker-Meier

Abstract

The paper deals with special definitions of phase and Doppler observations of electromagnetic signals and their modeling in four-dimensional space times of general relativity with an earth related metric. Particular attention is paid to phase measurements in GPS and, in addition, to possible contributions of clock transports. After a short explanation of the most essential fundamentals of physical theories and observations with standard clocks solutions of the direct and inverse geodetic problems in media are treated, which are directly or indirectly important for all observation equations concerning coordinate determinations. Then, based on the theory of electromagnetic wave propagation in a refractive medium the observation equations for phase measurements with standard clocks (oscillators) are derived. Finally, the observation equations of clock transports along terrestrial paths, directly resulting from the fundamental form of general relativity, are discussed. It can be shown that their application yields a valuable improvement in the accuracy of coordinate determinations.

1. Preliminary remarks

Producing of earth related observations and mapping them into physical models are the main topics of geodesy. Thereby, the objective functions consist of coordinates of points of the earth's surface and parameters for describing its extrinsic gravitation field. Now as before, classical <u>Newtonian mechanics</u> based on three-dimensional Euclidean space and absolute time dominates in geodetic modeling. Generally, we leave this physical concept when

- electromagnetic signals are used for observations,   (1-1)

- test bodies attain high relative velocities compared with   (1-2)
  the velocity of light   or/and

- strong gravitation fields exist.   (1-3)

Then we have to replace Newtonian mechanics by <u>general relativity</u>, which involves a consistent theory of mechanics and electrodynamics in four-dimensional (pseudo-) Riemannian space. Observations with so-called <u>standard clocks</u> play an important role in general relativity. They are the fundamental observables in the theory of relativity and can be used for measuring the travel time of electromagnetic signals as well as for phase and Doppler observations. In addition, <u>rigid measuring rods</u>, which are the fundamental observation instruments of <u>Euclidean geometry</u>, can still be used in relative small spacetimes.

At present, the presuppositions of (1-1) to (1-3) for leaving classical physics are especially fulfilled in VLBI and GPS brought about first of all by great accuracy in observations combined with (1-2). Formulations of <u>observation equations for phase measurements</u> based on general relativity is one of the two objectives of the following representations. The other concerns with problems of terrestrial <u>clock transport</u> as an essential procedure for the synchronization of standard clocks in general relativity.

2. Fundamentals of geometrical optics in general relativity

<u>Geometrical optics</u> is that part of electrodynamics dealing with the propagation of electromagnetic waves in the approximation for very short wave lenghts.

In general, electromagnetic waves are solutions of Maxwell's equations or the <u>wave equation of electrodynamics</u>. Let A be a component of an electric or a magnetic wave field, then

$$A = a\ e^{-i\phi}\ , \quad \phi = \phi_o + \int d\phi\ = \phi_o + \int k_\alpha\ dx^\alpha\ , \tag{2-1a}$$

$$a(x^\alpha) \qquad\qquad = amplitude, \qquad \phi(x^\alpha) = phase, \tag{2-1b}$$

$$x^\alpha = (ct\ ,\ x^i) = 4\text{-coordinates}, \quad c = velocity\ of\ light, \tag{2-1c}$$

$$k_\alpha = (\omega/c,\ k_i) = wave\ 4\text{-vector}, \quad \omega = angular\ frequency; \tag{2-1d}$$

$$\alpha,\ \beta,\ \gamma,\ \ldots\ \in \{0,\ 1,\ 2,\ 3\}, \quad i,\ j,\ k,\ \ldots \in \{1,\ 2,\ 3\} \tag{2-1e}$$

is true. The phase of the wave is an invariant, i.e., it is independent of the reference system caused by the fact that it results as the inner product of the wave 4-vector k_α and the coordinate differentials dx^α.

The wave 4-vector is normal to the wave front $\phi = $ constant which follows from

$$k_\alpha = k_\alpha(x^\alpha) = \phi_{,\alpha} \ . \tag{2-1f}$$

In case of geometrical optics amplitude and wave 4-vector are nearly constant. In a medium of refractive index n relativistic geometrical optics can be based on Fermat's principle:

$$\int_{ray} k_\alpha dx^\alpha = \text{extremal} \tag{2-2a}$$

under the side-condition:

$$k_\alpha k^\alpha = K_o^2(1-n^2) \ , \tag{2-2b}$$

$$K_o = k_\alpha u^\alpha/c \quad = \text{frequency in the rest frame of the medium} \ , \tag{2-2c}$$

$$u^\alpha = u^\alpha(x^\beta) \quad = \text{4-velocity of the medium} \ , \tag{2-2d}$$

$$n = n(x^\alpha, K_o) = \text{refractive index of the medium} \ . \tag{2-2e}$$

The characteristic equations of geometrical optics describe the propagation of an electromagnetic wave. They follow from the Euler equations of the variational principle (2-2a); see E. STÖCKER-MEIER:

$$dx^\alpha/dq = (k^\alpha - (u^\alpha/c)K_o(1-n^2-nn'K_o))\lambda \ , \tag{2-3a}$$

$$Dk_\alpha/dq =: dk_\alpha/dq - \Gamma^\beta_{\alpha\gamma}k_\beta(dx^\gamma/dq) \tag{2-3b}$$

$$= (-K_o^2 nn_{,\alpha} + (u^\beta_{;\alpha}k_\beta/c)K_o(1-n^2-nn'K_o))\lambda \ ,$$

$$\lambda =: C/(nK_o((n+ n'K_o)^2- 1)^{1/2}) \ , \tag{2-3c}$$

$$n' = \partial n/\partial K_o \ . \tag{2-3d}$$

If the refractive index is a function of frequency (2-3d), we call the medium dispersive. q is an affine coordinate, defined by

$$g_{\alpha\beta}(dx^\alpha/dq)(dx^\beta/dq) = C^2 = \text{constant} \ . \tag{2-3e}$$

The two ordinary differential equations (2-3a,b) determine not only the coordinates but also the wave 4-vector at each point of the ray. One equation depends on the other so that it is necessary to solve them together. It is also possible to eliminate the wave 4-vector by differentiation of (2-3a) with respect to q and insertion of (2-3b). Accordingly the characteristic equations (2-3a,b) reduce to an ordinary differential equation of second order:

$$D^2 x^\alpha/dq^2 =: d^2 x^\alpha/dq^2 + \Gamma^\alpha_{\beta\gamma}(dx^\beta/dq)(dx^\gamma/dq) = (dt/dq)^2 R^\alpha \ , \tag{2-4a}$$

$$R^\alpha = R^\alpha(n, n', n_{,\alpha}, u^\alpha, u^{\alpha;\beta}, g_{\alpha\beta}, dx^\alpha/dq) \ . \tag{2-4b}$$

R^α is called disturbing function. In vacuum we obtain:

$$R^\alpha = 0 \quad \text{for} \quad n = 1 = \text{constant} \ , \tag{2-4c}$$

so that the world lines of the signals in vacuum are <u>geodesic lines</u> in the four-dimensional Riemannian spaces. Considering the equations (2-4a) we have to eliminate the coordinate q and to write the equations as functions of coordinate time t:

$$d^2x^i/dt^2 = G^i + \bar{R}^i ,$$

(2-5a)

$$G^i = ((1/c)\Gamma^o_{\beta\gamma}(dx^i/dt) - \Gamma^i_{\beta\gamma})(dx^\beta/dt)(dx^\gamma/dt) ,$$

(2-5b)

$$\bar{R}^i = R^i - (1/c)R^o(dx^i/dt) .$$

(2-5c)

The 3-acceleration of the signal is composed of a gravitational part G^i and a disturbing part $\bar{R}^i$. We formulate initial-value and boundary-value problems to solve the differential equations by numerical integrations or series expansions. In case of the propagation of electromagnetic signals through the atmosphere the initial-value and boundary-value problems are called <u>direct or inverse geodetic problems</u>. They are directly or indirectly important for all observation equations concerning coordinate determinations. Without the effects of refraction and gravitation electromagnetic signals propagate along straight lines. With respect to the straight lines in Euclidean spaces we define the so-called <u>reduction of travel time</u> δt:

$$\delta t = \Delta\bar{t} - \Delta t ,$$

(2-6a)

$$\Delta t = \text{observed travel time} ,$$

(2-6b)

$$\Delta\bar{t} = \text{Euclidean travel time}$$

(2-6c)

$$= (\Delta_{ij}((x^i)_b - (x^i)_a)((x^j)_b - (x^j)_a))^{1/2}/c .$$

For a distance of 20000 km we obtain a reduction of 1-2 cm as a result of the earth gravitational field, which ought to be taken into account.

3. <u>Observations with standard clocks</u>

Every <u>arbitrary clock</u> $\tilde{U}_n$ is nothing else but an oscillator, that is a physical system, which changes its phase periodically:

$$\overset{n}{\phi}(\overset{n}{\tau}) = \overset{n}{\phi}(\overset{n}{\tau}_o) + \overset{n}{\omega}_o(\overset{n}{\tau} - \overset{n}{\tau}_o) = \overset{n}{\phi}(\overset{n}{\tau}_o) + 2\pi\overset{n}{N} + \overset{n}{\varphi}(\overset{n}{\tau}) ,$$

(3-1a)

$$\overset{n}{\tau} = \text{oscillator time} ,$$

(3-1b)

$$\overset{n}{\omega}_o = d\overset{n}{\phi}/d\overset{n}{\tau} = \text{constant angular frequency} ,$$

(3-1c)

$$\overset{n}{N} = \text{cycle count} , \quad \overset{n}{\varphi} = \text{fractional phase part} < 2\pi .$$

(3-1d)

As a rule the so-called <u>standard clocks U_m</u> are the observing instruments to measure travel times, phases and frequencies of electromagnetic signals. Standard clocks are able to measure proper time differentials, which correspond to the fundamental form of general relativity with the exception of constant scale factors $\overset{m}{M}$:

$$\overset{m}{M}d\overset{m}{\tau} =: d\overset{m}{\tau} = (g_{oo} + (2/c)g_{oi}v^i + (1/c^2)g_{ij}v^iv^j)^{1/2} dt , \qquad (3\text{-}2a)$$

$$\overset{m}{\tau} = \text{proper time} , \qquad \overset{m}{\omega_o} = \text{proper frequency} . \qquad (3\text{-}2b)$$

Recently the best approximations of standard clocks are atomic clocks. In general relativity it is only possible to compare two clocks, which are moving together on one and the same world line. In this case the physical

hypothesis of consistency: $\qquad\qquad\qquad\qquad\qquad\qquad (3\text{-}3)$

two standard clocks of the same type have same scale factors

$$\overset{1}{M} = \overset{2}{M} = \dots \overset{m}{M} \qquad\qquad\qquad\qquad\qquad (3\text{-}3a)$$

and for two standard clocks of different types the ratios of the scale factors are constant:

$$\overset{1}{M}/\overset{2}{M} = \text{constant} , \qquad \overset{2}{M}/\overset{3}{M} = \text{constant} , \dots \qquad (3\text{-}3b)$$

is assumed; see J.L. SYNGE. Arbitrary clocks $\tilde{U}_n$ have scale factors, which depend on time:

$$\overset{n}{M} = \overset{n}{M}(\overset{n}{\tau}) . \qquad\qquad\qquad\qquad\qquad\qquad (3\text{-}4)$$

The observables of standard clocks U_m are:

cycle counts $\qquad\qquad\qquad \overset{m}{N} , \qquad\qquad\qquad\qquad (3\text{-}5a)$

fractional phase parts $\quad \overset{m}{\varphi} < 2\pi \quad$ and $\qquad\qquad (3\text{-}5b)$

proper time intervals $\quad \overset{m}{\Delta\tau} . \qquad\qquad\qquad\qquad (3\text{-}5c)$

Compared to the phases of an arbitrary clock $\tilde{U}_n$ we obtain

phase differences $\qquad \overset{m}{\Delta\phi_{mn}}(\tau) = \overset{m\,m}{\phi(\tau)} - \overset{n\,m}{\phi(\tau)} . \qquad (3\text{-}5d)$

4. Phase and Doppler observations

Phase differences are used in NAVSTAR–GPS to measure electromagnetic waves by means of a local standard oscillator U_m (3-2a,b) in the terrestrial receiving station P_m. The transmitter is a standard clock U_n in the GPS-satellite P_n. It generates the electromagnetic wave, which travelles to the receiver and controls there an arbitrary oscillator $\tilde{U}_n$. The output of the GPS-receiver is the difference between the phases of U_m and $\tilde{U}_n$:

$$\Delta\phi_{mn}(t_m) = (\overset{m}{\phi}(t_m))_m - (\overset{n}{\phi}(t_m))_m + 2\pi N_{mn} . \qquad (4\text{-}1a)$$

The

integer cycle count N_{mn} (4-1b)

represents the number of full cycles in the observed phase. With respect to the phase variation in time and in space we yield the observation equations for phase measurements:

$$\Delta\phi_{mn}(t_m) = (\phi(t_m))_m^m - (\phi(t_m))_n^n + 2\pi N_{mn}$$
$$+ \omega_o(\Delta t_{nm} + G_{mn}(t_m) + R_{mn}(t_m)) \ ,$$ (4-2a)

$$G_{mn}(t_m) = \int_{t_n}^{t_m} d(\tau(t))^n - dt \qquad = \text{gravitational reduction} \ ,$$ (4-2b)

$$R_{mn}(t_m) = (1/\omega_o) \int_{(x^\alpha)_m}^{(x^\alpha)_n} k_\alpha dx^\alpha = \text{refractive} \qquad \text{reduction} \ ,$$ (4-2c)

$$\Delta t_{nm} = t_m - t_n = \text{coordinate travel time} \ .$$ (4-2d)

The gravitational reduction (4-2b) follows from the difference between proper time of the satellite-oscillator and coordinate time used in the problem. To see what influence gravitation has, the reduction (4-2b) is computed with an earth related metric. It results 1-2 cm due to the potential difference between satellite and receiver. With the observation equation (4-2a) it is possible to determine the time difference (4-2d). To calculate the

spatial coordinate differences $(x^i)_n - (x^i)_m$ (4-2e)

we have to solve a direct geodetic problem and to introduce (4-2d) as initial-values.

Because of inaccurate satellite orbits and refraction models we have to use relative coordinates. The observations which belong to relative positioning are the station differences $D\Delta\phi_{ab}$. They are defined as the differences between the phase measurements made by the two receivers U_a and U_b in case of simultaneously emitted signals:

$$D\Delta\phi_{ab} =: \Delta\phi_{an}(t_a) - \Delta\phi_{bn}(t_b) \ .$$ (4-3a)

Inserting (4-2a) into (4-3a) the observation equation can be expressed as:

$$D\Delta\phi_{ab} = (\phi(t_b))_a^a - (\phi(t_b))_b^b + 2\pi N_{ab} + \omega_o((t_a - t_b) + \Delta G_{ab} + \Delta R_{ab})$$ (4-3b)

$$\Delta G_{ab} = \text{gravitational anomaly} \ ,$$ (4-3c)
$$\Delta R_{ab} = \text{refractive} \qquad \text{anomaly} \ .$$ (4-3d)

ΔG_{ab} and ΔR_{ab} follow from the gravitational and refractive effects, which are different in both stations. The observation equation (4-3b) yields

$$N_{ab} =: \overset{a}{N_{an}} - \overset{b}{N_{bn}} \quad , \quad t_a - t_b \quad , \tag{4-4a}$$

$$(\phi(t_b))_a - (\phi(t_b))_b \quad , \tag{4-4b}$$

whereby the coordinate time difference $t_a - t_b$ can be transformed into the

baseline vector $\quad (x^i)_b - (x^i)_a \quad , \tag{4-4c}$

by solving a direct geodetic problem. (4-4b) depends on the

clock offset $\quad (\overset{a}{\phi(t_o)})_a - (\overset{b}{\phi(t_o)})_b \tag{4-4d}$

and the

clock scale rate $\quad \overset{a}{M}/\overset{b}{M} \tag{4-4e}$

between both receivers. t_o is a reference time. It is shown in chapter 5, that it is also possible to determine the phase difference (4-4b) by means of a clock transport. A transport of clocks is a good completion to simultaneous phase measurements and it should be carried out additionally, because this method reduces the unknowns.

The time derivatives of phase measurements (4-1b) are defined as Doppler displacements:

$$\Delta\overset{m}{\omega}_{mn}(\tau) = \overset{m}{\omega}_o - \overset{n\,m}{\omega}(\tau) = (d\Delta\overset{m}{\phi}_{mn}/d\tau)_{\overset{m}{\tau}} \quad . \tag{4-5a}$$

Differentiating equation (4-2a) with respect to $\overset{m}{\tau}$ gives in the local observing system of the receiver:

$$\Delta\overset{m}{\omega}_{mn}(\tau) \approx \overset{m}{\omega}_o - \overset{n}{\omega}_o(\overset{m\,n}{M/M})(1+(v_n/c)\cos\varphi \tag{4-5b}$$

$$+(1/2)(v_n/c)^2(\cos^2\varphi - \sin^2\varphi) - W_n/c^2 + R\omega) \quad ,$$

v_n = velocity of the satellite , $\tag{4-5c}$

φ = angle between wave propagation and direction of $\tag{4-5d}$
satellite motion ,

W_n = potential difference between satellite and receiver $\tag{4-5e}$

$R\omega$ = refractive reduction . $\tag{4-5f}$

In addition to the longitudinal and transversal Doppler effects the gravitational potential influences the Doppler observations with the order of 30 cm in case of measurements to a TRANSIT-satellite. The refractive reduction depends on the direction of the gradient of the refractive index relative to the velocity of the satellite.

5. Clock transports

The transport of clocks is a terrestrial method measuring phase differences to synchronize earth related clocks within the scope of general relativity. The basis of a clock transport reads as follows:

One standard clock is positioned in each of two terrestrial observation stations:

$$U_1 \quad \text{in} \quad P_a \;, \quad \text{measuring proper time} \; \overset{1}{\tau} \;, \qquad (5\text{-}1a)$$

$$U_2 \quad \text{in} \quad P_b \;, \quad \text{measuring proper time} \; \overset{2}{\tau} \;.$$

A third transportable standard clock

$$U_3 \quad \text{in} \quad P_c \;, \quad \text{measuring proper time} \; \overset{3}{\tau} \;, \qquad (5\text{-}1b)$$

is located in P_c moving along a path C_{ab} from P_a to P_b. The observations are the proper times of U_3:

$$\Delta\overset{3}{\tau}_{ab} = \overset{3}{\tau}_b - \overset{3}{\tau}_a \;, \quad \overset{3}{\tau}_p = \text{arbitrary times for } P_c \equiv P_p, \; p \in \{a,b\}, \qquad (5\text{-}2a)$$

and the phase differences between U_3 and U_m in P_a and P_b:

$$\Delta\phi_{m3} = \overset{m}{\phi}(\overset{m}{\tau}) - \overset{3}{\phi}(\overset{m}{\tau}) \;, \quad m \in \{1,2\} \;. \qquad (5\text{-}2b)$$

Integrating the fundamental form (3-2a) yields the observation equation for the transport of the clock U_3 (5-2a):

$$M\Delta\overset{3}{\tau}_{ab} = \int_{t_a}^{t_b} \left((g_{oo} + (2/c)g_{oi}v^i + (1/c^2)g_{ij}v^i v^j)^{1/2}dt\right)_{C_{ab}} \;. \qquad (5\text{-}3a)$$

If the gravitational field and the velocity v^i of U_3 are known sufficiently accurate along the path C_{ab} it is possible to determine the coordinate time difference:

$$\Delta t_{ab} = t_b - t_a \;. \qquad (5\text{-}3b)$$

In the earth related system S_E the difference between (5-2a) and (5-3b) can be splitted into three parts. K_1 depends on height h above a reference station P_B, K_2 is a function of velocity v and K_3 results from Coriolis's acceleration:

$$\Delta t_{ab} - \Delta\overset{3}{\tau}_{ab} =: K_1 + K_2 + K_3 \;, \qquad (5\text{-}4a)$$

$$K_1 = \int_{\tau_a}^{\overset{3}{\tau}_b} (1/c^2) \, (((gh)_B - gh) \, d\overset{3}{\tau})_{C_{ab}} \;, \qquad (5\text{-}4b)$$

54

$$K_2 = \int_{\tau_a}^{\tau_b^3} (1/2) \ ((v/c)^2 \ d\tau)_{C_{ab}}^3 \ , \tag{5-4c}$$

$$K_3 = \int_{\tau_a}^{\tau_b^3} -(1/c^2) \ (\omega \ \epsilon_{3jk} v^j x^k \ d\tau)_{C_{ab}}^3 \ , \tag{5-4d}$$

g = absolute value of gravity acceleration of the earth , (5-4e)

ω = absolute value of angular velocity   of the earth . (5-4f)

The values (5-4b,c,d) were calculated for a transport of clocks through the area of the Eifel with $v \approx 50$ km/h. The results are shown in table 1.

Comparing two standard clocks the <u>observation equations of phase measurements</u> (5-2b) reduce to:

$$\Delta\phi_{m3}^m(\tau) = \overset{m}{\phi}(\overset{m}{\tau_0}) - \overset{3}{\phi}(\overset{m}{\tau_0}) + (\overset{m}{\omega_0} - \overset{3}{\omega_0}\overset{3}{M}/\overset{m}{M})(\tau - \overset{m}{\tau_0}) \ . \tag{5-5a}$$

(5-5a) is called the clock equation, which allows to determine the clock offsets and the scale rates:

$$\overset{1}{\phi}(\overset{1}{\tau_0}) - \overset{3}{\phi}(\overset{1}{\tau_0}) \ , \quad \overset{1}{M}/\overset{3}{M} \ ,$$

$$\overset{2}{\phi}(\overset{2}{\tau_0}) - \overset{3}{\phi}(\overset{2}{\tau_0}) \ , \quad \overset{2}{M}/\overset{3}{M} \ . \tag{5-5b}$$

The objective of the <u>clock synchronisation of general relativity</u> is to compare coordinate times in different places. With the assumption

$$(v^i)_a = (v^i)_b = 0 \ , \quad g_{\alpha\beta,o} = 0 \tag{5-6a}$$

(3-1a) and (3-2a) gives in P_a and P_b:

$$(\overset{m}{M}/\overset{m}{\omega_0})(\overset{m}{\phi}(\tau) - \overset{m}{\phi}(\overset{m}{\tau_{oa}})) = (g_{oo})_a^{1/2}(t_a - t_{oa}) \ , \quad m \in \{1,3\} \ , \tag{5-6b}$$

$$(\overset{n}{M}/\overset{n}{\omega_0})(\overset{n}{\phi}(\tau) - \overset{n}{\phi}(\overset{n}{\tau_{ob}})) = (g_{oo})_b^{1/2}(t_b - t_{ob}) \ , \quad n \in \{2,3\} \ .$$

t_{op} are the coordinate times belonging to τ_{op}. If t_{oa} is a known quantity, we can determine

$$t_{ob} = t_{oa} + \Delta t_{ab} \tag{5-6c}$$

by means of (5-3b). t_a and t_b are given functions of the observed phases in P_a and P_b on the basis of (5-6b) provided that we have

$$\overset{m}{\phi}(\overset{m}{\tau_{oa}}) - \overset{n}{\phi}(\overset{n}{\tau_{ob}}) \ , \quad \overset{m}{M}/\overset{n}{M} \tag{5-6d}$$

additionally from the comparision of the standard clocks (5-5b). It is also possible to describe with (5-6b) the phase differences (4-4b) as functions of coordinate time.

path	compo-nent	reductions for the path: $C_1 \rightarrow C_2$ ns	$C_2 \rightarrow C_1$ ns
C_1	K1	-0.37218	-0.37218
	K2	0.00968	0.00968
	K3	-0.09894	0.09894
C_2	K1	-0.25502	-0.25502
	K2	0.01443	0.01443
	K3	0.08909	-0.08909
total reductions		-0.61294	-0.59322

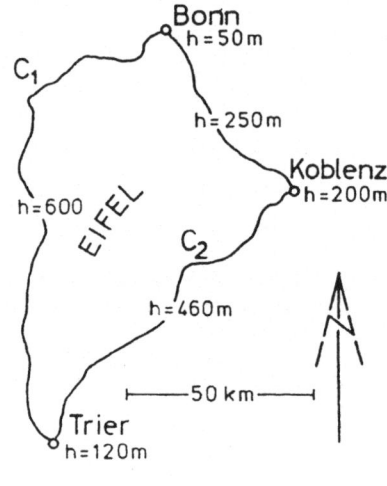

TABLE 1: clock transports: $C_1 \rightarrow C_2$ = Bonn-Trier-Koblenz-Bonn,
$C_2 \rightarrow C_1$ = Bonn-Koblenz-Trier-Bonn

6. References

BORN, M./WOLF, E. (1986): Principles of optics, Pergamon Press, Oxford/ New York/Toronto/Sydney/Frankfurt

FRIEDRICH, J. (1988): Relativistische Grundlagen geodätischer Modelle, Dissertation Universität Bonn, Mitt. aus den geod. Instituten der Universität Bonn Nr.76

HEITZ, S. (1985): Grundlagen relativistischer Modelle, Vorlesungsmanuskript Universität Bonn

JACKSON, J. D. (1975): Classical Electrodynamics, John Wiley & Sons, Inc. , New York/London/Sydney/Toronto

LANDAU, L. D./LIFSCHITZ, E. M. (1975): The classical theoriy of fields, Vol.2, Pergamon Press, Oxford/New York/Toronto/Sydney/Braunschweig

MØLLER, C. (1982): The theory of relativity, Oxford University Press, Delhi/Bombay/Calcutta/Madras

STÖCKER-MEIER, E. (1988): Grundlagen der Elektrodynamik in relativistischen geodätischen Modellen, Veröff. der Deutschen Geodätischen Kommission, Reihe C, München

SYNGE, J. L. (1976): Relativity: the general theory, North Holland Publishing Company, Amsterdam/New York/Oxford

WELLS, D. (1986): Guide to GPS-positioning, University of New Brunswick Graphic Services, Fredericton/New Brunswick Canada

Second session: Application of GPS

Chairman: Prof. Pâquet, Brussels

THE USE OF GPS AT IGN : GEODESY, GEOPHYSICS, ENGINEERING

by

Claude Boucher, Pascal Willis

Abstract

The Institut Géographique National has purchased since 1985 four GPS receivers (TR5S from SERCEL, 5 channels, single frequency type). For research and production purpose, a specific software, called GDVS, was then developed. This software is now operational and has been used with success to process several GPS campaigns.

Two major campaigns were performed in 1986 : one between France and England and the second one between France and Italy. One of the main topic of these campaigns was to connect tide gauges in the Channel area or in the Mediterranean Sea area, to a global reference frame.

GPS was also used to provide very shortly local geodetic network of a few tens of kilometers for geodetic or geophysical purposes. Some examples of such campaigns are the following : Djibouti, Pyrénées, Provence.

For the future, the IGN intends to continue these uses of GPS and also to continue the development of GPS for photogrammetry (airborne GPS) and kinematic GPS.

1. Introduction

Since 1985, the French Institut Géographique National has been involved in GPS and its possible applications. GPS receivers have been purchased and a specific research software called GDVS (*Géodésie par mesures de Distances et Variations de distances sur Satellites*) has been developed. Several applications of the GPS technology have been investigated for science or production purposes. For the Institut Géographique National, being both a geodetic Institute and a mapping Agency, these applications are very numerous : geodesy (small size specific geodetic network, national geodetic network), geophysics (active zones surveys, volcanoes surveys, tide gauges connections) or engineering (photogrammetry with airborne GPS receivers).

We summarize here the latest examples of use of GPS at IGN, giving at the same time a large range of possible applications of this highly accurate tool for geodesists. It should be noticed that the accuracy of GPS strongly depends on the way it is used : absolute or relative positioning, using the pseudo-ranges or the carrier beat phases [WELLS (1986)], type of receiver, real time or post-processing solution, software capabilities. Table 1 gives a summary of the possible use of GPS and the typical achievable accuracy.

Type	Accuracy
Navigation	20 m
Differential navigation	2 m
Trajectography	0,5 m
Geodesy (L1 receiver)	3 ppm
Geodesy (L1/L2 receiver)	1 ppm
Kinematic	0,01m
High accuracy	0,02 ppm

Classes of applications of GPS

Table 1

The Institut Géographique National has presently 4 TANS (TRIMBLE) for navigation, 4 TR5S (SERCEL) for geodesy (single frequency receivers) and plan to purchase in 1988 3 dual frequency receivers (to be determined) : 2 for surveying and one for satellite tracking.

2. Navigation

The initial purpose of the development of GPS for the U.S. military
agencies was to provide at any time and anywhere an absolute position
(in a precise global reference frame). This application only requires
low cost GPS receivers that can perform pseudo-range measurements on
the L1 frequency. The Institut Géographique National possesses four
receivers of this type : TANS from TRIMBLE. A simple software is
needed to obtain a real-time position from the pseudo-ranges measurements. The accuracy of this basic use of GPS is about 20 metres.
Figure 1 shows an example of this type of positioning using a TR5S
(SERCEL) equipment at Hasparren (Netherlands). Every point represents
an independent position estimated every 0,6 second. This figure shows
a good consistency of the result (internal precision at a few metres
level). The dots seem to scatter in two different populations. This
difference is due to the change of ephemerides.

In this type of positioning, the quality of the broadcast orbit is
essential and may be degraded in a near future (block II satellites).
This real time navigation can be used with profit for the fast determination of a large number of points, for instance to locate a
gravity survey. The I.G.N. is achieving such a large project in Africa
(Mali : 640 points, Guinea Bissau : 30 points, Guinea Konakry : 737
points, Togo : 74 points, Benin : 150 points, Ivory Coast : 167 points,
Central African Republic : 710 points, ...). GPS navigation can also be
used for very small scale stereopreparation, for example for the SPOT
images.

3. Differential navigation

In the navigation accuracy (20 metres), a large part comes from the
precision of the orbit. This effect is in fact a regional effect and
will cause a more or less common bias on the estimated position of the
receiver in the same area.

During the campaign presented in figure 1, another TR5S receiver was
present at Saint-Mandé (France). The results obtained were very similar,
showing the same skip due to the change of ephemerides. Figure 2
represents the result of the differential positioning (real-time
difference of estimated three-dimensional coordinates every 0,6s). In
this differential positioning, there is no more break if the achievable
accuracy is at the level of 2 metres. A very important aspect of the
method is that the receiver observes simultaneously the same configuration of satellite. Only L1 pseudo-range receivers and real-time
single point processing software are needed [NARD et al. (1987)]. The
final solution is obtained in a post processing mode by differentiating
the real time navigation solution.

When another system is available to transmit information from one
receiver to another, this relative positioning can also be obtained in
real time. Two solutions are possible : differentiating the estimated
position (as described before) or differentiating the raw pseudoranges.

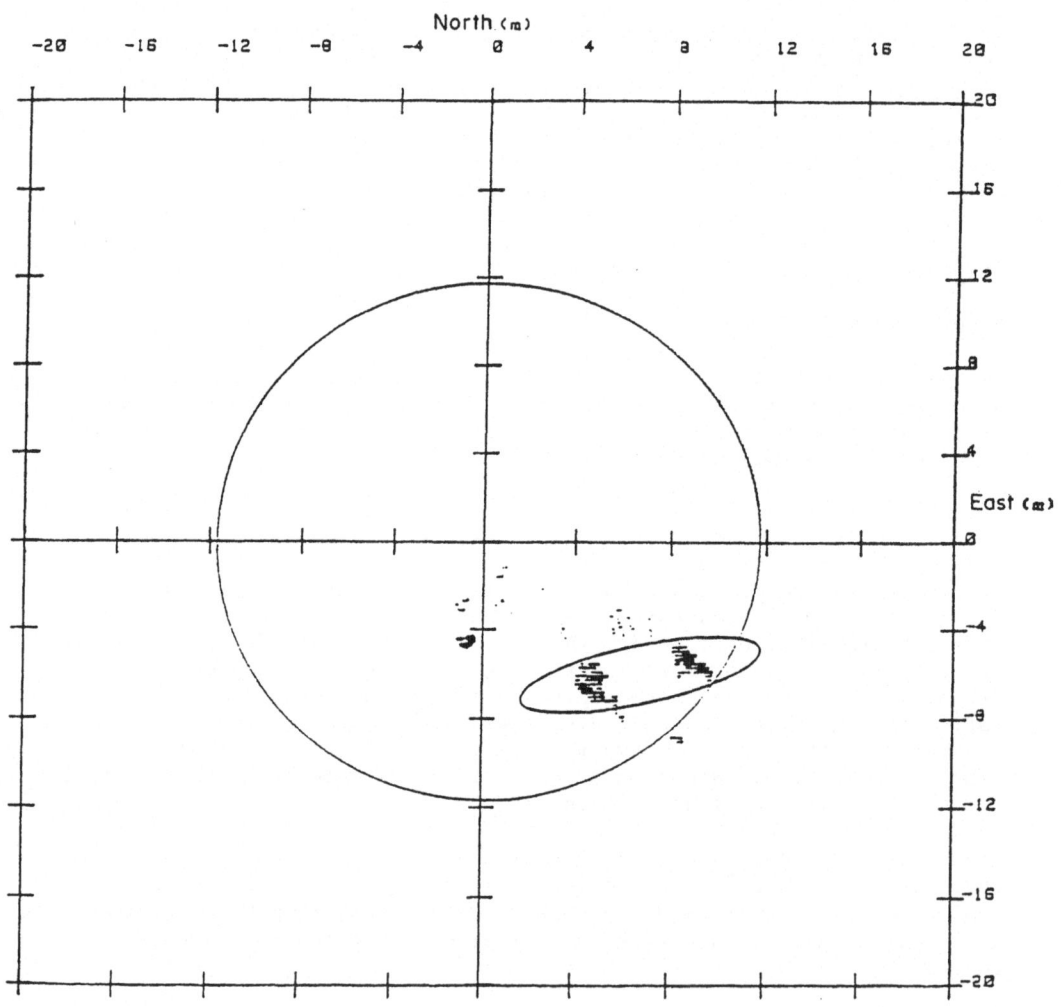

Figure 1

USE OF GPS FOR NAVIGATION

—∎—

JUNE 1986 SERCEL GPS CAMPAIGN
SITE "HASPARREN"

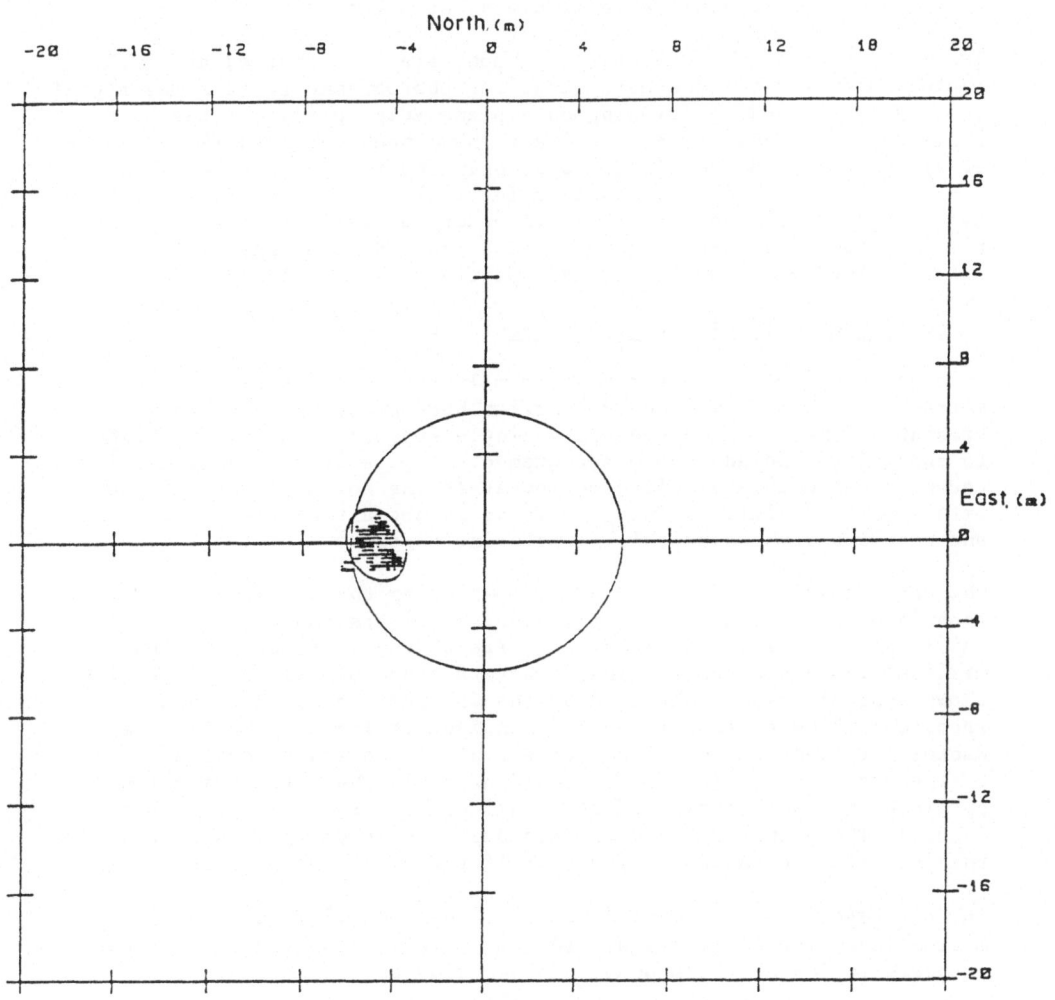

Figure 2
USE OF GPS FOR DIFFERENTIAL NAVIGATION
—
JUNE 1986 SERCEL GPS CAMPAIGN
RELATIVE POSITIONING SAINT MANDE-HASPARREN

4. Trajectography

At this point, the limit of accuracy is due to the noise of the raw pseudo-range measurement. When Doppler measurement exists (carrier beat phase or Doppler Integrated measurement), it is possible to use these measurements to smooth the pseudo-ranges and then decrease the noise level of this observable that we could call Doppler Integrated Pseudo-Ranges.

For this application, a L1 code and phase multichannel receiver is needed, and the data analysis is performed in an off-line mode. Figure 3 shows an example of this type of positioning applied to an airborne receiver. The data were recorded on an aircraft (Caravelle) and were processed, using a specific software developed by SERCEL. This figure shows the difference between the estimated position (every 0,6 second) and the STRADA estimation. Several tests on roads and on aircrafts have proved a 0,5 metre achievable accuracy, or better.

Two major uses of the GPS trajectography are investigated at I.G.N. : photogrammetry and roads data base. For photogrammetry, the major goal is to use this GPS positioning on a plane able to recover the photogrammetric perspective. Several tests have been realized [R. BROSSIER et al. (1986)] : Amiens (1986) with only one receiver, Lunel-Vichy (1987) with two receivers. Another interest for our Institute is to be able to precisely position a car in order to realize, in a very short time, a road data base. Several production GPS campaigns will take place in 1988 (metropolitan France, Martinique, Guadeloupe, ...).

5. Geodesy with single frequency receivers

In order to improve the accuracy from 1 or several metres to centimetres or millimetres, the GPS carrier beat phase must be used. Presently, the I.G.N. possesses 4 single frequency TR5S from SERCEL (5 channels, code and phase measurements). To solve ambiguities, it is necessary to obtain simultaneous observations for at least one hour with 4 or 5 satellites. The processing is not a real-time processing and requires a phase capability off-line software.

The data analysis software used is GDVS using the double difference technique. When using single frequency receivers and broadcast orbits, the relative positioning accuracies are at the level of 3ppm (part per million) for distances ranging to several tens of kilometres. Table 2 shows typical results obtained by the GDVS software. These data were recorded during a geodetic evaluation test realized by SERCEL near Nantes (Brittany). Every line corresponds to an independent estimation of the baseline (difference of coordinates in the three dimensions) obtained from one session of observations (5 GPS satellites during 1 hour). The mean value and the standard deviation were estimated from this set of independent estimated differences of coordinates.

In this case, the agreement (internal consistency) is at a 4 millimetres level (at 1σ) for a distance of 2,7km. This type of application of geodetic GPS has been used in production since 1986 and several campaigns already mentioned were carried out. Let us give some examples of such realizations.

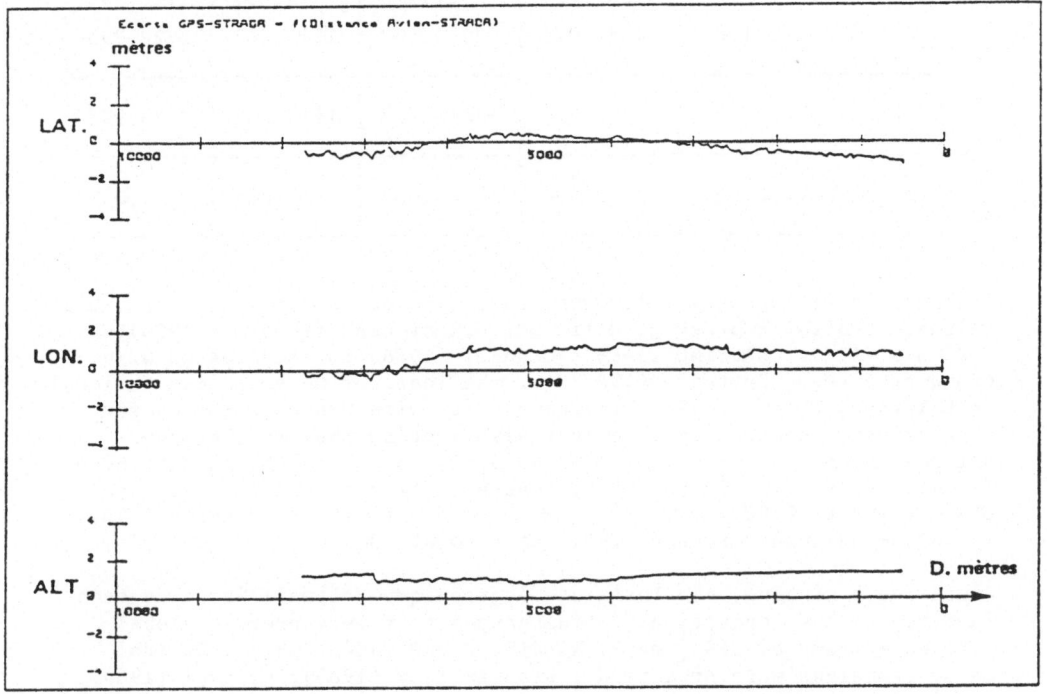

Figure 3

GPS profile of a landing aircraft

Comparison in meters with STRADA reference

SERCEL Results

GDVS results for the baseline SERCM001-AGFA (2.7km)

SERCEL test campaign (January-February 1988)

Session	Day	DX (m)	DY (m)	DZ (m)	D (m)
1	28-01-1988	-1385.034	-2011.471	1217.832	2729.002
2	28-01-1988	-1385.029	-2011.466	1217.839	2728.999
3	29-01-1988	-1385.027	-2011.471	1217.832	2728.998
4	29-01-1988	-1385.022	-2011.460	1217.846	2728.993
5	30-01-1988	-1385.026	-2011.470	1217.837	2728.999
6	30-01-1988	-1385.021	-2011.464	1217.847	2728.997
7	31-01-1988	-1385.024	-2011.470	1217.837	2728.998
8	31-01-1988	-1385.023	-2011.463	1217.843	2728.995
Mean value		-1385.026	-2011.467	1217.839	2728.998
Standard deviation		0.004	0.004	0.005	0.003

For small scale geodetic networks, this type of use of GPS is specially valuable : high accuracy geodetic network at CERN (1985 and 1986), TURTMANN (1985), CASSINO (1986), Channel (1986) in cooperation with Ordnance Survey, Zottegem (1987) for the Institut Géographique National de Belgique, Channel (1987) in cooperation with Ordnance Survey for Trans-Manche Construction. We must also mention that in 1988 the national geodetic network will be entirely done with GPS in Martinique Island. For geodynamic purposes, several networks were surveyed by GPS : Djibouti (1987) under the coordination of the IPGP, Pyrénées (1987) for GRGS-Toulouse, Provence 1987, as a joint IGN-IPGP project.

Since the beginning, the Institut Géographique National has also been involved in the connection of tide gauges to a very precise global reference frame by GPS [see P. WILLIS, C. BOUCHER (1987)]. Several major campaigns were organized : France-Italy (1986), Channel (1986). Figure 4 shows the geographic location of the tide gauges and of the SLR and VLBI sites in the area of the Channel. This activity will continue in the future and there are plans for such campaigns in Spain and Portugal in 1988 and in the Pacific zone in 1989.

6. Geodesy with dual frequency receivers

In order to improve the ionospheric correction, it is necessary for large distances (greater than several tens of km) to use dual-frequency equipment. Presently, the I.G.N. has no receivers of this type. During 1988, we plan to buy 3 of these receivers (two for surveying and one

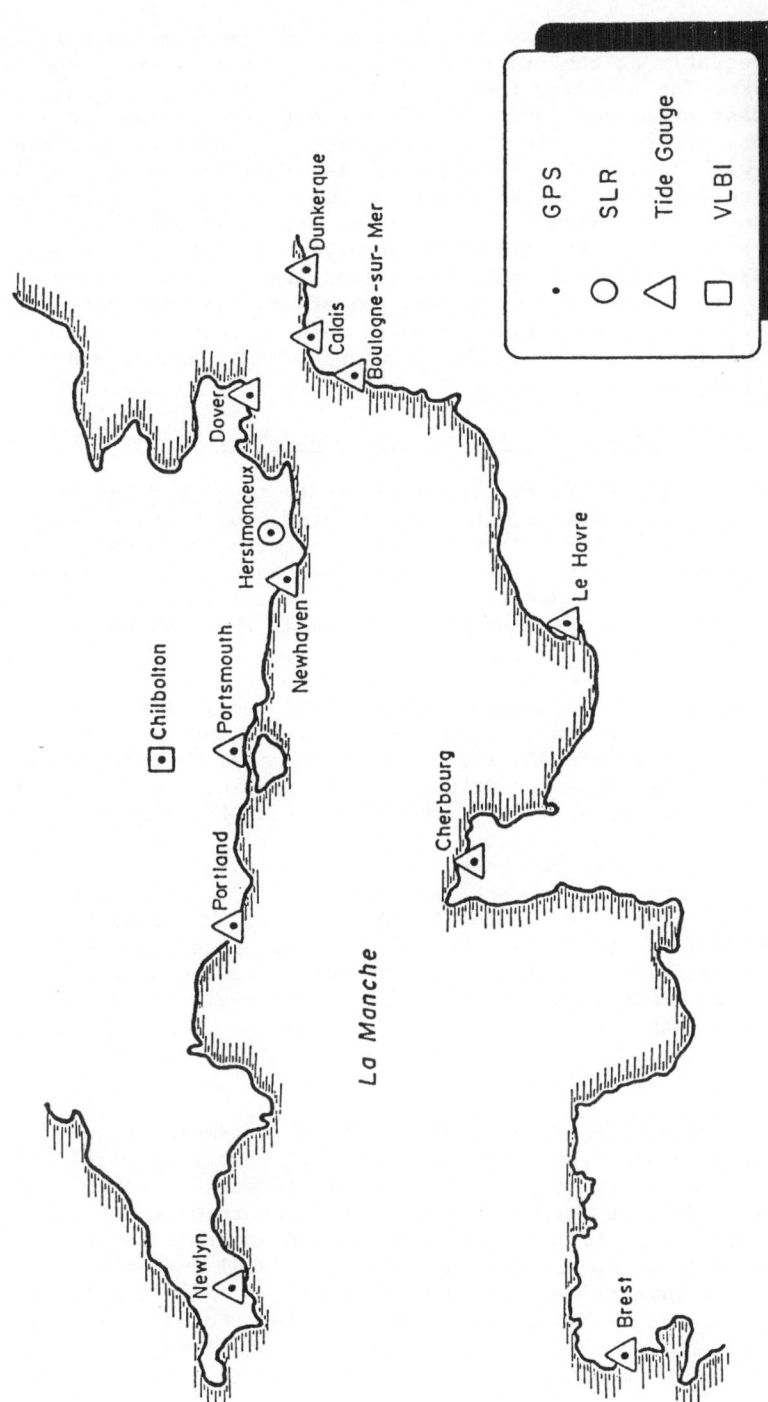

Legend:
- GPS — •
- SLR — ○
- Tide Gauge — △
- VLBI — □

Stations: Dunkerque, Calais, Boulogne-sur-Mer, Dover, Herstmonceux, Chilbolton, Portsmouth, Newhaven, Portland, Newlyn, Cherbourg, Le Havre, Brest

La Manche

Figure 4 - **IGN-OS Joint Campaign August 1986**

for satellite tracking. The expected accuracy of the relative positioning is presently at the order of 1 ppm for distances up to one or two hundred km. The limiting factor of this precision is the quality of the broadcast orbit when using a software package. It must be noted that it is possible to improve the single frequency results when they are combined with dual frequency measurements. In 1987, the I.G.N. has realized, in collaboration with Ordnance Survey, a specific geodetic network for the future Channel tunnel, using both dual frequency (4 TI 4100 from TRIMBLE) and single frequency (4 TRIMBLE 4000 S and 4 TR5S from SERCEL).Figure 5 shows the colocation between single and dual frequency instruments. The combination of all the observations, and also the previous terrestrial geodetic observations, allowed us to obtain an homogeneous network RTM 87 (Réseau Trans Manche 1987) of the quality of 1 ppm.

7. Kinematic positioning using the carrier beat phase

For the surveys described below, it is necessary to record one hour (at least) of GPS observations, in order to distinguish (in a least squares sense) the signature of the ambiguities with the signature of the position parameters. When these ambiguities are known, in a way or another, it is possible to use the phase observation in a non-ambiguous mode, as it is done with pseudo-range, as long as there is no cycle slip in the data.

Several techniques are possible to estimate the ambiguities : calibrate the ambiguities on a well known baseline, use the switching antenna technique [B.W. REMONDI (1986)], for example. The estimated accuracy is at the level of 2 centimetres for baselines smaller than a few kilometres [G. MADER (1986)]. This kinematic GPS using the phase must be distinguished from the trajectography (paragraph 4) when the observable is the smoothed pseudo-range and not the carrier beat phase. Some preliminary tests were performed at I.G.N. using the GDVS results and will be presented soon. In 1988, the I.G.N. plans to realize estimation tests for this method on a near production basis. This technique can be of great use for photogrammetry in providing a large number of control points at 1 centimetre level for large scale aerotriangulation.

8. High accuracy

Finally, the millimetric precision of the GPS measurement can be used, when using dual-frequency receivers and a specific software including orbit determination. This concept of "fiducial network", developed by J.M. DAVIDSON (1986), has proved to give excellent results by several authors [Y. BOCK et al. (1986)]. Presently, when an orbit determination is realized (using as fiducial points the VLBI and/or the SLR sites), it is possible to achieve accuracies of 2×10^{-8} in relative positioning for baselines of a hundred km to a few thousands km. The I.G.N. has not yet obtained this capability but is willing to improve its experience on the orbit determination problems in a near future.

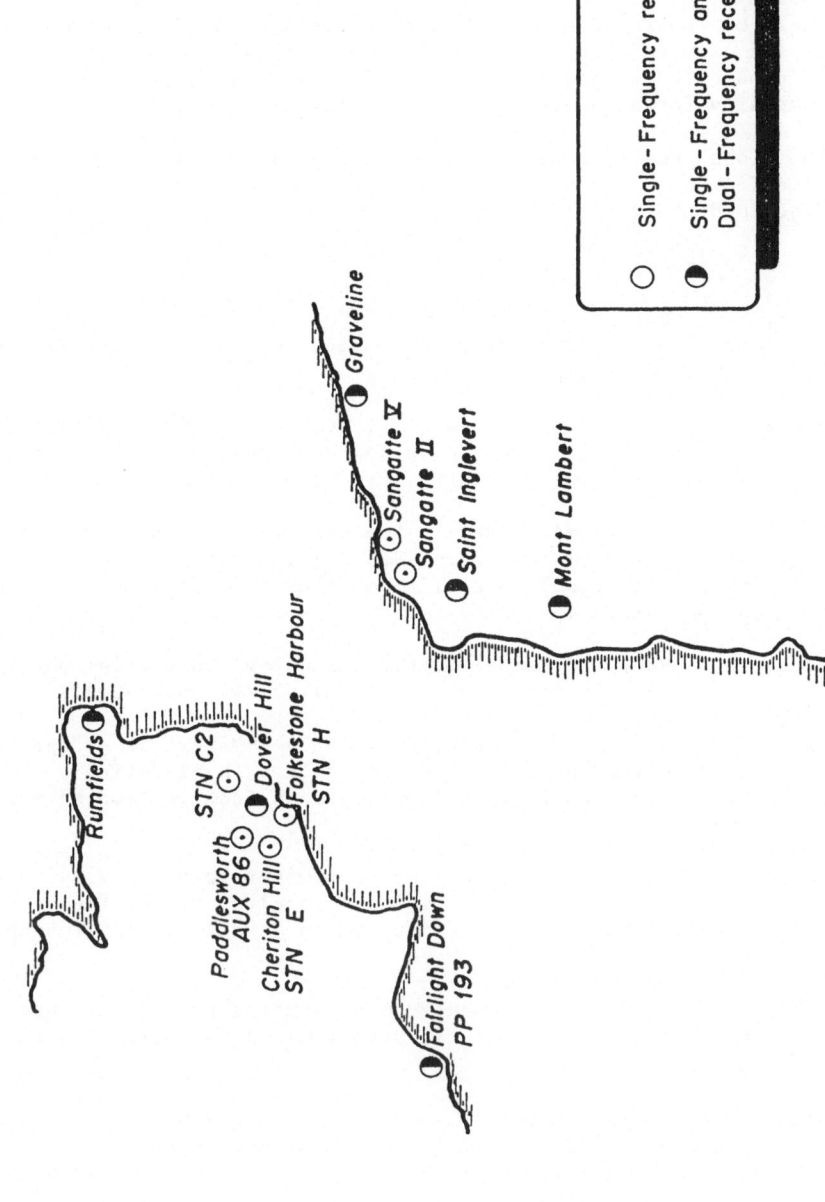

Figure 5 - IGN-OS Joint Campaign October 1987

Rumfields

Paddlesworth
AUX 86
Cheriton Hill
STN E
STN C2
Dover Hill
Folkestone Harbour
STN H

Fairlight Down
PP 193

Graveline

Sangatte V
Sangatte II
Saint Inglevert

Mont Lambert

○ Single-Frequency receiver

◑ Single-Frequency and
   Dual-Frequency receiver

69

9. Conclusions

Since 1985, the Institut Géographique National (France) has been involved in the GPS. Several receivers were bought and plans for a new equipment exist. A specific software (GDVS) has been developed. A large range of applications of the Global Positioning System have been tested or developed and actually demonstrated on the field for science or production purposes. Some investigations must be continued, specially for high accuracy GPS and for orbit determination.

The Global Positioning System will probably obtain a very large number of different users interested in one or several aspects of the above described applications of GPS.

10. References

Bock, Y., Abbot, R.I., Counselman C.C. III, King, R.W. (1986) : A demonstration of 1-2 parts in 10^{-7} accuracy using GPS, Bulletin Géodésique, pp. 241-254

Brossier, R., Nard, G., Rabian, J., Gounon, R. (1986) : Use of TR5S-B GPS receiver in airborne photography surveys, Proceedings of the ISPRS Symposium, Stuttgart, pp. 589-596

Davidson, J.M., et al. (1986) : Demonstration of the fiducial concept using data from the March 1985 GPS field test, Proceedings of the 4th International Geodetic Symposium on Satellite Positioning, pp. 603-612

Mader, G. (1986) : Dynamic positioning using GPS carrier phase measurements, Manuscripta Geodaetica, Vol. 11/4, pp. 272-277

Nard, G., Gounon, R., Broustal, J. (1987) : Matériels et applications du GPS différentiel de haute précision, hybridation avec d'autres moyens, 3ème Colloque National sur la Localisation en mer, pp. 196-208

Remondi, B.W. (1986) : Performing centimeter level surveys in seconds with GPS carrier phase : initial results, Proceedings of the 4th International Geodetic Symposium on Satellite Positioning, pp. 789-798

Wells, D.E. (1986) : Recommended GPS terminology, Proceedings of the 4th International Geodetic Symposium on Satellite Positioning, pp. 903-923

Willis, P. (1987) : Application de la technique GPS pour la localisation précise, 3ème Colloque National sur la Localisation en mer, published in Revue Géomètre, n° 3, Mars 1988, pp. 41-44

Willis, P., Boucher, C. (1987) : Tide gauge connection using GPS, XIXth IUGG Symposium, Vancouver, Canada

GPS APPLICATIONS OF CTS

by

Barbara Kolaczek

Abstract

An outline of present conventional terrestrial systems and posibility of applications of GPS to the CTS improvement are presented.

Introduction

GPS achieving a high accuracy in determinations of relative station positions and base lengths have been succesfully applied to regional Earth crust motions but it has not been applied yet to reference CTS determinations. Some suggestion of GPS applications to the improvement of the BTS is presented here.

The problem of definition and practical realization of the terrestrial and celestial coordinate systems was always the basic problem in geodetic and astronomic investigations. It became crucial when the new observational techniques, such as VLBI, lunar and satellite laser ranging and GPS achieved accuracies of the order of 1 mas and of a few centimeters in determinations of radiosource positions, of lengths and directions of terrestrial bases and of station positions, respectively. At this level of accuracy station positions can no longer be assumed motionless in relation to each other due to the Earth crust motions. Anyhow there seems to be a general agreement that the best way of the practical realization of a reference terrestrial coordinate system is the determination of a conventional terrestrial system – CTS consisting of a number of stations with accurately determined geocentric coordinates. Conceptionally the definition of CTSs presently used in the process of evaluation of the Earth rotation parameters is similar to the definition of the Conventional International Origin – CIO or BIH 1968 and BIH 1979 coordinate systems based on classical astrometric stations. The new CTSs consist of ten to several tens of stations carrying on observations by VLBI, laser and Doppler techniques, but their geocentric coordinates are not connected with local verticals as it was in the case of astrometric stations. Independent CTSs are determined in the process of ERP evaluations by different techniques but also in determinatiions of the Earth gravitational field. The homogeneous BIH reference terrestrial system BTS for all observational techniques was computed in 1984 and improved later. The problem of the reference terrestrial system has been discussed and reviewed at many meetings (B. Kolaczek, G. Weiffenbach, 1974; E. Gaposchkin, B. Kolaczek, 1981; I. I. Mueller, 1985; G. A. Wilkins, A. Babcock, 1987).

Outline of present CTSs and the BTS

CTSs consisting of sets of stations with the most accurately known geocentric positions which are presently determined in the evaluation process of the Earth rotation parameters – ERP depend not only on a set of participating stations and on the applied observational technique but also on the applied model of computations, adopted constants etc. Several different CTSs were determined from different sets of observational data by different analysing centers evaluating ERP series especially during the MERIT Campaign (BIH, 1986; Boucher, Altamimi,1986, 1987). The accuracy of station coordinates of the laser station network (73) determined for instance by the Center for Space Researches – CSR and the VLBI station network (13) determined by the National Geodetic Service - NGS range from several to 20 centimeters (BIH, 1986).

Since 1984, the BIH Terrestrial System - BTS, the reference CTS for all observational techniques participating in ERP observations has been computed (C. Boucher, M. Feissel, 1984). The BTS consists of stations at which observations by the use of at least two observational techniques (VLBI, laser, Doppler) were carried out. The BTS (1986) consists of 51 stations participating in evaluation of 6 ERP series, see Tables 1 and 2 (BIH, 1986; C. Boucher, Z. Altamimi, 1986).

Differences of station coordinates determined in BTS (1985) and BTS (1986) ranging from 1 to 20 cm (Table 1) and the transformation parameters of these two systems (Table 3) show that the accuracy of station coordinates of the BTS (1986) is of the order of a few centimeters. A higher accuracy of the reference terrestrial system is required.

The BTS (1986) has other drawbacks. Distribution of the BTS (1986) stations in longitudes and latitudes is very unhomogeneous. They are located mostly on the North American (23) and European-Asian plates (16). Only 12 stations are located outside Europe and North America. Nearly half of these stations are located in the vicinity of plate tectonic boundaries. 6 stations move with the velocity greater than 2cm/year due to the tectonic plate motions. 40 stations move with the velocity of the order of 1 - 2 cm/year (BIH, 1986). At the plate tectonic boundaries some irregular crust motions are possible.

There are only 16 VLBI stations in the BTS (1986). Therefore the present BTS (1986) is based mostly on laser and Doppler station networks. Transformation parameters between the individual terrestrial systems and the BTS (1986) given in Table 4 (BIH 1986) show the biggest discrepancies of the Doppler station coordinates.

Applications of GPS to CTS improvements

In this situation the application of the GPS technique to the improvement of the BTS ought to be considered. First of all the GPS technique could be applied as the additional observational technique at the laser nad VLBI sites of the BTS (1986), especially at its European and North American sites. Repetitions of these GPS measurements which are easier than the repetitions of laser or VLBI collocations could check the stability of the BTS network.

GPS mobile receivers achieving such a high accuracy in determinations of relative station positions and base lengths could be used for connections of the present BTS sites located in the vicinity of tectonic plate boundaries with well chosen sites located on the stable parts of the tectonic plates. It could allow to check motions of the BTS sites and to create consistent continental networks.

The application of the GPS technique to the ERP determinations is expected (Pâquet, Louis, 1987, Zelensky et al. 1987). In order to ensure the highest accuracy of CTS determinations in the process of evaluation of the ERP about 20 - 30 stations well distributed in latitude and longitude ought to be taken into account (Mueller et al., 1982). In the case of small number of stations participating in ERP determinations the geometric conditions of a solution and the influence of errors of station coordinates are important. It is well known fact in the case of IRIS-VLBI series of the ERP based on observations of several stations (IRIS, 1987). Bad geometric configuration of VLBI stations increases errors of ERP determinations or eliminates one of pole coordinates from determinations. Errors of station coordinates participating in ERP determinations in the order of a few centimeters can introduce similar errors in ERP determinations in the case of ten or smaller number of stations not well distributed in latitude and longitude.

In the near future new systems of satellite observations such as DORIS - Doppler Orbitography and Radiopositioning Integrated by Satellite (Doner et al., 1985) and PRARE - Precise Range and Range Rate Equipment (Hartl et al., 1985) will be introduced in practice. Some coordination of activities is needed in order to use all observational techniques for improvement of the reference CTS.

Table 1. BTS (1986) sites (BIH, 1986).

SITE	NETWORKS	COORDINATE DIFFERENCES BTS(1986) - BTS(1985)			PLATE
		dX (cm)	dY (cm)	dZ (cm)	
Algonquin	2 7	- 7	+ 4	- 7	NOAM
American Samoa	5 6 7	+ 3	-23	+ 3	PCFC
Arequipa	5 6 7	-10	+ 3	+ 7	SOAM
Bear Lake	5 6 7	-19	+14	- 5	NOAM
Bermuda	5 7	-12	+12	+ 4	NOAM
Canberra	3 5 7	-	-	-	INDI
Cerro Tololo	5 6	-	-	-	SOAM
Chilbolton	1 2 7	+ 2	-15	- 9	EURA
Dionysos	5 7	+58	-12	-18	EURA
Effelsberg	1 2 7	+ 3	-19	- 9	EURA
Flagstaff	5 6	-	-	-	NOAM
Fort Davis	1 2 4 5 6 7	-15	- 6	- 7	NOAM
Goldstone	1 2 3 5 6 7	-13	+11	- 4	NOAM
Grand Turk	5 7	-10	+ 5	- 6	NOAM
Grasse	4 5 6 7	- 8	- 3	0	EURA
Graz	5 6 7	- 2	- 3	-13	EURA
Greenbank	1 2 7	-18	0	- 8	NOAM
Haystack	1 2 3 5 6 7	-18	- 4	- 8	NOAM
Helwan	5 6	-	-	-	AFRC
Herstmonceux	5 6 7	- 3	+ 2	-11	EURA
Johannesburg	1 2 7	-	-	-	AFRC
Kootwijk	5 6 7	- 1	- 3	- 3	EURA
Kwajalein Atoll	2 5 6 7	+13	+43	+ 3	PCFC
Madrid	2 3 7	-	-	-	EURA
Maryland Point	1 2 7	-21	0	- 3	NOAM
Matera	5 6	-	-	-	EURA
Maui	4 5 6 7	+ 6	+ 6	-11	PCFC
Mazatlan	5 6	-	-	-	NOAM
Metsahovi	5 6 7	-174	-16	-74	EURA
Monument Peak	5 6	-	-	-	NOAM
Motu Hiumoo	5 6	-	-	-	PCFC
Mount Hopkins	5 7	-23	- 5	-12	NOAM
Natal	5 6 7	+ 1	-24	- 2	SOAM
Onsala	1 2 7	- 8	-14	-10	EURA
Owens Valley	1 2 3 5 6 7	-13	+11	- 9	NOAM
Pasadena	5 6 7	-12	+13	- 4	NOAM
Patrick Afb	5 7	-21	+ 5	- 2	NOAM
Platteville	5 6	-	-	-	NOAM
Potsdam	5 6	-	-	-	EURA
Quincy	5 6 7	- 5	+15	-12	NOAM

Table 1 cont.

| SITE | NETWORKS | COORDINATE DIFFERENCES BTS(1986) – BTS(1985) | | | PLATE |
		dX (cm)	dY (cm)	dZ (cm)	
Richmond	1 2 7	-19	+ 1	- 8	NOAM
San Diego	5 6 7	+12	+18	- 5	NOAM
Santiago	5 6	-	-	-	SOAM
Shanghai	5 6 7	-	-	-	EURA
Simosato	5 6	-	-	-	EURA
Vernal	5 6	-	-	-	NOAM
Washington	5 6 7	-12	+11	- 4	NOAM
Wettzell	1 2 5 6 7	- 7	+15	-11	EURA
Yarragadee	5 6 7	+19	-12	-14	INDI
Yuma	5	-	-	-	NOAM
Zimmerwald	5 6 7	+11	0	-24	EURA

Table   2. Networks   and   series   of   the   Earth   rotation   parameters
contributing   to   the   BTS (1986) (BIH,   1986).

	Contributing Networks				Series of ERP		
Nr	Set of station coordinates	Time span	Nb of coloc. sites		ERP Label		Time span
1	SSC (NGS) 87 R 01	1980–1986	12		NGS	87 R 01	1984.0–1987.0
2	SSC (GSFC) 87 R 01	1979–1986	15		NGS	87 R 01	1984.0–1987.0
3	SSC (JPL) 83 R 05	1971–1983	5		JPL	83 R 01	1984.0–1987.0
4	SSC (JPL) 87 M 01	1983–1986	3		JPL	87 M 01	1984.0–1987.0
5	SSC (CSR) 86 L 01	1976–1986	42		CSR	86 L 01	1984.0–1986.7
6	SSC (DGFII) 87 L 01	1980–1984	37		DGFII	87 L 01	1984.0–1985.0
7	SSC (DMA) 77 D 01	1975–1984	38				

Jet Propulsion Laboratory - JPL, Pasadena, USA
NASA Goddard Space Flight Center - GSFC, Greenbelt, USA
National Geodetic Survey - NGS, Rockville, USA
Center for Space Research - CSR, Austin, USA
Deutsches Geodätisches Forschunginstitut - DGFI, Munich, FRG
Defence Mapping Agency - DMA, Washington, USA.

Table 3. Transformation parameters between the three realizations of the BTS (the uncertainties are given in the second lines) (BIH, 1986).

BTS	T1 m	T2 m	T3 m	D 10^{-6}	R1 "	R2 "	R3 "
1984/1985	0.023	0.044	0.037	-0.0021	-0.0002	-0.0014	-0.0042
	0.020	0.020	0.020	0.0029	0.0008	0.0007	0.0007
1985/1986	0.051	-0.023	0.070	-0.0089	-0.0025	-0.0006	-0.0069
	0.022	0.022	0.022	0.0032	0.0009	0.0008	0.0008

Table 4. Transformation between the individual terrestrial systems and the BTS (BIH, 1986).

Network	SSC	T1 CM	T2 CM	T3 CM	D 10^{-8}	R1	R2 $0{.}''001$	R3	A3
SSC (NGS)	87 R 01	10.8	12.5	5.8	-2.4	-4.5	10.3	-1.8	0.6
SSC (GSFC)	87 R 01	153.5	-96.1	45.0	-2.7	-4.1	10.4	10.3	0.9
SSC (JPL)	83 R 05	5.5	-27.6	-15.8	-1.6	-4.8	-2.6	4.9	1.4
SSC (JPL)	87 M 01	.0	.0	.0	-2.2	1.9	3.7	-10.6	
SSC (CSR)	86 L 01	.0	.0	.0	.0	-4.5	3.1	-4.4	
SSC (DGFI)	87 L 01	.3	-3.3	6.2	1.3	8.0	-6.0	120.6	
SSC (DMA)	77 D 01	-16.7	-21.2	-431.4	57.1	36.1	8.4	-795.1	

BIH (1984 – 1986): BIH Annual Reports for 1984, 1985 and 1986, Paris, France,

Boucher C., Altamimi Z. (1986): Comparison of various Reference Systems with the BIH Terrestrial System (1985), BIH Annual Report for 1985, Paris, France,

Boucher C., Altamimi Z. (1987): Complement of Analysis of BIH Terrestrial System for 1986, BIH Annual Report for 1986, Paris, France,

Boucher C., Feissel M. (1984): Realization of the BIH Terrestrial System, Intern. Symposium on Space Techniques for Geodynamics, Sopron, Hungary,

Doner M., Lefebrve M. (1985): Doppler Orbitography and Radiopositioning Integrated by Satellite – DORIS, CSTG Bull. No 8,

Gaposchkin E. M., Kolaczek B. (eds) (1981): Proc. of the IAU Coll. No 56 on Reference Coordinate Systems for Earth Dynamics held in Warsaw, Reidel Publ. Company, Dordrecht, Netherlands,

Kolaczek B., Weiffenbach G. (1975): Proc. of the IAU Coll. No 26 on Reference Coordinate Systems for Earth Dynamics held in Torun, Warsaw Technical University, Warsaw, Poland,

Hartl Ph., Reigber Ch. (1985): Precise Range and Range Rate Equipment – PRARE, CSTG Bull. No 8,

IRIS (1987): IRIS Earth Rotation Bulletin, IRIS Subcommission, NGS, Rockville, USA,

Mueller I. I. (ed) (1985): Proc. of the Intern. Conference on Earth Rotation and the Terrestrial Frame held in Columbus, The Ohio State University, Columbus, Ohio, USA,

Mueller I. I., Zhu S. Y., Boch Y. (1982): Reference Frame Requirements and the MERIT Campaign, Report of the Department of Geodetic Science and Surveying of the Ohio State University, Columbus, USA,

Pâquet P., Louis L. (1987): Simulations to Recover Earth Rotation Parameters with GPS System, Proc. of the IUGG Symposium on Variations in Earth Rotation in Vancouver (in press),

Wilkins G., Babcock A. (1987): Proc. of the IAU Symposium No 128 on the Earth's Rotation and Reference Frame for Geodesy and Geodynamics held in Coolfont, USA, Reidel Publ. Company, Dordrecht,

Zelensky N., Ray J., Liebrecht P. (1987): Proc. of the IUGG Symposium on Variations in Earth Rotation held in Vancouver (in press).

AGEDEN - AN APPLICATION OF GPS FOR
GEODYNAMIC INVESTIGATIONS IN AUSTRIA

by

Karl Rinner   Peter Pesec   Günter Stangl
Bernhard Hofmann-Wellenhof   Herbert Lichtenegger

Abstract

In the opinion of geologists, there are at least three active tectonic structures in Austria. For the determination of possible relative movements or strain in these areas, small networks in the size of about 50 km times 50 km were established and are planned to be measured repeatedly by using GPS. In Fall 1987 the station positions of the networks were measured, for the first time, as one part of the Austrian Geodynamic Densification Network (AGEDEN) project. The tectonic situation, the planning and realization of the measurements as well as the computational methods and results are presented.

1. Introduction

More than one decade ago, Rinner (1974) proposed to start geodynamic investigations in Austria based on geodetic methods. Relative movements should be detected and proved in tectonically active structures. Presently, the realization of this proposal is in progress as one part of the Austrian Geodynamic Densification Network (AGEDEN). Three areas were selected where small networks in the size of about 50 km times 50 km were established. To determine the strain components in these areas, which is the most important goal of the AGEDEN project, repeated GPS measurements are planned. The results of the first set of GPS observations are now available. Therefore, apart from the Laser measurements at the observatory Graz-Lustbühel, the AGEDEN results can be regarded as another Austrian contribution to the European WEGENER-MEDLAS project where geodetic geophysical investigations in the transition zone between the African and the Eurasian plate are being performed.

2. General geologic, geodynamic, and geodetic aspects

2.1 Geologic aspects

Austria can be characterized by at least three geologic units namely the Bohemian Massif in the northeast, the centrally situated Eastern Alps, and some lowlands and basins.

The Bohemian Massif, consisting essentially of gneiss and granite, is a spur of the Variscian mountain which was folded up in the early Paleozoic era. In the tectonic sense, it can be regarded as the oldest part.

The Eastern Alps were mainly formed by various tectonic processes starting in the late Mesozoic period. They are divided into several east-west directed chains built-up by various minerals. A simplified presentation is given in Fig. 1. An important feature of the alpine tectonics is the movement of a huge mass of marine sediments to the north resulting in the formation of the Northern Limestone Alps. This process began some 60 million years ago and originated from the collision of the African plate against the Eurasian plate and the subsequent uplift of the ocean ground.

Finally, most of the basins were filled up in the Tertiary. They are therefore the youngest geologic formations. Roughly speaking, they separate the Eastern Alps from the Bohemian Massif and the Pannonian basin.

2.2 Geodynamic aspects

The alpine region is still tectonically active and the Austrian territory is covered with a number of geodynamic

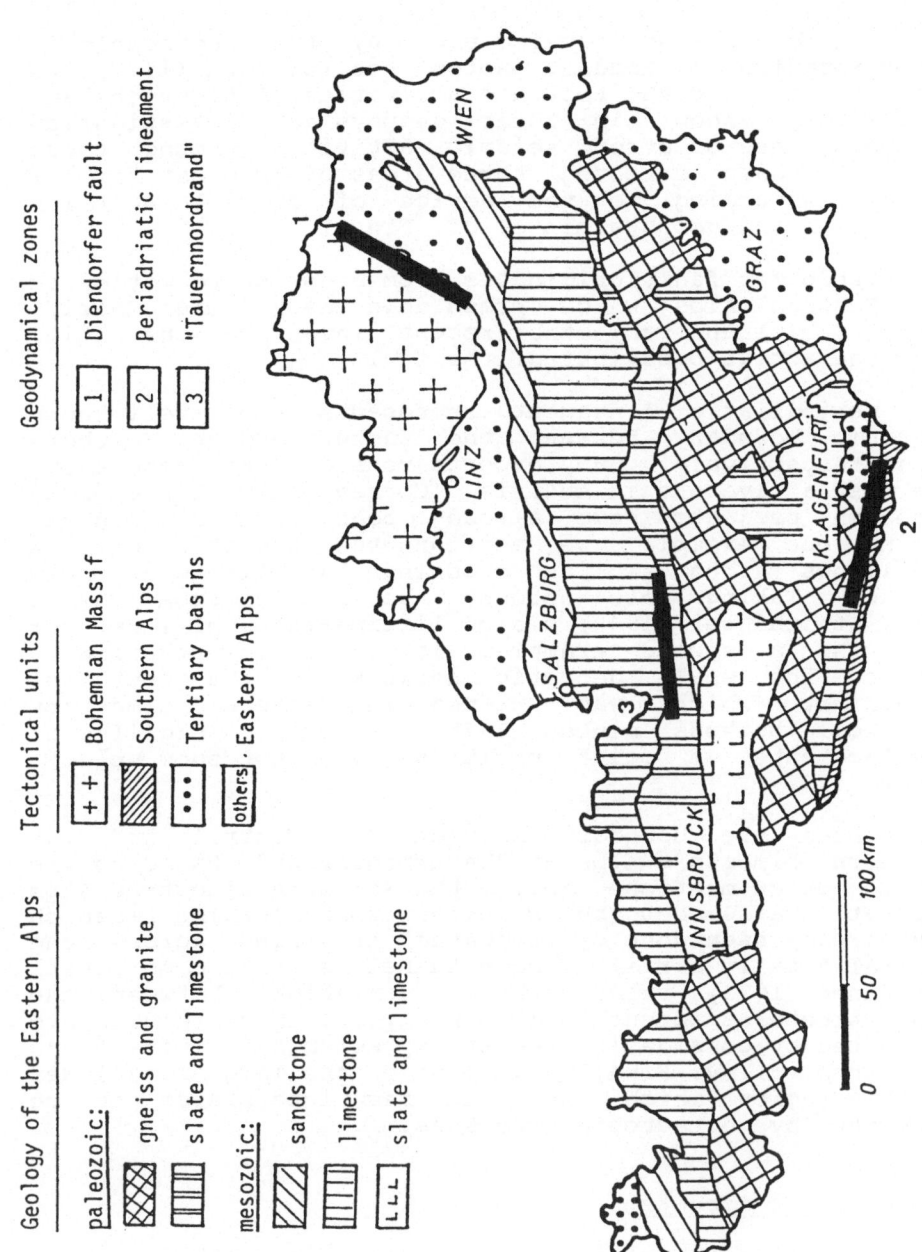

Figure 1: Geology and tectonics in Austria

Geology of the Eastern Alps

paleozoic:

- ⊠ gneiss and granite
- ⬚ slate and limestone

mesozoic:

- ⧄ sandstone
- ⊟ limestone
- ⌐ slate and limestone

Tectonical units

- + + Bohemian Massif
- ⧄ Southern Alps
- ∷ Tertiary basins
- others · Eastern Alps

Geodynamical zones

- 1 Diendorfer fault
- 2 Periadriatic lineament
- 3 "Tauernnordrand"

0   50   100 km

zones which has been revealed e.g. by the photogeologic interpretation of Landsat photos, see Tollmann (1977). As another proof for the tectonic activities, Gutdeutsch and Aric (1976) show in their geophysical investigation numerous zones of strong seismic activity. Among these zones we have selected three areas of interest for our project according to the advice of several Austrian geologists and geophysicists, cf. Fig. 1.

The Diendorfer fault originates from a northward motion of the Eastern Alps which press down the Bohemian Massif. Figdor (1976) and many other recent investigations still show the active behaviour of the fault.

The periadriatic lineament represents the well-known tectonic boundary between the Eastern and the Southern Alps, respectively. Many experts consider this zone being the most active one in Austria. Consequently, a number of research projects have already been performed there. Giving one example, we mention the repeated precise levellings along a line crossing the lineament near Villach. Taking into account the results from 1952 to 1964, Steinhauser (1980) claims lifting-rates of 1 mm per year. As a second example, it is also worthwhile to mention the small geodetic network in the south of Klagenfurt which was re-measured several times by classical methods since 1975. However, significant displacements of point positions did not show up, cf. Peters (1979).

We denote the boundary between the Central and the Northern Limestone Alps as "Tauernnordrand". Based on the alpine motion mentioned above, the Northern Limestone Alps are still assumed to be in motion. Corresponding research has already been done by repeated levelling across the Central Alps in a railway base-tunnel. According to Senftl and Exner (1973), relative height variations between the two structures amount to 1 mm per year. These results are supported by considering the strong seismic activity along the Enns valley. Finally, we mention the large landslides in this region which are sometimes interpreted to be generated by neotectonic processes.

2.3 Geodetic aspects

For monitoring the geodynamics by geodetic methods there should be established rectangularly-shaped networks in selected areas along the active structures. This configuration enables us to detect any possible tendency of motion. Consider a rectangle with diagonals as shown in Fig. 2. Baseline variations along and across the corresponding structure mainly indicate extensions and compressions. The diagonals may proof transforms. Moreover, the application of GPS for the 4 points of the basic network gives 9 coordinate differences and, therefore, 9 equations for the strain matrix whose elements can, consequently, be evaluated without any hypothesis. Using in addition Hooke's law, we may also derive the stress components of the region in consideration. This yields a more valuable information on e.g. imminent earthquakes.

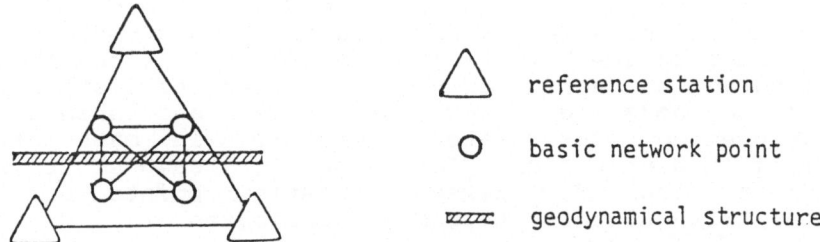

Figure 2: Idealized geodynamic network

We have added 3 more points outside the geodynamic zones of interest to enable a transformation into a fixed reference system. This is necessary for re-measurements in order to detect motions of the basic network relative to the reference system. Moreover, three reference points allow a check of the stability of the reference system.

Turning to GPS means that we need at least seven receivers for fulfilling geodetic and geodynamic requirements, namely 3 for the reference points and 4 for the points selected in the geodynamic zone. The idealized minimum configuration is presented in Fig. 2.

3. Site selection and field work

3.1 General remarks

The planning of AGEDEN was originally based on the availability of 10 TI-4100 GPS receivers to be placed at our disposal by the geodetic institutes of the Universities Hannover, Munich, Copenhagen, by the University of Federal Armed Forces Munich, and by the Alfred-Wegener Polar Institute Bremerhaven on the basis of free equipment exchange. In the early stage of preparations the Federal Office of Metrology and Surveying (BEV) in Vienna, Austria indicated its interest to add four additional observation sessions for strengthening the first order national triangulation network, checking levelled heights, establishing GPS reference sites, and applying GPS methods for small networks. Therefore, it was decided to organize a seven-day campaign by common efforts thus reducing the costs per site considerably. The sequence of sessions was arranged as economically as possible by minimizing driving.

Although, with respect to the organization, the campaign should be considered on the whole the following sections deal only with the "geodynamic part" of AGEDEN, i.e. the observation days 1, 4 and 5.

3.2 Site selection

As explained in section 2.3, the geometric form of the network should be selected rectangularly and symmetrically with respect to the investigated faults. The sites should be situated on solid ground in order to distinguish between exogenous and endogenous forces and should be well-monumented for later recovery in case of a superficial destruction. For economic reasons it was obvious to look for already existing points of the national triangulation network as candidates.

The reconnaissance for the Diendorfer fault (session day 1) and the Tauernnordrand (session day 5) was done by the BEV with geological assistance, for session day 4 (periadriatic lineament) by the Institute of Applied Geodesy and Photogrammetry of the Technical University Graz. All sites to be selected had to meet the following additional requirements:
- no obstruction above 10 degrees (with allowed exceptions in those directions where no satellites could be observed according to the present satellite configuration),
- no high-power lines nearby,
- easily accessible by car because of the somewhat bulky equipment used.

Naturally, it was not possible to select all sites according to all of the conditions mentioned above. In particular the ideal geometric form of Fig. 2 could not be maintained in all cases. However, as shown in Fig. 3, a reasonable station distribution could be attained.

For each session day the network configuration was planned in such a way that three DÖNAV sites, see Seeber et al. (1987) and Kirchner et al. (1987), could be observed simultaneously yielding the opportunity to compute the coordinates of the monitor stations in a fixed frame.

3.3 Field work

The GPS campaign AGEDEN was carried out during days 255 to 262 of 1987 (September 10 to September 17, 1987). Since, finally, only 8 GPS receivers were at our disposal one point per session had to be dropped, and we had to dispense with the originally planned spare receiver to be located at the observatory Graz-Lustbühel (Fig. 3). Six GPS satellites were available giving a reasonable geometry during the period 6 UT (Universal Time) and 10 UT. Days 255, 258, 259 were devoted to the geodynamic part of AGEDEN.

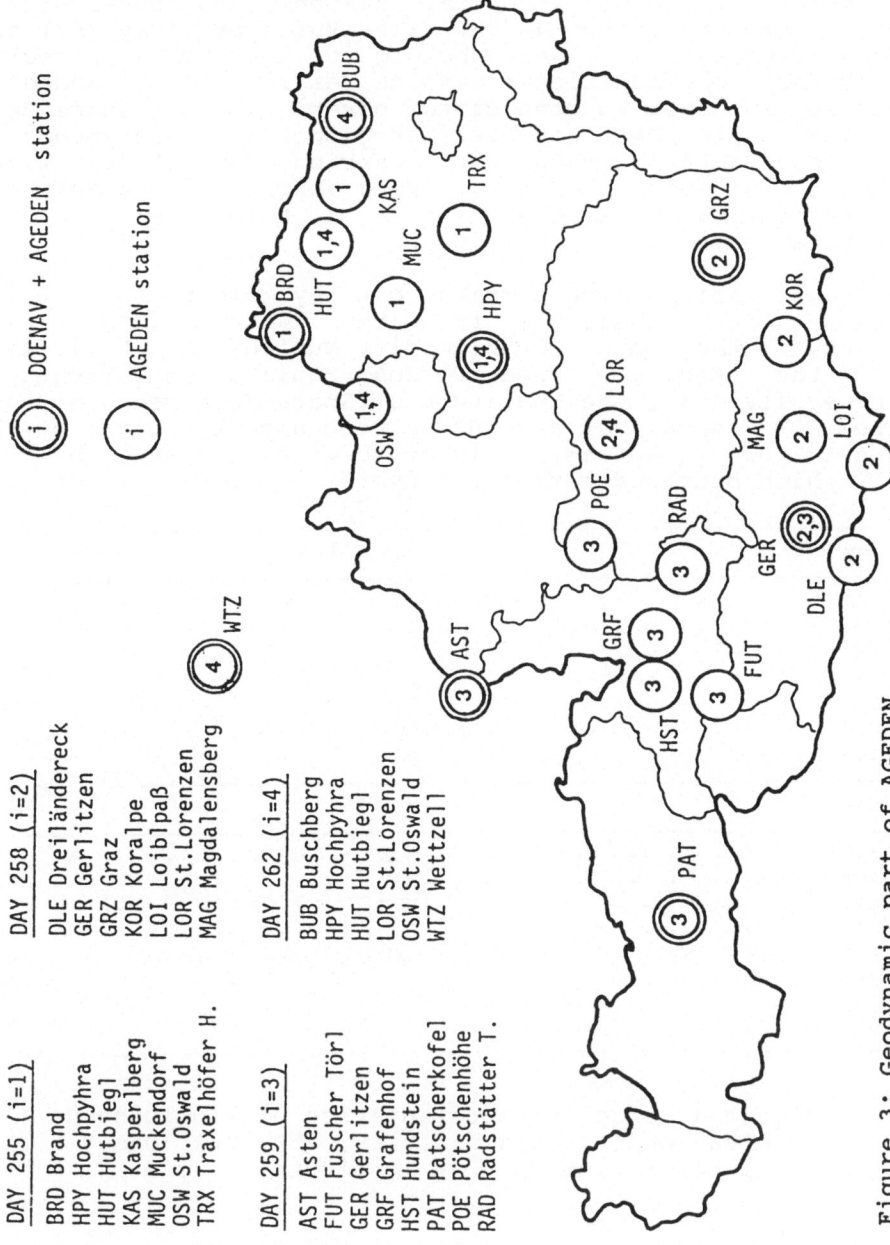

DAY 255 (i=1)

BRD Brand
HPY Hochpyhra
HUT Hutbiegl
KAS Kasperlberg
MUC Muckendorf
OSW St.Oswald
TRX Traxelhöfer H.

DAY 259 (i=3)

AST Asten
FUT Fuscher Törl
GER Gerlitzen
GRF Grafenhof
HST Hundstein
PAT Patscherkofel
POE Pötschenhöhe
RAD Radstätter T.

DAY 258 (i=2)

DLE Dreiländereck
GER Gerlitzen
GRZ Graz
KOR Koralpe
LOI Loiblpaß
LOR St.Lorenzen
MAG Magdalensberg

DAY 262 (i=4)

BUB Buschberg
HPY Hochpyhra
HUT Hutbiegl
LOR St.Lorenzen
OSW St.Oswald
WTZ Wettzell

(i) DOENAV + AGEDEN station

(i) AGEDEN station

Figure 3: Geodynamic part of AGEDEN

85

Unfortunately, for technical reasons, no observations could be carried out at the sites Buschberg (day 255) and Hochwurzen (day 258) which belong to the DÖNAV network. Therefore, one additional session was immediately added at day 262 in order to connect the observations of these days to the basic DÖNAV network. Altogether 60 measurements of 3.5 hours duration each were carried out at 46 sites giving a huge amount of data stored on about 420 cassettes. Half of the data have been used for the computation in this publication.

The observations were carried out by eight observation teams with observers from the Technical University Hannover, the Technical University Munich, the University of the Federal Armed Forces Munich, the Technical University Graz, the Institute of Space Research Graz, and the BEV Vienna. For day 262 we were also supported by the observatory at Wettzell. Six teams were assisted by the BEV which provided cars and drivers.

```
 day 255                              day 258

 BRD         --- --- --- --- ---      DLE    - --- --- --- --- --- ---
 HPY      --- --- --- --- --- ---     GER    - --- --- --- --- --- ---
 HUT  --- --- --- --- --- --- ---     GRZ    --- --- --- --- --- ---
 KAS           -- --- --- ---         KOR    -- --- --- --- --- --- ---
 MUC  --- --- --- --- --- --- ---     LOI                - --- --- ---
 OSW  --- --- --- --- --- --- ---     LOR    --- --- --- --- --- ---
 TRX  --- --- --- --- --- --- ---     MAG        --- --- --- --- ---

 day 259                              day 262

 AST  --- --- --- --- --- --- ---     BUB    --- --- --- --- --- --- ---
 FUT  --- --- --- --- --- --- ---     HPY    --- --- --- --- --- --- ---
 GER    - --- --- --- --- --- ---     HUT    - --- --- --- --- --- ---
 GRF  --- --- --- --- --- --- ---     LOR    -- --- --- --- --- --- ---
 HST  --- --- --- --- --- --- ---     OSW    --- --- --- --- --- --- ---
 PAT  --- --- --- --- --- --- ---     WTZ    -- --- --- --- --- --- ---
 POE    -- --- --- --- --- --- ---
 RAD    -- --- --- --- --- --- ---
```

Figure 4. Graphic representation of simultaneous observation for days 255, 258, 259, 262. One "-" indicates a time span of 10 minutes.

In Fig. 4 the time span of the recorded data is displayed for each station. It can be seen that, apart from initial delays, problems could be detected only at the stations KAS, LOI and MAG. For these stations the acquisition of 4 simultaneous satellites, which is necessary for clock synchronization, could not be performed in time, partly because of obstruction problems, partly because of signal interference. For the majority of stations, however, the observations could be carried out satisfactorily.

4. Computations

The software used is a modification of the Bernese Second Generation Software Package, see Gurtner et al. (1985), which had to be adapted for the UNIVAC/1100 computer in Graz. Most of the changes result from the memory restrictions (about 128 kByte). Parts of the data transfer programs and the whole main frame program (in menue technique) had to be re-written.

4.1 Data transfer and preprocessing

The data measured in intervals of 3 seconds were available on altogether 420 digital cassettes in a binary mode. In order to make these data readable they had to be transformed into hexadecimal code. Due to the lack of a direct connection between the digital cassette reader and the UNIVAC/1100 computer, 300 megabyte of data had to be stored on PC-floppies, transferred to a VAX-750, reformatted and stored again on magnetic tapes. These tapes were, finally, read into the UNIVAC system for further use.

The second step was to pick out from the hexadecimal strings 30 second data samples in a format readable by the Bernese software. This means that the original data set was reduced by a factor of 10. For this data reduction a modified version of the Bernese program TIREAD was used, which extracts the necessary informations like satellite orbital data, pseudoranges and the phase measurements for L1 and L2. All data were stored on so-called session files containing the data set of all simultaneously observing stations per day.

4.2 Orbit data

For the numerical integration we used broadcast ephemerides for the calculation of boundary values and as earth potential representation the GEM 10 model with maximum degree of 8, see Beutler et al. (1986). Concerning solar radiation, we only estimated the direct pressure parameter but not the others.

The observations of each day had to be processed separately. Consequently, the orbit computation is bounded by the time span of the corresponding session. In the actual case this leads to a short arc of about 5 hours.

4.3 Processing of the observables

The P-code for pseudoranges enables us in a first step to estimate preliminary station coordinates and receiver clock information in form of user defined polynomials. At some stations we got large residuals for the beginning epochs, indicating warm-up effects of the oscillators.

The second step can be denoted as a screening of the single differences. These, computed from the phase observables, were formed for each day referring to one reference location. The reference site selection depends on various criteria as e.g. maximum time span of observation, no gaps during the observation, no extreme large distances to the other stations ("minimum maximum distance"), small residuals for the P-code solutions. In fact, none of the selected sites could fulfill all requirements optimally but in general the decision was not really difficult. Considering the number of epochs and distances for each pair of stations, we had about 360 epochs (3 hours of shared observation) available with distances ranging from 30 km to 150 km.

In a third step by using the computed orbit data, the preliminary coordinates and the clock polynomial coefficients, the single differences were screened for cycle and half cycle slips. At first the L1 frequency and the L2 frequency were screened separately. The result was used for computing and screening L3. This somewhat sensitive step was performed by two independent teams. After fixing (repairing) cycle and half cycle slips, fitted polynomials of second degree showed rms in the range of 0.02 cycles and 0.05 cycles or 4 to 10 millimeters for L1.

Quite interesting was the following behaviour of some satellites: comparing the first and the second L1 epoch, a half cycle jump apparently occurred and was corrected. But checking the ambiguities, the previously performed application of the half cycle slip correction resulted in a non-integer behaviour of the ambiguities. The way out of this dilemma was simple: we retrieved the original data again, deleted the first epoch and the effect mentioned above vanished.

A simple single layer ionospheric model can be helpful for screening the L1 and L2 data. In our case we obtained a reduction of 40 percent for the rms of the polynomial fit.

4.4 Estimation of the results

According to our computer memory limitations we must impose the following restrictions during the computations: 1) One day only, 2) Six satellites, 3) 500 epochs, 4) 65 unknowns.

The orbit and the station coordinates of the reference site were always fixed (i.e., not estimated). The coordinates of the remaining stations as well as relative ambiguities were computed with L1, L2 and L3 data. For the ionosphere a single layer model was applied to L1 and L2 and the Saastamoinen model was taken into account for all frequencies. For the L1 and L2 data computations we tried to fix the ambiguities by rounding to nearest integers. As expected, this often leads to considerable biases.

5. Numerical results

Before presenting figures we summarize the specifications
met for the evaluation. The computations are based on the
WGS72. Apart from a fixed reference site, the coordinates
of the stations were estimated. The only other estimated
quantities were relative integer ambiguities. This means
that the final adjustment procedure did not contain clock
parameters, orbital elements, and parameters of the
atmosphere. However, the measured data were corrected for
the ionosphere with a simple single layer model and for
the troposphere with the Saastamoinen model.

The data and the interpretation presented below are mainly
based on the first day of the campaign. The results for
the six baselines referring to Muckendorf (MUC in Fig. 3)
are given in Table 5.1. There are shown results for L1,
L2 and L3, the linear combination of L1 and L2 which
eliminates the main influence of the ionosphere. Moreover,
we also present on the one hand the results for fixed
integer ambiguities and on the other hand for the
estimated ones. Note that for L3 ambiguities must always
be estimated.

Baseline	L1	L2	L3	Differences (L1-L2)	(L1-L3)	Ambiguities
MUC-BRD	58746.806	.823	.779	-.017	.027	estimated
	.756	.820		-.064		fixed
MUC-HUT	43717.614	.625	.601	-.011	.013	estimated
	.608	.611		-.003		fixed
MUC-HPY	43980.344	.324	.362	.020	-.018	estimated
	.399	.362		.037		fixed
MUC-TRX	42854.400	.410	.384	-.010	.016	estimated
	.434	.400		.034		fixed
MUC-KAS	62224.724	.742	.696	-.018	.028	estimated
	.736	.748		↴.012		fixed
MUC-OSW	47460.625	.645	.594	-.020	.031	estimated
	.644	.658		-.014		fixed

Table 5.1. Baseline estimations for L1, L2 and
L3 data of day 255. All numbers are given in
meters. The baseline length is shown only once
for L1 with estimated ambiguities. For L2 and L3
and for the computation with fixed ambiguities
only fractions of meters are given.

From Table 5.1 we can make the following statements. The reference site selected fulfills the criterion of baselines with about the same length: they range from 42 km to 62 km. Obviously, the strategy of fixing the ambiguities to the nearest integers often leads to worse results which can easily be seen in the difference L1-L2.

Baseline		\multicolumn{5}{c}{Satellites}				
		3	6	11	12	13
MUC-BRD	L1	3.95	0.24	2.63	-0.01	0.07
	L2	2.98	0.22	3.13	-0.03	0.08
MUC-HUT	L1	-1.20	-0.13	-0.13	-0.06	-0.03
	L2	-1.05	0.00	0.06	0.14	0.12
MUC-HPY	L1	96.19	0.37	0.90	17.94	71.44
	L2	17.27	0.82	0.88	0.98	46.96
MUC-TRX	L1	1.81	-0.10	-0.03	0.00	0.43
	L2	1.73	-0.05	-0.01	-0.01	-0.09
MUC-KAS	L1	-0.14		0.08	0.11	0.11
	L2	-1.12		-0.88	-0.82	-0.86
MUC-OSW	L1	1.46	-0.19	-1.12	-0.88	-0.76
	L2	2.41	-0.15	-0.09	0.07	0.19

Table 5.2. Relative integer ambiguities.

The estimated relative integer ambiguities in Table 5.2 were calculated by the adjustment of double differences. In all but one cases the double differences were formed with SV 9 as reference satellite. Only for baseline MUC-HUT and L1 carrier SV 12 was used as reference satellite. The gap for satellite SV 6 in baseline MUC-KAS results from technical problems in recording data of this satellite. The rms errors of the shown ambiguities are in all cases less than 0.10 m. Thus some of the ambiguities differ significantly from integer numbers, cf. e.g. baseline MUC-OSW for satellite SV 3 (both carriers).

Quite noteworthy are the results for baseline MUC-HPY for satellites SV 3, SV 12 and SV 13 where we get large numbers in contrast to all other results. There could be a correlation to the results of Table 5.1 where the behaviour of the same baseline differs from the others, compare e.g. the L1-L3 difference where the only negative sign appears for MUC-HPY. The same holds for L2-L3. This proof is left to the reader.

Why are the results of the remaining days not shown here? There are some reasons. One of them is the remarkable deterioration of the rms of single differences, cf. Table 5.3, which could not yet analyzed fully.

	day 255		day 258		day 259		day 262	
Observ's	6120		6230		8220		4790	
Unknowns	47	18	47	18	56	21	40	15
rms (L1)	.014	.027	.040	.047	.039	.046	.025	.042
rms (L2)	.020	.024	.048	.057	.044	.056	.037	.062
rms (L3)	.012	-	.035	-	.033	-	.015	-

Table 5.3. Statistics on the geodynamic part of AGEDEN.

Considering the rms of the single differences for day 255 and comparing it with the corresponding values of the days 258, 259 and 262, we recognize a factor of 2 to 3. An explanation for this could be an incomplete cycle slip elimination. This somewhat tedious step must be done again since it is the crucial point of the Bernese software.

Note that in Table 5.3 the two different numbers of unknowns (and therefore two different rms columns per day) result from the two calculations where either ambiguities were estimated or kept fixed.

Another reason for omitting more results is the preparation of a final report for a project sponsored by the Austrian Science Research Council to be finished in fall 1988.

6. Conclusion

The GPS measurement campaign presented in the previous sections is to be considered as "zero-measurement" for geodynamic investigations. Being aware of the accuracy requirements for expected motions along the geodynamic structures in the amount of millimeter/year we must state that the accuracy gained so far is not sufficient yet to recognize very slow motions in short time intervals (i.e., by repetition measurements). This is not really surprising since the present level of GPS ranges from 1 ppm to, under optimum conditions, 0.1 ppm. In our actual case with baselines of about 50 km we need at least an accuracy of 0.02 ppm for millimeter motions during one year. This means that in the present development status of GPS a repetition period of at least about 5 years is suggested. Nevertheless, GPS can be regarded as an important tool for geodynamics too especially when combined with other observation techniques like Satellite-Laser-Ranging. In addition, it is probably a matter of a short time span to pass when GPS will yield even better results.

7. References

Beutler, G.; W. Gurtner; M. Rothacher; T. Schildknecht; I. Bauersima (1986): "Determination of GPS orbits using double difference carrier phase observations from regional networks". Proceedings of the Fourth International Geodetic Symposium on Satellite Positioning. Austin, Texas, April 28 - May 2, Vol. 1, pp. 319-335.

Figdor, H. (1976): "Schwereanomalie und Geomechanik der Diendorfer Störung". Technical University of Vienna, Dissertation.

Gurtner, W.; G. Beutler; I. Bauersima; T. Schildknecht (1985): "Evaluation of GPS carrier difference observations: the Bernese second generation software package". Proceedings of the First International Symposium on Precise Positioning with the Global Positioning System. Rockville, Maryland, April 15-19, Vol. 1, pp. 363-372.

Gutdeutsch, R.; K. Aric (1976): "Erdbeben im ostalpinen Raum". Mitteilungen der Zentralanstalt für Meteorologie und Geodynamik, Vienna, Vol. 210.

Kirchner, G.; P. Pesec; G. Stangl ed. (1987): "Information on current Laser- and GPS-TI4100 activities in Austria". Paper presented at the XIX General Assembly of IUGG in Vancouver, Canada.

Peters, K. (1979): "Krustenbewegungsmessungen im Karawankenprofil und an der Torscharte". Geowissenschaftliche Mitteilungen, Technical University of Vienna, Vol. 15, pp. 257-307.

Rinner, K. (1974): "Der geodätische Beitrag zu geodynamischen Projekten". Zeitschrift für Vermessungswesen, Vol. 99, No. 8, pp. 325-335.

Seeber, G.; G. Wübbena; A. Schuchardt ed. (1987): "Status report on DÖNAV". Paper presented at XIX General Assembly of IUGG in Vancouver, Canada.

Senftl, E.; C. Exner (1973): "Rezente Hebung der Hohen Tauern und geologische Interpretation". Verhandlungen der Geologischen Bundesanstalt, Vienna, Vol. 2, pp. 209-234.

Steinhauser, P. (1980): "Rezente Krustenbewegungen an der Nivellement-Linie Villach - Thörl Maglern". Mitteilungen der österreichischen geologischen Gesellschaft, Vienna, Vol. 71/72, pp. 317-322.

Tollmann, A. (1977): "Die Bruchtektonik Österreichs im Satellitenbild". Neues Jahrbuch für Geologie und Paläontologie Abhandlungen, Stuttgart, Vol. 153, pp. 1-27.

INVESTIGATION OF AN ALTERNATE METHOD OF PROCESSING GLOBAL POSITIONING SURVEY DATA COLLECTED IN KINEMATIC MODE

by

Clyde C. Goad

Abstract

Kinematic survey phase data are reduced using a standard double difference formulation when the initial integer values of the biases are known or determined. Two experimental results are presented. A byproduct of the reduction was the generation of a measure of success similar to the Geometrical Dilution of Precision (GDOP) familiar to those using measurements of pseudo ranges. This new measure is called RDOP for Relative Dilution of Precision.

1. Introduction

With the advent of Global Positioning System (GPS) phase tracking receivers in early 1983 (Bock et al., 1984; Goad and Remondi, 1984), the survey community immediately recognized the increased productivity and superior precision which had not been available, economically, prior to this time. Increases in productivity of a factor of 6 over traditional terrestrial techniques were common. In 1985, Remondi (1985a, 1985b) proposed the use of the same GPS survey quality receivers which could maintain lock on the satellite signals during periods of motion, in kinematic mode spending only a very short time (the order of seconds) at each survey mark visited. This offers another major increase in the number of locations which can be determined. Also shown by Remondi (1985b) was that the path of the antenna could be determined even while the receiver (and antenna) were in motion between survey marks at a resolution coincident with the noise characteristics of the receiver's collected data. Mader (1986) applied the Remondi mathematical model to show that a precise path of a rather rapidly moving platform (aircraft) also could be determined from GPS phase measurements.

The mathematical model used for these investigations was to determine the Doppler change due to motion of the roving antenna by differencing double differences (of phase measurements) in time with the same measurements collected earlier before any movement of the antennas. In the work reported here, an alternate technique is reported which allows one to perform and reduce data obtained in kinematic mode with a traditional double difference algorithm when a precise initial baseline is available. Thus organizations already in possession of GPS phase tracking receivers which can also maintain phase lock on satellite signals during motion, possibly can use vendor-supplied or in-house software to process data collected kinematically which was originally developed only for use in processing data from static surveys.

2. Clinton Lake Dam Survey

During July 1987, the U.S. Army Engineer Topographic Laboratories (ETL) used four ETL-owned Global Positioning System (GPS) receivers for a traditional GPS survey operation at Clinton Lake Dam near Lawrence, Kansas, USA. In addition to the traditional data collection, Corps personnel, along with the author, spent two days collecting survey data in kinematic mode. It was the purpose of this investigation to test the feasibility of utilizing such data for dam monitoring purposes. If successful, three-dimensional surveys can be performed at much lower cost when compared to a conventional terrestrial or static GPS data collection and reduction. Kinematic surveying requires the use of GPS receivers which maintain "lock" on the GPS satellite being tracked during time of movement from one survey marker to another. If lock is not lost, initialization procedures do not have to be redone. This savings in time enables one to spend only a very short time (theoretically just seconds) to recover sufficient data for baseline vector determination. Of particular interest to this experiment on 15 July 1987 using 4 receivers was to measure all 42 dam survey marks using two receivers in fixed positions and two as rovers. All points were recovered from the collected data.

3. The Math Model

The mathematical model is based on the double difference formulation (Bossler et al., 1980). In equation form it is given in simplified form as follows:

$$\phi_{ij}^{km}(t) = -\frac{f}{c}\left[\rho_i^k - \rho_i^m - \rho_j^k + \rho_j^m\right] + N_{ij}^{km}$$

where ϕ_{ij}^{km} is the double difference measurement generated from individual phase observables

$$\left(\phi_i^k - \phi_i^m - \phi_j^k + \phi_j^m \right) \ ,$$

ϕ_i^m is the phase observable from station i to satellite m,

ρ_i^m is the slant range or distance from station i to satellite m,

N_{ij}^{km} is a bias which should be an integer ,

$\quad$ f $\quad$ is carrier frequency, c is speed of light.

If the integer N can be identified, then the double difference is a function only of differences of ranges (in this simplified model), thus allowing very precise determination of relative positions. These biases are easily determined when starting from known stations, since the range combinations are easily computed, or by collecting data for a long period (30 mins. to 2 hrs.) so that both vector and biases can be determined simultaneously using the changing geometry of the station/satellite configurations.

In fact, the biases do not even have to be known in order to obtain baselines. Subtracting the initial double difference from all that follow allows one to eliminate the biases from these new (Doppler) measurements and express them in terms of the change in position. This technique has been used for the case of precise determination of a moving platform, but its strength is also its weakness. That is, once the bias is removed it cannot then be used even if it is known since it is differenced away.

4.0 The Experiment

4.1 Survey Description

The experimental surveys were conducted on two days, 14-15 July 1987. Four GPS phase tracking receivers were used. One occupied a mark just north of the Clinton Lake Dam (North Point) and another receiver occupied a point just south of the dam (South Point). The remaining two receivers were used in kinematic mode. Forty-two survey marks are located on the top of the earthern dam labeled AA-1 to AA-42. AA-1 is the southern most survey marker. The number is then sequential from the south to north. The main idea was to discover if, collecting only 5 minutes of data, subsequent processing could yield cm-level precision if a known initial baseline vector could be used to obtain the integer biases. With known biases and no losses of lock by the receivers, theoretically the baselines can be determined if there are a sufficient number of linearly independent measurements.

4.2 Day 195

The first day of the two day experiment occurred on day 195 of 1987 (14 July 1987). Since neither the Corps nor Ohio State had ever performed a kinematic survey before, it was decided to be very conservative on this first day. No analysis software was available to use in survey planning (more about this later).

As can be seen from Tables 1 and 2, fairly long visit times were used for this first day. Rover #2 remained at AA-5 for the entire second session. The long visit times at the

Table 1. Day 195 Schedule for Rover #1

Mark	UTC Start (H:M:S)	Stop (H:M:S)	GPS Start (Sec)	Stop (Sec)
AA1	21:28:45	22:33:30	250125	254010
AA2	22:35:30	22:47:45	254130	254865
AA3	22:49:15	22:58:45	254955	255525
AA4	23:00:00	23:05:30	255600	255930
AA6	23:07:15	23:12:15	256035	256335
AA7	23:15:45	23:18:30	256545	256710
AA7	23:30:15	23:36:00	257415	257760
AA8	23:39:00	23:49:30	254940	258570
AA7	23:51:00	23:52:45	258675	258765
AA9	23:54:30	23:59:00	258870	259140
AA10	00:00:15	00:05:15	259215	259515
AA11	00:07:00	00:12:00	259620	259920
AA12	00:13:45	00:18:00	260025	260280
AA13	00:19:30	00:23:30	260370	260610
AA14	00:24:45	00:28:45	260685	260925
AA15	00:30:15	00:34:15	261015	261255
AA16	00:36:00	00:41:15	261360	261675
AA17	00:42:30	00:47:30	261125	262365
AA18	00:48:45	00:52:45	262125	262365
AA1	01:00:00	01:20:00	262800	263985

Table 2. Day 195 Schedule for Rover #2

Mark	UTC Start (H:M:S)	Stop (H:M:S)	GPS Start (Sec)	Stop (Sec)
AA2	21:45:30	22:30:00	251130	253800
AA3	22:32:30	22:45:00	253950	254700
AA4	22:46:15	22:55:15	254775	255315
AA5	23:00:00	23:05:30	255420	256920

initial points were chosen so as to provide initial starting vectors and integer biases both for the first and the second day. What was not known at the time of the survey, but discovered during the data processing, was just how weak the solutions were due to poor distribution of the first 4 satellites. One major conclusion of the study is to wait for a minimum of 5 satellites before initiating a kinematic survey. Nevertheless, initial values were obtained and processing was successful (and laborious).

4.3 Day 196

The second day of the two day experiment occurred on day 196 of 1987 (15 July 1987). It was decided to proceed rather blindly on this day moving to a new survey marker every 6 minutes. It usually took about 1.5 minutes to move, thus useful data were recorded for approximately 4.5 minutes per mark. One rover started at marker AA-1 (southern end) and the other stated at marker AA-42 (northern end). The receivers then moved toward each other at the rate of 6 minutes per mark. After they met in the center of the dam, larger steps were taken to provide a better duplicative sampling across the dam. When the visibility dropped to 4 satellites, longer observation times were chosen. All 42 marks were visited on this day. The visit times are given in tables 3 and 4.

5.0 Data Reduction

The reduction of the phase proceeded differently than the way normal static GPS survey data are usually processed. As mentioned earlier, knowledge of the integer biases was a requirement for successful reductions. These bias values were obtained either by using a previously determined precise starting baseline vector or by occupying the initial (or some other) point long enough to recover unambiguously the integer value of the biases. Once determined, the processing only requires the data start/stop times be known or determined in order to isolate the data collected at the desired mark. Using only data collected while stopped at a mark and constraining the biases to their previously determined integer values, the data are processed in double difference mode. The processing proceeds in this manner from mark to mark unless a cycle slip is discovered. Discovery is rather simple due to an obvious inability to obtain an adjustment which reduces the measurement residuals to the small level of 0.05 cycle.

There is not a set algorithm for fixing a cycle slip. Obviously, the more satellites that are tracked the easier it is to isolate the satellite with the new integer bias (loss of lock). A minimum of 5 satellites is suggested. In this study, some data were collected using only four satellites, but the strength of the results is usually lower and cycle slips were much harder to identify and repair.

6. RDOP

It was determined very early in the data processing that some simple quantifiable characteristic was needed in order to assess the quality of the reductions. Because most analysts are very familiar with the concept of Geometric Dilution of Precision (GDOP), some value similar to this should be generated for the relative baseline reductions. Thus Relative Dilution of Precision (RDOP), or the GDOP for the relative case, is generated. This is given as follows:

$$RDOP = [TRACE[A^T \Sigma^{-1} A]^{-1} / \sigma_\phi^2]^{1/2}$$

where

Table 3. Day 196 Schedule for Rover #1

Mark	UTC Start (H:M:S)	UTC Stop (H:M:S)	GPS Start (Sec)	GPS Stop (Sec)
AA42	20:54:00	21:18:00	334440	335880
AA41	21:19:45	21:26:00	335985	336360
AA40	21:27:45	21:36:00	336465	336960
AA39	21:37:30	21:42:00	337050	337320
AA38	21:43:30	21:48:00	337410	337680
AA37	21:49:30	21:54:00	337770	338040
AA36	21:55:30	32:00:00	338130	338370
AA35	22:00:45	22:12:00	338445	339120
AA34	22:13:45	22:18:45	339225	339525
AA33	22:20:15	22:30:00	339615	340200
AA32	22:37:30	22:42:00	340650	340920
AA31	22:43:15	22:48:00	340995	341280
AA30	22:43:15	22:48:00	340995	341280
AA29	22:49:15	22:54:00	341355	341640
AA28	22:55:15	23:00:00	341715	342000
AA28	23:08:45	23:36:00	342525	344160
AA27	23:37:15	23:42:00	344235	344520
AA26	23:43:00	23:48:00	344580	344880
AA25	23:49:15	23:53:00	344955	345180
AA24	23:54:00	24:00:00	345240	345600
AA22	00:01:15	00:10:00	345675	346200
AA19	00:11:45	00:20:00	346305	346800
AA16	00:21:45	00:30:00	346905	347400
AA12	00:32:15	00:00:00	347535	349185

Table 4. Day 196 Schedule for Rover #2

Mark	UTC Start (H:M:S)	UTC Stop (H:M:S)	GPS Start (Sec)	GPS Stop (Sec)
AA1	20:56:15	21:06:00	334575	335160
AA2	21:07:15	21:12:00	335235	335520
AA3	21:13:15	21:19:00	335595	335880
AA4	21:19:15	21:24:00	335955	336240
AA5	21:25:15	21:30:00	336315	336600
AA6	21:31:15	21:36:00	336675	336960
AA7	21:37:15	21:42:00	337035	337320
AA8	21:43:15	21:48:00	337395	337680
AA9	21:49:15	21:54:00	337755	338040
AA10	21:55:15	22:12:00	338115	339120
AA11	22:13:15	22:18:00	339195	339480
AA12	22:19:15	22:24:00	339555	339840
AA13	22:25:15	22:30:00	339915	340200
AA14	22:31:30	27:36:00	340290	340560
AA15	22:37:30	22:42:00	340650	340920
AA16	22:43:30	22:48:00	341010	341280
AA17	22:49:30	22:48:00	341010	341280
AA18	22:55:30	23:00:00	341730	342000
AA18	23:11:30	23:36:00	342690	344160
AA19	23:37:30	23:42:00	344250	344520
AA20	23:43:30	23:48:00	344610	344880
AA21	23:49:30	23:53:00	344970	345180
AA22	23:54:30	24:00:00	345270	345600
AA23	00:01:30	00:10:00	345690	346200
AA26	00:22:15	00:20:00	346335	346800
AA29	00:22:15	00:30:00	346935	347400
AA30	00:32:15	00:40:00	347535	348000
AA37	00:42:15	01:00:00	348135	349200

$(A^T\Sigma^{-1}A)^{-1}$ is the least squares covariance matrix

A is the design matrix (matrix of partial derivatives of the double difference with respect to the unknown baseline elements),

σ_ϕ is the uncertainty in a double difference measurement,

Σ is the double difference covariance matrix (diagonal elements are equal to σ_ϕ^2).

The units of RDOP used in this study were meters/cycle. Thus, theoretically, multiplying RDOP by the uncertainty of a double difference measurement yields the (spherical) relative position error. It was discovered that RDOP values of 0.1 m/cycle were of acceptable quality.

Of particular interest is that RDOP values can be generated days prior to a survey. Only approximate knowledge of the survey location needs to be known along with anticipated survey start/stop times. Precise orbits of the GPS satellites are not required for RDOP calculations; the broadcast almanacs are ideal for these calculations.

It should also be emphasized that the RDOP generation must model the normal matrix (vector of unknowns) as closely as possible. In the analysis here the initial baseline vector was known which, in turn, allowed for determination of the (integer) biases. Thus, for the baseline, recoveries no biases were estimated. This is sometimes called the "fixed" solution. RDOP values for the fixed case will be substantially smaller than corresponding RDOP values for the case when biases are not known apriori and are part of the parameter set being estimated (the "float" case).

7.0 Results

Table 5 gives the comparisons for all points visited two or more times along with the computed RDOP values (for the fixed case). The table reveals that when RDOP values of 0.10 meters/cycle were achieved, excellent recoveries were obtained. It should be emphasized again that the visits to the dam marks were done "in the blind." That is, the movement from one mark to the other was done on a regular bases without regard to satellite geometry. Had RDOP values been available before the survey, then more reliable results would have been achieved. Not too surprising is that low RDOP values were computed when five satellites were available. Since our surveys started and ended with four satellites above the cutoff mask angle, some of the early and later visits have higher RDOP values and therefore cannot be trusted even though the integer biases were known. Centimeter repeatability was achieved much of the time even when RDOP values were 0.40 meters/cycle.

8. Herndon Test Results

Subsequent to the Clinton Lake Dam survey by the U.S. Army Corps of Engineers, the U.S. National Geodetic Survey (NGS) performed a similar kinematic test at the Defense Mapping Agency facility near Herndon, Virginia. Based on recommendations from the author, NGS personnel waited until five satellites were available. Fewer sites were used which enabled many more visits per site. The test spanned two days. The results of the data analysis are given in table 6. Time of mark occupation is given along with vector recoveries and associated RDOP values.

Because tracking from five satellites was always available, the data reductions were performed with much more ease. Although cycle ships were encountered, they were much easier to repair with the availability of the redundant fifth satellite.

Table 5.
Kinematic Survey Comparisons
in WGS 84 Δx, Δy, Δz coordinates

Day	Session	Rover	Δx (m)	Δy (m)	Δz (m)	RDOP (m/cycle)
			C-2		C-3	
			North Pt. to South Pt.			
197	A		276.670	-1961.927	-2358.472	0.09
197	B		276.665	-1961.929	-2358.470	0.04
197	C		276.664	-1961.931	-2358.461	0.04
197	D		276.671	-1961.929	-2358.470	
			North Pt. to AA1			
197	A		353.544	-1737.754	-2116.166	0.09
195	B	1	353.545	-1737.736	-2116.174	0.17
195	A	1	353.545	-1737.755	-2116.162	0.03
196	A	2	353.528	-1737.742	-2116.149	0.18
			North Pt. to AA2			
195	A	1	349.856	-1699.199	-2068.478	0.04
195	A	2	349.854	-1699.202	-2068.485	0.06
196	A	2	349.841	-1699.201	-2068.455	0.27
			North Pt. to AA3			
195	A	1	342.461	-1660.206	-2021.954	0.06
195	A	2	342.457	-1660.205	-2021.960	0.05
196	A	2	342.462	-1660.191	-2021.966	0.30
			North Pt. to AA4			
195	A	1	335.288	-1621.826	-1975.540	0.08
195	A	2	335.291	-1621.826	-1975.549	0.06
196	A	2	335.272	-1621.831	-1975.513	0.36
			North Pt. to AA5			
195	B	2	327.903	-1583.402	-1928.969	0.08
196	A	2	327.879	-1583.408	-1928.936	0.46
			North Pt. to AA6			
195	A	1	320.681	-1545.015	-1882.307	0.08
196	B	2	320.660	-1545.001	-1882.281	0.68

North Pt. to AA7

197	A		313.137	-1506.359	-1835.526	0.07
195	A	1	313.134	-1506.369	-1835.507	0.12
195	B	1	313.136	-1506.372	-1835.512	0.19
196	A	2	313.064	-1506.430	-1835.338	1.48

North Pt. to AA8

| 195 | B | 1 | 305.946 | -1467.860 | -1788.910 | 0.08 |
| 196 | A | 2 | 305.947 | -1467.803 | -1788.970 | 1.88 |

North Pt. to AA9

| 195 | B | 1 | 298.558 | -1429.379 | -1742.299 | 0.13 |
| 196 | A | 2 | 298.553 | -1429.296 | -1742.395 | 0.81 |

North Pt. to AA10

| 195 | B | 1 | 291.124 | -1390.861 | -1695.533 | 0.46 |
| 196 | A | 2 | 291.114 | -1390.818 | -1695.559 | 0.05 |

North Pt. to AA11

| 195 | B | 1 | 283.828 | -1352.484 | -1648.905 | 0.41 |
| 196 | A | 2 | 283.813 | -1352.458 | -1648.900 | 0.08 |

North Pt. to AA12

195	B	1	276.534	-1314.010	-1602.305	0.41
196	B	1	276.533	-1314.021	-1602.293	0.13
196	A	2	276.525	-1313.997	-1602.305	0.08

North Pt. to AA13

| 195 | B | 1 | 269.034 | -1275.109 | -1555.309 | 0.41 |
| 196 | A | 2 | 269.023 | -1275.108 | -1555.294 | 0.08 |

North Pt. to AA14

| 195 | B | 1 | 261.620 | -1236.679 | -1508.716 | 0.39 |
| 196 | A | 2 | 261.595 | -1236.658 | -1508.712 | 0.08 |

North Pt. to AA15

| 195 | B | 1 | 254.249 | −1198.217 | −1462.084 | 0.38 |
| 196 | A | 2 | 254.228 | −1198.220 | −1462.062 | 0.08 |

North Pt. to AA16

197	B		246.889	−1159.664	−1415.449	0.07
195	B	1	246.885	−1159.637	−1415.455	0.33
196	A	2	246.877	−1159.652	−1415.449	0.08
196	B	1	246.884	−1159.641	−1415.448	0.27

North Pt. to AA17

| 195 | B | 1 | 239.501 | −1121.277 | −1368.840 | 0.33 |
| 196 | A | 2 | 239.500 | −1121.285 | −1368.830 | 0.08 |

North Pt. to AA18

195	B	1	232.247	−1082.735	−1322.092	0.37
196	A	2	232.247	−1082.734	−1322.090	0.09
196	B	2	232.245	−1082.749	−1322.078	0.11

North Pt. to AA19

| 196 | B | 1 | 224.806 | −1043.972 | −1275.199 | 0.29 |
| 196 | B | 2 | 224.808 | −1043.989 | −1275.190 | 0.11 |

North Pt. to AA21

| 197 | B | | 210.059 | −967.080 | −1181.936 | 0.04 |
| 196 | B | 2 | 210.056 | −967.084 | −1181.920 | 0.13 |

North Pt. to AA22

| 196 | B | 1 | 202.717 | −928.623 | −1135.282 | 0.31 |
| 196 | B | 2 | 202.717 | −928.632 | −1135.285 | 0.27 |

North Pt. to AA26

197	C		173.284	−774.425	−948.531	0.04
196	B	1	173.288	−774.435	−948.531	0.11
196	B	2	173.284	−774.426	−948.531	0.30

North Pt. to AA28

| 196 | A | 1 | 158.584 | −697.335 | −855.134 | 0.08 |
| 196 | B | 2 | 158.585 | −697.371 | −855.104 | 0.11 |

North Pt. to AA29

| 196 | A | 1 | 151.145 | −658.904 | −808.614 | 0.08 |
| 196 | B | 2 | 151.144 | −658.904 | −808.603 | 0.28 |

North Pt. to AA31

| 197 | C | | 136.409 | −581.632 | −715.074 | 0.04 |
| 196 | A | 1 | 136.412 | −518.630 | −715.072 | 0.08 |

North Pt. to AA36

| 197 | C | | 92.391 | −350.581 | −435.185 | 0.08 |
| 196 | A | 1 | 92.402 | −350.544 | −435.222 | 0.11 |

North Pt. to AA37

| 196 | A | 1 | 85.016 | −312.172 | −388.650 | 0.82 |
| 196 | B | 2 | 85.134 | −312.971 | −388.594 | 0.17 |

North Pt. to AA42

| 197 | D | | 9.780 | −122.823 | −163.893 | 0.07 |
| 196 | A | 1 | 9.776 | −122.818 | −163.891 | 0.11 |

Table 6.
Herndon Kinematic Results

Baseline	Day/ Session	dx (m)	dy (m)	dz (m)	dh (m)	Rdop (m/cycle)	Time (min)
optk-31309	343a	14.936	-58.372	-74.008	0.238	0.08	7
	343b	14.925	-58.351	-73.996	0.228	0.12	5
	343c	14.930	-58.360	-74.001	0.232	0.08	21
	344a	14.935	-58.360	-74.005	0.230	0.03	11
	344b	14.926	-58.351	-73.987	0.234	0.03	10
	344c	14.929	-58.366	-73.984	0.247	0.03	14.5
optk-w83rma	343a	-80.930	-62.216	-49.319	2.325	0.12	5
	343b	-80.920	-62.222	-49.316	2.333	0.12	6
	344a	-80.933	-62.208	-49.320	2.318	0.07	2
	344b	-80.931	-62.205	-49.307	2.324	0.07	2
optk-31313	343a	-54.174	-26.967	-13.710	2.572	0.12	5
	343b	-54.162	-26.964	-13.711	2.572	0.12	5.5
	344a	-54.173	-26.957	-13.707	2.567	0.07	2
	344b	-54.167	-26.957	-13.703	2.571	0.07	2
	344c	-54.169	-26.972	-13.699	2.585	0.07	2
	349a	-54.169	-26.959	-13.728	2.556	0.15	5
	349b	-54.169	-26.957	-13.707	2.568	0.13	5
	349c	-54.167	-26.967	-13.712	2.573	0.14	4
optk-w83 ecc	343a	59.724	0.795	-16.515	-0.800	0.12	6
	343b	59.727	0.782	-16.512	-0.788	0.13	5
	344a	59.719	0.804	-16.508	-0.804	0.07	2
	349a	59.718	0.795	-16.513	-0.800	0.13	5
	349b	59.718	0.795	-16.514	-0.801	0.12	5
optk-rm1 hern	349a	1.967	-9.182	-11.168	0.271	0.09	9
	349b	1.973	-9.183	-11.158	0.279	0.13	5

Also of interest is to compare the occupation times with their corresponding RDOP values. Note that in some cases the satellite geometry was so good that shorter occupation times yielded smaller RDOP values. It is clear that time on site is only one of several components which contribute to the overall success of the survey.

9. Summary

Table 5 shows clearly that the experiment was successful. The data were processed with vendor supplied software. Less than the desired 5 minutes of data per site were collected with better than expected results. A new measure (RDOP) was also devised which can be used even prior to the survey for planning purposes. Even better, the RDOP values can be generated for use in realtime to determine just how long one needs to remain at a mark to achieve centimeter-level results. The second day required four hours of GPS data collection in kinematic mode for all 42 points to achieve the same level of results as a traditional terrestrial survey at many less points, and also achieved a three dimensional (geometric) reduction.

10. Acknowledgements

These excellent results could not have been achieved without the diligent cooperation of the Army Corps of Engineers survey team. Any new or untried exercise such as this is stressful on participants; the cooperative efforts created an environment which contributed significantly to the success of the mission. The author is indebted to them for their efforts.

References

Bock, Y., C.C. Counselman, S.A. Gourevitch, R.W. King and A.R. Paradis (1984): Geodetic accuracy of the Macrometer^{TM} model V-1000, *Bulletin Geodesique,* Vol. 58, pp. 211-221

Bossler, J.D., C.C. Goad, P.L. Bender (1980): Using the Global Positioning System (GPS) for geodetic surveying, *Bulletin Geodesique,* Vol. 54, pp. 553-563

Goad, C.C., B.W. Remondi (1984): Initial relative positioning results using the Global Positioning System, *Bulletin Geodesique,* Vol. 58, pp. 193-210

Mader, G.L. (1986): Dynamic positioning using GPS carrier phase measurements, *Manuscripta Geodaetica,* Vol. II, pp. 272-277

Remondi, B.W. (1985a): Global Positioning System carrier phase: description and use, *Bulletin Geodesique,* vol. 59, pp. 361-377

Remondi, R.W. (1985b): Performing centimeter-level surveys in seconds with GPS carrier phase: initial results, *Navigation: Journal of the Institute of Navigation,* Vol. 32, No. 4, pp. 386-400

TMMacrometer is a registered trademark of Aero Service Division, Western Atlas International, Inc., Houston, Texas, USA.

EXPERIENCES WITH THE WM 101 GPS RECEIVER

by

Lars E Sjöberg

Abstract

Since June 1987 the Royal Institute of Technology (RIT) has three WM 101 GPS receivers at its disposal. The equipment has been used together with the PoPS software in some campaigns and pilot projects, such as the GPS test network at Gävle, a densification network in the Båstad region, the triangle RIT-Lovön-Uppsala, height control for photogrammetric mapping and the Åland campaign. The experience gained until now is summarized in the presentation. In general the equipment meets the specifications. However, several steps of the first version of PoPS were more or less malfunctioning. The newly delivered second version appears to run more satisfactory.

1. Background

Early in 1985 the Swedish GPS group was formed by members from the Department of Geodesy at the Royal Institute of Technology (RIT), the Onsala Space Observatory at Chalmers Institute of Technology, the Department of Geodesy at the University of Uppsala, the National Land Survey (NLS), the Board of Road Administration (BRA) and the Swedish Hydrographic Department. Since that time the informal group has grown to the order of 25 persons at 10 universities or other departments. The goal of the group is to stimulate the development of GPS in the country, by arranging seminars and starting projects, and by creating funds for purchasing equipment. At present the group is finalizing a research plan for GPS in Sweden.

In Spring 1986 test measurements with various GPS receivers started at a test network at Gävle, central Sweden. At about the same time an application for the purchase of two GPS receivers plus accessories submitted by the universities of the group, was approved by the Swedish Council for Planning and Coordinating of Research and the Wallenberg Foundation. The NLS provided funds for an additional receiver. Guided by the test measurements and other sources the "group-of-owners" decided to order three Wild-Magnavox 101 receivers and the PoPS software. The equipment was delivered in June 1987.

2. Experiences in the field

The WM 101, until now a pure L1-unit for tracking up to six satellites, is a robust, easily transportable GPS receiver. According to the manufacturer the unit can operate in the temperature range -25° C to $+55^{\circ}$C. We have not been able to check these limits. (One reason was the unusually mild winter.) The sealed construction of the unit is not only weather proof but even buoyant. Wild states also, that in most cases it is not necessary to carry external batteries into the field, but one can rather rely on a (fully charged) internal battery. Our experience is the opposite: The standard rule is not to trust the internal battery for the daily observation session! A new, fully charged battery is said to operate for 4.5 hours. Used batteries do not! After a few months all battery indicators are malfunctioning, and even if they were operating properly, one cannot change from internal to external battery (or vice-versa) without a break in the session.

The WM 101 has an advanced control panel allowing sophisticated as well as simple operations. As the receiver may be pre-programmed the field measurements can be performed by "any" personnel after a few instructions. However, the field work should gain from an improved field manual.

The satellite data is normally recorded as so-called compressed data. Assuming that this data is being recorded at one minute intervals, a standard casette tape can record 6 satellites for about 14 hours.

The criteria for selecting GPS control points differ in some respects
from classical geodetic work. In the latter case one essential criterion
for optimum point location is the intervisability of points. In the
satellite method the visibility of satellites in all azimuts is a
dominating criterion. In addition signal reflecting surfaces in the
neighbourhood of control points should be avoided. These different
demands cause a problem in combining GPS networks and terrestrial networks. Furthermore, the field experience of the traditionally working,
local surveyor cannot automatically be deployed for the planning of a
GPS' network. In other words, traditional control points can hardly be
adopted as GPS points without field reconnaissance. In a forrestry
country like Sweden traditional control points cannot generally be
used for GPS without placing the GPS antenna on a special mast above
the point. This fact deteriorate both speed and accuracy.

The overall experience of the receiver and antenna units is that they
are both reliable and easy to operate. The only hardware problems
occuring since the delivery of the equipment in June 1987 are the above
mentioned battery problems, some broken battery cable locking
nails and a few unexplained power-cuts during measurement leading to
interruped observation sessions.

3. Experiences with PoPS

The PoPS (Post Processing Software) from Wild-Magnavox is developed to
run on an IBM PC XT equipment with a coprocessor. A MEMTEC cassette terminal
is used for reading the cassettes with WM 101 data into the computer.
Special attention has to be paid to the fact that newest desk top
computers are not always compatible with the MEMTEC terminal. This
creats a problem to read the cassettes. One good solution is the Compaq 360
portable PC.
Although the PoPS is written in FORTRAN 77 the user is not permitted
access to the source code. This is naturally an important problem for a
research organisation.

3.1 The PoPS version 1.04

The version 1.04 of the software was delivered together with the
receivers. It includes "user friendly", data-management system modules
with the components field preparation, data transfer, data pre-proces-
sing, baseline definitions, repair of the cycle-slips and computation.
The computation module uses the principle of double differences of
baseline measurements. It includes tropospheric and ionospheric modelling
of the signal propagation as well as clock parameter modelling.

In starting up the computations the program allows the observation interval to be subdivided into shorter sessions and meteorological data to be
updated. However, there is no possibility to change observation data.

The PoPS manual describes the software installation, programs and a numerical test example. The various software modules are described in detail chapter by chapter. However, there is no useful overview of the program, but the user is forced to read the main part of the manual to, hopefully, grasp it. Subsequently there is a great need for a software overview and a short version of the manual. In addition, detailed instructions are missing in the manual. Appendices describing the computer algorithms are also missing.

In certain modules logfiles are created. These often huge files contain a lot of information,but do not always include other essential information. It should be a clear advantage if the user could choose what to be stored.

The preprocessing module includes the repair of cycle-slips by polynomial fit of differences between successive baseline differences. This processing is often rather critical and the computed result depends to a high degree on the operator's experience. In practise it seems that outliers should be rejected more restrictively than advised in the manual. More research is needed to define optimum rejection limits! (Cf. Jivall and Jakobsson, 1987).

The computing module allows processing of up to 10 network points and 78 unknowns. However, there is an additional limitation of 60 ambiguites to be solved for. If there are many ambiguities, several of these have to be resolved before entering the computing module,possibly implying the introduction of gross errors to the data.
In Frei et al. (1986) some approximate times are given to process various modules of PoPS. In an independent study Jivall and Jakobsson (1987) conclude that the total necessary operator time is longer. In particular the processing of the computing module is much longer.

The PoPS routine for tropospheric correction from actual data was malfunctioning in this PoPS version . In addition PoPS did not run for most sessions belonging to the RIT-Lovön-Uppsala triangle. The reason is not known. See section 4.4.

Finally, several sections of the PoPS were missing, e.g. with no information on error codes.

3.2 The PoPS version 2.01

The updated version 2.01 of PoPS arrived only at the beginning of March 1988. According to WM Satellite Survey Company "PoPS now fulfils all specifications as originally planned. It contains the transformation, results and data-editing components." A most well-come new feature is the option "AutoPoPS", allowing the software to run in a time-saving automatic mode. Other features are

   * more flexible use of log-files

* rigorous treatment of correlations between zero- and double-
     differences

and
   * "computation windows" allowing the use of only a part of the
     collected data in session

Due to the late receipt of the new PoPS version there was little time to
fully investigate it. However, it appears that most defects pointed out
for PoPS version 1.04 have been corrected. The same holds for the manual.
See also Section 4, in particular the iteration problem discussed in
Section 4.4.

4.   Results

In this section we discuss results and other experiences gained during
specific campaigns and projects.

4.1   The Gävle test network

The GPS test network outside Gävle consists of six triangle stations,
four out of which are located in forest. Two of these (Västerbro and
Mårtens klack) are equipped with permanent towers,while at the other
two (Sälgsjön and Nyhammar) transportable masts are used. The final two
points (Hille and Mårtsbo) are also located on permanent towers. See
Fig. 1. The baselines, ranging between 8 and 40 km, were observed by
laser Geodimeter to a standard error of adjusted distances of about 1 ppm.
Three of the station heights have been determined by spirit levelling,
while the other three are so far determined only by trigonometric
levelling.

During April 1986 the test network was measured by the NLS and BRA
in corporation with GPS Survey Services, England using four Trimble
4000S GPS receivers. The rather cold weather (-7°C in the evenings)
revealed some shortages of the equipment and it is most likely that
these defects affected results. The data was processed with the BASE-
LINER Software. More details are given in Jonsson (1986).

In June 1986 two Texas TI 4100 were used by the NLS and RIT in the test
network. The aim was to keep the observation sessions to one hour each
and to observe three baselines per day. In contrast to the April campaign the signal propagation was now disturbed by birch trees having come
into leaf   at one station. The data was processed with the Magnet-
4100 software by Texas Instruments, Dallas. See also Jonsson (1986).

In November 1986 the test network was observed with three WM 101 receivers
by Wild in cooperation with RIT and NLS. The data was processed by Wild
Heerbrugg using the PoPS software.

In Table 1 we compile the result of the 1986 test measurements as Geo-
dimeter-GPS differences. The table suggests that there are no systematic

differences (because the mean differences are within the standard errors). The manufacturers provide the following baseline accuracies to there instruments:

Trimble 4000S	10 mm + 2 mm/km
Texas TI 4100	5 mm + 1 mm/km
Wild WM 101	10 mm + 2 mm/km

The differences from four Trimble 4000 S baselines are exessive to these numbers. This is the case for all TI 4100 baselines.However,it should be kept in mind that the differences include Geodimeter errors. Nevertheless the WM 101-differences to Geodimeter are all within the 10 mm + 2 ppm limit.

Summarizing, the test measurements in Gävle gave a good insight to the practical work with various GPS receivers and softwares. The tests certainly guided the group to choose the WM 101 and PoPS. For more details we refer to Jonsson (1986) and (1987). The WM 101 data was further analysed by Jivall and Jakobsson (1987). By statistical testing they came to the conclusions that

a) one cannot reject the hypothesis that there is no bias in the (weighted) GPS-Geodimeter differences (Δ), and

b) the apriori standard error for Δ is too big.

Based on the WM 101 ellipsoidal heights and levelled orthometric heights geoid undulations were determined for all points. The RMS variation of these undulations with respect to there mean is 0.50 m. The geoid undulations were also estimated from astro-geodetic levelling (Bo Jonsson, private communication). The standard deviation of the difference between the two types of undulations is 0.13 m. At the RIT a computer program has been developed to determine gravimetric geoid undulations from gravity data and potential coefficients combined in a least squares sense. For details on this method the reader is referred to Sjöberg (1986) and (1987). Preliminary results on undulation differences derived from GPS and the OSU 86 C potential coefficients complete through degree 250 yielded a standard deviation of 0.26 m. A slight reduction of this deviation to 0.22 m was achieved when including a (rather incomplete) set of gravity data within a 2° cap around each computation point. Work is underway to improve the terrestrial data.

Table 1. Comparison of Geodimeter (T) and GPS (G) baselines (T-G)
From Jonsson (1987).

Baseline	Length km	No. of obs.	TRIMBLE 4000S BASE-LINER Difference		Texas TI4100 MAGNET 4100 Differ-ence		WM101 PoPS Differ-ence	
			mm	ppm	mm	ppm	mm	ppm
Mårtens klack-Västerbo	37	1	59	1.6			-37	-1.0
-Mårtsbo	40	4	13	0.3			-37	-0.9
-Nyhammar	18	2	5	0.3			-16	-0.9
-Hille vt	29	3	-23	-0.8			-16	-0.6
Sälgsjön-Västerbo	13						-17	-1.3
-Nyhammar	16						-17	-1.3
-Hille vt	12	1	75	6.3	19	1.6	10	0.8
Mårtsbo-Västerbo	16	1	94	5.8	25	1.5	0	-0.0
-Sälgsjön	8	4	35	4.2			25	3.0
-Hille vt	15	5	89	5.2	20	1.3	19	1.2
Nyhammar-Hille vt	11	1	-31	-2.8	-30	-2.7	-23	-2.0
Mean Difference				+2.2		0.42		-0.29
Standard deviation				3.2		2.1		1.45
RMS Difference				3.8		1.9		1.4

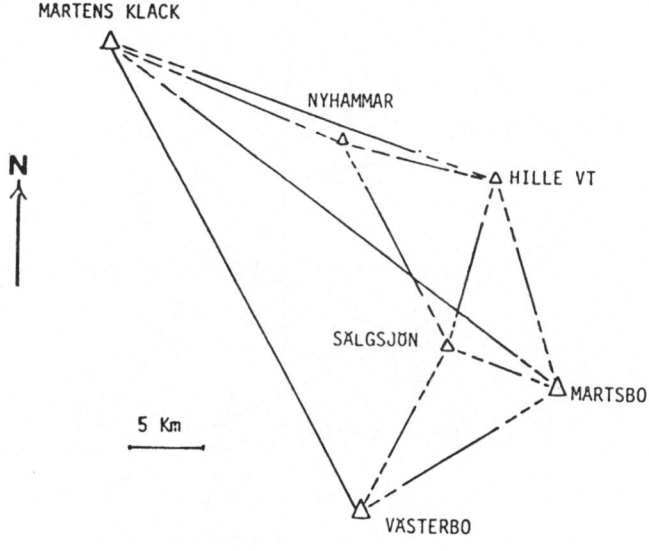

Fig 1   The GPS test network at Gävle

4.2 The Båstad region densification network

In 1986 ten triangulation points in the densification network of the Båstad community, southern Sweden, were selected for a GPS test. The purpose of the study is to investigate the practical and economical feasibility of GPS to this type of control net. A preliminary report on the study was presented by Håkansson (1987). The network configuration is shown in Fig 2. The observed baselines were measured during eight days with 2-4 stations occupied each day. The number of stations (10) was selected to fit the PoPS (version 1.04) software. However, due to software restrictions on ambiguities (see Section 3.1) the network had to be adjusted in two steps. (An alternative would have been to solve for ambiguities prior to the adjustment. However, this might lead to biased solutions.) The <u>internal</u> standard error of coordinates of the adjusted network is in the order of 1-2 mm. Including external factors the accuracy is estimated to one order of magnitude larger.

The stations 1, 2 and 9 were used as fitting points in a 4 parameters Helmert adjustment in x and y to fit GPS coordinates to regional, municipal coordinates. The resulting scale factor was 1.66 ppm and the RMS residuals in the seven new points were 1.8 cm and 1.9 cm in x and y, respectively.

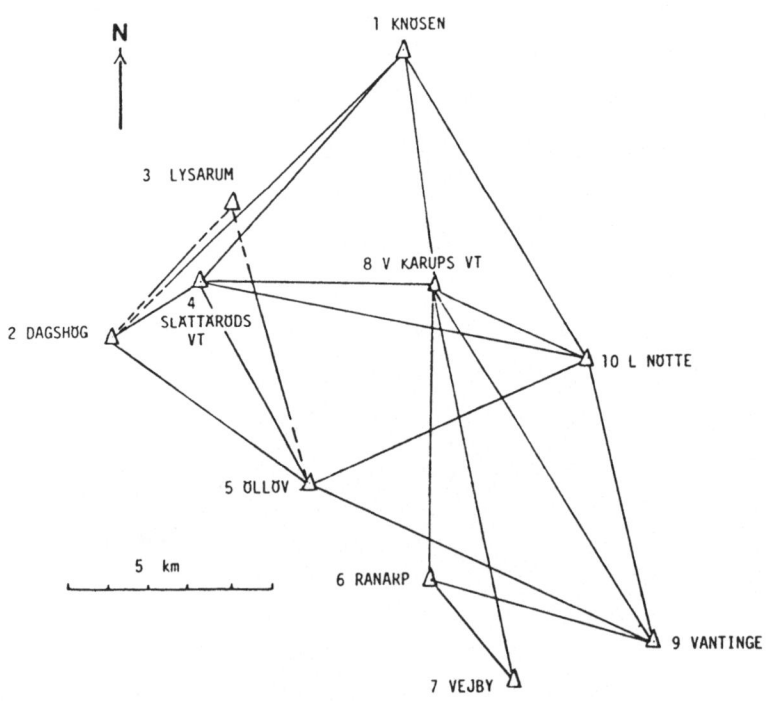

Fig 2 The GPS network at Båstad

The orthometric heights of the GPS points were generally determined by
spirit levelling connection to the levelling network of the community.
In a few cases this connection was carried out by trigonometric levelling.
A one parameter bias fit was performed between GPS ellipsoidal heights
and the terrestrial (orthometric) heights for the stations 1, 2 and 9.
This type of fit could be valid if the geoidal height is practically
constant in the area. However, the actual fit indicates a slope of the
geoid of about 1.5 dm between the points 9 and 2, or +1.0 cm/km. No
detailed geoid undulations are known in the area as yet, but the astro-
geodetic geoid for Sweden, GEOID RAK 1970 (Ussisoo, 1975), confirms a
slope of that order. If the bias fit is replaced by a three parameter fit
(including slopes) to the same points (1, 2 and 9) a residual of 5.0 cm
still exists for the central point L Nötte (cf. Fig. 2). Obviously the
(orthometric) height determination by GPS could benefit from a detailed
geoid mapping already for this small size area.

Based on the above study it appears (Håkansson, 1987), that GPS is
economically motivated today for triangulation down to municipal densification networks. Technical development and cost reductions are likely
to improve the cost benefit ratio GPS to traditional technique.

4.3 The Åland-campaign

The idea to connect the national height systems in Sweden and Finland
by an Åland link is an old dream within the Nordic Geodetic Commission.
Inspired and economically supported by professor Kakkuri, the RIT and
the University of Stuttgart decided to cooperate with the Finnish Geodetic Institute on such a project based on GPS techniques. Any height
connection better than, say, 10 centimeters of accuracy is a good check
on the quality of the existing networks. In addition there is a possibility to investigate the accuracy of the ED 87 horisontal coordinates in
the region. Although the ionospheric effects might be critical for the
solution, and the WM 101 GPS units will not be updated for the L2 frequency
until the end of 1988 most essential experiences are expected from such
a project. More details of this project will be presented elsewhere (see
e.g. Kakkuri and Sjöberg, 1988). Here we   give a brief, preliminary
report on the observations and computations.

The observation campaign was carried out during 4 days in October, 1987
with 5 WM 101 receivers. One receiver was permanently located at station
Märket, while the other four receivers altered in the network in such a
way that each baseline was observed during two days. (See Fig 3.)

Some preliminary statistics and computing results are given in Tables 2
and 3. The duration of each session was about 2 hours.

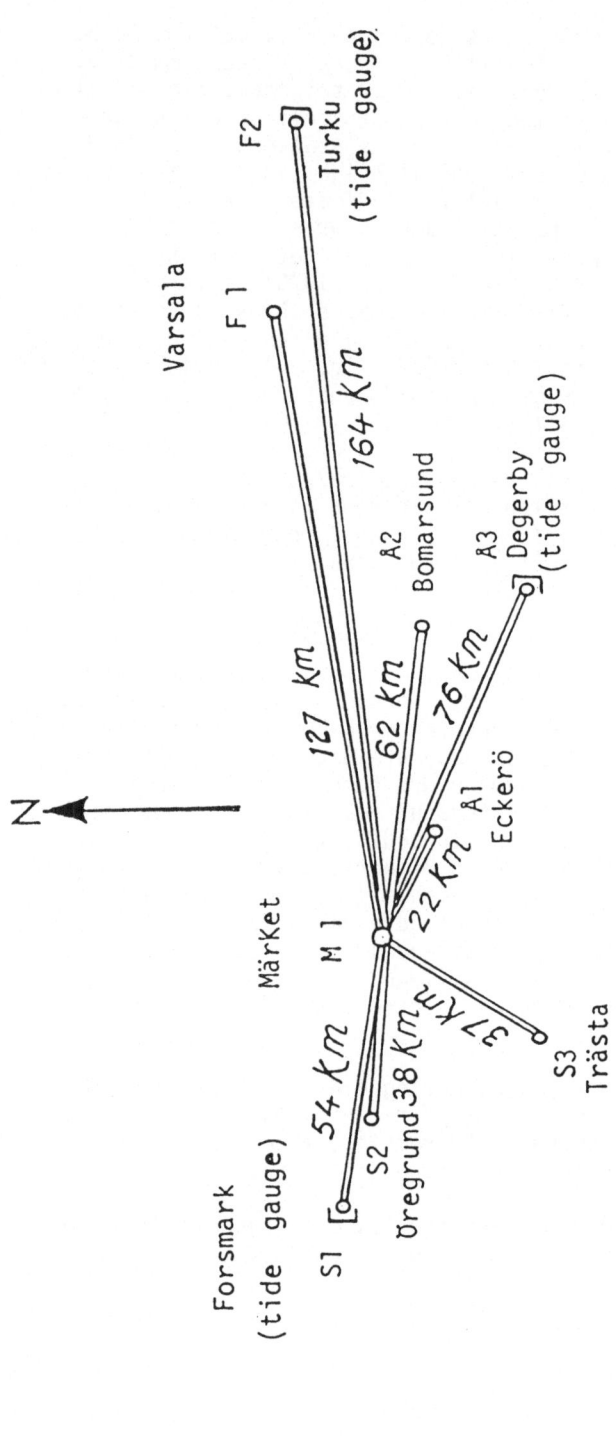

Fig 3 The net configuration of the Åland campaign

Days	Receiver locations					
	230	239	241	G1	G2	
1	S1	S2	M1	A2	A3	
2	S2	S3	M1	A1	A3	
3	S1	A1	M1	F1	F2	
4	S3	A2	M1	F1	F2	

116

Table 2. Computation statistics on the Åland GPS experiment.
No. of parameters. RMS differences in meters.

Session no. Date	1+2 87-10-12	3+4 87-10-13	5+6 87-10-14	7+8 87-10-15	All
Observations	3595	3009	2842	2651	12097
Parameters	52	47	44	45	35
Ambiguities	40	35	32	33	11*
Resolved ambig.	40	31	23	31	3
RMS of $\}\{$ before	0.014	0.016	0.031	0.021	0.031
D.diff. after	0.019	0.021	0.032	0.025	0.031

*Option used: Ambiguities of previous solutions.

Table 3. Preliminary slope distances and ellipsoidal heights from net
adjustment. No tropospheric corrections. Standard errors from
two daily sessions.

Site pairs	d $|m|$	Δh $|m|$
M1 -S1	53722.087 $\pm$ 0.035	-1.962 $\pm$ 0.018
-S2	38185.542 $\pm$ 0.015	-0.712 $\pm$ 0.005
-S3	36580.840 $\pm$ 0.003	-7.666 $\pm$ 0.008
-A1	22071.045 $\pm$ 0.003	3.263 $\pm$ 0.001
-A3	75771.000 $\pm$ 0.011	2.274 $\pm$ 0.052
-A2	62362.170 $\pm$ 0.003	-1.754 $\pm$ 0.023
-F1	122617.093 $\pm$ 0.007	-17.172 $\pm$ 0.057
-F2	164445.633 $\pm$ 0.115	1.810 $\pm$ 0.005

4.4 Other results

A local GPS network in the Stockholm region is under development. The network will be connected to several trig. points of the community. Six baseline measurements, (ranging from 3.1 to 14.7 km) have so far been successfully completed. The standard deviations of the computed slope distances
and ellipsoidal heights among sessions spanning a time period of about
two hours are included in Figs 5 and 6.

In connection with the above measurements the (60 km) baselines from the
Uppsala first order triangulation point to Lovön and to RIT were measured
during 5 days. The Uppsala data could not be read with the PoPS version
1.04. With the version 2.01 the data can be read, but no meaningful single
point positioning results are obtained. The reason is under investigation.

Finally we report some preliminary results from a GPS campaign aiming
at a three-dimensional photogrammetric ground control. The network, shown
in Fig 4, was observed during 4 days and data was divided in less than two
our sessions. In Tables 4 and 5 we present some computation statistics
based on the PoPS. All operators (one assistant and four students) started
from the same data set. The differences are due to different approximate
coordinates (!) and different results of data screening. Table 5 shows
that the final coordinates may vary several centimeters for baselines
within 2.1 km merely due to these two processing factors.

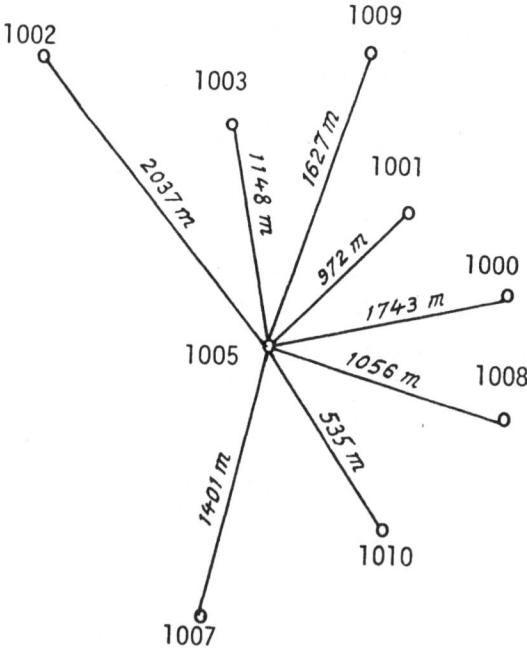

Fig 4 The local photo control GPS network

118

Table 4. Computation statistics for various PoPS operators
RMS double differences in meters

Operator No. of	Assistant	Stud. 1	Stud. 2	Stud. 3	Stud. 4
Observations	1775	1733	1609	1636	1566
Parameters	48	45	43	43	43
Ambiguities	24	24	22	22	22
Resolved amb.	24	24	22	22	21
RMS of double (I)	.009	.023	.009	.013	.010
diff. (II)	.009	.024	.010	.014	.011

Table 5 Statistics on students' versus assistant's estimated coordinate
differences for the points 1001-1003, 1007, 1009 and 1010
(JP=assistant). Units: mm.

	Stud 1-JP	Stud 2-JP	Stud 3-JP	Stud 4-JP
No. of differences	18	18	18	18
Max. difference	36	31	80	83
Mean difference	-1.7	-2.3	-20.5	18.6
Standard deviation of differences	15	17	42	51

We have found that the coordinates computed with the PoPS might be
significantly dependent on the choice of approximate coordinates of the
unknown points. In one case a variation of the approximate coordinates
by about 40 meters changed the final coordinates by millimeters. This
effect, probably due to a critical linearization of the observation
equations, usually converges when iterated.

In Figs 5 and 6 we summarize the standard deviations of some session
results for slope distances and ellipsoidal heights, respectively. Generally
the standard deviations are well within 1 ppm even for baselines ranging
to 150 km. It is noteworthy that this result was obtained with the WM
101 as a single-frequency receiver and with the PoPS software. The use
of two-frequency receivers and a software allowing satellite orbit
improvements might significantly reduce possible systematic errors.

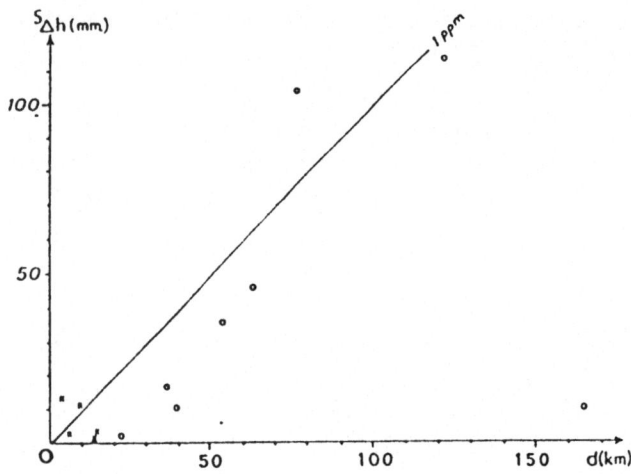

Fig 5 The standard deviation of GPS height differences versus
distance. For x and o, see Fig 6

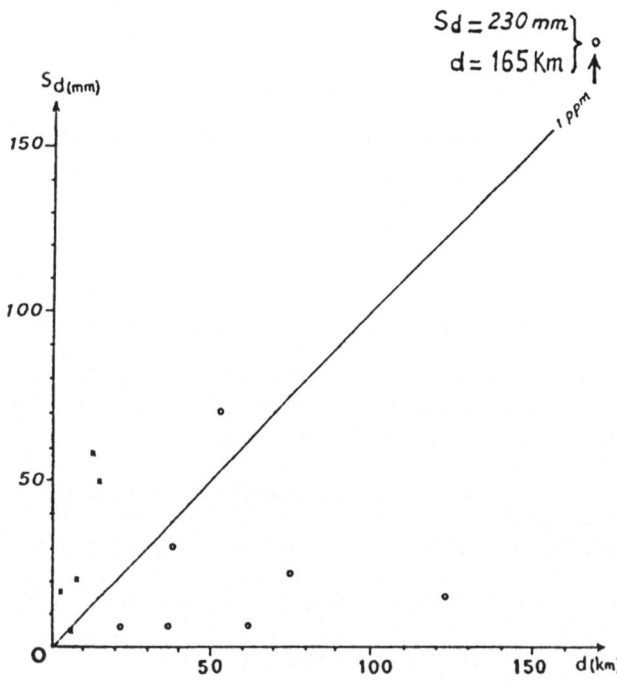

Fig 6 The standard deviation of GPS slope distance from session
means versus distance
x = the local network in the Stockholm region (2 hours sessions)
o = the Åland campaign with one day sessions

120

5. Conclusions and final remarks

Except for some minor problems our WM 101 receivers have proved to be reliable and easy to operate. In the near future the receivers will be updated with a new firmware allowing pre-programming for automatic time control of measuring sessions. In addition this installation is expected to solve the aforementioned battery problems.

The PoPS version 1.04 was incomplete and malfunctioning in several respects. Hopefully most of these defects have been repaired in the recently received new version of the software. However, there appears to be a linearization problem in it.

The computing results are generally satisfactory and well within the 10 mm + 2 ppm of accuracy as prescribed by Wild Magnavox Survey Co. Improvements are expected after the L2 frequency update at the turn of the year.

Acknowledgements. The computations were performed by Mr A. Forsberg and Dr. J. Piechocinska. Valuable comments were given by them and Dr. K. Hayling. All this support is cordially acknowledged.

References

Frei, E., R. Gough and F. Brunner (1986): PoPS^{TM}: A new generation of GPS post processing software. Proceedings of the 4th International Geodetic Symposium on Satellite Positioning, 28 April-4 May 1986, Austin, Texas

Håkansson, A. (1987): GPS i Bjäre. The Royal Institute of Technology, Department of Geodesy, Stockholm

Jivall, L. and L. Jakobsson (1987): Mäta med GPS. Tekniska Skrifter - Professional Papers. LMV-Rapport 1987:18. (Diploma Theses RIT.)

Jonsson, B. (1986): GPS-observations with Trimble 4000 S and Texas TI 4100 - Status Report Sept. 1986, Proceedings of the 10th General Meeting of the Nordic Geodetic Commission, Helsinki 29 Sept - 30 Oct. 1986, 366-379

Jonsson, B. (1987): Några svenska erfarenheter av positionsbestämning med Global Positioning System (GPS). Svensk Lantmäteritidskrift, 1987, No. 1, 11-18

Kakkuri, J. and L.E. Sjöberg (1988): Åland GPS levelling experiment in 1987: preliminary report on the results. (This volume.)

Sjöberg, L.E. (1986): Comparison of Some Methods of Modifying Stokes' Formula. Bolletino di Geodesia e Scienze Affini, 45, 3,229-248

Sjöberg, L.E. (1987): Refined Least-Squares Modification of Stokes' Formula. Presented at the IAG Scientific Meeting Gsm3 "The Challenge of the cm-Geoid-Strategies and State of the Art", the XIX General Assembly of the IUGG, Vancouver, August 9-22, 1987

Third session: GPS-Campaigns

Chairman: Prof. Sjöberg, Stockholm

ÅLAND GPS LEVELLING EXPERIMENT IN 1987
PRELIMINARY REPORT ON THE RESULTS

by

Juhani Kakkuri, Erik W.Grafarend, and Lars E.Sjöberg

1. Introduction

A GPS levelling experiment from the Turku tide gauge, Finland, over the Åland archipelago to the Forsmark tide gauge, Sweden, was performed 12 to 15 Oct., 1987 as a joint work of the Finnish Geodetic Institute, Helsinki, the Royal Institute of Technology, Stockholm, and the University of Stuttgart. This experiment was originally outlined in Vancouver, Canada, on the occasion of the XIX I.U.G.G. General Assembly, and its realization became possible through the agreements Professor Juhani Kakkuri made with Professors Lars E. Sjöberg of the Royal Institute of Technology and Erik W. Grafarend of the University of Stuttgart. Practical measurements were performed with five Wild/Magnavox GPS receivers. Three of them were provided by Professor Sjöberg and the remaining two by Professor Grafarend. The Finnish Geodetic Institute took care of expences caused.

The first preliminary computation was performed by Professor Sjöberg using the PoPS Program version 1.04. Geocentric coordinates, which he delivered to Professor Kakkuri, were converted to orthometric heights in the Finnish Geodetic Institute using the available geoidal information. Orthometric heights thus obtained were further compared with the orthometric heights of the levelling bench marks, and results of the comparison are shown in the following paragraphs.

2. Geodetic information

The following sites were included in the experiment:

Table 1. Sites of the Åland GPS levelling experiment.

Site	No	Name	Lat.	Long.	N60 Height [m]
S1	1380604	Forsmark tide gauge	60°24'1	18°10'8	-
S2	1288092	Öregrund	60 20.4	18 26.7	-
S3		Trästa	60 02.2	18 44.4	-
M1	SF72	Märket	60 18.1	19 07.9	3.1521
Å1	67116	Eckerö	60 15.2	19 31.1	0.4382
Å2	62138	Bomarsund	60 12.7	20 14.6	5.9983
Å3	75400	Degerby tide gauge	60 01.9	20 23.1	2.1685
F1	66139	Vartsala	60 31.5	21 18.6	21.2596
F2	257F	Turku tide gauge	60 25.7	22 06.0	1.9473

A precise levelling line (Fig. 1), which connects the Turku tide gauge through the levelling bench marks at Vartsala (66139), Bomarsund (62138) and Eckerö (67116) directly to the Degerby tide gauge as well as to the Märket lighthouse on the boundary between Finland and Sweden, was levelled in 1963 to 1975 over the Åland archipelago, cf. **KAKKURI and KÄÄRIÄINEN (1977)**. Between Märket and Sweden no levellings have been performed as yet.

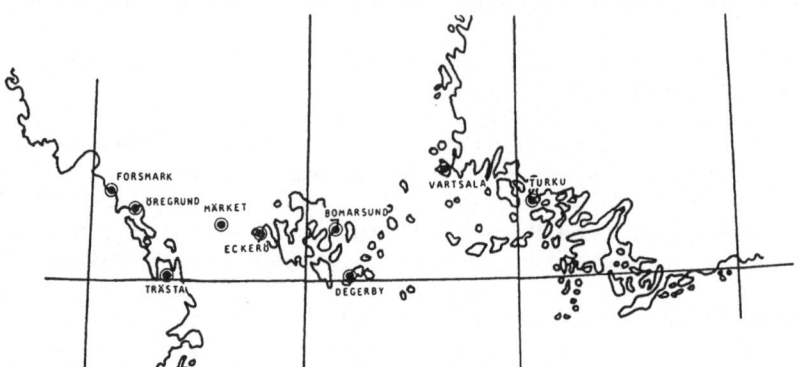

Fig. 1. The GPS levelling experiment area.

Three different geoids are available for the area: 1) a Bomford 70 geoid obtained with astrogeodetic levelling, 2) a SEASAT geoid derived from satellite altimetric observations (**VERMEER, 1983**), and 3) a gravimetric geoid computed with the mass point technique from worldwide gravimetric data, cf. **HEIKKINEN (1981)**, **VERMEER (1984)**. The reference system of the Bomford geoid is the International Reference Ellipsoid 1924, and the Geodetic Reference System 1980 serves as the reference system for the SEASAT and gravimetric geoids. All the above mentioned geoids are translated and biased for various reasons, and á priori precision is estimated to vary from 20 to 30 cm for each of them.

The geoidal heights of the sites which are on the Finnish side, are as follows:

Table 2. Geoidal heights at the GPS levelling area.

Site	Geoidal heights in metres		
	Bomford 70	SEASAT	Gravimetric
Märket	-7.190	19.665	21.904
Eckerö	-7.811	19.119	21.252
Bomarsund	-8.177	18.630	20.573
Degerby	-8.489	18.446	20.346
Vartsala	-7.236	19.009	20.696
Turku	-6.881	19.239	20.831

3. Observational information

The site "Märket" was kept permanently occupied during the whole GPS campaign. The other sites were occupied on two occasions during the campaign:

Table 3. Occupation of the sites.

Day	Sites occupied								
Oct. 12.	S1	S2		M		Å2	Å3		
-"- 13.		S2	S3	M	Å1		Å3		
" 14.	S1			M	Å1			F1	F2
" 15.			S3	M		Å2		F1	F2

Antennas of the GPS receivers were set up directly above the bench marks at the sites. That was not possible at Turku nor at Märket, and there precise eccentricity ties of the antennas to the bench marks were performed. The height differences between the electrical centers of the antennas and the bench marks were as follows (Finnish sites only):

Table 4. Heights of the GPS antennas in metres from the bench marks of the Finnish sites.

Site	12.Oct.	13.Oct.	14.Oct.	15.Oct.
Märket	18.4065	18.4065	18.4065	18.4065
Eckerö		1.263	1.653	
Bomarsund	1.287			1.668
Degerby	1.049	1.082		
Vartsala			1.216	1.219
Turku			1.085	1.162

On 12th to 15th Oct., 1987 the orbits of the GPS satellites were rather unfavourable, because useable "windows" were open during late nights. In spite of this inconvenience measurements were performed successfully.

4. Results obtained with PoPS 1.04

The first preliminary computation was performed as already mentioned by Professor Sjöberg using the PoPS Program version 1.04. Due to software problems (SJÖBERG, 1988) the preliminary computations were limited to day by day results. As each bench mark was observed during two days the computations gave two sets of geocentric X,Y,Z coordinates, and the averages of the sets were brought into the following analysis at the Finnish Geodetic Institute:

Step 1: Convert (X,Y,Z) into $(\emptyset,\cap,H_E)$, where $\emptyset$ = latitude, $\cap$ = longitude, and H_E = ellipsoidal height.

Step 2: Compute the orthometric height $H_O = H_E - N$, where N = geoidal height.

Step 3: Compute the orthometric height differences h_O of the sites in relation to Turku (for GPS and levelling).

Step 4: Compute the difference $\delta h_O = h_O(\text{GPS}) - h_O(\text{levelling})$.

Step 5: Remove the bias and translation with a fit by a plane described by $\delta h_O = aS_x + bS_y + c$, where a, b and c are the bias and translation parameters, resp., S_x is the WE component and S_y the SN component of the normed distance of a site from Turku.

The above procedure was applied to each available geoid (Table 2) and results obtained are shown in the following table:

Table 4. Values of δh_O in centimetres for different geoids; PoPS 1.04 Program.

Site	S_x	S_y	Geoid applied		
			Bomford 70	SEASAT	Gravimetric
Turku	0.000	0.000	0.0	0.0	0.0
Vartsala	-0.269	+0.088	+25.5	+ 4.1	- 5.4
Bomarsund	-0.626	-0.113	+72.0	+33.3	- 1.8
Degerby	-0.582	-0.237	+58.8	+33.9	+ 3.1
Eckerö	-0.870	-0.076	+98.6	+43.7	-10.4
Märket	-1.000	-0.039	+90.1	+36.0	-28.7

The plane formulae were applied with the following results:

Bomford 70 geoid: $\delta h_O = -97.87S_x - 22.33S_y + 1.50$; $\sigma_O = \pm 9.1$ cm
SEASAT geoid: $\delta h_O = -38.54S_x - 59.67S_y - 0.08$; $\sigma_O = \pm 4.2$ cm
Gravimetric geoid: $\delta h_O = +28.65S_x - 67.97S_y + 4.51$; $\sigma_O = \pm 6.5$ cm

5. Discussion

As can be seen, the geoids applied to the Åland GPS levelling experiment are strongly biased. E.g., the bias of the Bomford 70 geoid in relation to the gravimetric geoid is 1.5 sec of arc, partly due to different ellipsoids, and the SEASAT geoid lies between them. The shape of each geoid is, in general, rather similar, and, therefore, anyone of them can be applied to GPS levelling when converting ellipsoidal heights into the orthometric height system. The best accuracy was provided by the SEASAT geoid, and in that case the accuracy of the orthometric height of a single point is $\sigma_0/\sqrt{2} = \pm3.0$ cm. The "worst" result given by the Bomford 70 geoid can be explained by the sparsity of the astrogeodetic observations used.

The above results must be considered sufficient, because the levelling net used is rather sparse. Namely, the distances between consecutive bench marks vary from 20 to 70 kilometres, and the total distance of the Märket lighthouse from the Turku tide gauge is 165 km.

The accuracy of the GPS orthometric heights definitely depends on the accuracy of the geoid. New treatment of the data with more precise ephemeris and advanced program versions will doubtlessly bring some improvement, but any sub-centimeter accuracy cannot probably be reached without more accurate geoid determinations. Such works are under way.

6. Epilogue

After completing the manuscript of this report Juhani Kakkuri received new results from L.E.Sjöberg and E.W.Grafarend, obtained with the PoPS 2.01 Program version. These results, indicating the superiority of the gravimetric mass point geoid, are shown in the following Tables 5 and 6.

Table 5. Values of δh_0 in centimetres for different geoids; Sjöberg's solution with PoPS 2.01 Program.

Site	S_x	S_y	Geoid applied		
			Bomford 70	SEASAT	Gravimetric
Turku	0.000	0.000	0.0	0.0	0.0
Vartsala	-0.269	+0.088	+16.2	- 9.6	-19.1
Bomarsund	-0.626	-0.113	+51.2	+12.3	-22.8
Degerby	-0.582	-0.237	+38.4	+11.6	-19.2
Eckerö	-0.870	-0.076	+76.0	+19.0	-35.1
Märket	-1.000	-0.039	+74.4	+18.4	-46.4

The plane formulae were applied with the following results:

Bomford 70 geoid: $\delta h_0 = -82.37S_x + 8.86S_y - 2.69$; $\sigma_0 = \pm6.3$ cm
SEASAT geoid: $\delta h_0 = -21.80S_x - 35.78S_y - 5.79$; $\sigma_0 = \pm5.6$ cm
Gravimetric geoid: $\delta h_0 = +45.48S_x - 44.23S_y - 1.18$; $\sigma_0 = \pm3.1$ cm

Table 6. Values of δh_o in centimetres for
different geoids; Grafarend's solution
with PoPS 2.01 Program.

Site	S_x	S_y	Geoid applied		
			Bomford 70	SEASAT	Gravimetric
Turku	0.000	0.000	0.0	0.0	0.0
Vartsala	-0.269	+0.088	+16.8	- 8.9	-18.4
Bomarsund	-0.626	-0.113	+52.5	+13.6	-21.5
Degerby	-0.582	-0.237	+41.4	+14.6	-16.2
Eckerö	-0.870	-0.076	+78.2	+21.3	-32.8
Märket	-1.000	-0.039	+78.6	+22.6	-42.1

The plane formulae were applied with the following results:

Bomford 70 geoid: $\delta h_o = -85.73 S_x - 6.23 S_y - 2.85$; $\sigma_o = \pm 5.4$ cm

SEASAT geoid: $\delta h_o = -25.19 S_x - 38.13 S_y - 5.91$; $\sigma_o = \pm 5.4$ cm

Gravimetric geoid: $\delta h_o = +42.01 S_x - 46.44 S_y + 1.32$; $\sigma_o = \pm 2.5$ cm

7. Acknowledgements

The authors want to thank Messrs M. Ollikainen, M. Takalo and Dr M. Vermeer of the Finnish Geodetic Institute, Dr J. Piechocinska and Messrs E. Asenjo, H.Fan and A.Forsberg of the Royal Institute of Technology, and Messrs H. Kremers and K. Rösch of the Stuttgart University, for making the practical observation work as well as for taking part in the computations. Thanks are addressed also to Mr R. Chen, who took part in computations. A joint report will be published in the near future on the final results of the Campaign.

7. References

Heikkinen,M. 1981. Solving the Shape of the Earth by using Digital Density Models. Rep. Finn. Geod. Inst. 81:2. Helsinki.

Kakkuri,J. and J.Kääriäinen, 1977. The Second Levelling of Finland for the Åland Archipelago. Publ. Finn. Geod. Inst. 82. Helsinki.

Sjöberg,L.E., 1988. Experiences with the WM 101 GPS receiver. (This volume).

Vermeer,M.,1983. A New Seasat Altimetric Geoid for the Baltic. Rep. Finn. Geod. Inst. 83:4. Helsinki.

Vermeer,M., 1984. Geoid Studies on Finland and the Baltic. Rep. Finn. Geod. Inst. 84:3. Helsinki.

Experiences with TRIMBLE receivers in the control network of the F.R.G.

by

Wolfgang Augath

Abstract

Since 1987 the State Survey Office of Lower Saxony works with a set of TRIMBLE 4000 SL receivers for the establishment of the 3rd and 4th order densification network.

In this report first experiences are given with investigations about the accuracy of short distances (< 10 km), the organisation of field work, the design of the network, the costs and the organisation of the evaluations.

1. Introduction

Since 1972 the whole horizontal control network of Lower Saxony is
renewed in two steps by the State Survey Office; first: 1st and 2nd
order net and second: 3rd and 4th net with a final point density of
one point/2 km². In addition the local cadastral authorities are
establishing a minor control point net with a density of about 10
points/km² only for cadastral purposes.

The observations of the fundamental 1st and 2nd order net were finished
in 1982 (microwave measurements with SIAL MD60 on observation platforms). A neighbourhood accuracy of less than +/- 2 cm for distances
up to 200 km was obtained (Augath, 1988).

At the same time about 40% of the second step, the renewal of the
3rd and 4th order (traverse nets) were finished. In this network
a neighbourhood accuracy of less than +/- 1 cm for distances up to
10 km is reached. To shorten the time for the termination of this
work, additionally a TRIMBLE 4000 SL-GPS equipment was purchased in
1987.

2. Description of the TRIMBLE 4000 SL-receiver

The TRIMBLE 4000 SL-receiver is an one frequency receiver (L1 =
1575,42 MHZ, $\lambda \sim 19$ cm) with five channels, which works by the
multichannel principle (see figure 1). While the five channels
work simultaneously, the oscillator must not be so stable (that
means also not so sensitive against external effects in the field
work) as in other receivers, working in a sequential mode. For geo-
dectic purposes the equipment also contains a microstrip antenna,
which has only small variations of the antenna phase center and a
low sensitivity against multipath effects (see chapter 3). The power
supply works with external batteries. It is also possible to put an
additional internal battery into the receiver, which can work for
one hour in case of an emergency during the measurements. The data
are stored in a 1-megabyte-chip in 4 files:

1. Ionospheric model and clock parameters   (ION)

2. Tracking data of the satellites   (EPH)

3. Local data of the station   (MES)

4. Measurements (pseudoranges, dopplercounts,   (DAT)
   phase differences) all 15 sec.

If necessary, p.e. for permanent stations, the beginning of the
measurements and the choice of the satellites can be programmed
in advance. The internal computer also determines coordinates
in WGS 84, velocities and directions in real time mode.

The data transfer form the receiver to an external computer needs
6 min for 4 hour of measurements. The TRIMBLE Nav. Inc. delivers
software for all calculations belonging to the receiver, as

Data transfer   (TRIM4000)

Baseline computation   (TRIM640)

Multistation adjustment (preliminary release)   (TRIMVEC)

For this work the computers have to use MS DOS and should have a
mathematic coprocessor. The COMPAC Portable III p.e. needs 7
minutes for a baseline calculation.

The necessary additional software as

data management
3 D-adjustments
transformation in a local datum and combination with terrestrial
observations

has to be done by the user (see chapter 4.5).

Table 1: Description of the GPS receiver TRIMBLE 4000 SL
```
+-----------------------------------------------------------------+
| L1 - C/A-code receiver                                          |
| 5 channels (multichannel mode) = 5 satellites simultaneously   |
| microstrip antenna for geodetic purposes                       |
| max. cable length 60 m                                         |
| 1-Megabyte-chip as data storage (max. 18.5 h of 5 satellites) |
| RS 232C-interface, 9600 bit/s                                  |
| power supply: external batteries 10 - 30 V                     |
|               additional internal battery possible             |
| weight      : 14.5 kg                                          |
| water resistant                                                |
+-----------------------------------------------------------------+
```


3. Test measurements

Before using the receivers in the field, some tests were made to
get experiences about the stochastic model of the observations
and possible systematic effects. For this purpose and for the
training of the staff measurements in local configurations and
in typical testnets were made.

3.1 Internal accuracy and antenna phase center

For the determination of the internal accuracy a receiver configuration as shown in figure 1 was used. The distance between the antennas should be just so large, that there is no influence caused by
the neighbour antenna. On the other hand it should be possible to determine the threedimensional relations with simple methods. A typical
result of this test in given in figure 1.

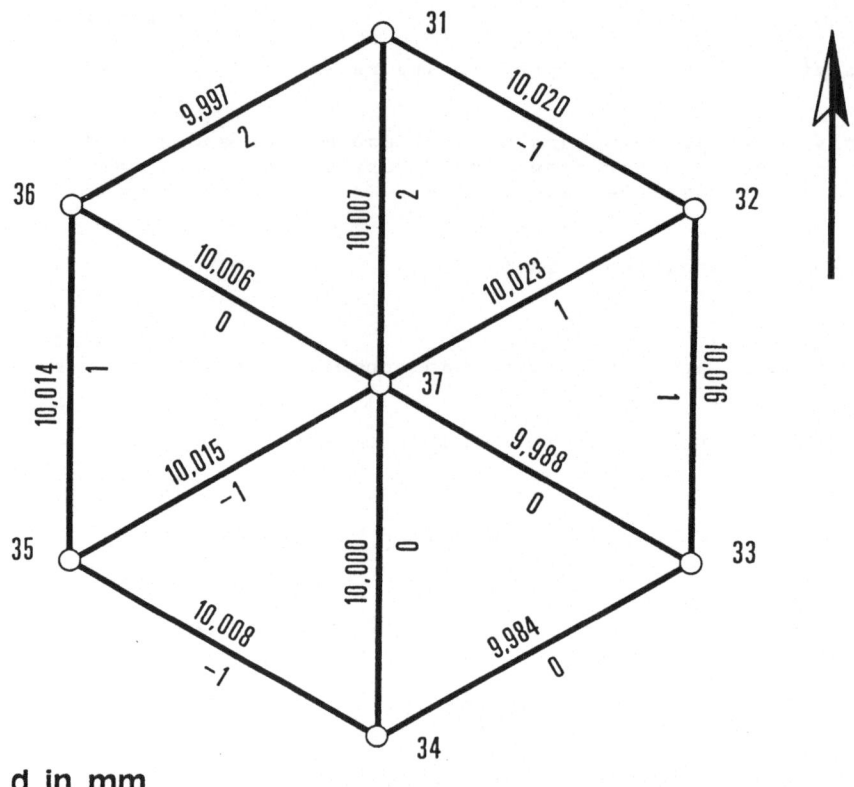

d in mm

figure 1: Internal accuracy in a local configuration

While in all field measurements the antenna is always orientated to north, for test purposes some antennas were turned off. In this way possible excentricities or variations of the antenna phase center should be discovered. Summerizing we received the following results:

- The internal accuracy of observations between antennas with the same orientation is about +/- 1 mm.

- It seems that some antenna phase centers are not in the geodetic center. The maximum difference of one antenna was 3 mm.

- If you turn an antenna (90° or 180°), there are differences of 1 - 2 mm, with a maximum value of 7 mm.

3.2 Multipath-effects

Mulitpath-effects were not recognized, but on the other hand, the field work was organized in such a way, that they were hardly to expect (distances between the antenna and cars or houses greater than 20 m).

3.3 Test nets

For the planned activities in the 3rd and 4th order (distances from
1 - 10 km, medium 2 - 3 km) in 1987 measurements were done in GPS-
testnets at Hannover (Seeber, a.o., 1986: internal accuracy +/- 3
mm) and Turtman (L + T, 1988: internal accuracy +/- 1 mm). The Turtman-testnet was realized by the SWISS geodetic commission. A comparison between the TRIMBLE-values and the known values is given as "baseline difference" in figure 2 and 3. They don't differ in

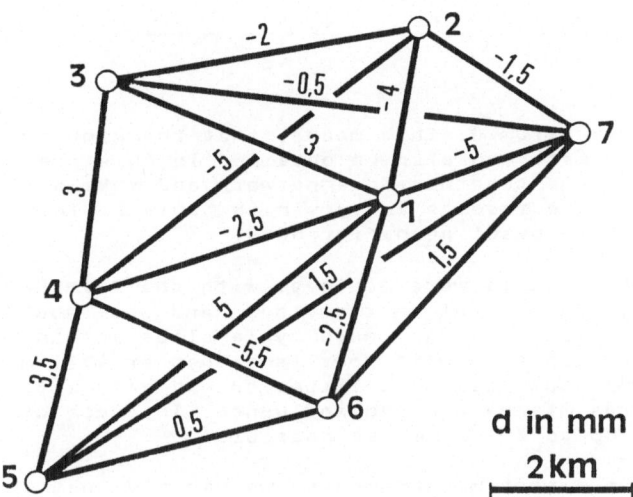

figure 2: Baseline differences in the Hannover-testnet

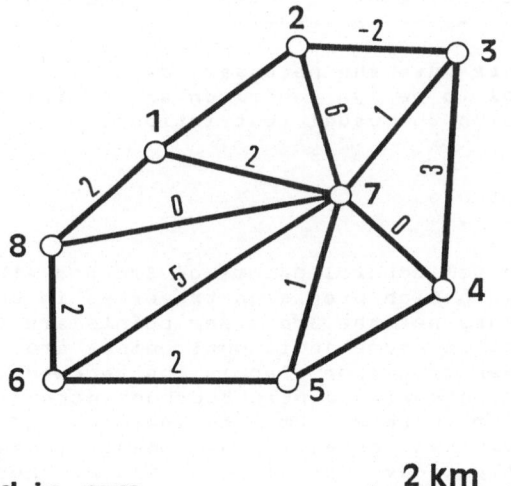

figure 3: Baseline differences in the Turtman-testnet (prelimary
          results)

Table 2: Residuals errors in the Hannover-testnet

NR	NAME	RESIDUALS IN METERS		
		X	Y	Z
1	LIN	−0.000	0.001	0.002
2	MSD	0.005	−0.004	0.008
3	VEL	−0.008	0.002	−0.003
4	BEN	0.008	0.012	0.016
5	RON	−0.005	−0.004	−0.015
6	MHL	0.010	0.003	0.010
7	LVA	−0.004	0.000	−0.010

a significant way from 0. This means, that the high internal accuracy
of the antenna tests can also be obtained in this area. If you adjust
the single sessions together in a network and make a 7-parameter-
transformation, the results are given in table 2. They are of the same
size as the simple baseline differences.

These very good results were achieved with the TRIMVEC-software without manual assistance. But on the other hand one should not forget,
that simple distructions can cause cycle slips and in this case the
minimum observation time of 1 hour (see chapter 4.2) may not be sufficient. In addition "ill" satellites or bad values for the approximate coordinates may have a bad influence. In these cases errors in
the baselines up to 2 - 3 cm are possible.

As summary of the test measurements, we can say that the TRIMBLE 4000
SL GPS-receiver produces an accuracy of better than +/- 5 mm, if the
distances are up to 5 km (or some more). Unexpected effects make it
recommendable to design the network not only by accuracy restrictions
but also as usual by reliability restrictions.

4. Use in the horizontal control network of 3rd and 4th order
----

The test measurements gave the necessary data for the functional
and stochastic model to design a horizontal control network of
3rd and 4th order with the usual restrictions.

4.1 General circumstances

Till now the horizontal control points of 3rd and 4th order are determined together with high precision traverses in one step (design
see figure 4). In this net the 3rd order points are the nodal points,
the 4th order points and some additional points are in the traverses.
The frame of this densification step is the renewed fundamental net of
1st and 2nd order points with a neighbourhood accuracy of +/- 1 cm up
to 10 km and +/- 2 cm up to 200 km. The locations of these points are
selected under "classical" aspects. That means, these point are often
located on the mostly woody tops of mountains or churchtowers has been
used. The visibility of satellites on these points is often not possible without signalization towers for the GPS-antenna. The location of

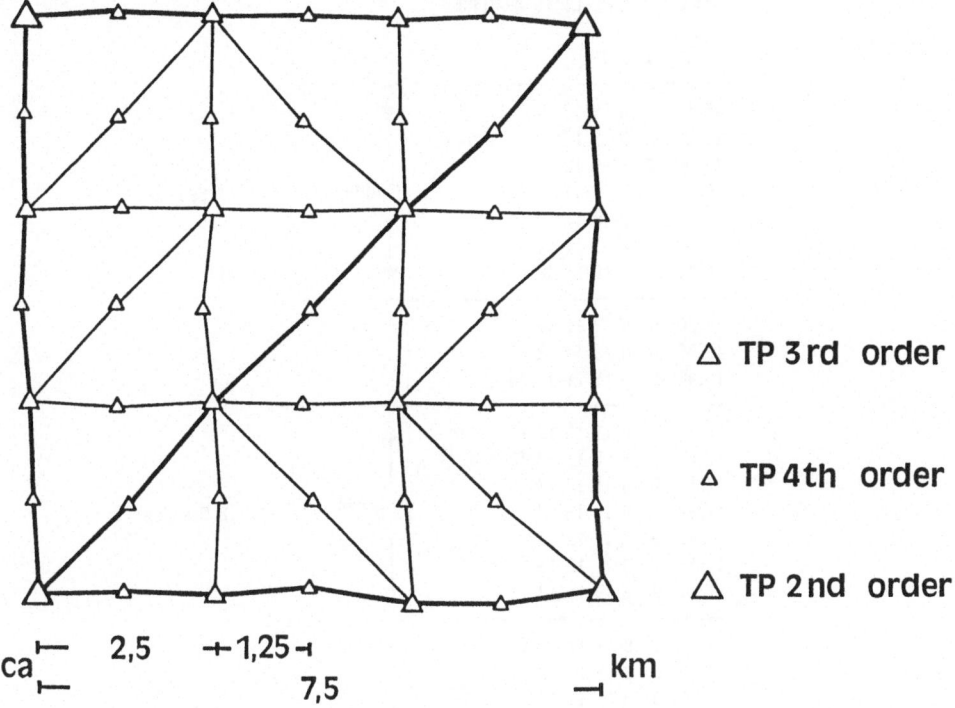

figure 4: Design of high precision traverse nets of 3rd and 4th order

the new 3rd and 4th order points is always suitable for GPS-measurements:

- visibility > 15 degrees, if necessary on eccentric marks

- points near to the road, cable length less than 60 m.

4.2 Organisation of the field work

Because of the limited time with 4 and more satellites in 87/88 (~4 hours), only 3 sessions are possible in 4th order nets (see figure 5). The minimum observation time per session is 1 hour. The moving time from one point to the next is up to half an hour. This short time is only sufficient in the 3rd and 4th order nets.

```
TIME    SATELLITES AVAILABLE

15.00  6  9 12
-------------------------------------------
15.10  6  8  9 12
15.20  6  8  9 12
15.30  6  8  9 12
15.40  6  8  9 11 12        ↑
15.50  6  8  9 11 12      1ʰ 10ᵐⁱⁿ        1.Session
16.00  6  8  9 11 12
16.10  6  8  9 11 12        ↓
-------------------------------------------
16.20  6  8  9 11 12
16.30  6  8  9 11 12
16.40  6  8  9 11 12 13
-------------------------------------------
16.50  6  8  9 11 12 13     ↑
17.00  6  8  9 11 12 13
17.10     8  9 11 12 13
17.20     8  9 11 12 13   1ʰ 10ᵐⁱⁿ        2.Session
17.30     8  9 11 12 13
17.40     8  9 11 12 13     ↓
17.50  3  8  9 11 12 13
-------------------------------------------
18.00  3  8  9 11 12 13
18.10  3  8  9 11 12 13
18.20  3     9 11 12 13
-------------------------------------------
18.30  3    11 12 13        ↑
18.40  3    11 12 13
18.50  3    11 12 13
19.00  3    11 12 13      1ʰ 10ᵐⁱⁿ        3.Session
19.10  3    11 12 13
19.20  3    11 12 13        ↓
19.30  3    11 12 13
-------------------------------------------
19.40  3    12 13
```

figure 5: GPS-timetable in 3rd and 4th order nets

4.3 Design of the network

Till now no final investigations about the design of 3rd and 4th
order networks are available. The result of the test measurements
was a stochastic model of the observations and some practical rules
for the design with accuracy and reliability criteria. Because of
the special conditions in Lower Saxony (1st and 2nd order net renewed, limited personal staff) it was useful; p.e. with 6 receivers
to move only 4 receivers and to put the remaining 2 receivers on
the next 1st and 2nd order points as reference stations. In this
way the following connections are given in three sessions as shown
in figure 6.

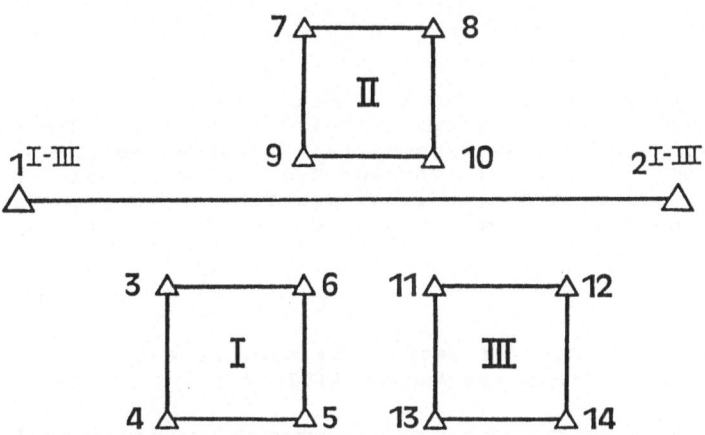

I:number of session

figure 6: Design with accuracy criteria in 3rd and 4th order GPS-nets
using 2 reference stations and 4 mobile receivers.

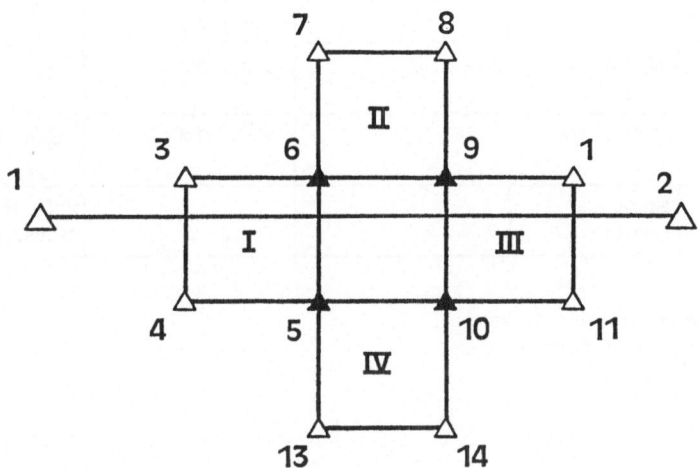

I: number of session

figure 7: Design with reliability criteria in GPS-nets
(two sessions | point --> reduncy ～0,5)

To reach only accuracy criteria a simple polar determination from
the 1st and 2nd order points would be sufficient. This configuration also has the advantage, that the wellknown distance between
these points can be compared with the GPS-values just after the
measurements.

For reliability purposes each point should be used in two sessions
as shown in figure 7.

4.4 Costs

The cost of the existing methods of measurement were put together in Augath (1976). Table 4 shows an enlargement with GPS-techniques (state 1988), which allows a comparison after a reduction to the price level of 1976 for 1st, 2nd and 3rd/4th order point determinations. In all orders GPS-techniques are less expensive. For that reason only these techniques should be used for control point determinations in future.

Table 4: Costs per point of methods of measurement
(specifications see Augath (1976))

measurement methods distances between the points	angle measurements	distance measurement with microwaves	electro-optical distance measurement	combinations	GPS
> 30 KM	63.000 DM	7.000 DM	33.600 DM •) 21.600 DM ••)	-	1.600 DM
5 - 15 KM	10.300 DM	2.600 DM	5.000 DM	-	1.100 DM
2 - 5 KM	2.000 DM	-	2.000 DM	1.000 DM	600 DM
remarks	single-points			nets of traverses	State: 1988

•) Signalization with wooden towers
••) Signalization with iron towers

4.5 Concept of the evaluations

The concept of the evaluations of GPS-measurements uses the special EDV-conditions in Lower Saxony (data lines from the State Survey Office to all cadastrial offices, distributed processing with local SIEMENS MX2/MX500 workstations (PC), identical procedures on all PCs available). It is divided into two parts:

1. Data transfer and preliminary evaluations (see figure 8)

2. Common adjustment (see figure 9)

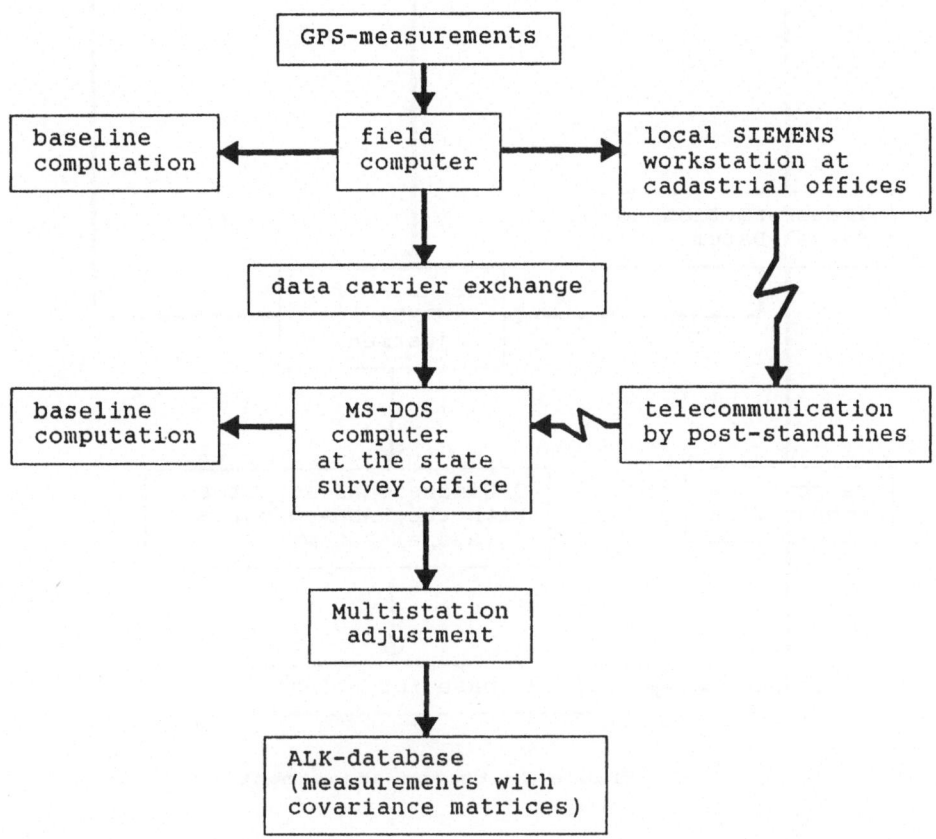

figure 8: Data transfer and preliminary evaluation

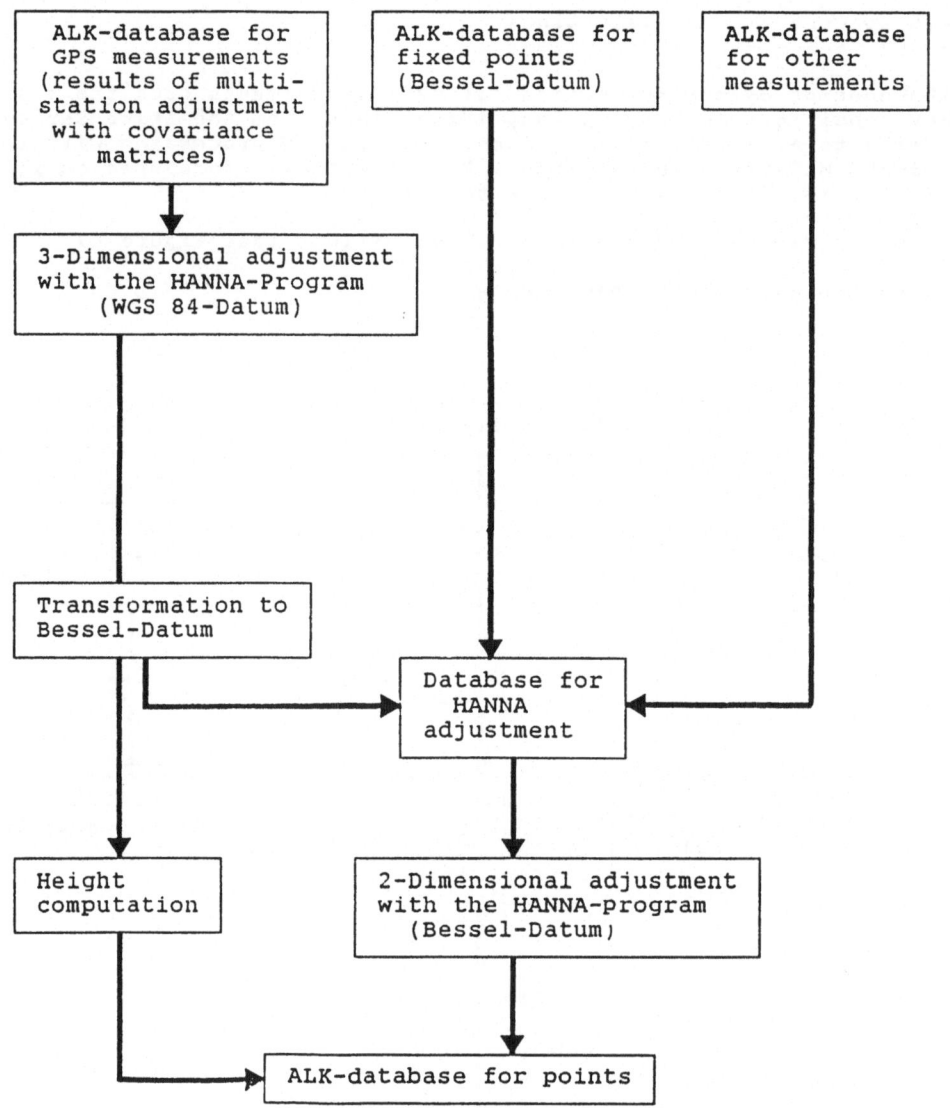

figure 9: Common adjustment

5. Conclusion

The TRIMBLE 4000 SL GPS-receiver is an efficient instrument working in the field without problems. The handling is not complicated. The accuracy of connections up to 5 km is +/- 5 mm and better. GPS-nets should be designed using accuracy and reliability criteria. GPS-techniques are less expensive than all other methods of measurement used for horizontal control point determinations. The TRIMVEC-software is working efficiently and orientated on practical purposes. The necessary integration of this MSDOS-software with the existing adjustment and data base software is solved at Hannover with the HANNA (1982) - and ALK (Sellge 1986) - systems.

6.Literature

Augath, W. (1976) : Untersuchungen zum Aufbau geodätischer Lage-
netze. Wissenschaftliche Arbeiten der Lehr-
stühle für Geodäsie, usw. an der Technischen
Universität Hannover Nr. 72, Hannover 1976.

Augath, W. (1988) : 50 Jahre dezentrale Grundlagenvermessung in
Hannover.
Nachrichten aus der Nds. Vermessungs- und
Katasterverwaltung (in press), 1988

Brouwer, F. and   First experiences with TRIMBLE 4000 SX receivers
Husti, L. (1987) : in the Netherlands.
Szentendre, 1987

HANNA (1982)   : Handbuch für die Benutzung des Hannoverschen
Ausgleichungsprogramms HANNA
(unpublished), 1982

L + T (1988)   : Bundesamt für Landestopographie, Bern
(personal information), 1988

MI (1988)   : Einrichtung, Nachweis und Erhaltung der Fest-
punktfelder (Festpunktfelderlaß). Herausgege-
ben vom Niedersächsischen Minister des Innern,
Hannover, 1988

Seeber, G.,   Anlage eines örtlichen 3-D-Netzes mit GPS-TI 4100-
Wübbena, G.,   Beobachtungen.
Augath, W. (1986) : Zeitschrift für Vermessungswesen, S. 481 - 489,
1986

Sellge, H. (1986) : General View on the ALK-system.
in Pelzer, H. and Niemeyer, W. (editors):
Determination of Heights and Height changes
and recent vertical crustal movements in
Western Europe
Dümmler Verlag Bonn, 1986

CAMPAIGNS WITH WM101 IN AUSTRIA 1987

by

Günter Stangl

Abstract

Three campaigns with receivers WM-101 took place in Austria 1987 .
They were used in networks with average distances from 1 km to 50 km .
We hoped to answer mainly three questions :

1) Can we compose precise threedimensional networks more economically
   and more precise than the conventional methods ?
2) How good are receivers with one frequency compared with two-frequency ones at distances of some tens of kilometers ?
3) Is it possible to integrate GPS-measurements into the cadastrial
   network and the Austrian height system ?

The three campaigns have shown that in most cases we can answer these
questions in a positive way . The precision of WM-101 seems to lie at
the 1 - 2 ppm level for distances ranging from 5 km to 50 km .

1. Introduction

In 1987 a group of interested institutions planned and executed three campaigns with GPS-receivers WM-101 for testing purposes . These institutions were :

Technical University of Vienna (H. Kahmen)
Technical University of Graz (H. Sünkel)
Federal Bureau of Metrology and Surveying (J. Zeger , E. Erker)
University of Innsbruck (G. Chesi)
Austrian Academy of Sciences (K. Rinner)
R&A Rost (Wild representative in Austria)
ETH Zürich (H. Kahle)
UBW München (G. Hein)
Chamber of civil engineers in Styria and Carinthia
Österreichische Draukraftwerke AG
Bleiberger Bergwerksunion

2. Projects

Table 1 shows the dimensions of the projects , figures 1 and 2 their distributions

No.	receivers	points	max. diameter	average point distance
1	5	8	10 km	5 km
2	5	5	70 km	40 km
3	5	7	30 km	10 km
4	5	7	20 km	8 km
5	2	4	4 km	2 km
6	3	7	2 km	1 km
7	3	3	450 km	200 km
8	3	9	5 km	2 km
9	4	4	60 km	40 km
10	4	7	7 km	2 km
11	4	5	20 km	10 km
12	4	7	2 km	1 km
13	4	12	7 km	2 km

Table 1   Size of projects measured with WM-101 Austria 1987

Project 1 was a network measured by Macrometer V-1000 in 1985
(Rinner et al. 1986)   plus a baseline of 1 km .
Project 2 used some first order triangulation points , partially observed by V-1000 , and tested the equipment in high mountains (2000 m)
Projects 3 and 4 were designed for comparing large orthometric height differences with GPS-derived geometric heights .
Projects 5 and 6 were performed at the same day , holding one receiver fixed in project 5 and moving the others . The distance between both projects was 60 km . The points of project 6 form parts of a network observing a dam in altitude of 1800 m . Project 5 consists of three points in a V-shaped valley plus the fixed point on a mountain in a mining district .

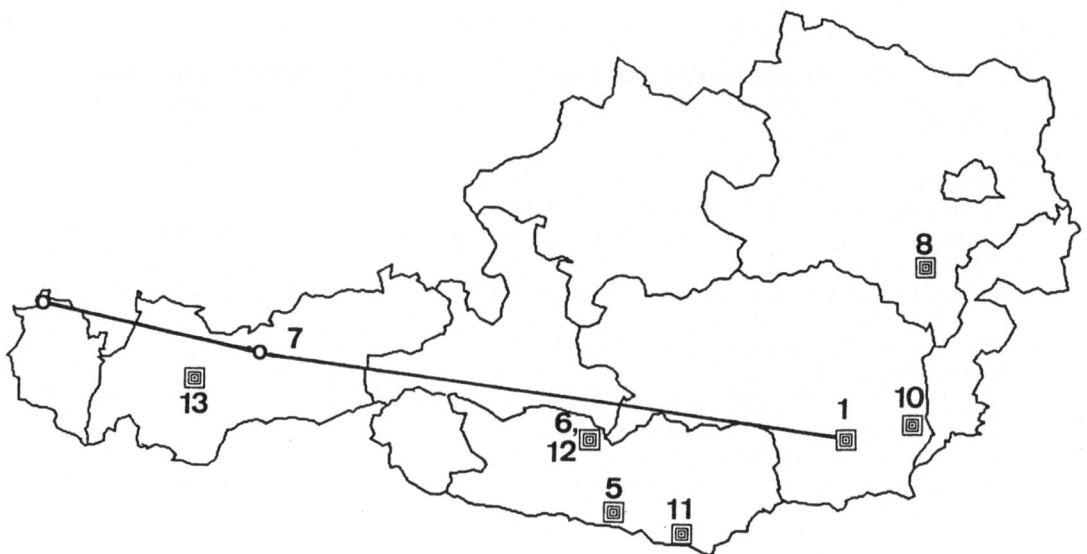

Figure 1 WM-101 measurements : The long baseline project and local networks

Figure 2 WM-101 measurements in major triangulation networks

Project 7 consists of two baselines bridging almost whole Austria .
It was tested if the software were able to get reasonable results and
how they were to compare with Doppler and TI-4100 results .

For details of projects 1 to 7 see (Höggerl et al. 1988) .

Project 8 covered a test net of the Technical University of Vienna ,
for details see   (Kahmen et al. 1987) .

Project 9 was a comparison between TI-4100 and WM-101 at the same time ;
the receivers of different types had been separated by only some meters .
Project 10 was designed for creating new points in a triangulation network . Three receivers had been situated on reference points , the
fourth one had to measure 4 points in 4 hours . The average time for
one point was 45 minutes . It should be tested if there were an economic
way to get coordinates with good accuracy .
Project 11 observed parts of a high precision network of the Technical
University of Vienna .
Project 12 repeated project 5 (because it partially failed) , now using
two fixed points in the area itself .
Project 13 lasted for two days , a controlling network for a tunnel ,
repeating itself by observing the 6 points again .

3. Results

Till now there have been no results of projects 11 , 12 and 13 . All
other projects have been computed with PoPS (Frei et al. 1986)   ,
most of them also with the Bernese Second Generation Software Package
(Gurtner et al. 1985)   at the Technical University of Vienna . Coordinates are published by H. Kahmen and J. Schwarz from the Institut für
Landesvermessung und Ingenieurgeodäsie , Abteilung Ingenieurgeodäsie in
(Höggerl et al. 1988) .

3.1 Local networks

The smallest network (Project 6) delivered the best results . Comparing
the distances , the differences in general are below 10 mm . A similar
result yields project 8 . On the other side , project 5 showed huge differences of more than 100 mm in some distances . Unfortunately the quality of the connection of points is not known .
Projects 1 and 10 were only tested with the cadastrial net , but showed
quite good results . The differences in project 1 range from 10 to 50 mm
after a 7-parameters-Helmert-transformation , project 10 delivers them
a little bit higher . These two projects show the possibility of bringing
together cadastrial and GPS-coordinates in a local area , provided the
heights of the cadastrial net are good .

3.2 Major triangulation networks

Projects 2 , 3 , 4 and 9 contain points of first to third order triangulation networks . They have cadastrial and ED-79 coordinates , but not
all of them have orthometric heights .
As expected , residuals after transformation were smaller for coordinates
in ED 79 (better than 100 mm) than the cadastrial ones .

3.3 Comparing GPS networks

There were two possibilities of testing outcomes of GPS receivers of different type .
First , parts of projects 1 and 2 were measured by Macrometer V-1000 in 1985 (Rinner et al. 1986) . The points LUS , PLB , SLB , PLT and FUR are members of project 1 , the others , SKL , REN and KOR , of project 2 . Taking LUS as reference point , table 2 shows the differences in coordinates X , Y , Z together with the distance references .

Station	dX (mm)	dY (mm)	dZ (mm)	ds (mm)	ds (ppm)
LUS	0	0	0	–	–
PLB	86	-5	29	17	2
SLB	47	18	-11	-15	4
PLT	11	-21	-20	-9	2
FUR	9	-9	-18	-18	8
SKL	99	-28	72	1	0.1
REN	232	-131	201	55	1.5
KOR	24	-123	-306	216	4

Table 2   Differences WM-101 minus Macrometer V-1000

Up to now there is no reasonable explanation for the quite big difference in stations REN and KOR . It should be mentioned that REN is identical for both receivers .
The second comparison between a one-frequency receiver and one with two frequencies promised a more stringent proof of the quality of the WM-101 . Unfortunately one of three comparing points showed bad measurements in both receivers (but not of the same source) . Therefore point FRA in table 3 must be considered with great caution .

Station		dX (mm)	dY (mm)	dZ (mm)	ds (mm)	ds (ppm)
KOE		0	0	0	–	–
FRA	(L1 - L1)	-98	-166	-242	-37	1
	(L1 - L3)	244	-201	-203	42	1
ROS	(L1 - L1)	48	30	85	-21	0.5
	(L1 - L3)	-26	53	53	45	1

Table 3   Differences WM-101 minus TI-4100

The moderate differences in ROS can be explained by different satellite geometries . WM-101 used SV 8 and the other satellites during their whole path , where as TI-4100 was obliged to change the configuration of its 4 satellites .

4.   Conclusions

As projects 6 and 8 have shown , WM-101 can create threedimensional coordinates with an accuracy between 5 mm and 10 mm . If we take the longest distances in these local networks of high precision an accuracy of 1 - 2 ppm results . Since we are lacking a network with longer lines in the same precision we must rely on the result of project 9 which might confirm this assumption . Concerning the large differences between V-1000 and WM-101 we must take in mind that we do not know how the co-ordinates of V-1000 points have been computed .

Therefore we can use WM-101 in every case to improve and increase our triangulation network which has an accuracy of 2 to 10 ppm .

As can be seen in tables 2 and 3 caution must be taken by comparing distances , the results seem too optimistic .

Measuring 3 to 4 points for each receiver and for each observation window would be a good chance for lowering the costs .

Taking into account speed and accuracy , receivers like WM-101 seem to be a very useful tool for institutions which have to install networks with a lot of points at distances from 1 to 50 km .

References

Frei , E. ; R. Gough ; F. K. Brunner (1986) : PoPS[TM] : A new generation of GPS post-processing software . Proceedings of the Fourth International Geodetic Symposium on Satellite Positioning . Austin , Texas , April 28 - May 2 , Vol. 1 , pp. 455 - 473 .

Gurtner , W. ; G. Beutler ; I. Bauersima ; T. Schildknecht (1985) : Evaluation of GPS carrier difference observations : the Bernese second generation software package . Proceedings of the First International Symposium on Precise Positioning with the Global Positioning System . Rockville , Maryland , April 15 - 19 , Vol. 1, pp. 363 - 372 .

Höggerl , N. ; H. Kahmen ; G. Kienast ; J. Schwarz ; G. Stangl ; H. Sünkel ; J. Zeger (1988) : Die WM 101 GPS-Kampagne 1987 in Österreich . Österreichische Zeitschrift für Vermessungswesen und Photogrammetrie , 75. Jahrgang/1987 , Heft 4 , pp. 167 - 201 .

Kahmen , H. ; J. Schwarz ; T. Wunderlich (1987) : GPS-Messungen im Test-netz "Neue Welt" . Österreichische Zeitschrift für Vermessungswesen und Photogrammetrie , 75. Jahrgang/1987 , Heft 3 , pp. 123 - 134 .

Rinner , K. ; J. Zeger ; B. Hofmann-Wellenhof ; E. Erker (1986) : Über die Macrometer-Kampagne 1985 in Österreich . Österreichische Zeit-schrift für Vermessungswesen und Photogrammetrie , 74. Jahrgang/1986 , Heft 1 , pp. 1 - 25 .

RESULTS ON LONG BASELINES IN EUROPE

Preliminary comparisons with Laser and VLBI solutions

by

Bernd Breuer

Hermann Seeger

Abstract

In November 1986 simultaneous GPS observations at 10 european VLBI and LASER tracking stations were performed within the GINFEST project. The observations were recorded during three days (322-324) using TI4100 receivers.
This paper presents some adjustment results of this network evaluated at the Geodetic Institute in Bonn. The computations have been performed using submitted Broadcast (BE) and Precise (PE) Ephemeris kindly provided by the Defence Mapping Agency in Washington. In order to keep the GPS internal scale factor unchanged no orbit improvement has been used in the solutions.
First comparisons between the different GPS solutions and LASER as well as VLBI results are presented.

1. Introduction

Within the European GINFEST project (Geodetic Intercomparison Network for Evaluating Space Techniques) simultaneous GPS observations were performed between 18th and 20th of November 1986. Participating stations together with their comparison Space techniques are shown in figure 1.

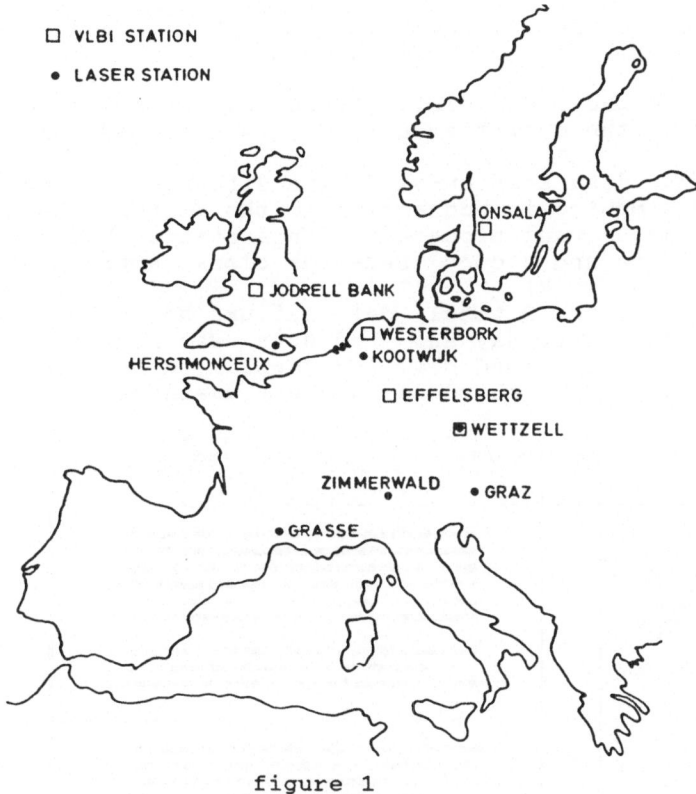

figure 1

Several computations were performed to prove the repeatability and stability of the GPS solution. The influence of weather data and the usage of so called Precise Ephemeris provided by the Defence Mapping Agency were tested with the given GPS data sets.

Comparisons between GPS and other coordinate sets were made by testing the scale factors and transforming the coordinate sets using a 7 parameter Helmert transformation.

During the comparisons we found out, that at station Grasse and at station Westerbork no already published and therefore comparable site was used for the GPS measurements. We will try to get the correct excentricity later on.

2. Details on the GPS solutions

2.1 Evaluation of the GPS baselines

For this adjustment a modified version of the Second Generation Bernese Software package /1/, installed at the RHRZ 3081 IBM mainframe computer at the University of Bonn was used.

To perform the computations simultaneous P-code and carrier phase of two frequencies (L1 and L2) was used.

For correcting the ionospheric refraction the so called L3 linear combination of the dual frequency data was formed /2/.

With this raw data station-satellite differences (double differences) were computed in connection to the station Westerbork and used in the adjustment procedure. (This station has a high amount of recorded data and a good receiver clock approximately in the middle of the network)
As shown in figure 2 total data failure has occured at Station Wettzell at the first day and bad data was recorded at station Effelsberg at the second day. Additionally at station Graz about 1 hour of data of the third day was not available.

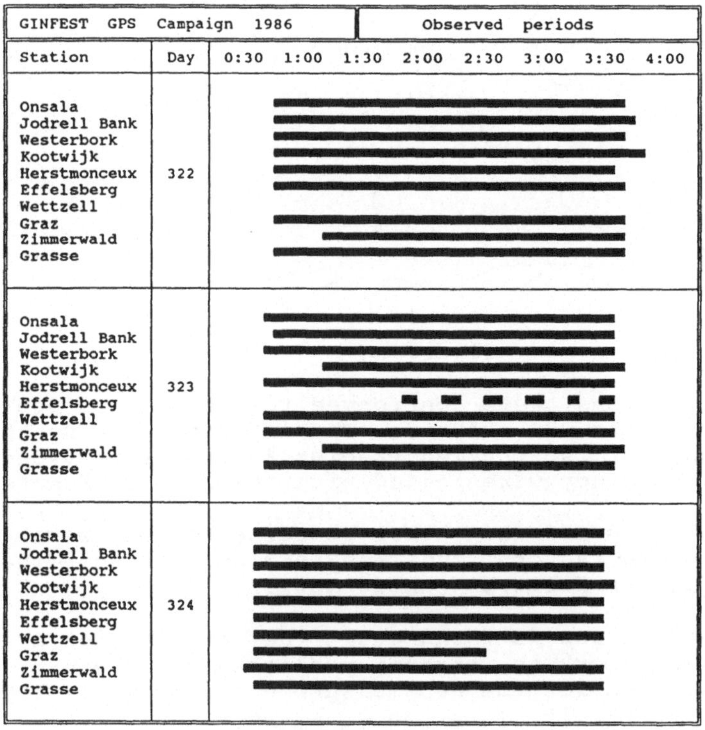

figure 2

The overall quality of the recorded data was good - the receivers at station Herstmonceux and Jodrell Bank showed large clock corrections and also more significant half and even some quarter cycle slips in the data. This was the reason for not using the data set recorded at Herstmonceux at the third day.

Weather information (pressure, wet and dry temperature) was not provided at station Graz for the whole campaign. The given data was added to the phase data files and used as input values for the Lanyi-Davis tropospheric model. A comparison of the weather data during the whole campaign shows larger regional weather disturbances (relative humidity) at the third day and mainly in the southern region (ZIM,GRS). At the first two days a smooth weather situation at all stations provided a good correction of the tropospheric error by forming double differences. Due to the missing weather data at station Graz extrapolated values of the weather information recorded at Wettzell and Zimmerwald was used to correct the tropospheric delays.

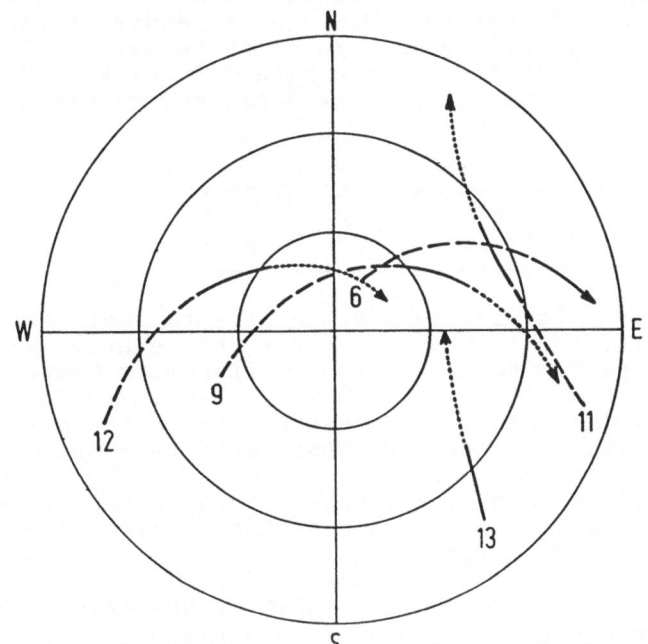

station: Effelsberg   date: 19.11.1986

figure 3

Orbit information was firstly evaluated with the transmitted Broadcast Ephemeris (BE) and secondly as 15 minute coordinate sets provided by the NSWC-DMA orbit computation center at Dahlgreen, Virginia (PE). In a preliminary step this orbit data was used to compute smoothed single arc coefficients for each day for each satellite with the computation of 6 Keplerian elements and the solar radiation pressure parameters (although the observations were taken at 2:00 o'clock in the morning all satellites were in the sunlight during the whole campaign). This orbit information was then used without orbit improvement during the network adjustments in order to keep the GPS internal system scale factor(s) unchanged.

Approximately 90 baseline components and satellite ambiguities were estimated by the use of about 9000 double difference observables each day. The final solution used for comparisons with LASER and VLBI results is a three day adjustment using either the Precise or the Broadcast Ephemeris.

Finally one should mention that the satellites geometry given in November 1986 was not optimal (compare figure 3). Nearly all satellites show more or less a moving direction from west to east over middle Europe. This produces a statistically large variance of the x-baseline components in comparison with the y- and z-values.

2.2 Results of the Precise (PE) and Broadcast (BE) solutions

To get a view in the repeatability and stability three single day solution distances were compared with the final three day values using BE and PE orbits (compare appendix A1/A2). The mean absolute value difference in case of PE is with 3.2 cm not that much smaller than in the BE case with 4.8 cm. With the first look at the total satellite coordinate differences (PE-BE) this may be surprising:

```
delta x  =  app.   5 Meter
delta y  =  app.  25 Meter
delta z  =  app.  10 Meter
```

The fact, that this differences are repeated each day in the same range and direction is the reason for the relatively good repeatability of the BE solution. The differences between both solutions are obvious in comparisons of absolute coordinate differences. Herin mainly two systematic effects related to the fixed base station Westerbork (WES) could be seen:

- all stations which are located south of WES are shifted to the west in longitude - the stations north of WES are not moved or in the opposite direction.
- all stations east of WES are drifting to the south in the latitude - the stations west of WES are changed in the opposite direction with approximately the same values.

The changes are approximately related to the baseline lenghts which would properly be expected.

The effect on the baseline lenght will be analysed later on in chapter 3.3.

The remaining residuals of the BE solution are in the range of 5 - 9 centimeters. Smooth systematic waves can be seen which are now identified as orbit error effects. With the PE solution this systematics are gone - the residuals are in the range of 2 - 4 centimeters.

solution typ	$sigma_x$	$sigma_y$	$sigma_z$	
Broadcast	4.0	1.9	1.0	[cm]
Precise	2.5	1.2	0.7	[cm]

The special satellite geometry above Europe can be seen in the variances of the PE and the BE solutions. In both cases the mean x-component is twice the y-value, the y-component approximately twice the z component.

3. Comparisons with other Space techniques

The LASER solution we are using for the presented comparison was evaluated by the IFAG analysis center and taken from reference /3/. It includes station coordinates of 5 LASER sites excluding station Kootwijk in the Netherlands. The solution was adjusted by use of the Jan/Oct. 1986 ranging data to LAGEOS and shows internal coordinate accuracies of maximal ± 3 centimeters (station Grasse).
Additionally a LASER solution evaluated at the Delft computer center was used for comparisons with GPS which showed more or less the same results as seen with the IFAG solution. This results are therefore not presented here in detail /4/.

The VLBI solution used here is a combined set of coordinates. The coordinates of the stations Jodrell Bank and Westerbork are computed using the 12 hours of data observed during the GINF-2 VLBI experiment (October 1987) /5/. It is a single band (6cm-5GHz) preliminary solution with an internal accuracy of about ± 30 centimeters. The coordinates of the stations Effelsberg, Onsala and Wettzell are evaluated in the NASA Crustal Dynamics Project and published 1987 /6/. These station coordinates are presented with an internal accuracy of about ± 2 centimeters.

The given LASER and VLBI coordinates have been centered to the GPS ground marks by using mainly ED79 excentricity vectors. This could cause an additional station error of approximately 2cm in the comparable coordinates.

3.1 Comparison with LASER coordinates

Routinely used 7 parameter transformations were used to compare the GPS with the described LASER solution. Appendix B1 contains post transformation residuals of a solution with 4 stations (HRS,WTZ,GRZ,ZIM) using the PE-GPS coordinates. The scale factor can be seen with $3.0 * 10^{-7}$ meters. Looking at the raw range differences one can find a systematic offset at all distances in connection with Herstmonceux. If one assumes that there could be a 35 cm excentricity error a total agreement of both solutions of ± 5 cm could be found.

A comparison without station Herstmonceux (appendix B2) leeds to a scale factor of $1.4 * 10^{-8}$ meters. Using this factor the three baselines are identical with the LASER distances within 3 centimeters.

3.2 Comparison with VLBI results

Using the same procedure as in the LASER case a similar problem with station Jodrell Bank is found (compare appendix C1 and C2). If one remembers the a-priori accuracy of the used VLBI solution at Jodrell Bank it is a possible result of the comparison. A second transformation without Jodrell Bank shows with small redundancy a 10 cm discrepancy at station Effelsberg. This could be influenced by the large excentricity vector of about 6.5 km, which was taken from a local network adjustment. A second reason may be the GPS solution with only two days of observations instead of three as on the other sites.
The reached scale factor within the 3 station solution is with $1.8 * 10^{-8}$ in the same region as found in the LASER case.

3.3 Broadcast and Precise solution transformation

To give an idea in which range the BE and PE solutions are comparable in terms of baseline lenghts the following table shows all scale factors from the LASER/VLBI transformations.
The correct excentricities of the 3 station solutions so far (VLBI: EFB-WTZ-ONS - LASER: WTZ-GRZ-ZIM) show a factor of 10 between the PE and BE baseline lenght in both cases.

stations used for scale	LASER		VLBI	
	BE	PE	BE	PE
3	$1.9 * 10^{-7}$	$1.4 * 10^{-8}$	$2.3 * 10^{-7}$	$1.8 * 10^{-8}$

This could be interpreted as external accuracy of the BE and PE baseline lenght solutions. Due to the small redundancy no comments on rotation or translation parameters are possible until now.

4. Conclusion

Using precise ephemeris in the computations of long baselines the internal and external accuracy is in the range of $1-2 * 10^{-8}$ without orbit improvement in the adjustment procedure. A systematic change between BE and PE solutions was found and has to be further analysed in the future.

Eccentricity vectors of some stations have to be checked again with the background of the now available computations. Additional observations to connect VLBI and LASER sites should be done with a better satellite geometry and perhaps over a longer period.

5. <u>References</u>

/1/ Gurtner, W.;   Evaluation of GPS Carrier
    Beutler, G.;   Difference Observations: The
    Bauersima, I.;   Bernese Second Generation Software
    Schildknecht, T.   Package
                          in: Proceedings of the First
                          International Symposium on Precise
                          Positioning with GPS
                          Rockville 1985

/2/ Goad, C.:   Precise Positioning with the
                          Global Positioning System
                          CERN Accelerator School,
                          Genf, 14.-18.04.1986

/3/ Hauck, H.;   Station coordinates 1986 (Jan/Oct)
    Ehlert, D.:   from LAGEOS laser ranging
                          evaluated by the IFAG analysis
                          center
                          Bologna Mai 1987

/4/ Ambrosius, B.A.C.;   Recent results of LASER station
    Noomen, R.;   Positioning
    Wakker, K.F.;   Paper presented at the 12th General
    Papazissi, E.:   Assembly of the European Geophysi-
                          cal Society, Strassbourg, 1987

/5/ Campbell, J.:   GINFEST - VLBI Report
                          presented at:
                          Meeting of the EVN-Technical
                          Working Group
                          Jodrell Bank 17-18.03.1988

/6/ Ryan, J.W.;   Crustal Dynamics Project
    Ma, C.:   Data Analysis-1987
                          Volume 1 - Fixed Station
                          VLBI Geodetic Results
                          1979-86
                          GSFC, Greenbelt 1987

6. Appendix

GINFEST GPS Campaign	322-324 1986		Baseline differences			A 1
From *)	To *)	length (322-324)	difference [cm] 322	323	324	
Westerbork	Jodrell Bank	597703.999	2.2	− 1.1	5.1	
	Onsala	602617.445	1.1	0.1	− 0.1	
	Herstmonceux	487197.078	1.5	3.0	#	
	Kootwijk	98042.686	5.7	0.3	− 2.9	
	Effelsberg	261382.879	5.0	+	− 3.1	
	Wettzell	607402.148	+	1.0	− 1.2	
	Graz	908705.067	− 1.4	1.6	8.0	
	Zimmerwald	674109.718	0.5	− 3.3	7.1	
	Grasse	1017885.823	− 1.0	4.9	2.0	
Jodrell Bank	Onsala	1011707.969	5.5	− 0.3	7.4	
	Herstmonceux	319858.563	− 1.4	0.6	#	
	Kootwijk	560625.996	0.7	− 2.9	5.2	
	Effelsberg	701153.980	2.1	+	1.8	
	Wettzell	1150683.801	+	− 0.2	− 0.2	
	Graz	1436917.783	1.3	− 0.4	9.2	
	Zimmerwald	992321.233	1.2	− 2.5	1.5	
	Grasse	1251482.654	− 0.6	1.7	− 0.6	
Onsala	Herstmonceux	1045590.115	2.6	3.8	#	
	Kootwijk	700502.424	6.9	− 0.3	− 4.2	
	Effelsberg	825390.835	7.0	+	− 3.7	
	Wettzell	919659.333	+	0.4	5.0	
	Graz	1172773.762	− 3.9	− 1.7	14.9	
	Zimmerwald	1207160.071	0.5	− 3.9	9.9	
	Grasse	1553502.822	− 1.4	6.1	2.5	
Herstmonceux	Kootwijk	406374.508	− 2.4	1.8	#	
	Effelsberg	467174.804	0.0	+	#	
	Wettzell	917014.933	+	2.6	#	
	Graz	1182909.938	1.2	2.5	#	
	Zimmerwald	684818.546	− 1.6	− 1.3	#	
	Grasse	932779.453	− 0.2	1.2	#	
Kootwijk	Effelsberg	195093.609	0.3	+	− 1.8	
	Wettzell	602457.486	+	2.8	− 3.1	
	Graz	899498.852	− 1.6	3.4	6.2	
	Zimmerwald	601438.752	− 3.6	− 1.5	7.2	
	Grasse	939543.703	− 5.5	5.8	2.9	
Effelsberg	Wettzell	455167.377	+	+	− 4.8	
	Graz	738061.761	0.0	+	4.4	
	Zimmerwald	412959.000	− 3.8	+	9.4	
	Grasse	757921.306	− 5.3	+	4.4	
Wettzell	Graz	302087.045	+	− 0.5	7.8	
	Zimmerwald	475895.397	+	− 3.1	3.1	
	Grasse	753186.737	+	6.1	− 5.4	
Graz	Zimmerwald	610851.027	1.1	1.9	4.5	
	Grasse	764478.435	1.6	5.9	− 5.8	
Zimmerwald	Grasse	349684.090	− 1.3	8.3	− 5.7	

*) referred to TI 4100 antenna phase center

PE - solutions   (without orbit improvement, incl. weather data)

GINFEST GPS Campaign	322-324 1986		Baseline differences		A 2

From *)	To *)	length (322-324)	difference [cm]		
			322	323	324
Westerbork	Jodrell Bank	597703.999	− 3.7	0.4	8.1
	Onsala	602617.321	− 2.2	− 3.1	5.3
	Herstmonceux	487196.975	− 2.6	− 8.6	#
	Kootwijk	98042.688	7.0	− 2.8	0.2
	Effelsberg	261382.809	4.0	+	− 3.2
	Wettzell	607402.074	+	1.4	− 1.9
	Graz	908704.960	− 3.4	3.6	0.6
	Zimmerwald	674109.509	− 4.0	−11.6	5.9
	Grasse	1017885.561	− 3.9	− 1.0	4.4
Jodrell Bank	Onsala	1011707.921	− 5.0	− 2.4	21.4
	Herstmonceux	319858.471	0.2	− 1.4	#
	Kootwijk	560625.977	− 4.6	− 0.7	5.8
	Effelsberg	701153.922	− 2.9	+	− 0.3
	Wettzell	1150683.736	+	1.8	2.5
	Graz	1436917.680	− 2.4	− 3.8	− 3.1
	Zimmerwald	992321.069	− 0.8	− 1.5	− 3.2
	Grasse	1251482.390	− 1.3	1.8	− 5.1
Onsala	Herstmonceux	1045589.890	− 4.9	−12.0	#
	Kootwijk	700502.303	5.1	− 6.0	5.5
	Effelsberg	825390.639	3.6	+	1.7
	Wettzell	919659.099	+	− 0.4	− 0.1
	Graz	1172773.500	− 8.0	1.1	0.5
	Zimmerwald	1207159.713	− 7.6	−17.1	13.2
	Grasse	1553502.425	− 7.0	− 4.1	10.3
Herstmonceux	Kootwijk	406374.404	− 7.6	− 7.4	#
	Effelsberg	467174.730	− 5.3	+	#
	Wettzell	917014.850	+	− 4.7	#
	Graz	1182909.831	− 3.9	− 1.1	#
	Zimmerwald	684818.444	− 2.2	− 1.4	#
	Grasse	932779.271	− 3.5	2.3	#
Kootwijk	Effelsberg	195093.552	− 1.7	+	− 3.6
	Wettzell	602457.437	+	3.2	− 1.7
	Graz	899498.767	3.3	5.5	1.3
	Zimmerwald	601438.559	− 8.7	− 6.9	4.8
	Grasse	939543.450	− 9.3	2.4	3.8
Effelsberg	Wettzell	455167.364	+	+	0.5
	Graz	738061.716	0.0	+	4.4
	Zimmerwald	412958.862	− 7.4	+	6.3
	Grasse	757921.114	− 7.3	+	7.1
Wettzell	Graz	302087.016	+	1.6	1.2
	Zimmerwald	475895.286	+	17.5	−12.2
	Grasse	753186.595	+	− 4.2	9.2
Graz	Zimmerwald	610850.949	− 2.2	−10.5	15.1
	Grasse	764478.341	1.4	− 2.0	15.6
Zimmerwald	Grasse	349684.044	0.8	11.4	2.1

*) referred to TI 4100 antenna phase center

BCE - solutions   (without orbit improvement, incl. weather data)

Transformation GPS-PE-solution to LASER (4 sites)		B 1

Translation	dX:	− 2.324	± 0.06	Meter		
	dY:	− .618	± 0.06	Meter		
	dZ:	.914	± 0.06	Meter		
Scale factor	dm:	1.0000003	± 0.0000001			
Rotation around	X-Axis :	1.112	± 0.08	seconds		
	Y-Axis :	.998	± 0.04	seconds		
	Z-Axis :	.184	± 0.02	seconds		

Station	vE	vN	vH	c m	vX	vY	vZ
HRS	5.2	−5.9	6.0		7.4	6.4	0.6
WTZ	2.7	9.8	−12.0		15.5	0.2	−2.5
GRZ	3.8	−5.7	12.3		11.6	5.7	5.5
ZIM	−11.7	1.8	−6.3		−3.6	−12.4	−3.5

From site	to site	LASER distance	GPS distance	length diff.	within sc.fact.
HRS	WTZ	917015.036	917014.691	0.345	0.068
HRS	GRZ	1182910.149	1182909.779	0.370	0.012
HRS	ZIM	684818.840	684818.453	0.387	0.179
WTZ	GRZ	302086.973	302086.989	−0.016	−0.108
WTZ	ZIM	475895.297	475895.318	−0.020	−0.164
GRZ	ZIM	610851.070	610851.030	0.040	−0.145

Transformation GPS-PE-solution to LASER (3 sites)		B 2

Translation	dX:	− 2.125	± 0.02	Meter		
	dY:	− .451	± 0.02	Meter		
	dZ:	− .635	± 0.02	Meter		
Scale factor	dm:	1.00000001	± 0.00000007			
Rotation around	X-Axis :	1.306	± 0.04	seconds		
	Y-Axis :	1.030	± 0.02	seconds		
	Z-Axis :	.232	± 0.01	seconds		

Station	vE	vN	vH	c m	vX	vY	vZ
WTZ	1.1	3.6	0.1		−2.8	0.5	2.5
GRZ	− 2.1	−2.0	−0.1		1.9	−1.8	−1.4
ZIM	1.0	−1.6	0.0		0.9	1.2	−1.1

From site	to site	LASER distance	GPS distance	lenght diff.	within sc.fact.
WTZ	GRZ	302086.973	302086.989	−0.016	−0.021
WTZ	ZIM	475895.297	475895.318	−0.020	−0.027
GRZ	ZIM	610851.070	610851.030	0.040	−0.031

Transformation GPS–PE-solution to VLBI (4 sites)						C 1		

	dX:	− 3.711 ± 0.05	Meter
Translation	dY:	− 2.231 ± 0.05	Meter
	dZ:	1.713 ± 0.05	Meter
Scale factor	dm:	1.0000001 ± 0.0000001	
	X-Axis :	0.022 ± 0.03	seconds
Rotation around	Y-Axis :	0.212 ± 0.03	seconds
	Z-Axis :	0.226 ± 0.02	seconds

Station	vE	vN	vH	c m	vX	vY	vZ
ONS	7.6	4.6	−0.4		−4.8	7.0	2.5
JOB	6.7	−8.7	2.1		7.2	7.7	−3.6
EFB	−18.0	2.1	−4.0		−1.7	−18.4	−1.9
WTZ	3.7	1.9	2.2		−0.7	3.6	2.9

From site	to site	VLBI distance	GPS distance	lenght diff.	within sc.fact.
ONS	JOB	1011707.903	1011707.827	0.076	−0.064
ONS	EFB	825390.642	825390.644	−0.002	−0.116
ONS	WTZ	919659.212	919659.106	0.106	−0.021
JOB	EFB	701154.233	701153.865	0.368	0.271
JOB	WTZ	1150683.824	1150683.593	0.231	0.072
EFB	WTZ	455167.109	455167.250	−0.141	−0.203

Transformation GPS–PE-solution to VLBI (3 sites)						C 2		

	dX:	− 3.753 ± 0.06	Meter
Translation	dY:	− 1.859 ± 0.06	Meter
	dZ:	1.385 ± 0.06	Meter
Scale factor	dm:	1.00000002 ± 0.0000001	
	X-Axis :	0.026 ± 0.03	seconds
Rotation around	Y-Axis :	0.246 ± 0.07	seconds
	Z-Axis :	0.248 ± 0.03	seconds

Station	vE	vN	vH	c m	vX	vY	vZ
ONS	2.5	−4.4	0.0		0.3	3.1	−2.7
EFB	−10.3	−0.6	0.2		2.4	−10.0	−0.2
WTZ	7.8	5.0	−0.2		−5.4	6.9	2.9

From site	to site	VLBI distance	GPS distance	lenght diff.	within sc.fact.
ONS	EFB	825390.642	825390.644	−0.002	−0.017
ONS	WTZ	919659.212	919659.106	0.106	−0.089
EFB	WTZ	455167.109	455167.250	−0.141	−0.149

SUPERVISION OF THE CONTROL NETWORK
OF THE FEDERAL REPUBLIC OF GERMANY
WITH MACROMETER 1983 - 1985
- KONMAC -

by

Jürgen Kremers

1. General Aspects

During the years 1983 to 1985 Macrometer measurements had been done in
the Federal Republic of Germany in especially selected areas. Exclusively Macrometer I with only one frequency had been used. The
measurements, executed by the state land-survey agencies and the
university of the Federal Armed Forces, had first of all intended to
confirm the indicated accuracy of the Macrometer. Finally, however,
after having measured some combining elements between the twelve
partial networks, it was possible to form one entire network. The
structure of this network is because of the first intentions rather
inhomogeneous, but this network, covering great areas of the Federal
Republic of Germany and containing connections to the Netherlands,
Belgium and Austria, forms some sort of zero order network for the
German Primary Triangulation Network. This control network as a result
of Macrometer measurements is called KONMAC-network (figure 1).

The methods of calculating KONMAC had been very new for the state
survey agencies. The first time they had to handle in a great size with
three-dimensional geocentric coordinates. Up to now it was nearly
exclusively usual to use coordinates concerned to two-dimensional
reference surfaces. Questions had to be answered concerning transformations in the three-dimensional space, converting of observations to
station-centres, and the positions of the tradionally used reference
surfaces in a geocentric coordinate system.

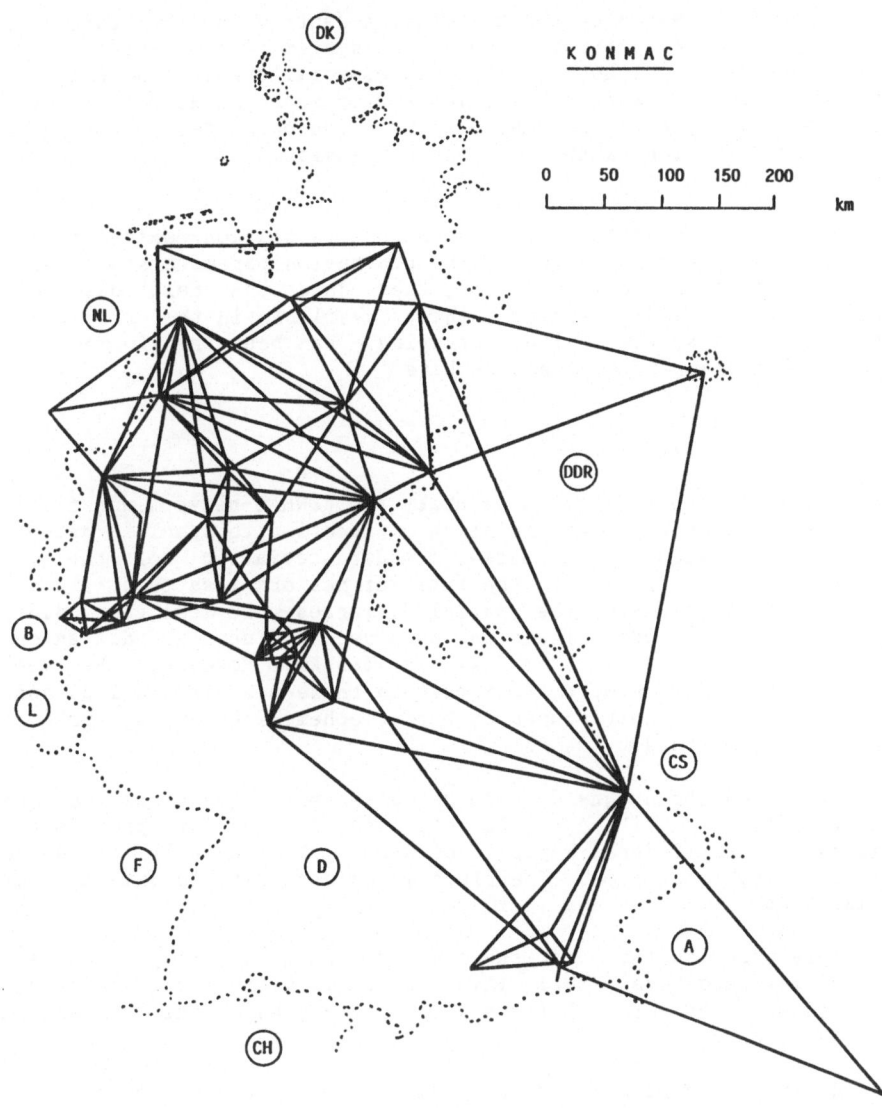

Figure 1

2. Adjustment

The space vectors between the different Macrometer-stations were introduced into the adjustment of the KONMAC-network as observations. The variance-covariance matrices of the three components of each vector were taken into consideration. These data had been the result of a pre-adjustment of the original measured Macrometer data, done in the U.S.A. The different space vectors had to be regarded to be respective independent as their correlations had not been known, neither mathematically nor physically. This neglect was not without any problems, because sometimes there had measured four Macrometers at the same time, and the so called "trivial vectors" were used for adjusting.

The adjustment was done for the station-centres in the three-dimensional geocentric cartesian coordinate system of the space vectors. Free adjustments without location to especially selected points had been done as well as free adjustments with such locations. They had been done partly with only three translation-parameters, partly with seven transformation parameters (three translations, three rotations, one scale-unknown).

For the spatial converting of observations to station-centres there had to be used stepwise got a-priori transformation parameters and differences in geoid-undulations. Using these data the relations between station-centres and measuring points, available in the traditionally used coordinate system of land surveying, had been transformed to the coordinate system of the space vectors.

3. Practical results

To get informations about its accuracy the KONMAC-network was first of all freely adjusted by renunciation of location to special groups of points. This lead to a spatial standard deviation for the points between 8 cm and 32 cm, in the interior network less than 2 dm. The relative corrections of the spatial distances were less than 4.6 ppm, on an average 0.83 ppm. There was no dependence upon the distances to be seen, which varied from 5 km to 466 km. However, the vectors measured in 1985 were more accurate than those measured in 1983 and in 1984. This was probably caused by better ephemerides and by technically improved Macrometers.

A further adjustment was done to get answers to questions concerning the respective position of the reference surfaces for positions and heights in the Federal Republic of Germany by the help of geodetic methods using satellites. Therefore it is necessary to give a special explanation.

Reference surface for the field of fixed height points is the Normal-Zero-Level (NN). This surface is defined by connecting the endpoints of vertical lines with the lengths of the corresponding NN-heights.

Reference surface for the position point field, based on the German Primary Triangulation Network (DHDN), is the Bessel-ellipsoid. Calculating of the DHDN started with the measured base-lines, continued with triangulation chains and lead finally to completing networks. The base-lines have to be regarded now some more intensively.

They had been measured at the real surface of the earth. Because the determination of coordinates had to be done on the Bessel-ellipsoid, they had to be reduced there. This was done by using the heights of the base-lines, determined in the actual height system of that time. The height differences between these old heights and the present actual NN-heights show there immediately the respective position of Bessel-ellipsoid and NN-level.

The height-differences between the surface of the so located Bessel-ellipsoid and NN-level are called "NN-undulations". Thus

the NN-undulations in the points of the base-lines are well known.
NN-undulations of further points are the differences between their
ellipsoidal heights and their NN-heights.

In connection with calculating the Macrometer measurements it was
planned to find out the NN-undulations for all points of the network.
Therefore the KONMAC-network, containing points of the base-lines
Meppen, Goettingen and Bonn, was freely adjusted and located by a 7-
parameter-transformation to these three points. The remaining differences in heights in these points were about +/- 2 cm.

The NN-undulations obtained for the points of the KONMAC-network fit to
the geoid-undulations evaluated by S. Heitz. These geoid-undulations,
astrogeodetically found out and concerned to the Bessel-ellipsoid of
the DHDN, show differences to the NN-undulations of about 4 dm.
Differences are about 1.5 dm if the comparison is taken with the
quasigeoid, astrogravimetrically found out by D. Lelgemann and transformed into the same respective position of geoid and Bessel-ellipsoid
as choiced by S. Heitz.

After that the KONMAC network was adjusted without constraint and
located by rotation-parameters and a scale-unknown to 22 points of the
DHDN. The resulting two-dimensional coordinates were compared with the
Gauss-Krueger coordinates of those 22 points. The differences in the x-
and in the y-axes were both about 4 dm to 5 dm. The vectors show the
well known two whirls that are always found if freely adjusted large
networks are compared with the coordinates of the DHDN.

To eliminate this effect, the KONMAC-network was transformed to diagnosis-coordinates of the DHDN. These diagnosis-coordinates had been
received by adjusting all angels and all distances of DHDN-measure-
ments. They are more homogeneous than the normally employed coordinates, for the scale is uniform and not varying as in the DHDN. This
lead to two-dimensional differences in the x- and in the y-coordinates
from both less than 1 dm. So by the help of geodetic methods using
satellites it was verified independently, that the outer accuracy of
the diagnosis-coordinates is much better than those of the usually
employed coordinates.

4. Comparison with results of Doppler-campaigns

A further analysis was done by comparing the results of Macrometer
measurements with coordinates by Doppler-campaigns. Along great distances Doppler-coordinates give an independent control to the results
of KONMAC. For this comparison were used

- the Doppler-campaign "Deutschland-Österreich" DOEDOC, observed in
  early summer 1977 and in summer 1979 in the Federal Republic of
  Germany and in Austria,
- the Doppler-campaign "Niedersachsen" NIEDOC, observed in late autumn
  1981 in Lower Saxon and
- the Doppler-campaign "Rheinland-Pfalz" RPDOC, observed in summer 1983
  in Rhineland-Palatinate.

From these campaigns there exist Doppler-coordinates for 13 points identical to the KONMAC-network. Two points have Doppler-coordinates from two campaigns.

For a first analysis the Doppler-coordinates of all identical points were taken to repair the rank-defect in translation of the KONMAC-network. The differences between Doppler- and KONMAC-coordinates reached about 5 dm for each direction of the coordinate-axes.

A final best fit location of the KONMAC-network to coordinates of each of the three Doppler-campaigns by using three times three transformation parameters and one unknown of scale lead to a reduction of the mentioned coordinate-differences of about 50%. As it was expected these differences were minimal in the points of NIEDOC (about 0.5 dm), for NIEDOC had taken place in a very small area. In the points of DOEDOC, spread over the whole KONMAC-network, the differences were about 3.5 dm and gave a more realistic result than the comparison with NIEDOC.

5. Effects to practical surveying in North Rhine - Westphalia

Meanwhile in North Rhine - Westphalia the use of GPS by Macrometer measurements has become routine. For all points, that have been determined by satellite methods, not only the traditionally used coordinates are offered to the users, but also three-dimensional geocentric cartesian coordinates.

To get the traditional Gauss-Krueger-coordinates as results of Macrometer measurements, NN-undulations are necessary, and they are offered as well in the lists of coordinates. Future Macrometer measurements shall stepwise lead to a condensation of the point-field with geocentric coordinates and NN-undulations.

Finally a special aspect, which is of importance for regions of land sinking, typical for the area of North Rhine - Westphalia, has to be mentioned. The interpretation of NN-undulations, obtained by condensation of satellite-geodetic networks, allows conclusions concerning the actuality of the NN-heights at the moment of Macrometer measurement. If there exists an inhomogeneous result in the NN-undulations, than it is possible to reconstruct actual NN-heights by interpolation of NN-undulations of neighbouring points.

GPS has met the expectations of the state survey agencies. It presents a lot of facilities to practical surveying. To exhaust these facilities, satellite-geodetic activities shall quickly increase their importance.

Bibliography

Schmidt, R. (publ.) 1986: Kontrolle des Deutschen Hauptdreiecksnetzes durch Macrometer-Messungen 1983-1985 -KONMAC-, Deutsche Geodätische Kommission, München, Reihe B, Heft 282

with references to further publications

Simulations with the Software Package of Darmstadt for Kinematic
Applications
- Some Numerical Results with a WM 101 -

by

Hans-Jürgen Euler

Abstract

The paper provides an overview about phase measurements recorded with
a Wild Magnavox WM 101 receiver. The problems of so-called raw data
measurements and determination of connected phases are outlined for
static and kinematic applications. Test measurements using WM 101 in a
weakly accelerated mode are presented.

0. Introduction

Static GPS measurements using WM 101 need at least 60 minutes of observation time to achieve the announced accuracy of 10 mm plus 2 ppm (WM 1986). For shorter baselines up to few kilometers length this time is too long, because only 3 to 4 baselines can be observed due to the daily GPS observation window of 4 hours. It is well-known that the ambiguity solution in the least squares algorithm is the critical component in view of shorter observation time. If the ambiguity is known, the solution of the baseline needs only minutes to achieve a convenient accuracy. The recent direction of investigation is to measure in kinematic mode. As described by Remondi (1986), the antenna will be moved while tracking satellites. This method gives the possibility to transfer the solved ambiguity to the next station, and the estimation of baseline components requires a shorter span of observation time. The first tests were done using TI4100 receivers. The receiver design of the WM 101 is different from the concept of TI4100, but we were interested to see what happens, if the WM 101 is moved during the time of observation.

1. Formulae for Phase Measurements

1.1 Phase Measurement

The formula for a observation of a single phase can be expressed like:

$$\phi_R^S(t) = \frac{f}{c} \, \rho_R^S(t) + \frac{f}{c} \, \Delta\rho_R^S(t) +$$

$$(f_R - f_S + \frac{f}{c} \, \dot{\rho}_R^S(t)) \, \delta t_R(t) + f \, \Delta T_R^S(t) + n_R^S$$

where

f	= nominal frequency
f_R	= receiver frequency
f_S	= satellite frequency
c	= speed of light
$\rho_R^S(t)$	= range between satellite and receiver at time t
$\Delta\rho_R^S(t)$	= ionospheric and tropospheric correction
$\delta t_R(t)$	= receiver clock correction at t

$\dot{\rho}_R^S(t)$ = range rate

$\Delta T_R^S(t)$ = clock offset between satellite and receiver clock

n_R^S = integer ambiguity

It is well known that the desired coordinates are hidden in the range vector ρ_r^S between the receiver and the satellite. In the formula the terms in relation to the receiver clock and especially the integer ambiguity are not known besides the coordinates of the receiver. The term considering the ionospheric and tropospheric effect can be replaced by a model.

1.2 Between Time Single Difference

The difference of two consecutive measurements at epoch t and t + Δt gives the between time single difference.

$$\phi_R^S(t+\Delta t) - \phi_R^S(t) = \frac{f}{c} (\rho_R^S(t+\Delta t) - \rho_R^S(t)) +$$

$$\frac{f}{c} (\Delta\rho_R^S(t+\Delta t) - \Delta\rho_R^S(t)) +$$

$$\frac{f}{c}(\dot{\rho}_R^S(t+\Delta t) \, \delta t_R(t+\Delta t) - \dot{\rho}_R^S(t) \, \delta t_R(t)) +$$

$$(f_R - f_S) \, (\delta t_R(t+\Delta t) - \delta t_R(t)) +$$

$$f \, (\Delta T_R^S(t+\Delta t) - \Delta T_R^S(t))$$

The integer ambiguity, which is not time dependent, is removed and if the receiver clock offset is known, it is possible to compute three position changes with three phase measurements to the same satellite in consecutive epochs.

In contrast to the phase formula we have to take into account the changes of corrections with respect to time.

Compared to the other terms, the correction due to the offset between the receiver clock and the satellite clock is the most prominent one. Assuming a clock drift of 1 ppm and a measurement interval of two seconds the correction for the Ll frequency is :

$$f \cdot (\Delta T_R^S(t+\Delta t) - \Delta T_R^S) \approx 3150 \text{ Cycles}$$

The correction term for the clock epoch gives us with the range rate of 900 m/s and a maximum receiver clock offset of 0.001 seconds :

$$\frac{f}{c}(\dot{\rho}_R^S(t+\Delta t) \, \delta t_R(t+\Delta t) - \dot{\rho}_R^S(t) \, \delta t_R(t)) \approx 0.1 \text{ Cycles}$$

The frequency offset correction with the presupposed offset for clock and frequency leads us to :

$$(f_R - f_S)(\delta t_R(t+\Delta t) - \delta t_R(t)) \approx 0.003 \text{ Cycles}$$

This is a sort of maximum estimation. In most cases the parameters will be only fractions of the values introduced in this estimation and corrections will be even smaller.

Noticing that the accuracy of a phase measurement is about 0.025 to 0.01 cycles we can rewrite the formula for the between time single difference:

$$\phi_R^S(t+\Delta t) - \phi_R^S(t) = \frac{f}{c}(\rho_R^S(t+\Delta t) - \rho_R^S(t)) +$$

$$\frac{f}{c}(\dot{\rho}_R^S(t+\Delta t) \, \delta t_R(t+\Delta t) - \dot{\rho}_R^S(t) \, \delta t_R(t)) +$$

$$f(\Delta T_R^S(t+\Delta t) - \Delta T_R^S(t))$$

Since changes of the elevation angle are very small, we may also neglect the correction due to weather for lower accuracies.

The between time single difference gives the possibility to compute a single receiver solution. Using three phase measurements to three different satellites at two consecutive epochs and introducing the clock correction, a direct solution of the horizontal and vertical position changes is practible. Since differences of consecutive observations are formed, only these errors are removed which are not time dependent. But orbital errors, satellite clock errors and effects of the propagation delay will corrupt the accuracy of the changes essentially.

1.3 Time-Receiver Double Difference

The difference of two between time single differences from two different receivers measured at the same time interval will be called time-receiver double difference :

$$\phi_{R1}^{S}(t+\Delta t) - \phi_{R1}^{S}(t) - \phi_{R2}^{S}(t+\Delta t) + \phi_{R2}^{S}(t) =$$

$$\frac{f}{c}(\rho_{R1}^{S}(t+\Delta t) - \rho_{R1}^{S}(t) - \rho_{R2}^{S}(t+\Delta t) + \rho_{R2}^{S}(t)) +$$

$$\frac{f}{c}(\dot{\rho}_{R1}^{S}(t+\Delta t)\,\delta t_{R1}(t+\Delta t) - \dot{\rho}_{R1}^{S}(t)\,\delta t_{R1}(t) -$$

$$\dot{\rho}_{R2}^{S}(t+\Delta t)\,\delta t_{R2}(t+\Delta t) + \dot{\rho}_{R2}^{S}(t)\,\delta t_{R2}(t)) +$$

$$f\,(\Delta T_{R1R2}(t+\Delta t) - \Delta T_{R1R2}(t))$$

With this difference we remove or reduce the satellite dependent terms of the observation equation. The solution associated with this equation gives position changes of one receiver relative to the second one.

2.   Phase data collection

2.1 Raw data measurements of WM 101

The WM 101 measures every 0.1 seconds the actual phase of a specific tracked satellite (WM 1986). After collecting data during a time interval of two seconds a curve fit is evaluated. The values of the phase and the ascent of the fitted curve (phase rate) at the mean of interval together with the propagation time could be written, if selected, as a raw data record on tape.

Since the WM 101 has only three tracking channels for phase measuring, additional satellites will be distributed, so that there are up to two satellites on each channel. After the two seconds for raw data measurements the receiver tracks on this channel a second satellite, if available; otherwise the measurements for the first will be continued.

If data compression is enabled, the receiver computes an additional curve fit for the selected time interval (up to 60 seconds) and records single values for phase, phase rate and propagation time on tape. The name of records with compressed data is "satellite data record".

As stated in the technical reference manual (WM 1986) the satellite data record and the raw data record have a similar structure (number of bytes and positions for output values). The satellite data record contains the continuous phase and other values as described above for one satellite at a specific epoch (e.g. full minute for 60 seconds compression interval and an offset of zero).

All satellite data records are synchronized; there is no offset between epochs of two measurements at a specific time interval.

In the raw data record a non-continuous phase is embedded and there is no synchronization to any other tracked satellite. The non-continuous phase has to be connected with the previous phase observation using the phase rates. Since the phase rate is the ascent of the fitted curve (derivative of phase measurements with respect to time), an integration of it gives a good approximation almost within one cycle for the continuous phase.

At first we tested an integration of a least squares polynomial of degree 3 for 10 measurements around the two seconds interval to be spanned. The second method was an integration of a cubic spline through the phase rates. Both approaches deliver valid results using static observations. The integer part of the connected phase is given by the integrated phase rate, the fraction by the observed phase of the epoch. The approximations with phase rates represent the continuous phase with deviations smaller than 0.1 cycles. In about 15 percent of the computations we found deviations of about half a cycle. In most cases (80 percent) the receiver indicated a half cycle slip and we are able to correct them. But sometimes one of the available receivers indicated half cycles where none occured and vice versa.

2.2 Problems with Raw Phase Measurements in a Weakly Accelerated Mode

By using the WM 101 with a moving antenna several problems arise. At first the internal curve fit of the 0.1 seconds data is performed. Since the receiver records only the phase and phase rate at the midpoint of the two seconds interval, these computed values represent the whole interval. With stronger accelerations within the interval the algorithm might not be able to follow them. Between observations while moving the antenna with constant velocity and measurements without motion for the receiver, there is no difference except bandwidth problems with the receipt of signal. Since it is not possible to accelerate the antenna between consecutive two seconds time intervals, only a weakly accelerated movement is appropriate. Numerical tests showed that with an acceleration of about 0.2 m/s^2 the reconstruction of the continuous phase using two seconds phase and phase rate does not end in several cycle slips. But due to the requirement of measurements in consecutive epochs, no interruption while tracking another satellite is allowed. This implies that only measurements with three satellites are applicable because of switching channels in the current design of WM 101.

The phase connection using the integration of a least squares polynomial was not as successful as for static observations. The solution with an integrated spline is much more flexible when stronger changes of phase rates occured. In case of moving antennas the latter method is preferable.

3. Kinematic measurements with WM 101

In December 1987 we made kinematic test measurements in Darmstadt using two WM 101. The reference frame for the observations was provided by several pillars at the roof of our building at outside campus. The antennas were set up on two pillars and three satellites were tracked. The satellites were selected in a configuration whose distribution was as good as possible with respect to different azimuths and good elevations above horizon. One of the antennas was moved between several pillars and back to the starting place. Polynomials for the clock correction of the receivers were determined using pseudo ranges. In the case of the moving antenna the observations when occupying the starting place were used.

Figure 1 shows the position changes in the case of the static antenna. The most prominent effect is the large drift in the x-component of about 4 meter. In the y-component the deviation is only about 0.1 m and in the z-component up to 1 m.

The coordinate changes of the moving antenna are given in figure 2. It is obvious that both solutions are corrupted by the same errors. The shapes of corresponding curves are similar when the antennas are held fixed. The sources of the drifts are satellite dependent like orbit errors, unmodelled atmospheric delays and satellite clock errors. In the dual receiver solution the drift is substantially reduced (see figure 3). The sources of the smaller drift are the receiver clocks which are not well modeled using pseudo ranges. The steep slope represents the position changes within few centimeters. A better accuracy could not be expected with only three measurements at one epoch.

4. Conclusions

The WM 101 is not designed for a kinematic application using the phase observable. Nevertheless, it seems to be possible to use this receiver in motions with low accelerations and no high restrictions concerning the accuracy of the solutions. For higher precision some additional continuous tracking channels are required in order to eliminate the receiver clock errors. A kinematic survey with higher accelerations needs by all means a modified output of the receiver. Shorter sampling intervals for phase and phase rates result in an unconvenient amount of data during the session. The better solution would be to introduce additional and more suitable values in the raw data record for reconstruction of the continuous phase.

5. References

Remondi, B. W. (1986): Performing Centimeter-Level Surveys in Seconds with GPS Carrier Phase: Initial Results, Proceedings of the Fourth International Geodetic Symposium on Satellite Positioning, University of Texas, pp 1229 - 1250

Wells, D.(1986): Guide to GPS Positioning, Canadian GPS Associates

WM (1986) : WM 101 GPS Satellite Surveying Equipment, Technical Reference Manual, WM Satellite Survey Company

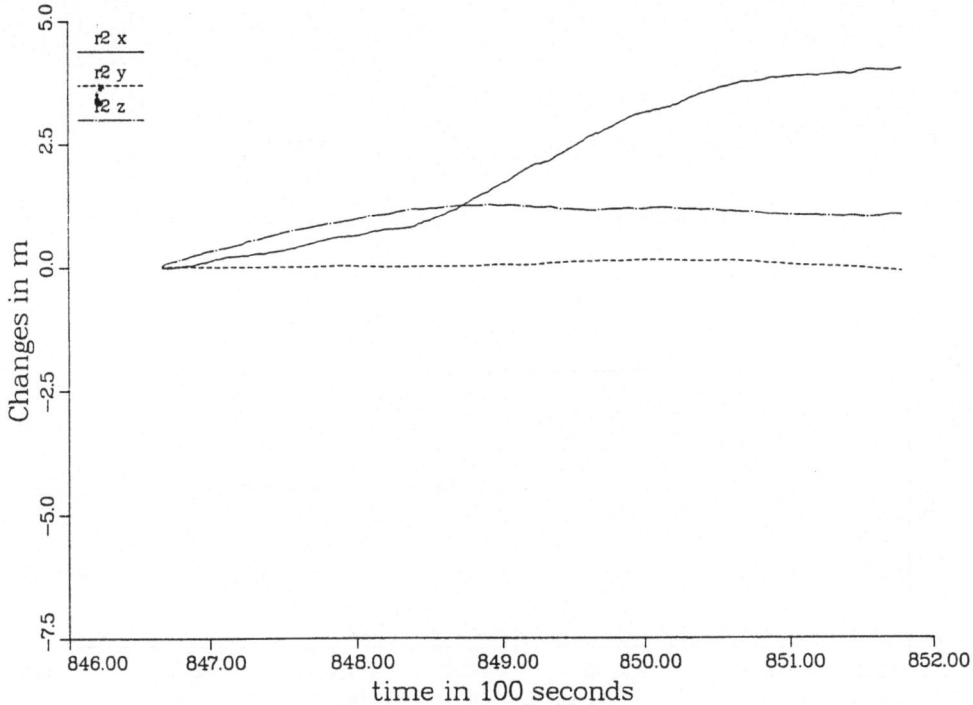

Figure 1   Changes of Coordinates (Static Antenna)

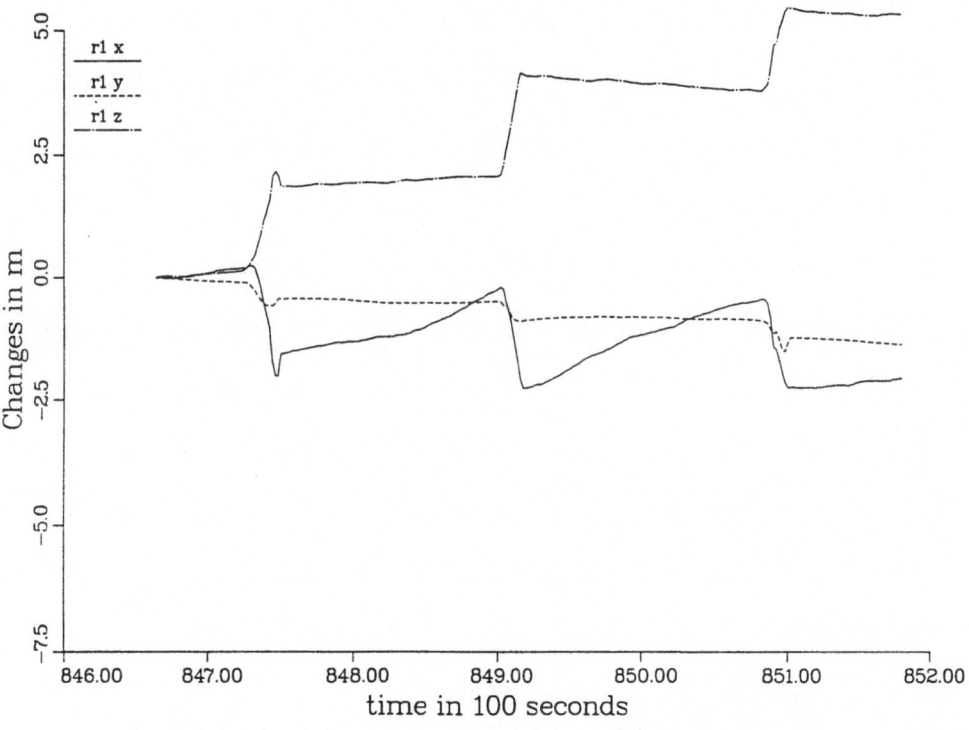

Figure 2   Changes of Coordinates (Moving Antenna)

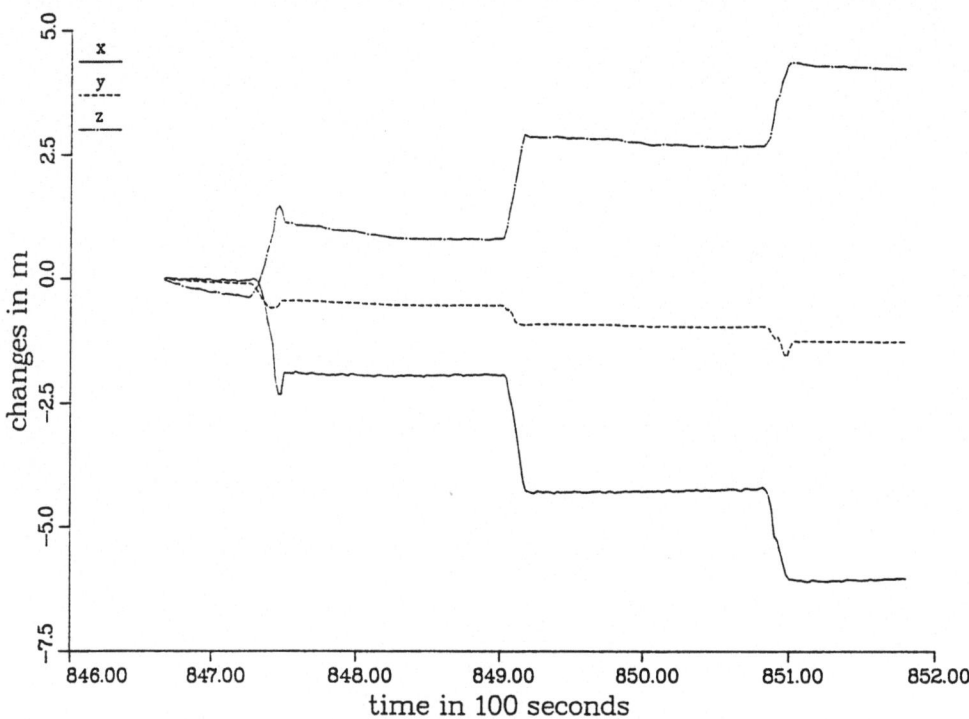

Figure 3   Changes of Relative Coordinates

Fourth session: Campaigns (continued) and Instruments

Chairman: Prof. Boucher, Paris

AN INTRODUCTION TO GPS AND ITS GEODETIC APPLICATIONS

by

Vidal Ashkenazi

Abstract

Although Block II satellites have yet to be launched, GPS has already made a very significant impact on navigation, geodesy, surveying and geophysics. This introduction to GPS is aimed at the non-specialist who wants to understand the basic principles of the system, the way measurements are made, the level of accuracies achieved and potential applications. The lecture uses frequent analogies in order to simplify concept visualization. Nevertheless, it covers the GPS signal structure, the nature of the two main observables and the manner in which they are handled by the different types of hardware, data processing strategies, and the likely influence of the technique on classical triangulation and levelling. The lecture is concluded with a brief summary of other satellite systems which present potential alternatives to GPS for navigation and positioning.

SYSTEM AND HARDWARE ORIENTED ASPECTS OF GPS- APPLICATIONS

by

Philipp Hartl

Abstract

The present situation of the GPS- implementation is reviewed. Some conclusions are drawn from the fact that up to
24 satellites might become operational. This recommends
modifications of the receiver hardware and software and may
lead to considerable system improvements. The present developments in the receiver market as well as the practical
applications are reviewed.

1. Introduction

The Global Positioning System GPS- NAVSTAR has been tested
during the last 5 years with great success. After the Challen-
ger disaster it was, however, not possible to install the
operational system according to the original schedule. Four of
the Block II satellites were destroyed during the last Shuttle
launch and there has not been an opportunity up to now for the
deployment of spacecrafts of the operational system.

But, according to the most recent plans, the first Block II
satellite will be launched before the end of this year by a
Delta rocket and in 1989 several Navstars will be launched,
some of them again by expendable launchers and others by
Shuttle flights. It seems that it will last only a few years
until the operational system is set up completely and will
become fully operational.

At least 21 satellites and most probably even 24 satellites
(including the hot spares) will then be available for practical use /1/.

There might be some intentional performance degradation for
the C/A - Code, but -apart from this- one can hope that most
of the time the Block-II satellites will be operated in the
same manner as the presently available ones. The P- code will
be encyphered additionally either very rarely, or there will
only be one of the satellites which will operate in this "more
classified" mode. For differential measurements and a posteriori data processing the GPS- system can be expected to be
fully available.

If this would be really the case then one could expect that
GPS would become an extraordinary powerful system with signals
generally available of at least 6 satellites at any time and
at any place in the world. Very often one could even apply 8
or sometimes more satellites. For spaceborne applications,
i.e. a GPS- receiver installed in low- orbiting satellites for
its own orbit determination one will have even situations,
where up to 14 GPS- satellites' signals could be received. From
the fact of more than four satellites simultaneous in view
one can derive several interesting consequences as will be
seen in the following.

2. Comparison of system performance

The previously planned 18 satellite version would have had
serious problems from the GDOP point of view, which will not
any more show up in the 24 satellite case; fig.1. This was
certainly the main reason for DOD to add the 6 spare units as
hot spares to the system, because only under this condition it
is possible to avoid worldwide the degradation problem. With
the 18 satellite configuration the Department of Transportation (DOT) would not be able to introduce GPS for Commercial

Aviation. But under the given circumstances one can expect that on a long turn GPS will be introduced for both, Commercial and General Aviation.

It is, indeed, unacceptable for the Commercial Airlines, where, worldwide, at least 2000 planes are permanently in the air that with the 18 satellite configurations various areas of a size like Spain and France together could not be served all day long with the required accuracy: Along the latitudes 30 to 50 degrees N and S there are many such areas where more than 15 minutes and at least once per day the PDOP is unacceptably bad in the 18 satellite case.

With the new policy one could expect many more "customers" and applications. There are many ten thousands of small aircrafts not equipped, yet, for instrument flight and landing (radio location aids) and thousands of small airfields cannot yet provide the proper instrumental support either. GPS - differential measurements could be introduced and will certainly be used as soon as the GPS- satellite system will be operating, provided the receivers are sufficiently low in price. Taking into account that there many ten thousands of customers in this field of aviation one can assume receiver costs for the airborne applications being less than 10 thousand Dollars per unit. The same is true for marine applications. The cost reduction can be expected during the next five to seven years.

Combined navigational systems consisting of INS plus GPS installations are even now in test. One can assume that future INS- instruments for airplanes will contain GPS -receivers as integral part without extra cost, weight and volume, because of the advances in technology.

Instead of going now to the many other applications of GPS, such as its use in millions of cars, we just conclude that the various types and large numbers of users can help to reduce the equipment cost drastically and to make the operation of the system very secure.

What is the technical and practical consequence for the geodetic applications, apart from the fact that one can expect a considerable reduction of the receiver costs?

- First of all, it would be possible to perform the measurements considerably faster and/or with better accuracy, provided one would make use of the advantage of receiving many more satellite signals simultaneously.

- This would be possible by two ways: One could use the 4 satellites with best geometry during the time of measurements.

- One could also use more than 4 channels and apply the weighted pseudoranges and carrier signals of all the satellites. This would have the advantage of making also benefits of the fact that the local geometry/ topography might recom-

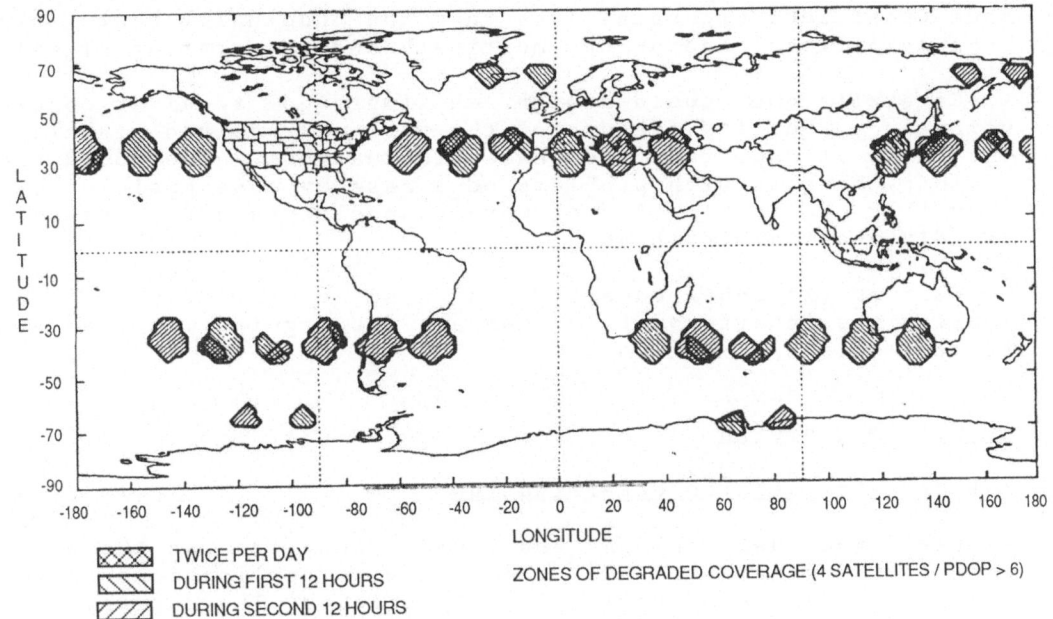

TWICE PER DAY
DURING FIRST 12 HOURS
DURING SECOND 12 HOURS

ZONES OF DEGRADED COVERAGE (4 SATELLITES / PDOP > 6)

Fig.1: Critical areas for the 18 GPS- configuration according to /1/

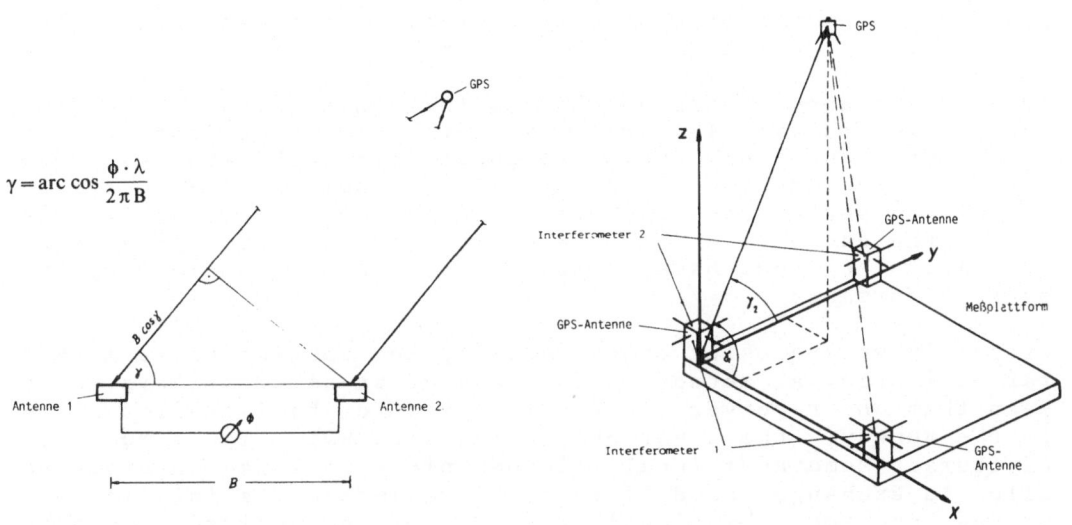

$$\gamma = \text{arc cos} \frac{\phi \cdot \lambda}{2\pi B}$$

Fig.2: GPS- interferometry for attitude measurement

mend other sets of satellites than the "nominally best ones" because of local shadowing, multipath and interfering signals.

- Generally one could design an improved receiver concept using as a quality measure of the various received signals on the basis of amplitude and phase fluctuations. One could then exclude channels with problems on a case by case basis. Apart from that the combination of more channels could help to solve rapidly phase ambiguities.

- This multichannel capacity is also of benefit, if one applies the interferometric concept for angular measurements; fig.2.

3. Receiver design

3.1 Dynamic position determination

So far, the use of C/A and P code technique was the normal code in the kinematic and the dynamic applications of GPS, i.e. in the field of navigation of vehicles. The carrier phase measurement technique has been applied, so far, almost exclusively in the stationary baseline measurements. But in a few publications /2/, /3/, /4/, /5/, it is described that one could also use the carrier phase measurement techniques for very precise position determination in the kinematik case. So far, this is a matter of post processing and of differential phase measurements to static receivers. The question is, how this could be realized reliably in the future in real time.

In order to answer this one has to take into account the following two facts: This requires at least some indicator of phase loss and/ or cycle slippage for each channel and a method to exclude the "bad" channel. In addition one must then be able to perform the measurement by "good" channels. If this is not the case, i.e. if there are less than four good channels left then must have at least some warning indicating the poor performance.

Presently we are using experimentally the carrier phase method for differential methods. In case of an aircraft this would mean that one receiver could, for example, be installed close to the runway, the other one in the aircraft. There must be, of course some additional telecommunication link in order to allow to exchange the differential measurement's information, or one can do the signal processing a posteriori. In both cases one can achieve accuracies in the order of 5 to 10 cm in the "kinematic mode". In the a posteriori case use the "double difference" method is advisable.

Another way to overcome the ambiguity problem it the use of both informations, the pseudorange and the carrier phase. With pseudorange one determines the coarse position and counts in parallel the phase carrier. In small time intervals, say 1 second samples, the results are improved by a recursive filtering technique.

There are also receiver developments going on, for example at JPL, which promise to achieve pseudorange accuracies in the order of 15 cm. If this is assured, then the phase ambiguity problem is solved and then it is possible to determine unambiguously the carrier phase range. But, even if a 15 cm pseudorange determination becomes not possible reliably enough, then at least with the help of a pseudorange accuracy of 30 to 50 cm the carrier phase measurement can be improved considerably in the future: The pseudorange information could at least reduce the total phase ambiguity problem to two to three wavelengths. It is expected to solve for the ambiguities in real time and to achieve the accuracy of the carrier phase measurement with, for example, 10 cm tolerance over larger distances (a few hundred km), provided that two- frequency systems are used.

This will be the case even with 4 to 5 satellites in view, provided the multipath problems can be overcome. In this aspect the fact that more than 4 satellites can be used, will be very important.

Kinematic position determination with aircrafts and automobiles as well as ships will become a common application in geodesy, photogrammetry etc. in the future. This requires for the receiver adjustable bandwidth in the control loop. Wider bandwidth causes additional receiver noise. One must find an optimal compromise between the signal/ noise ratio and the dynamic features. The control loops should be wide, if the one wants to avoid the loss of track and it should be low for good signal to noise ratio. An optimum is a function of the vehicles dynamic.

One other important application of GPS measurement technique is the determination of direction by means of interferometry. The simplest way of measurement would be possible by direct comparison of the carrier phases. This implies some modifications in the receivers. For a three antenna interferometer it must be possible to have 3 sets of receivers (or multiplexed channels) per satellite signal and corresponding phasemeters; fig.2; /4/.

As a conclusion of all these aspects it is quite obvious that for future developments much more than just 4 receiver channels should be included into the equipment. There are tendencies to build modular receivers which can be equipped with up to 12 channels per P- and C/A - code and which have additionally the capacity for the L2- code.

3.2 Receiver developments

Due to the very promising aspects many companies in the world have or will develop GPS- receivers. One can observe various tendencies. There is the military line, the dominant line in the civilian sector, namely for the use in cars, the civilian aircraft line, the ship line and finally the geodetic line.

As regards the military line one can see that in the United States advanced microprocessors and LSI-chips are going into production now in robot assisted factories. Some standardization is taking place in particular for reasons of economy and simplified logistics. The receivers use either one-, two-, or five- channel versions of the same hardware. For a next generation of receivers there are not only developments on their way for the US- DOD, which include among other versions a kingsize cigarette package one- channel receiver , but also some which are contracted by the European NATO partners to a US- European industrial consortium. In addition some national Defense Agencies in Europe have their own activities in this field /6/, /7/.

Controlled reception- pattern antennas (phased array concepts) are under design which minimize sensivity in the direction of jammers, if the enemy tries to disturb the navigation.

The military seems to have numerous applications, even in the field of artillery guiding, parachuting, bomb guiding etc. apart from the well- known plans in the area of aircraft, rocket, satellite, marine and infantry as well as land vehicle navigation.

As regards civil users the most important future market will be for auto cars, another big market will be in the area of aviation and marine.

For car navigation and the envisaged mass production many companies are active in developing own systems and chips. Many Japanese companies are preparing themselves for this market. According to Aviation Week /5/ Trimble contracted with the People's Republic of China to sell hybrid LORAN-C/GPS for integration and to have them finally produced in China.

For the civilian aeronautical applications hybrid systems like Inertial/ GPS, Laser Inertial/ GPS, TACAN/GPS etc. are under development.

In the geodetic receiver developments some uncertainty seems to exist due to the present bad and uncertain operational conditions of the GPS satellite system. But one can assume that this will change as soon as the new satellites are launched. Considerable advances can be seen in the software developments, and as regards the kinematic applications, in the development of more "dynamic" receivers.

As a final remark, it is to state that roughly 70 different types of GPS- receivers exist worldwide presently and much more are about to come. Competition will become tough for the industry and the costs will considerably drop in the future.

4. Practical applications

The practical applications of the carrier phase measurements in the civilian market increase. The applications will include

- Kinematic position determination with land vehicles (surveying) and aircrafts (photogrammetry)

- use for digital road mapping,

- attitude control,

- photogrammetric application (position and attitude),

- application to radar mapping and geomagnetic as gravimetric measurements,

- vehicle dynamic performance tests, etc.

This is certainly an incomplete list, but shows already the general tendency of the market of GPS to expand.

5. Conclusions

There are many interesting developments which make the GPS-technique even more promising in the future. It seems to be doubtful that the degradation due to the military aspects will become a problem during "normal" times. If this is the case and civilian applications remain possible with very high accuracy then GPS will have a tremendous market and variety of uses.

6. References

/1/ Ph. J. Klass : Defense Department will seek funds to expand Navstar Constellation.

Aviation Week and Space Technology /Oct. 5, 1987

/2/ J.R. Lucas, G.Mader: Verification of airborne positioning using GPS Carrier Phase measurements

Photogrammetric Engineering and Remote Sensing, to be published in 1988

/3/ Ph. Hartl, A.Wehr: Chancen der GPS- Satellitennavigation für die Luftphotogrammetrie

Bildmessung und Luftbildwesen 54, Heft 6, 1986

/4/ Ph. J. Klass: Industry devising GPS receivers with hybrid navigation aids

Aviation Week and Space Technology/ Dec.14, 1987

/5/ Ph. J. Klass: First production GPS receiver delivered ahead of schedule

Aviation Week and Space Technology/ Sep. 21, 1987

GPS AS LOCAL GEOLOGICAL CONTROL AND NATIONAL GEODETIC CONTROL

by

P. A. Cross and P. C. Sellers

Abstract

In June 1987 GPS surveys were carried out around six Greek SLR sites. These sites are part of the WEGENER/MEDLAS network in the Eastern Mediterranean region.

In a three week campaign six small fiducial networks, typically 5-10 km, were successfully surveyed. Results show that the local geological stability of a SLR site can easily be monitored at the (sub)centimetre level. Since the fiducial networks included Greek national triangulation pillars, GPS also proved to be an efficient tool for tying the zero-order SLR (baseline) network to the national control network.

1. INTRODUCTION

WEGENER is an acronym for Working group of European Geo-scientists for the Establishment of Networks for Earthquake Research, a voluntary co-ordinating body promoting research directed towards achieving a better understanding of crustal dynamics, kinematics and processes leading to earthquakes. One of the major components of WEGENER is MEDLAS, which is a satellite laser ranging project with the specific objective to establish reference positions for the long-term monitoring of tectonic motions in the eastern and central Mediterranean. The determination of rates of motion and the development and verification of kinematic models will lead to a better understanding of regional plate tectonic mechanisms and the physical processes which drive them.

Of particular interest is the Aegean region, which is believed to have the largest strain rates associated with continental deformation in the world. The region is deforming in a very complicated manner. The African plate is subducting under the Eurasian plate, and the Arabian plate is colliding with the Eurasian plate. Greece and Western Turkey are situated in a region where three major plates are converging. Much of the continental deformation, as indicated by earthquake activity, appears to be occurring in a zone about 300 km wide within a larger region being stretched horizontally by the southward (relative to the Eurasia plate) movement of the Hellenic Trench. The trench, where the African plate is subducting under the Eurasian plate, is believed to be moving at about 6 cm yr^{-1}.

To address the problem a number of permanent and mobile satellite laser ranging sites have been established at key locations in Turkey, Greece, Switzerland, Israel, Italy, and the islands of Sardinia and Lampedusa, see Fig. 1. The mobile sites will be re-occupied in turn for periods of about ten weeks by one of the three mobile satellite laser ranging units which have been developed, and are operated, by West German (MTLRS-1), Dutch (MTLRS-2), and U.S.A. (TLRS-1) groups. Since the MEDLAS campaign began in March 1986 all the mobile sites have been occupied once, and over half have been occupied a second time. A first comparison of repeated baseline measurements will be possible as the observation data collected in 1987 is made available over the next six months.

Furthermore, in a series of meetings (Frankfurt am Main, 7-7-8 October 1986 and Athens, 3-4 March 1987) GPS surveying has been explored as a technique that could be used in a complementary manner to these ongoing SLR activities in the Eastern Mediterranean area.

In particular two items have been considered:

(i) The densification of the WEGENER/MEDLAS network by the use of GPS.

(ii) The use of GPS for the measurement of local Laser Site Control Networks (LSCN).

The first item is being addressed by a number of groups. It is considered essential to determine the partitioning of strain geographically within the region. One of the fundamental problems in plate tectonics today is the relationship between deformation on the scale of plates (i.e. 100's of km) and deformation on the scale of individual faults (10's of km). At the large scale deformation can be modelled using the mathematics of fluid dynamics. However, at a smaller scale the deformation is brittle, or discontinuous, and the fluid model is no longer valid. By monitoring crustal deformation in the Aegean down to a scale of about 10 km it is hoped to identify the scale at which deformation becomes discontinuous.

The second item involves checking whether the SLR sites are geologically stable with respect to their immediate surroundings (say 10-20 km). These measurements are essential in order to separate very local effects from strain derived from the repeated determination of SLR baseline lengths (first priority). It also provides ties to the existing first through fourth order networks of any national geodetic control.

This paper reports on a first test with GPS in 1987 to measure six LSCN's in Greece, i.e. on item (ii) foregoing.

2. GPS MEASUREMENTS IN GREECE

In a multi-national effort (Greece, Great Britain and the Netherlands) preparations for the design and measurement of six LSCN's were made in the first half of 1987. The Greek sites were chosen in preference to other MEDLAS sites because, as already mentioned, they are the first to be reoccupied in the MEDLAS campaign and it was therefore of paramount importance that their local stability be assessed.

After an initial design effort at the March meeting in Athens the Hellenic Military Geographical Service (HMGS) checked the availability of and accessibility of the proposed LSCN sites and provided monumentation and documentation where necessary. Geologists from the University of Athens advised on the extent of the local geological features and whether or not the selected monuments were representative of those features.

Then between June 3 and 19 six LSCN's were measured by five Trimble 4000S receivers (two made available by the Netherlands' Goedetic Commission and three by the

University of Newcastle upon Tyne). Fig. 2 shows the
location of the sites and the order in which they were
visited and Fig 3. shows the LSCN's at each site. Table 1
gives details of the occupation times. Spirit levelling was
used at all stations to measure the height difference
between the antenna and the survey mark on the triangulation
pillar.

ASKITES	June 3 and 4
KARITSA	June 6 and 7
CHRISOUKELLARIA	June 9 and 10
DIONYSOS	June 12 and 13
ROUMELI	June 15 and 16
KATTAVIA	June 18 and 19

Table 1. Measurement schedule

In general the LSCN networks consisted of six stations. The
central station was on the the SLR pad and another four were
within a radius which varied typically from 5km to 10km.
The sixth station was the nearest Greek first order control
network which could be many tens of kilometres away. The
networks were measured in two days each of two sessions:
the first using satellites 6, 8, 9, 11 and 12 between
approximately 12hr and 15hr UT and the second using
satellites 3, 9, 11, 12 and 13 between approximately 15hr
and 18hr UT, see Fig. 4 for the satellite availability
during the relevant periods. Usually on the first day all
five of the five receivers were deployed in the region of
the SLR pad (i.e. one on the pad and the other four on the
other LSCN stations) and on the second day one receiver was
moved from one of the LSCN stations to the distant first
order station. In this way there were nominally either four
or two measures of all LSCN vectors and four measures of the
connection between the first order point and four of the
LSCN stations. Fig. 5, the Askites LSCN, gives a typical
example but note that here there were power problems at
stations B and D on day 2.

3. ANALYSIS OF THE GPS DATA

The measurements were processed independently at Delft
University of Technology and the University of Newcastle
upon Tyne. Essentially the vectors between simultaneously
observing pairs of stations were computed using the TRIMVEC
software package, and the vectors adjusted as a 3-D network
using software developed at the two institutions. The
computations are not yet complete and a number of detailed
decisions have yet to be made regarding the use of the
various software, especially with regard to the inclusion of
the observed meteorological and some fine details on the
selection of starting coordinates in the use of TRIMVEC. A
full technical report is planned which will give details of
all final coordinate estimates and measures of their quality
as well as descriptions of the stations used.

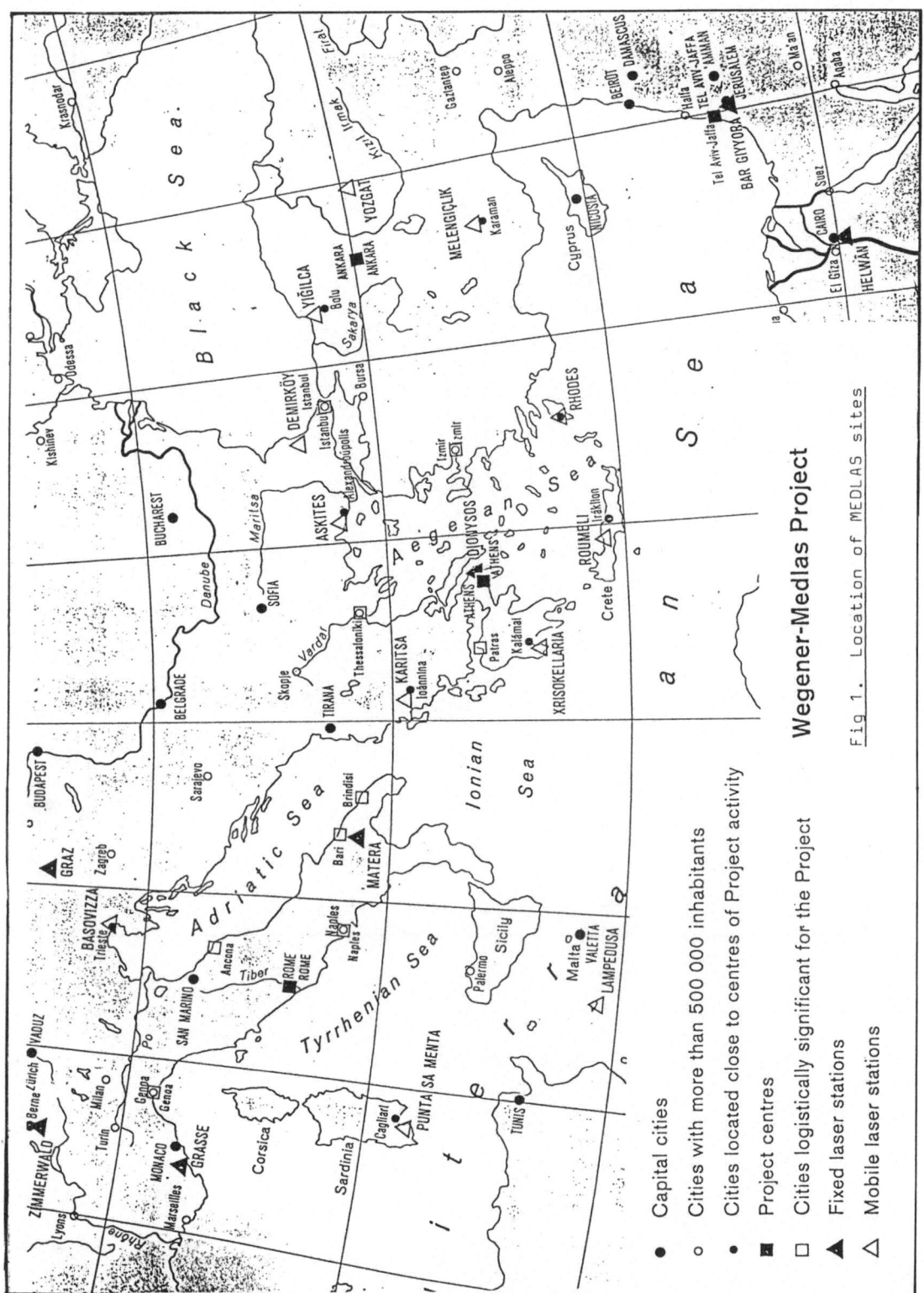

Wegener-Medlas Project

Fig 1. Location of MEDLAS sites

- Capital cities
- Cities with more than 500 000 inhabitants
- Cities located close to centres of Project activity
- Project centres
- Cities logistically significant for the Project
- Fixed laser stations
- Mobile laser stations

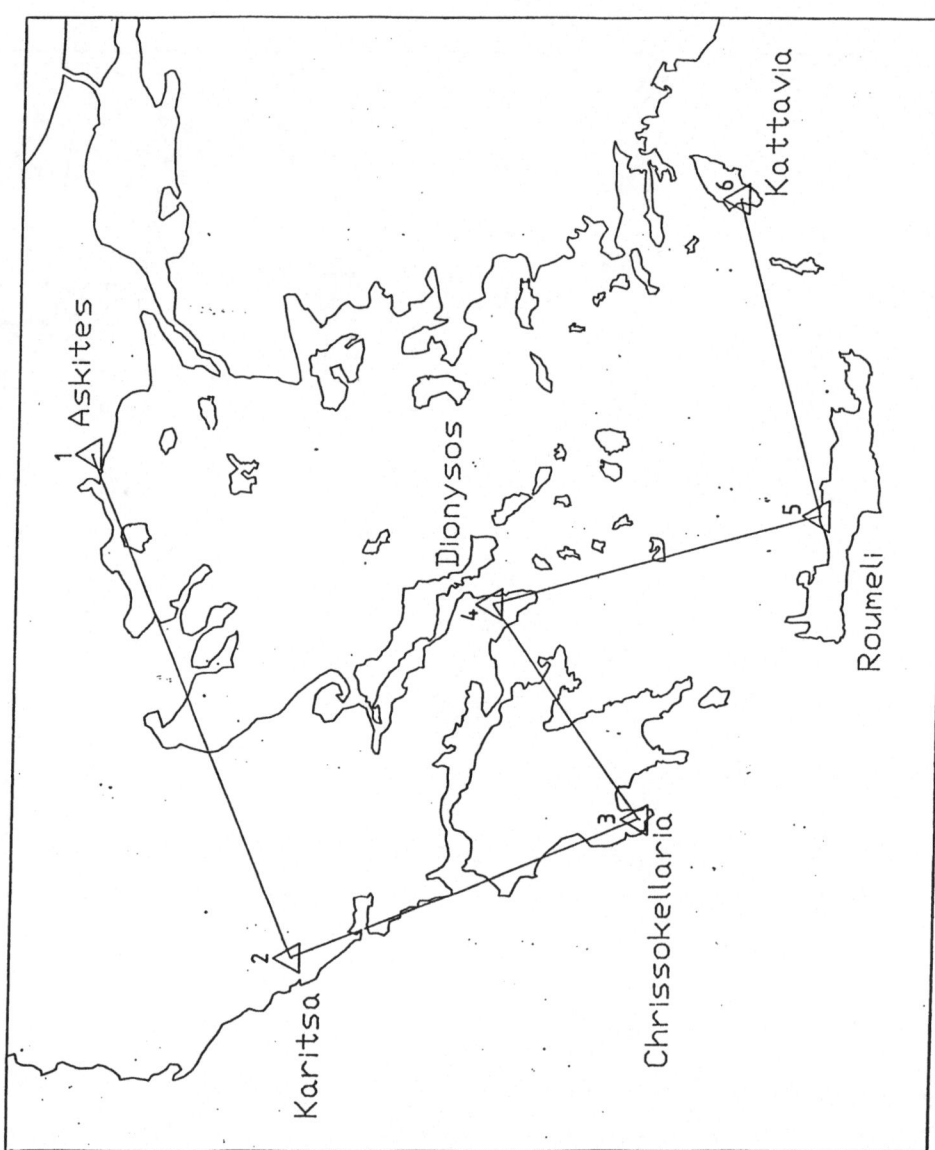

Fig 2. LSCN observing schedule

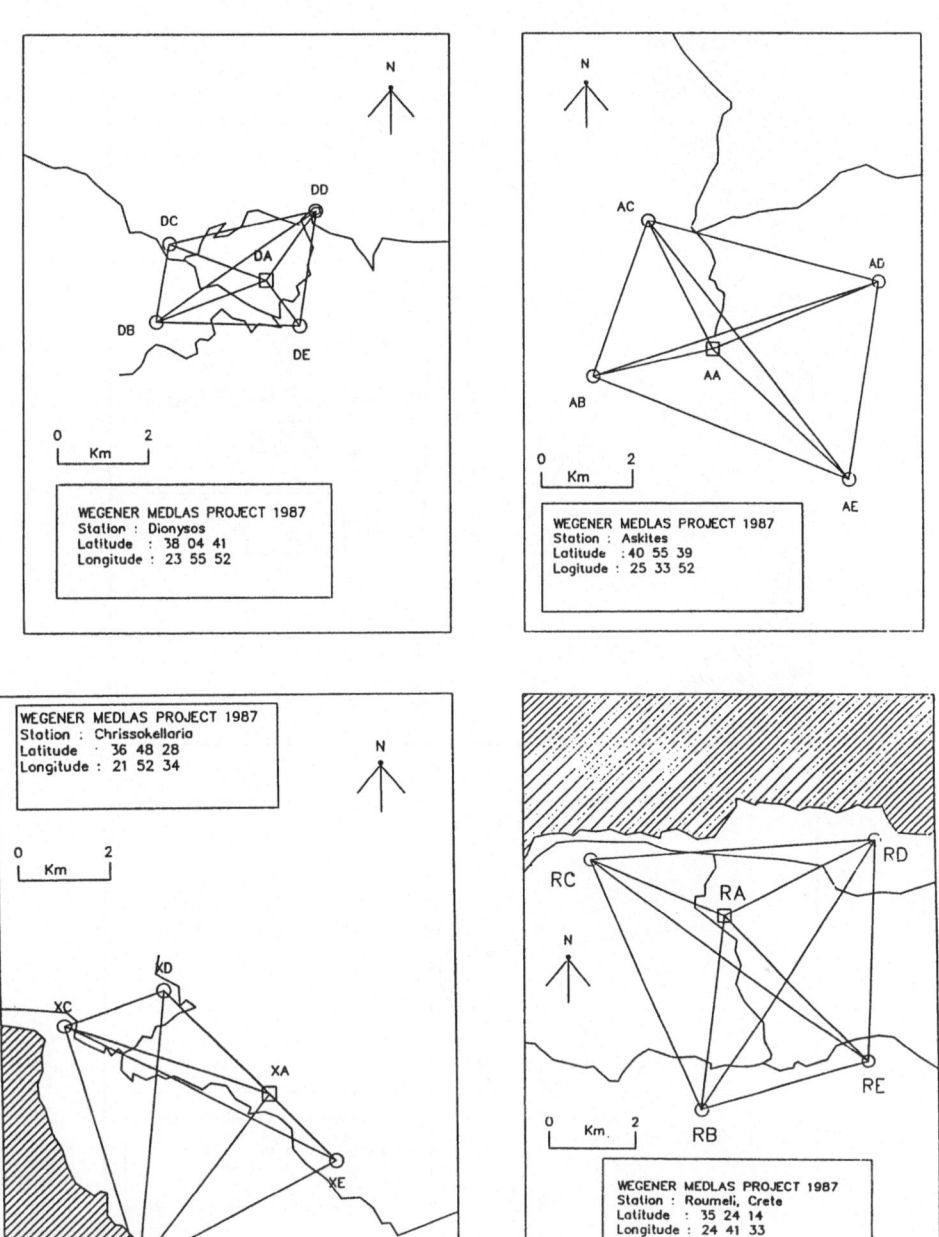

Fig. 3   LSCN's at MEDLAS sites in Greece

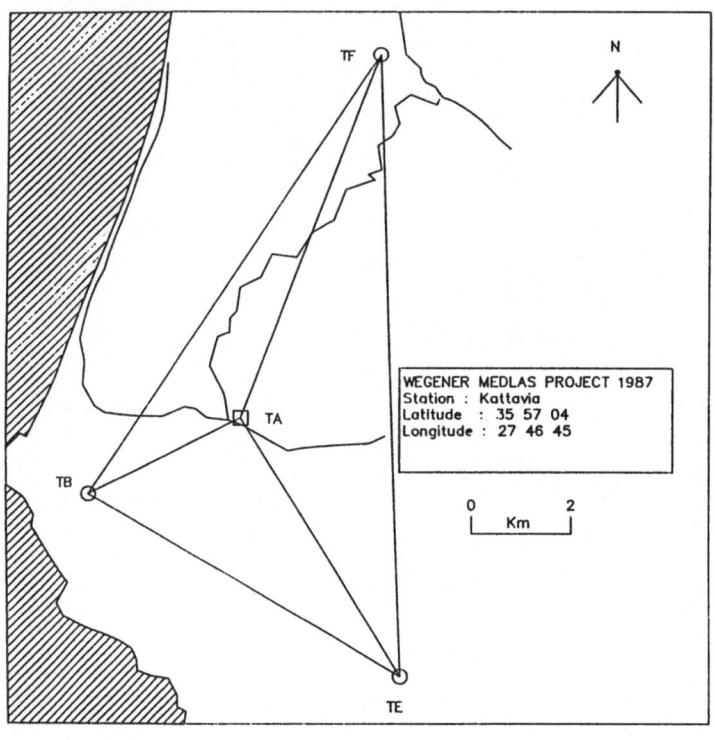

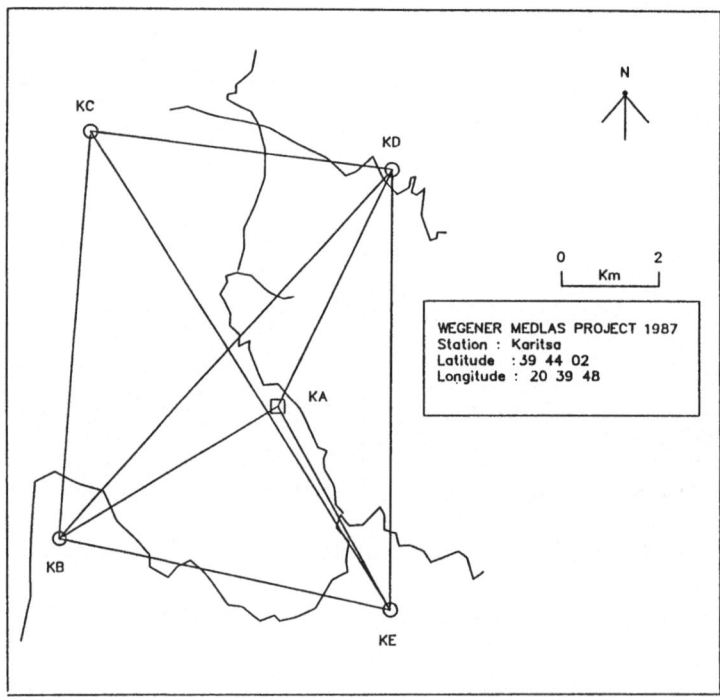

Fig.3 (continued) LSCN's at MEDLAS sites in Greece

ALERT SUMMARY ATHENS JUNE 1st 1987

GDOP satellite choice left to computer

TIME	SATELLITES AVAILABLE	CHOSEN	Pdop
12.30	6		
13.00	6 9		
13.30	6 8 9		
14.00	6 8 9 11	6 8 9 11	8
14.30	6 8 9 11	6 8 9 11	9
15.00	6 8 9 11 12	8 9 11 12	3
15.30	6 8 9 11 12 13	8 9 12 13	3
16.00	6 9 11 12 13	6 9 12 13	5
16.30	6 9 11 12 13	6 9 12 13	4
17.00	3 9 11 12 13	3 11 12 13	3
17.30	3 9 11 12 13	3 9 11 13	3
18.00	3 9 12 13	3 9 12 13	69
18.30	3 9 12 13	3 9 12 13	8
19.00	3 12 13		
19.30	3 12 13		

Fig 4. GPS satellite availability

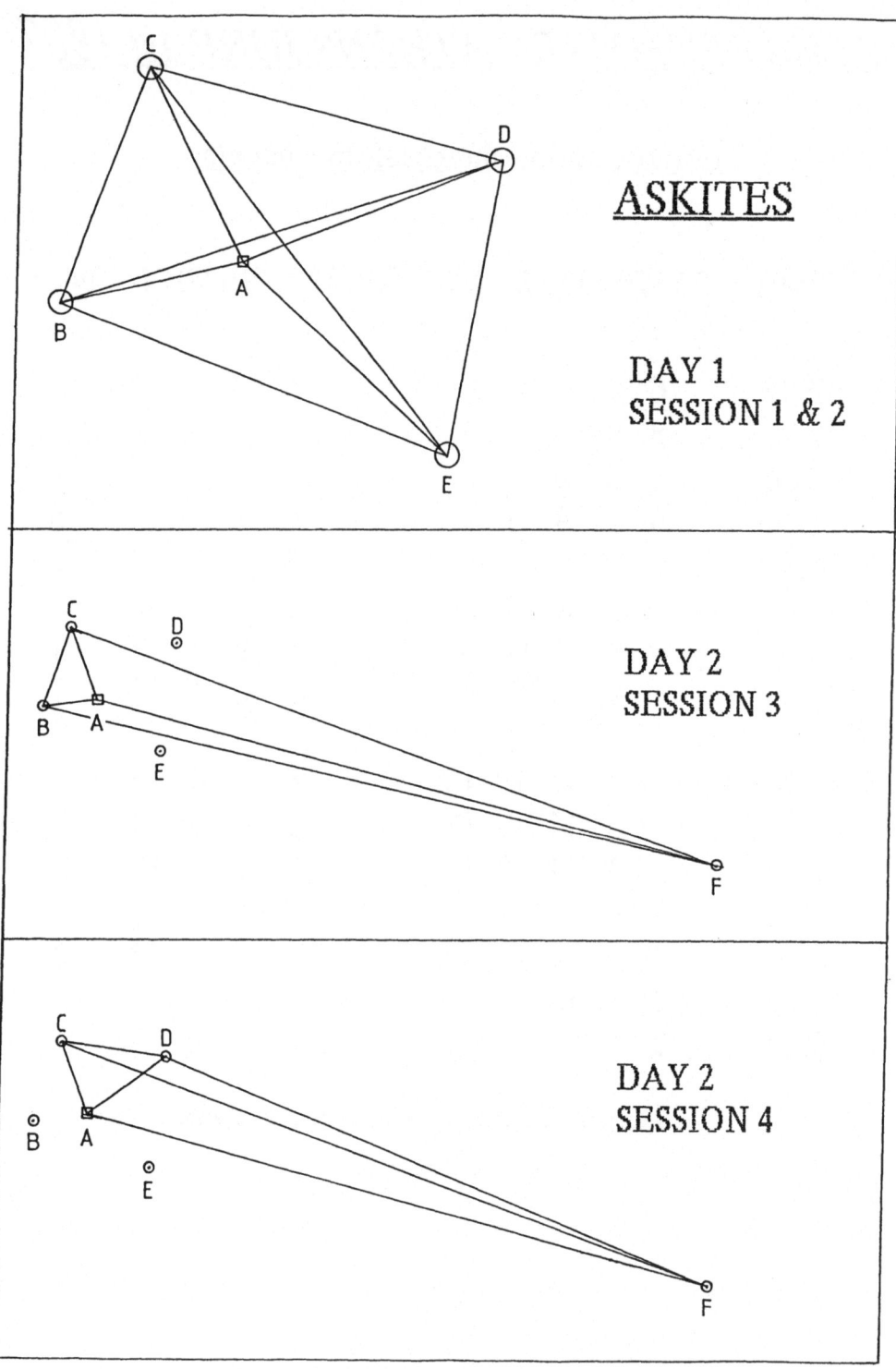

ASKITES

DAY 1
SESSION 1 & 2

DAY 2
SESSION 3

DAY 2
SESSION 4

Fig 5.   Observing sequence at Askites

198

So far it appears that from the pseudo-range and double-difference TRIMVEC solutions (mostly with fixed integer biases) that the LSCN baseline networks are determined with a precision of the order of 5mm to 15mm. In those cases where ties had to be made to remote higher order control sites (typically at distances above 20 kilometres) the precision degraded to the 10cm level. As a more detailed example of the internal precision of the networks, Table 2 shows the a typical subset of the network adjustment corrections (residuals), and Table 3 gives some of the baseline standard errors (after network adjustment), for the Askites LSCN (see Fig. 5). For the same network an external reliability analysis shows that the maximum effect of a marginally detectable error is only 10mm for the stations in the vicinity of the pad (with probabilities of type 1 and type 2 errors set at 0.05 and 0.10 respectively). For the distant point (30km from the pad) the corresponding figure is 87mm.

obs	dX	dY	dZ	slope-distance
A-B	-0.0052	-0.0023	-0.0050	2763.456
A-C	-0.0065	-0.0023	-0.0090	3128.322
A-D	0.0056	0.0042	-0.0024	4210.934
A-E	-0.0056	0.0004	0.0041	4522.685
A-F	0.1057	0.0049	-0.0400	29762.908
B-C	0.0036	0.0050	0.0039	3789.048
B-F	0.1918	0.0192	-0.0321	32090.631

Table 2   Corrections to baseline obs (m)

From	to	X	Y	Z	Distance
A	- B	0.0017	0.0011	0.0014	0.0014
A	- C	0.0017	0.0011	0.0014	0.0013
A	- D	0.0022	0.0018	0.0019	0.0020
A	- E	0.0026	0.0122	0.0023	0.0020
A	- F	0.0151	0.0122	0.0116	0.0123
B	- C	0.0024	0.0015	0.0020	0.0022
B	- F	0.0152	0.0122	0.0116	0.0125

Table 3.   S.E. of derived baselines (m)

Further investigations are taking place both at Delft and Newcastle regarding the detailed strategy for network adjustment, especially regarding the exact functional and stochastic models to adopt. At present the covariance matrices output by TRIMVEC are being used for the stochastic model and the functional model assumes all vectors to be in an identical coordinate system.

4. CONCLUSIONS

If one takes care of (the classical) problem areas such as

- transportation,
- communication,
- power supply (charged batteries!), and
- scrupulous antenna offset measurements

then GPS proves not only to be an easy tool to control the geological stability of SLR sites the (sub)centimetre level, but also is an efficient tool to provide ties between zero-order space geodetic networks and the classical goedetic first order control.

ACKNOWLEDGEMENTS

The authors thank the Hellenic National Committee on Geodesy and Geophysics and the National Technical University of Athens for their overall support and the Hellenic Military Geographical Service for the excellent logistical support.

Also GPS Survey Services Ltd (United Kingdom), especially B. Hogarth, who kindly made available to the University of Newcastle upon Tyne two Trimble 4000S receivers for the duration of the campaign.

J. W. Smit (DUT), M. W. Rayson and P. N. Rands (UNuT) participated in the measurement campaign. W. H. van Ooijen (DUT), M. W. Rayson, P. N. Rands, S Sutisna, A. Sharif, M. Khalid and N. Ahmad (UNuT) helped to process and analyze a substantial stack of diskettes.

PRESENT STATE OF THE DÖNAV CAMPAIGN

by

Günter Seeber

Abstract

Within this report a short review on the performance of the DÖNAV field campaign, the scientific aims of the project and the present status of the data evaluation are given.

1. Project DÖNAV

A more comprehensive report was given during the last General Assembly of the IUGG in Vancouver (Seeber et al. 1987). In the sequel the main items of that report are cited; the present state of the work is reflected.

In August 1986 a group of geodesists from Austria, Denmark and the Federal Republic of Germany discussed the possibility to observe - as early as possible - a multi-station GPS network with two-frequency receivers on stations of the corresponding national networks. The main impetus to this approach was

- the availability of a rather high number of two-frequency P-code TI 4100 GPS receivers within the groups
- the availability of 7 operational prototype satellites at that time
- the intention to dispose rather early on a suitable and comprehensive set of GPS data for manifold investigations.

The cooperating agencies and institutes were

Universität Bonn
Alfred-Wegener-Institut, Bremerhaven
Amt für Militärisches Geowesen, Euskirchen
Institut für Angewandte Geodäsie, Frankfurt
Institut für Weltraumforschung, Graz
Universität Hannover
Geodätisches Institut, Kopenhagen
Technische Universität München
Universität der Bundeswehr München, Neubiberg.

It was decided to perform the project DÖNAV already in November 1986. DÖNAV stands for "Deutsch-Österreichische Navstarkampagne" because the anticipated network covers the complete area of both Austria and the Federal Republic of Germany. Approximately 50 stations were selected with station-to-station distances of about 100 to 150 km. The distribution of stations is indicated in Figure 1.

Considering the number of available instruments it was decided to observe in a combination of fixed and mobile receivers. Six receivers were installed at reference stations, being partly Laser- and VLBI-stations or partly observing stations within the Doppler campaigns EDOC and DÖDOC, namely

Graz (Austria, part time)	(Laser, EDOC, DÖDOC)
Zimmerwald (Switzerland)	(Laser, EDOC, DÖDOC)
Wettzell (FRG)	(Laser, VLBI, EDOC, DÖDOC)

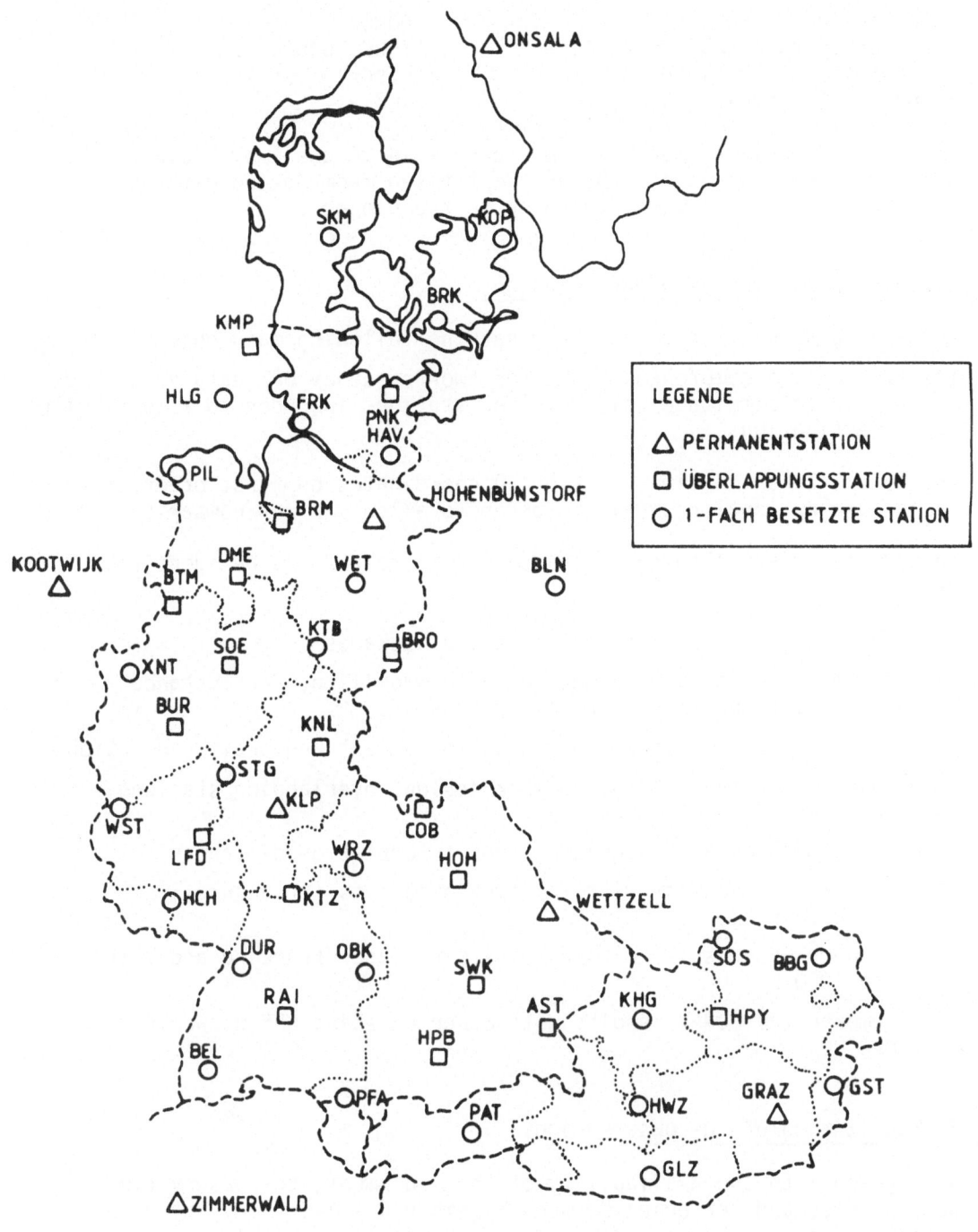

Fig. 1: Distribution of DÖNAV stations

Kloppenheim (FRG)	(DÖDOC)
Hohenbünstorf (FRG),part time	(EDOC, DÖDOC)
Kootwijk (Netherlands)	(Laser)
Onsala (Sweden)	(VLBI).

Six mobile receivers were planned to rove in groups from South to the North with two overlapping points kept fixed from day to day. In total, 12 TI 4100 receivers were available to the project.

2. Main objectives of DÖNAV campaign

DÖNAV is a scientific project and has the following main objectives

(a) to create a comprehensive set of two-frequency GPS data on a large number of simultaneously occupied stations in order to allow various investigations

(b) to create an "optimum" set of 3D coordinates on first order stations of the national networks in order to allow various comparisons.

In addition the following individual aims are and can be identified as

- to test and compare different software packages
- to perform special investigations into modelling, for instance ionosphere, troposphere, orbits, clocks ...
- to investigate techniques on orbit improvement and orbit generation
- to investigate the role of network design, overlapping stations and redundancies
- to test the accuracy dependency from interstation distances
- to compare GPS results with existing data sets (triangulation, DÖDOC, ED79 etc.)
- to compare GPS results with results from Laser tracking and VLBI (up to 1000 km)
- to compare the DÖNAV results with other existing GPS networks like KONMAC.

3. Realization of the observations

With respect to the availability of the equipment, the observations were planned and performed between November 4 - 14, 1986. The observation window was between 1:00 and 4:30 hours in the early morning.

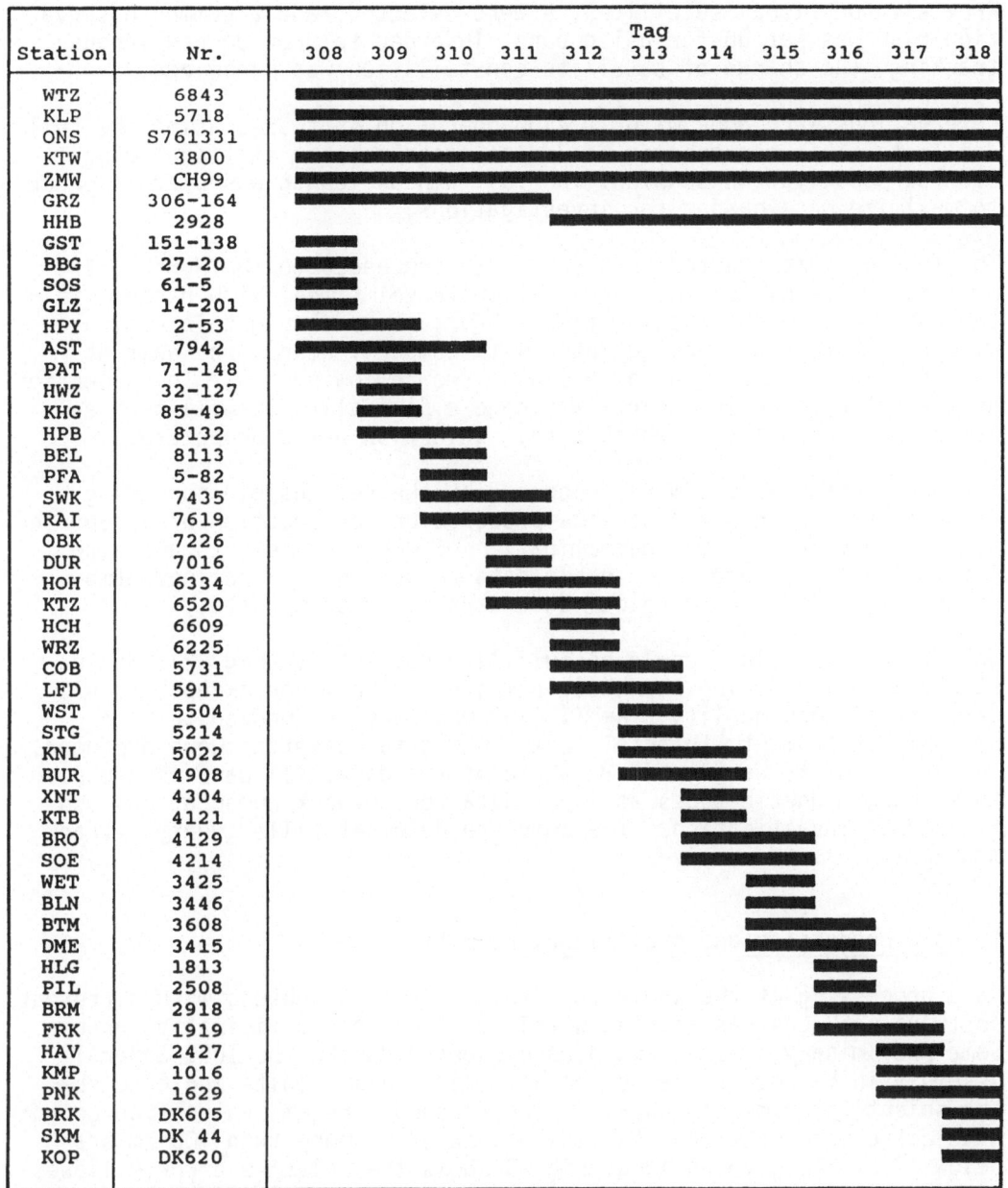

Station	Nr.	308	309	310	311	312	Tag 313	314	315	316	317	318
WTZ	6843											
KLP	5718											
ONS	S761331											
KTW	3800											
ZMW	CH99											
GRZ	306-164											
HHB	2928											
GST	151-138											
BBG	27-20											
SOS	61-5											
GLZ	14-201											
HPY	2-53											
AST	7942											
PAT	71-148											
HWZ	32-127											
KHG	85-49											
HPB	8132											
BEL	8113											
PFA	5-82											
SWK	7435											
RAI	7619											
OBK	7226											
DUR	7016											
HOH	6334											
KTZ	6520											
HCH	6609											
WRZ	6225											
COB	5731											
LFD	5911											
WST	5504											
STG	5214											
KNL	5022											
BUR	4908											
XNT	4304											
KTB	4121											
BRO	4129											
SOE	4214											
WET	3425											
BLN	3446											
BTM	3608											
DME	3415											
HLG	1813											
PIL	2508											
BRM	2918											
FRK	1919											
HAV	2427											
KMP	1016											
PNK	1629											
BRK	DK605											
SKM	DK 44											
KOP	DK620											

Fig. 2: DÖNAV observation plan

From six functional satellites, 5 were selected, and a common observation plan was set up for all groups. In order to provide continuous tracking, the change of satellite constellation was minimized.

Observations were done with TI 4100 operating software, using the full data rate. This corresponds to 7 data cassettes per station during a 3.5 hour observation session. The full window length was used in order to maximize data output for investigations.

The mobile teams started in Austria and proceeded to the North. It was tried to minimize the interstation travel times. With respect to the unfavourable observation time - short after midnight in beginning winter - most crews were equipped with campers. Nearly one half of the "mobile" stations was observed on two following days. The selection of the interconnection points was done empirically. It will be seen later from the analysis whether this selection was appropriate.

The local observations were supported by the responsible surveying authorities with respect to site preparation, demarcation of eccentric marks (where necessary), determination of centric elements and also logistic help. In order to enable data reduction with refined atmospheric models, meteorological data were recorded on site.

The observation plan could be fulfilled completely (Fig. 2). No instrument failed during the whole campaign. With minor exceptions all data are of good quality. The TI 4100 GPS receiver proved to be a reliable instrument. In total, some 1000 data cassettes were produced, corresponding to nearly 600 Megabyte of raw data. All dara were copied to 9-track magnetic tapes as input data for network computations and individual investigations. The complete data set fills twenty 9-track magtapes.

4. Data processing and preliminary results

Data processing of the whole data set or of some subsets with different software packages has started within most of the participating groups. Some preliminary results could be presented during the IUGG General Assembly in Vancouver (Seeber et al. 1987). All results indicate that the scientific aims of DÖNAV will be reached. The RMS values for coordinate differences between stations separated by more than 100 km are below 10 cm. To give an idea, Fig. 3 shows the relative error ellipses for the distances between the reference stations. Table 1 reflects the corresponding RMS values for the distances, coming from the complete adjustment with the Hannover TIPOSIT software, using mean meteorological data and no orbit adjustment techniques. The final adjustment with more elaborated models should provide even better results.

Table 1: RMS values for distances between reference stations from
complete adjustment

Line		number of simultaneous sessions	distance [km]	RMS distance [m]
ONS	ZMW	11	1207	± 0.031
ONS	WTZ	11	920	± 0.030
ONS	HHB	7	492	± 0.035
ONS	KTW	11	700	± 0.028
KTW	ZMW	11	601	± 0.030
KTW	KLP	11	298	± 0.028
KTW	WTZ	11	602	± 0.022
KTW	HHB	7	331	± 0.024
ZMW	KLP	11	383	± 0.033
ZMW	WTZ	11	476	± 0.024
ZMW	GRZ	4	611	± 0.028
GRZ	WTZ	4	302	± 0.032
GRZ	KTW	4	899	± 0.032
GRZ	ONS	4	1.173	± 0.041
GRZ	HHB	-	755	± 0.041
KPL	WTZ	11	322	± 0.022
KPL	HHB	7	337	± 0.035
WTZ	HHB	7	466	± 0.031

6. Further plans

It is planned to finish the individual computations by the end of June
1988 and to meet at Graz in order to compare the results and to agree
on a final solution. A final comprehensive joint publication on the
results is scheduled for the second half of 1988.

6. Acknowledgements

The realization of DÖNAV campaign was only possible through the support,
help and enthusiasm of various individuals and agencies. The participating groups, within a cooperative mind, brought in their own equipment,
staff and necessary funds. A large number of people participated in the
field parties and carried out   observations, partly under very rough
conditions. Most valuable support was given by the "Landesvermessungs-
ämter" in the Federal Republic of Germany and the "Bundesamt für Eich-
und Vermessungswesen" in Austria during the preparation and the observation period of the project. Observations in Zimmerwald, Kootwijk and

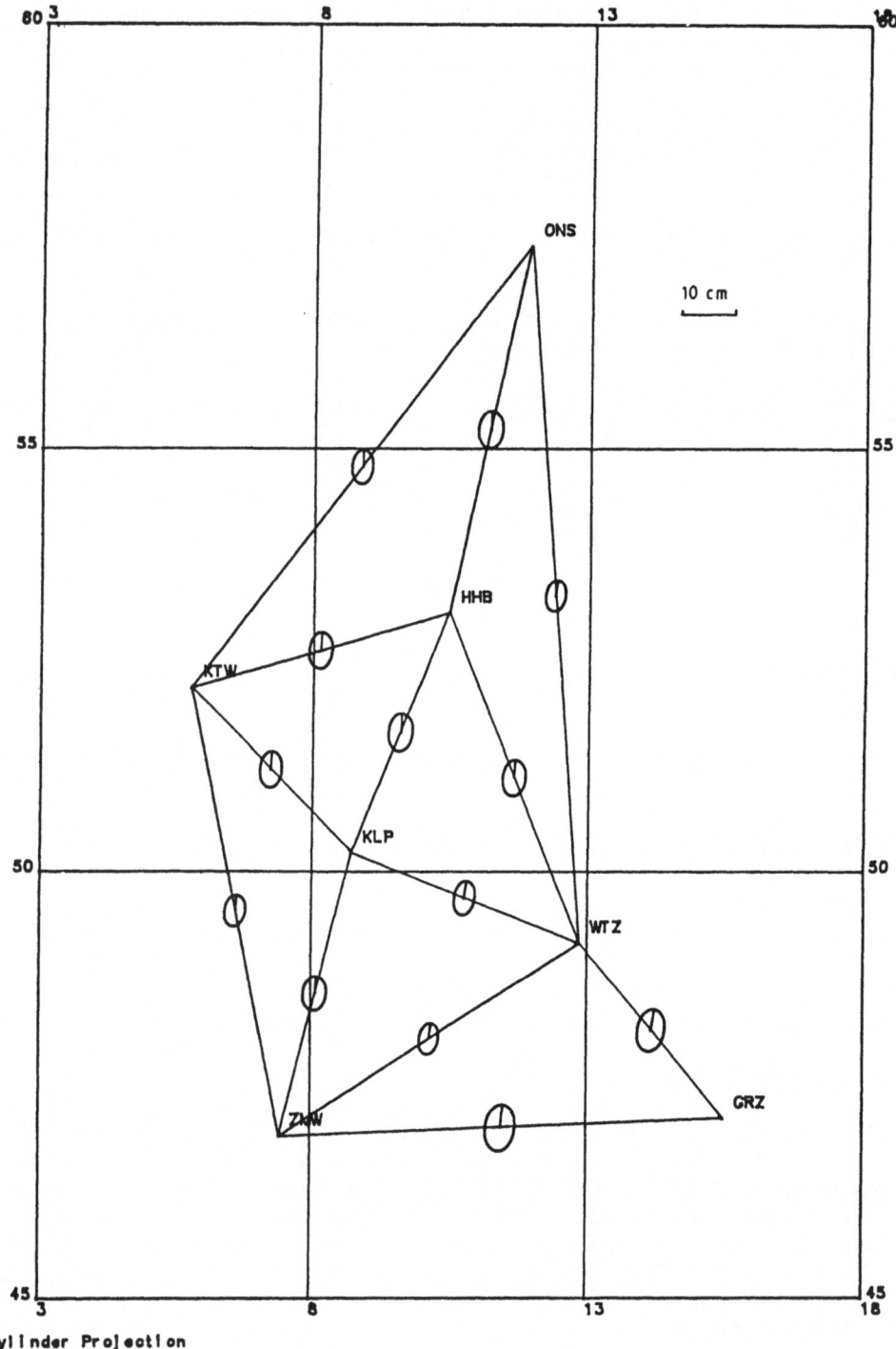

Zylinder Projection

Fig. 3: Relative error ellipses between permanent stations

Onsala were made possible through the support of the responsible bodies. Two TI 4100 receivers as well as the precise ephemeris were provided by the Defense Mapping Agency, Washington. This support is thankfully acknowledged.

7. References

Seeber, G., G. Wübbena, A. Schuchardt, H. Seeger, A. Müller, H. Schenke, F. Lohmar, G. Soltau, W. Schlüter, K. Rinner, P. Pesec, H. Pelzer, F. Madsen, R. Sigl, K. Deichl, G. Hein, H. Landau, A. Schödlbauer, H. Glasmacher, K. Krack (1987) Status Report on DÖNAV, IUGG Assembly, Vancouver.

Modeling of Phase Center Variation and its Influence on GPS-Positioning

by

Alain Geiger

Abstract: In this paper it is shown that a simple algorithm for modeling phase center variations could be helpfull to reduce the noise in GPS measurements. The more pragmatic the approach the simpler the correction algorithm. The pragmatic approach consists in a description of the effect of phase center variation and not of the variation itself. An error function derived for a certain antenna can directly be implemented as a distance correction. The effect on positions of different kinds of antennas, the problem of dual-frequency antennas and the reduction of error by relative positioning will be discussed.

0. Introduction

Since positioning for geodetic puposes has reached a very high degree of accuracy, many effects which could be treated in the sense of noise affecting the measurement become more and more significant for the systematic error reduction. One of these effects is the influence of phase center variation onto geodetic GPS-measurements. In this paper an approach will be made to describe the phase center variation and its influence. One of the basic ideas is to model the effect of variation on the range or phase measurement rather than the variation of the center itself. The variation of the center is difficult to determine. Since the procedure consists in principal in a differentiating algorithm (e.g. Tranquilla, 1986) the determination will be very unstable. For geodesy the true location of the 'phase center', which ever its definition may be, is of no importance. What geodesists are interested in is the direct range error due to phase center variation. This effect can directly be measured with respect to a 'mean center'. The procedure for measuring these phase shifts is described in the literature relevant to this subject. For GPS-antennas work has been carried out by e.g. Sims (1985), Tallqvist (1986), Schneck (1987). By taking a continuous satellite distribution into account we are able to describe a general behaviour of different antenna types in an explicit form. For other application of this method the reader is refered to Geiger (1987). A similar approach has been made in Beutler et al. (1986,1987) for modeling atmospheric effects.

1. Basic Equations and Satellite Density Distribution

We consider single-point solutions as well as differential observations. The method of observation is of minor interest for our purposes, since pseudorange-measurements anyway and phase observations after small modifications can be reduced to the simple well known equation:

$$r = \quad | x - s | - | y - s | + x_4 \tag{1}$$

Putting for single point and differential positioning

	single point	and differential positioning
s	Position of satellite	Position of satellite
x	Unknown position	Unknown differential position with respect to y
y	Reference station $y = s$	Reference station $y \neq s$
x_4	Fourth unknown for synchronisation etc. between satellite and receiver x	Fourth unknown for synchronisation etc. between receiver y and receiver x
r	Pseudorange- observation to satellite s	Difference of pseudorange- observations between $x-s$ und $y-s$

To avoid singularity problems we assume known positions of the satellite and reference station. Thus the variation of satellite and reference position will be held fixed:

$$\delta s = \delta y = 0$$

Linearization results in

$$v = A x - (r - r_0) \tag{2}$$

and in the solution of the unknown x

$$x = N^{-1} A^T P (r - r_0) \tag{3}$$

where

$$N = A^T P A \tag{4}$$

We define our coordinate system with the z-axis in zenithal direction. The y-axis is directed towards the center of the azimuthal sector of observations. θ = zenithal angle and λ = azimuth (from x-axis). In this local system the satellite-to-user line-of-sight vector e reads:

$$|e|^2 = e_1{}^2 + e_2{}^2 + e_3{}^2 = 1 \tag{5}$$

$$\begin{array}{rcl} e_1 & = -x & = -\sin\theta\cdot\cos\lambda \\ e_2 & = -y & = -\sin\theta\cdot\sin\lambda \\ e_3 & = -z & = -\cos\theta \end{array} \tag{6}$$

For **A** we find the known system matrix: (e = satellite-to-user line-of-sight vector)

$$\partial r\,/\,\partial x = (e_1, e_2, e_3, e_4) = (e, 1) \tag{7}$$

Assuming uncorrelated and uniformly weighted observations allows us to write: $P = E$
Introducing errors $r - r_0 = \delta r$ depending on the direction of observed range (satellite position) formula 3) can be rewritten:

$$\delta x\,(\theta,\lambda) = N^{-1}\,A^T\,\delta r\,(\theta,\lambda) \tag{8}$$

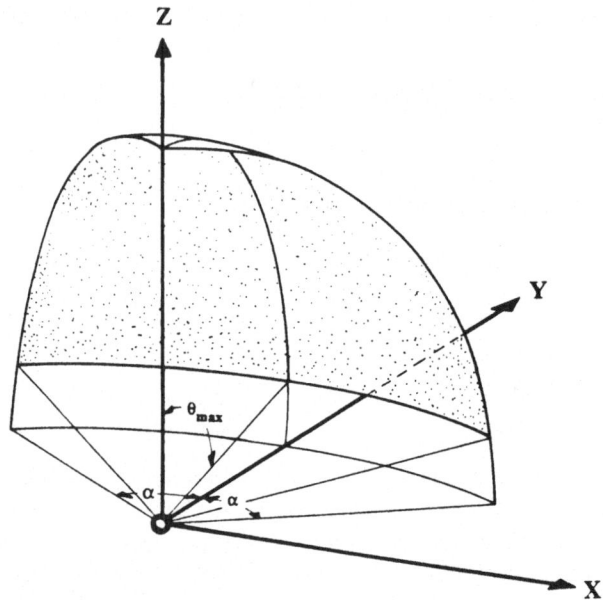

Fig.1:   Coordinate system and integration boundaries

All observations are made in an interval of zenithal angle of $[\,0,\theta_{max}]$ and in an azimuthal sector of $[90-\alpha\,,\,90+\alpha]$, α is in the range $[0,\pi]$. Normally a minimal elevation angle of 15^c to $20\,°$ will be respected. This corresponds to $70\,° < \theta_{max} < 75\,°$. Assuming a distribution $\beta(\theta)$ (independent on azimuth) of the total amount n of observations we can write

$$n = \int_{90-\alpha}^{90+\alpha}\int_0^{\theta_{max}} \beta\,\sin\theta\,d\theta\,d\lambda = 2\,\alpha\int_0^{\theta_{max}} \beta\,\sin\theta\,d\theta = 2\,\alpha\,I\beta_{10} \tag{9}$$

with the abbrevation for integrating a function β:

$$I\beta_{ij} = \int \beta \sin^i\phi \cos^j\phi \, d\phi \qquad (10)$$

In an analogous way we consider continuous matrix operations. Sums will be changed into integration and discrete observations will be replaced by the density function of observations. Assuming equal weighting the N-Matrix is formed by $\mathbf{A^T A}$ and it looks like:

$$\mathbf{N} = \begin{bmatrix} [e_1^2] & [e_1 e_2] & [e_1 e_3] & [e_1 e_4] \\ [e_2 e_1] & [e_2^2] & [e_2 e_3] & [e_2 e_4] \\ [e_3 e_1] & [e_3 e_2] & [e_3^2] & [e_3 e_4] \\ [e_4 e_1] & [e_4 e_2] & [e_4 e_3] & [e_4^2] \end{bmatrix} \qquad (11)$$

wherein [..] means sum over all observations. The sum will be replaced by the integraloperator

$$\int_0^{\theta_{max}} \int_{90-\alpha}^{90+\alpha} \dots \beta \sin\theta \, d\theta \, d\lambda \qquad (12)$$

If we assume a homogeneous λ distribution the integration over λ can be carried out immediately resulting in (N symmetric):

$$\mathbf{N} = 2 \begin{bmatrix} \frac{1}{2}(\alpha - \frac{1}{2}\sin 2\alpha)\, I\beta_{30} & 0 & 0 & 0 \\ 0 & \frac{1}{2}(\alpha + \frac{1}{2}\sin 2\alpha)\, I\beta_{30} & \sin\alpha\, I\beta_{21} & -\sin\alpha\, I\beta_{20} \\ 0 & + & \alpha\, I\beta_{12} & -\alpha\, I\beta_{11} \\ 0 & + & + & \alpha\, I\beta_{10} \end{bmatrix} \qquad (13)$$

Exactly the same procedure is applied to the term giving the error in position (equation 8)

$$\delta x\,(\theta,\lambda) = \mathbf{N}^{-1}\mathbf{A^T}\,\delta r\,(\theta,\lambda) \qquad (14)$$

The product $\mathbf{A^T}\,\delta r$ becomes:

$$
\mathbf{A}^T \, \delta\mathbf{r} \;=\;
\begin{bmatrix}
[\,e_1\,\delta\mathbf{r}\,] \\[4pt]
[\,e_2\,\delta\mathbf{r}\,] \\[4pt]
[\,e_3\,\delta\mathbf{r}\,] \\[4pt]
[\,1\,\delta\mathbf{r}\,]
\end{bmatrix}
\;=\; -\iint \delta r\,(\theta,\lambda)\cdot\beta(\theta)
\begin{bmatrix}
\sin^2\theta \cos\lambda \\[4pt]
\sin^2\theta \sin\lambda \\[4pt]
\sin\theta \cos\theta \\[4pt]
\sin\theta
\end{bmatrix}
d\theta\,d\lambda
\tag{15}
$$

If we observe satellites over the whole azimuthal range, $\alpha=\pi$, we see immediately that the x and y position-components of the error estimates $\delta\mathbf{x}$ are not affected.

Until now we deduced only formulas for single point positioning. The results can still be applied to differential positioning. Looking at the observation equation (1) $r = |\,x - s\,| - |\,y - s\,| + x_4$ it is seen that the error δr corresponds to the difference of errors δr_x made at station x and δr_y at station y resp. $\delta r = \delta r_x - \delta r_y$. The error estimates for the solution becomes:

$$
\delta\mathbf{x}\,(\theta_x,\lambda_x;\theta_y,\lambda_y) \;=\; \mathbf{N}^{-1}\,\mathbf{A}^T\,[\,\delta r_x\,(\theta_x,\lambda_x) - \delta r_y\,(\theta_y,\lambda_y)\,]
\tag{16}
$$

We introduced different sets of angles because of the parallaxe. For small (several 10 km) baselines this effect can be neglected. In this case the error becomes:

$$
\delta\mathbf{x}\,(\theta,\lambda) \;=\; \mathbf{N}^{-1}\,\mathbf{A}^T\,[\,\delta r_x\,(\theta,\lambda) - \delta r_y\,(\theta,\lambda)\,] \;=\; \mathbf{N}^{-1}\,\mathbf{A}^T\,\delta r\,(\theta,\lambda)
\tag{17}
$$

This equation is completely analogous to the formula for single point solutions. We only have to introduce δr as the difference between the ranging errors at different sites.

2. Modeling Phase Center Variation and its Effect on Positioning

For modeling the phase center variation we consider the following separation of the offset: we assume a constant, antenna fixed offset $\mathbf{p}$ which could be called 'mean phase center'. To this mean value a term $\delta r\,(\theta,\lambda)$ dependent on constellation is added. It is obvious that the effect in in range measurements will be:

$$
\Delta r\,(\theta,\lambda) \;=\; \mathbf{p}^T\mathbf{e} + \delta r\,(\theta,\lambda)
\tag{18}
$$

where $\mathbf{p}^T\mathbf{e}$ is the projection of the fixed phase center vector onto the line of sight to the satellite. With equation (8) we find immediately:

$$
\Delta\mathbf{x} \;=\; \mathbf{N}^{-1}\,\mathbf{A}^T\,\Delta r = \mathbf{p} + \mathbf{N}^{-1}\,\mathbf{A}^T\,\delta r\,(\theta,\lambda)
\tag{19}
$$

Since $\mathbf{p}$ is a constant term we ommit it in the examples to be discussed. We only treat the remaining, more interesting term

$$
\delta\mathbf{x} \;=\; \mathbf{N}^{-1}\,\mathbf{A}^T\,\delta r\,(\theta,\lambda)
\tag{20}
$$

which corresponds to the eqation (8). The problem is now reduced to the problem of determining a function $\delta r\,(\theta,\lambda)$ for a certain antenna. This may be solved either theoretically or empirically by measurement of the phase difference with respect to a 'mean phase center'. In this paper simple model functions have been used to show some principal results. Real functions may depart several mm from such curves. A stochastic approach has been used to account for departures from rotational symmetries of antenna characteristics.

To fit measured phase errors we could use a sum of

$$\delta r_{nm}(\theta,\lambda) = \cos^n\theta \cdot (cc_{nm} \cdot \cos^m\lambda + cs_{nm} \cdot \sin^m\lambda)$$
$$+ \sin^n\theta \cdot (sc_{nm} \cdot \cos^m\lambda + ss_{nm} \cdot \sin^m\lambda) \tag{21}$$

Spherical harmonics would also do the job. If the error function can be separated as follows $\delta r(\theta,\lambda) = f(\theta) \cdot g(\lambda)$ several equations are considerably simplified.

2.1 Directional Antenna

Directional means, that the antenna has a constant offset A from the rotational axis. In other words the errorfunction reads:

$$\delta r(\theta,\lambda) = A \tag{22}$$

We introduce this Function into equation (15). After a multiplication by the inverse of N (14) the following error in position results:

$$
\begin{aligned}
\delta x &= 0 \\
\delta y &= 0 \\
\delta z &= 0 \\
\delta(ct) &= A
\end{aligned}
\tag{23}
$$

This means that a directional antenna shows no systematic phase center induced errors in position. The whole error will show up in the synchronisation term. This fact is well known for VLBI-observations. It is to say that the error is not dependent on the satellite constellation.

2.2 Turnstile

The turnstile type antenna (crossed dipole) produces in first approximation a cosinus-shaped error function like:

$$\delta r(\theta,\lambda) = A \cdot \cos\theta \tag{24}$$

If we again introduce this assumption into equation (15) the following error in position will yield

$$
\begin{aligned}
\delta x &= 0 \\
\delta y &= 0 \\
\delta z &= -A \\
\delta(ct) &= 0
\end{aligned}
\tag{25}
$$

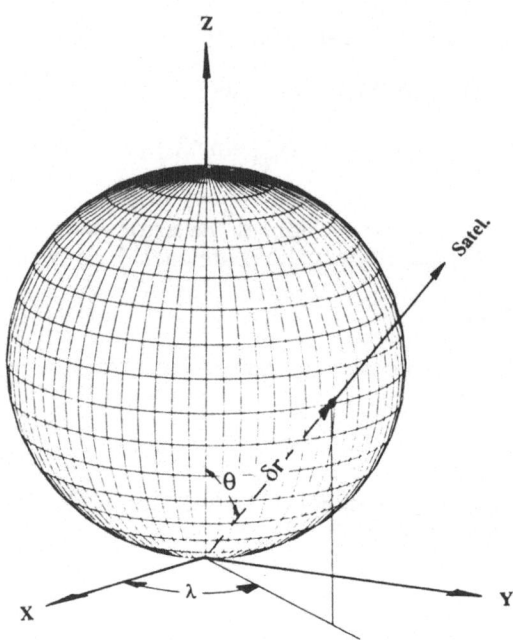

Fig. 2: Polar representation of the error function

This means that a cos-type phase-center errorfunction will produce an error in height only. The amplitude A could reach e.g. +3cm. This would lead to an error of 3 cm in absolute height while the position remains completely unaffected. Again it is important to say that the error doesn't depend on the satellite constellation. Hence it is possible to calibrate an antenna of this type once and to use the determined offset for all subsequent measurements. This advantage of certain antennas has already been pointed out by Counselman and Shapiro (1979).

2.3 Conical Spiral

Measurements by Sims (1985) of spiral antennas show a strong (not cos-shaped) dependence of the error function on the direction of the incoming raypath. This results in an positioning error which depends on the observed satellite constellation. The phase error measurement can be fitted by the expansion:
sum over n and m ($n \leq 4$; $m \leq 1$)

$$\delta r_{nm}(\theta,\lambda) = (cc_{nm} \cdot \cos^m \lambda + cs_{nm} \cdot \sin^m \lambda) \cdot \cos^n \theta . \tag{26}$$

The coefficients of influence on the position show quite a strong influence of the $\cos^2$-term.

2.4 Quadrifilar Helix

The same effects could occur with helical structures. Some test measurements for determinig constant antenna offsets (Cocard et al.,1988) revealed the necessity to assume such constellation dependent offsets. Following this work the constant part **p** ranges from 0.0 to 5.0 mm with a remaining stochastical part of 0.5 to 1.5 mm.

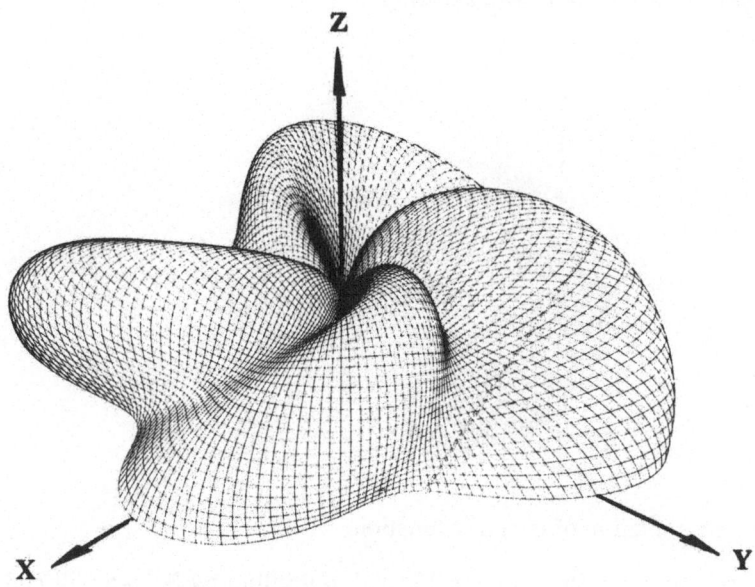

Fig. 3: Polar representation of a (hypothetical) error function of a 4-wire helix

Since the qudrifilar helix shows a higher symmetry than asymmetric antennas like 1-wire helix etc., the dependency on satellite constellation is slightly diminished compared to the conical spiral (one arm). Asymmetric antennas show an dependence on constellation (cut-off angle), even if the observations are homogeneously distributed over the whole hemisphere.

2.5 Micro Strip

As last example of antenna we consider a micro-strip type structure. There exist different testmeasurements with GPS-antennas of this kind, e.g. Tallqvist (1985), Schneck (1987).

$$\delta r(\theta, \lambda) = A \cdot \sin \theta \cdot \cos(\lambda - a_0) \qquad (27)$$

a_0 gives an azimuthal orientation of the antenna. A is a constant amplitude. This function may not be typical for all micro-strip antennas.

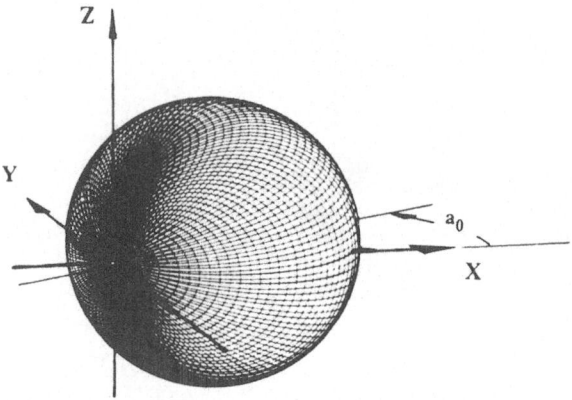

Fig. 4: Polar representation of the error function

Introducing this function into eqation 15) and multipling by N^{-1} we will end up with an error for the position of:

$$\delta x = -A \begin{bmatrix} \cos a_0 \\ \sin a_0 \\ 0 \\ 0 \end{bmatrix} \qquad (28)$$

The only assumption which has been made to reach this result is to take into consideration a homogeneous satellite distribution. However no restriction on the constellation has been introduced. Looking at the equation above, we see, that this error-type is completely independent on constellation during the observation. The antenna could be calibrated once. Since the offset appears in x and y (without affecting height and time), the antenna should always be oriented in the same direction a_0. If more or less identical antennas with the same orientation are used for baseline determination the error will be reduced significantly.

2.6 2-Frequencies antenna

Many GPS equipement use dual-frequency antennas. The phase centers of the two antennas have not necessarily to coincide. Thus two different errors have to be considered (the subscript denotes the frequencies L1 and L2 resp.) (see equation 18)

$$\Delta r_1(\theta, \lambda) = p_1^T e + \delta r_1(\theta, \lambda) \qquad (29)$$
$$\Delta r_2(\theta, \lambda) = p_2^T e + \delta r_2(\theta, \lambda)$$

and with equation (20)

$$\delta x_1 = N^{-1} A^T \delta r_1 = p_1 + N^{-1} A^T \delta r_1 (\theta,\lambda) \qquad (30)$$

$$\delta x_2 = N^{-1} A^T \delta r_2 = p_2 + N^{-1} A^T \delta r_2 (\theta,\lambda)$$

The movement of the two phase centers are not completely uncorrelated. In a first approximation a linear connection can be assumed:

$$\delta r_2 (\theta,\lambda) = \rho + (1+\mu) \, \delta r_1 (\theta,\lambda) \qquad (31)$$

Where ρ and μ are fixed numbers. It follows:

$$\delta x_2 = p_2 + N^{-1} A^T \delta r_2 (\theta,\lambda)$$
$$= p_2 + \rho N^{-1} A^T 1 + N^{-1} A^T \delta r_1 + \mu N^{-1} A^T \delta r_1 \qquad (32)$$

Finally we find the difference in position by:

$$\delta x_2 - \delta x_1 = p_2 - p_1 + \rho \, e_{ct} + \mu N^{-1} A^T \delta r_1 \qquad (33)$$

e_{ct} denotes the forth coordinate of the system (time axis) (0,0,0,1). This equation tells us that apart from the vector $p_2 - p_1$ which normally is accounted for, a synchronisation term and a $\mu N^{-1} A^T \delta r_1$ term dependent on the constellation should be taken into consideration. The value of μ could reach the order of 0.5 to 1.0. This error propagates into the solution where an effect of several mm would appear. The time term seems to be smaller than 5 mm. We see that even if the two 'mean phase centers' $(p_2 - p_1 = 0$) coincide there remains a small effect due to the different behaviour of the two frequencies.

2.7 Relative Positioning

By observing baselines with nearly identical antennas a significant amelioration of the situation can be achieved. Considering the difference in position due to the phase center variation (see also equation 17)

$$\delta x = \delta x_2 - \delta x_1 = p_2 - p_1 + N^{-1} A^T (\delta r_2 - \delta r_1) \qquad (34)$$

The subscripts 1 and 2 indicate now the stations 1 and 2 resp.

and considering two antennas with similar stochastical behaviour $\varepsilon(\lambda)$ of their departures from symmetrical characteristics

$$E[\, \varepsilon_i(\lambda) \, \varepsilon_i(\Lambda) \,] = \sigma^2_{\varepsilon i} \, \delta (\lambda - \Lambda) \qquad (35)$$
$$E[\, \varepsilon_i(\lambda) \, \varepsilon_j(\Lambda) \,] = \sigma_{\varepsilon ij} \, \delta (\lambda - \Lambda) \qquad i \neq j$$

$\sigma_{\varepsilon ij}$ is given by the correlation r of the two random functions ε_1 and ε_2:

$$\sigma_{\varepsilon ij} = r \, \sigma_{\varepsilon i} \, \sigma_{\varepsilon j} \qquad (36)$$

we can show (Geiger, 1988) that the resulting covariance matrix (the part which accounts for the phase center error) of the baseline will be

$$C = 2 (1-r) C_{11} \qquad (37)$$

where C_{11} is the covariance for single station solution e.g. station 1. r is the correlation between the two antennas. If the antennas, their surroundings and their orientations are completely identical the covariance would reduce to zero. However, a decrease of the covariance by a factor of 2 or more seems to be realistic.

Conclusions

If we arrange different antenna types according to their phase center behaviour the following table results:

Table 1:   Comparison of different antenna types

Type	Position		Height		Time	
	gen	ideal	gen	ideal	gen	ideal
Directional	-	-	-	-	fix	fix
Turnstile	-	-	fix	fix	-	-
Micro-Strip	fix	fix	-	-	-	-
4-wire Helix	dep	-	dep	dep	dep	dep
2-wire Helix	dep	-	dep	dep	dep	dep
1-wire Helix	dep	dep	dep	dep	dep	dep
Conical Spiral (one arm)	dep	dep	dep	dep	dep	dep

gen	Observations are not dispersed over the whole hemisphere.
ideal	Observations over the whole hemisphere.(360° in azimuth)
-	no Offset
fix	Offset fixed. Not dependent on constellation.
dep	Offset dependent on constellation. If dep appears in the column **ideal** it means that the offset still depends on the cut-off angle θ_{max}.

This very simple method corresponds very well to testmeasurements. Therefore the formulas can be used to construct simple but realistic error models without any complicate formalism. Applications can be extended to other problems as inclined tropospheric models, standard correction formulas (Saastamoinen), multipath, asymmetric satellite distribution (e.g. Geiger,1987). The formulas can easily be modified to satisfy generalized density functions. However, the formalism would get much more cumbersome. It is evident that in this case an a priori calculus or simulation directly by an GPS software package would be preferable and also superior to the complicated analytical solution.

Appendix I - References

Beutler, G., I. Bauersima, W. Gurtner, M. Rothacher, T. Schildknecht, A.Geiger (1987): Atmospheric Refraction and Important Biases in GPS Carrier Phase Observations. *Paper presented at IUGG Meeting, Vancouver, August 1987.*

Beutler, G., W.Gurtner (1986): Influence of Tropospheric Refraction on the Evaluation of GPS Phase Observation. *University of Berne. Preprint.*

Cocard, M., A.Geiger, B.Wirth, B.Bürki (1988): Versuche zur Kalibrierung von WM101 Antennen. *Institut für Geodäsie und Photogrammetrie, Zürich, Bericht 142, p 33.*

Counselman, C.C., I.I. Shapiro (1979): Miniature Interferometer Terminals for Earth Surveying. *Bull. Geod. 53, pp139-163.*

Geiger, A. (1987): Einfluss richtungsabhängiger Fehler bei Satellitenmessungen. *Institut für Geodäsie und Photogrammetrie, Zürich, Bericht Nr.130, p 40.*

Geiger, A. (1987): Simplified error estimation of Satellite positioning. *Paper pres. at GPS Technology Workshop, Jet Propulsion Laboratory, Pasadena.*

Geiger, A. (1988): Einfluss und Bestimmung der Variabilität des Phasenzentrums von GPS-Antennen. *Institut für Geodäsie und Photogrammetrie, Zürich, Bericht , in press.*

Gurtner, W., G. Beutler, S. Botton, M.Rothacher, A. Geiger, H.-G. Kahle, D. Schneider, A. Wiget (1987): The Use of GPS in Mountainous Areas. *Paper presented at Int. IUGG Meeting, Vancouver, August, 1987.*

Lee, K.F., (1984): Principles of Antenna Theory. *John Wiley & Sons .ISBN 0471901679.*

Rothacher,M., G.Beutler, W.Gurtner, A.Geiger, H.-G. Kahle, D.Schneider (1986): The 1985 Swiss GPS-Campaign. In: *Proceedings, Symp. on Satellite Positioning, Austin, Texas.*

Schneck (1987): Personal Communication, *Institut für Navigation, University of Stuttgart.*

Sims, M.L. (1985): Phase Center Variation in the Geodetic TI4100 GPS Receiver System's Conical Spiral Antenna. *Proceedings of the First International Symposium on Precise Positioning with the Global Positioning System. Rockville, Maryland. U.S. Department of Commerce, May 1985.*

Tallqvist, St. (1986): The GPS Microstrip Antenna Properties; Reduction of Multipath Contamination and other Interference by an RF Absorbent Ground Plane. *Paper presented at SATRAPE Meeting, Paris, 1986.*

Tranquilla, J.M. (1986): Multipath and Imaging Problems in GPS Receiver Antennas. In: *Proceedings, 4th International Geodetic Symp. on Satellite Positioning, Austin, Texas.*

Appendix II - Notation

In this paper bold symbols denote vectors or matrices. The following symbols are used:

r	pseudorange
r	vector of observations (range observations)
r$_0$	approximation of **r**
v	
N	Normalmatrix
A	systemmatrix
P	weighting matrix
e	satellite-to-user line-of-sight vector (unit lenght)
E	unity matrix
x	unknown position of observer
y	reference station vector
s	position of satellite
θ	zenithal angle of satellite position
λ	azimuthal angle of satellite position
β	function of satellite distribution depending on θ and λ
ϕ	integrable variable
α	azimuthal limit for density distribution ($90-\alpha < \lambda < 90+\alpha$)
θ_{max}	maximal zenithal angle of satellite position
n	number of observations
e_1	x-component of **e**
e_2	y-component of **e**
e_3	z-component of **e**
δ**r**	vector of range errors
δr	error function of ranging depending on θ and λ
N	refractivity
T	temperature
δ**x**	error vector of position **x**
δx	x- component of error vector of position **x**
δy	y- component of error vector of position **x**
δz	z- component of error vector of position **x**
cδt	t- component of error vector of position **x**
c	speed of light
t	time

Fifth session: Kinematic Applications

Chairman: Prof. Hartl, Stuttgart

HIKING AND BIKING WITH GPS: THE CANADIAN PERSPECTIVE

by

Petr Vaníček

Abstract

The paper attempts to review and put into perspective the findings of a recent by-invitation-only workshop on GPS held in Ottawa. The workshop brought together both GPS-information suppliers and consumers coming from all walks of life. Some unique exchanges of views took place which are worth bringing to the attention of the geodetic community.

Introduction

Let us begin, at the cost of sounding flippant, with an assertion that the technical problems of positioning with GPS have now been successfully solved. Most GPS researchers will probably agree with this statement if it is qualified by an accuracy limit of a few parts per million (ppm) in both the point positioning as well as relative positioning modes. What is understood by one ppm in the point positioning mode is that the average standard deviation in the 3 coordinates is equal to one millionth of the length of the geocentric position vector, i.e.,

$$\bar{\sigma} = r \times 10^{-6} \times \simeq 6.4 \text{ m}.$$

The concept of ppm in the context of relative positioning is well established and does not require any elaboration here.

It seems to us that GPS is now well established as a positioning tool in many application areas. This is the case in spite of only a partial satellite configuration having been available until now. Many routine position surveys are being conducted daily by means of GPS; ship fleets are being instrumented with GPS receivers, and readied for a switch over to GPS navigation. At last count, 60 makes of GPS receiver are now commercially available. Specifications are being written for 'GPS surveying' in rural and urban environments. More and more surveyors and navigators are becoming familiar with this system. Software of the fourth generation is now being written by various research groups. What else is there to do?

GPS positioning research now branches out in two directions:
(i) technical, of which most of the papers presented and discussed at this Workshop are examples, and
(ii) applicatory, which we have chosen to address in this contribution.
The aim of technical research is to improve further the accuracy of GPS-determined positions and position differences and the speed of their determination. The aim of the applicatory research is to facilitate and speed up the transition to a 'position oriented society.'

Five years ago, we predicted a sharp decline in the size and price of GPS receivers [Vaníček et al., 1983]. We also predicted the slow demise of terrestrial networks arguing that an easy, cheap, and fast positioning capability would make the monumenting of points of known positions economically untenable. Lastly, we predicted a spectacular rise of the numbers and kinds of consumers of position information. It is this last phenomenon that we would identify as an onset of 'position oriented society,' a society where easy availability of position information will revolutionize as many aspects of life as the ready availability of accurate time has done in the past.

Our first prediction is rapidly materializing: the new Rockwell-Collins DARPA GPS receiver now being designed will have a size of 9 x 7 x 1 cm with a strip antenna, noticeably smaller than a packet of 20 cigarettes, not too far away from the ultimate 'wrist locator.' Two and a half years ago, already an American manufacturer was offering GPS receivers for US$4000 (in orders of 1000 or more). We shall talk about our second prediction in the next paragraph, while the GPS User/Supplier Workshop, treated in the rest of this paper, has been a direct consequence of increasing interest in GPS among many diverse potential customer groups.

Active Control System

The Canadian Active Control System (ACS) has been under development for about two and a half years [Steeves et al., 1986]. It will consist of a number (50?) of Active Control Points (ACPs) spaced strategically throughout Canada. Eventually, it will replace conventional position control points, i.e., all geodetic networks.

An ACP will consist of a multichannel, two-frequency carrier tracking receiver controlled by a microcomputer which will continuously record carrier phase observations to all visible satellites. The collected data will be used for real-time orbit improvement, computing broadcast corrections to measured code-ranges for navigation purposes and for the verification of the integrity of the whole GPS satellite constellation. At some later stage, the individual (or regional master) ACPs may also broadcast electronic charts such as bathymetry, aeronautical, roads, and specialized maps of various scales.

The role the ACS is going to play in positioning will be two fold. Besides increasing significantly the accuracy and reach of differential positioning through the orbit improvement and superior accuracy of ACP's positions, it will have a very positive impact on land, sea, and air navigation. The fact that wherever one goes on, or in, the vicinity of Canadian territory, there will be a stationary reference GPS receiver nearby, will make for easier broadcasting and reception of differential corrections [Kalafus et al., 1983] rendering navigation more reliable, accurate, and impervious to both natural as well as artificial vagaries of the system. Both differential modes above call for powerful and reliable data links between the ACPs and the user. Different options exist and are now being vigorously investigated.

For the ACS to become a truly universal positioning tool, it must also be able to cater to two disparate user groups: those interested in knowing where they are themselves, and those interested in knowing where someone else is. We call them 'where-am-I' and 'where-are-you' users. Clearly a dispatcher of an ambulance fleet would not be interested in where he is but would wish to know where his ambulances are. On the other hand, a pilot of a ship navigating a narrow strait would be much helped by knowing where 'his' ship was and would not care much about the location of other ships unless they are likely to interfere with his own sailing.

It is the 'where-are-you' user group that brings a new dimension into positioning. Their applications consist of positioning of more or less sophisticated passive beacons whose design and communication links may vary widely from application to application. It is these applications that may force us to consider using other satellite systems, in addition to GPS, in the ACS scheme.

The enhancement of the ACS power by adding the electronic chart transmission capability would be immeasurable. One does not have to have much imagination to see that the whole present concept of mapping would have to undergo a fundamental change. In this context, the Geographical Information Systems will become a very important tool.

The Ottawa User/Supplier Workshop

In our recent book [Wells et al., 1987], we have identified many potential GPS users ranging from geophysicists interested in studying the earth's deformations, through legal surveyors, fishermen, and commercial aircraft pilots, to the more exotic hikers, bikers, and truck dispatchers. Over 100 such potential users from different walks of life and position suppliers had been invited to participate in a two-day workshop held in Ottawa on 21 and 22 January 1988 under the auspices of the

Geodetic Survey Division of the Survey and Mapping Sector of the federal Department of Energy Mines and Resources.

Four issues were addressed in four working sessions: human interface, impact, control, and applications.

(i) **Human interface**. In this session, the following topics were discussed: government-industry partnership in the coordination of effort; marketing/education/communication; general interface. The most interesting findings included different kinds of desired result displays by different users (CRT, voice, within/without permitted corridors), reliability versus accuracy for air navigation.

(ii) **Impact**. Social, legal, technological, and economic impacts were discusses in this session. Improved safety resulting from different applications of GPS positioning was thought would have a significant social impact. The issue of legal culpability in case of failure and the assurance of continuing availability dominated the discussions of legal considerations. The technological and economic considerations evolved around Canadian realities and would not be of general interest.

(iii) **Control**. Two topics were discussed in this session: political control, and administrative control. Interestingly, the absolute U.S. military control over GPS was not considered to be overly harmful to civilian applications. More concern was shown by the participants about the question of administrative control over ACS, i.e., government control versus privatization, levying of user fees.

(iv) **Applications**. Land, marine, and air applications were the topics for discussion. The equipment requirements by the different user groups were compiled in tables that will appear in the Proceedings of the Workshop. No major surprises surfaced during this session except, perhaps, for the identification of some fairly unorthodox users, such as robotic vehicles in agriculture and manufacturing, forest spraying, and surveillance used in the enforcement of national sovereignty at sea.

Conclusions

Instead of trying to summarize the lessons we have learned in Canada in the recent past, let us try to answer the following question, that many GPS positioning and navigation enthusiasts may have uppermost on their minds at this stage of GPS development: Should we, the position suppliers, i.e., navigators and surveyors, become involved in the non-technical issues the applications of GPS bring about? It appears to us that we should. For one simple reason: If we want to continue being professionally viable, we have no other choice. We have to demonstrate our ability to change, or rather our ability to lead the change in the society, not only in the technical sense but also in the application sense. Otherwise, technological progress, and the commercial evolution that goes with it, will soon overtake us and we will end in the 'professional dustbin' joining other extinct species.

References

Kalafus, R.M., J. Vilcans and N. Knable (1983). Differential operation of NAVSTAR GPS. *Navigation*, 30, pp. 187-204.

Steeves, R.R., D. Delikaraoglou and N. Beck (1986). Development of a Canadian Active Control System using GPS. *Proceedings of the Fourth International Geodetic Symposium on Satellite Positioning*, University of Texas at Austin, Austin, Tx, U.S.A., April, pp. 1189-1203.

Vaníček, P., D.E. Wells, A. Chrzanowski, A.C. Hamilton, R.B. Langley, J.D. McLaughlin and B.G. Nickerson (1983). The future of geodetic networks. *Proceedings of the International Association of Geodesy Symposia*, Symposium d, Vol. 2, Hamburg, FRG, August, pp. 372-379.

Wells, D., N. Beck, D. Delikaraoglou, A. Kleusberg, E.J. Krakiwsky, G. Lachapelle, R.B. Langley, M. Nawkiboglu, K.-P. Schwarz, J.M. Tranquilla and P. Vaníček (1987). *Guide to GPS Positioning*. 2nd revised ed., Canadian GPS Associates, Fredericton, N.B., Canada.

STATUS OF DUAL FREQUENCY
GPS DEVELOPMENT
AT TRIMBLE NAVIGATION

by

Timo Allison, Brian Westfall, Ralph Eschenbach
and Ron Hyatt

1.   Introduction

Trimble Navigation introduced its first GPS products in 1984 based on technology developed at Trimble Navigation and Hewlett-Packard.   The first products were the 5000A GPS Time/Frequency Monitor and the 4000A GPS Locator.   These products provide 100 nanosecond timing accuracy, one part in ten to the eleventh frequency resolution, 25 meter positioning, and 10 cm/sec velocity. In late 1985, Trimble introduced the 4000S GPS Surveyor to provide 1 cm $\pm$ 2ppm baseline measurement for the land surveyor. The 4000 series utilizes the L1 C/A code, carrier signals, and the broadcast ephemeris.   This product has recently been expanded to a 10-channel L1 C/A code and carrier tracking receiver with a one megabyte (14 hours) solid state data logger.   The post-processing software (TRIMVEC)^{TM} provides pseudo-range, triple difference, and double difference baseline solutions utilizing an IBM PC or compatible computer (Goad & Remondi, 1984).

There are nearly 700 of the 4000 and 5000 series units being utilized today in precision time and frequency systems, static and dynamic positioning, and land survey applications.

2.   Dual Frequency Design Goals

In September 1987, development work started on dual frequency capability for the 4000S product line. The design approach taken was to minimize changes to the current product to allow the upgrade of units already in use throughout the world. A very high quality L2 carrier phase measurement was the major goal in order to remove the effects of the ionosphere on very long lines and allow easier integer wavelength ambiguity resolution on intermediate length lines.

The dual frequency capability will be available in the 4000SX and the 4000SL during 1988. The products will have 5 channels of L1 C/A code and carrier tracking and 5 channels of L2 carrier tracking. Upgrades to existing units will require the addition of the L2 downconverter, 5 channels of L2 tracking, internal data logger, and a new dual frequency micro-strip antenna.

These dual frequency units can be operated as 10 channel L1 receivers also through keyboard control.

The system architecture is outlined in Figure 1. Dual down-converters are used for the L1 and L2 signals. Following down-conversion, all processing is fully digital. Acquisition and tracking of the L2 signals is made possible by using information from the L1 C/A-code-correlating channels.

The use of very narrow bandwidth carrier tracking loops, together with a sophisticated carrier phase measurement algorithm based on curve-fitting, has resulted in high accuracy L2 carrier phase measurements. L2 data integrity is enhanced by procedures which compare L1 and L2 carrier phase, instantaneous Doppler, and integrated Doppler observables.

Survey data is logged to a memory board which utilizes battery backed-up RAM. This has a capacity for 14 hours of continuous L1 and L2 data logging when tracking 5 satellites with a measurement epoch of 15 seconds.

2.   Current Status of Development

A special build of twenty-five dual frequency receivers were delivered to three universities in Japan in March. These units will be used in earthquake research activities throughout Japan. Automated monitoring stations will transfer the GPS data to a central computer for processing and analysis. These systems should be operational in the fall of 1988.

Regular production of the dual frequency receivers at Trimble Navigation will begin in the summer of 1988 with upgrades to existing receivers starting later in the year.

Post-processing software provides multi-station processing of L1 and L2 data separately as well as two L1, L2 combinations. These two combinations consist of an L1, L2 combination that provides for easier integer wavelength ambiguity resolution and a second combination that provides an ionospheric-free solution.

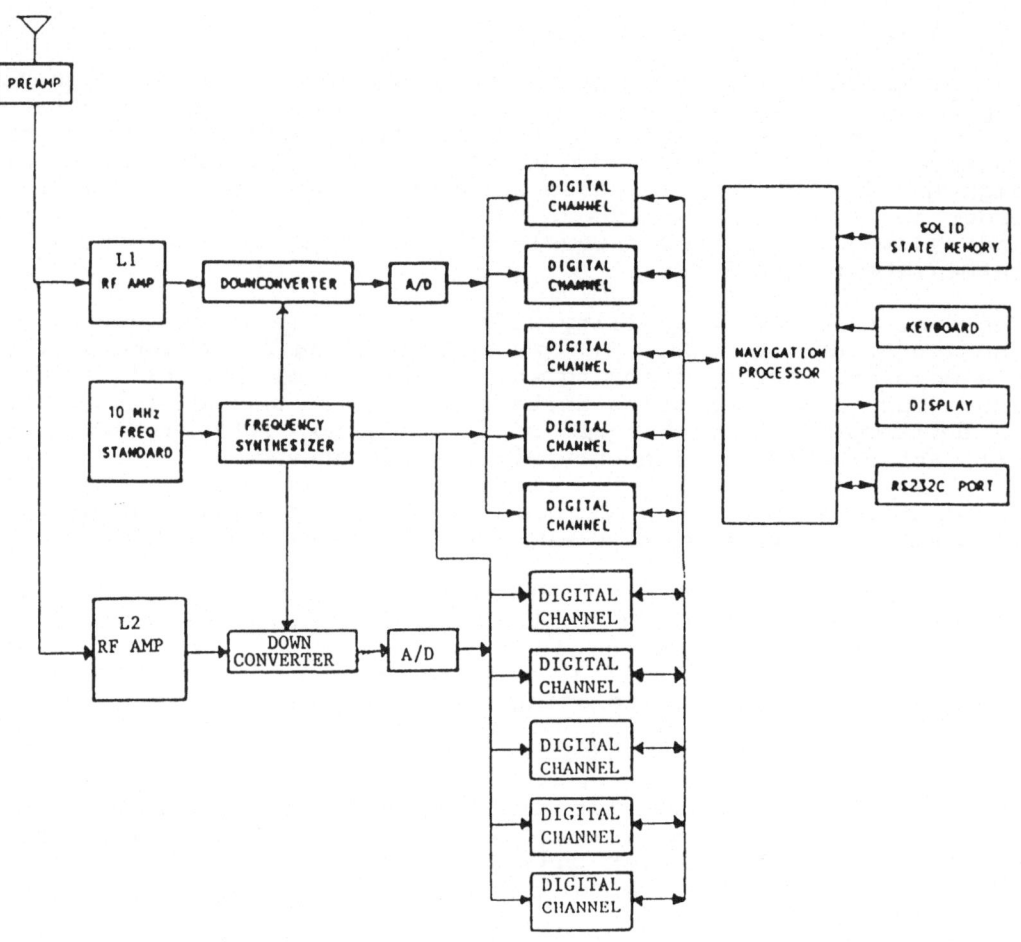

System Architecture for the Dual Frequency
4000SX and 4000SL

Figure 1

3.   Test Results to Date

Tracking of the L2 signal first occurred in November 1987, with
baseline testing beginning in February 1988.   Baselines ranging
from 6 to 146 kilometers have been measured since February.   The
table below compares the results of the L1 and L2 observations.

Date	L1 (meters)	L2 (meters)	L1/L2 (meters) (inon-free)
February 20	6248.4375	6248.4349	6248.4422
February 22	6248.4376	6248.4354	6248.4420
February 25	6248.4380	6248.4356	6248.4404
March 8	146167.30	146167.33	14167.32
April 2	53727.37	53727.39	53727.33
April 2	58706.80	58706.83	58706.77

This data indicates that L2 measurements are providing baselines essentially equivalent to the L1 measurements for the 6 kilometer line. The longer lines show differences below the 1 ppm level. These two to three hour data sets were taken at least one hour after sunset.

On April 2, two portable VLBI stations were occupied with dual frequency receivers. The table below shows the distances determined by VLBI in 1983 and 1985 as well as the GPS observation on April 2, 1988.

August	1983 VLBI	53727.199 meters
March	1985 VLBI	53727.238 meters
October	1985 VLBI	53727.231 meters
April	1988 TRIMBLE GPS	53727.277 meters

The GPS value is the L1, L2 combination that resolved the ambiguities with high confidence. Since the effects of troposphere and orbits are unknown, the repeatability of this result is unknown.

4.   Future Testing and Development Plans

Additional VLBI stations will be occupied during the summer of 1988, when the satellite constellation is visible during the daylight hours in California. Baseline distances of 100 kilometers to 500 kilometers will be measured to verify the removal of ionospheric effects. Extensive testing of 20 to 100 kilometer lines will be done to verify the ability to resolve ambiguities with dual frequency data that otherwise would be impossible.

Multi-station processing will be further developed for the new 32-bit personal computers to provide much faster, simultaneous processing of the L1, L2 data.

5.   Conclusions

High quality L2 carrier phase measurements have been demonstrated as a result of extensive laboratory and field tests. Verification of the removal of ionospheric effects will require additional testing during peak periods of ionospheric activity.

6.   References

[1] Goad, C.C., B.W. Remondi (1984):   Initial Relative positioning Results Using The Global Positioning System, *Bulletin Geodesique*, Vol. 58, pp. 192-210.

TABLE 1

DAY	L1 DISTANCE (meters)	L2 DISTANCE (meters)	L1/L2 DISTANCE (meters)
051	6248.4375	6248.4349	6248.4422
052	6248.4377	6248.4352	6248.4413
053	6248.4376	6248.4354	6248.4420
055	6248.4347	6248.4339	6248.4374
056	6248.4380	6248.4356	6248.4404
064	6248.4353	6248.4327	6248.4397

L1/L2 =   Ionosphere-Free Solution

TABLE 2

DAY 051	DISTANCE (meters)	DELTAX (meters)	DELTAY (meters)	DELTAZ (meters)
L1	6248.4377	-4221.110	-1395.637	-4390.604
L2	6248.4352	-4221.109	-1395.641	-4390.600
L1/L2	6248.4413	-4221.110	-1395.631	-4390.611
L1-L2	6248.4306	-4221.108	-1395.649	-4390.592

L1/L2 = Ionosphere-Free Solution

L1-L2 = 34 cm Wavelength Solution

TABLE 3

VLBI DATA - METERS

AUGUST	1983	53727.199
MARCH	1985	53727.238
OCTOBER	1985	53727.231
OCTOBER	1987	53727.251

TNL GPS DATA-METERS

APRIL	1988	53727.277

BASED ON BROADCAST EPHEMERIS AND STANDARD SURFACE METS.

KINEMATIC SURVEYING

by

V Ashkenazi, C de la Fuente and P J Summerfield

Abstract

The principles of GPS Kinematic Surveying are discussed, starting with the hardware requirements and software considerations. Details are given of kinematic tests carried out at Nottingham University, and preliminary results summarised.

1   INTRODUCTION

In the context of this paper, kinematic surveying is specifically defined as surveying using interferometric carrier phase observations in a kinematic mode. The survey begins with two receivers i and j positioned over two starting points A and B respectively. After an initial period, with both receivers static over these two points, the antenna of receiver j is lifted and moved on to a new point whose coordinates are unknown. It then remains over this new point for no more than about a minute, before it is picked up and moved again to another point. This process continues until all the unknown stations in the survey network have been occupied by receiver j. The data, recorded by receivers i and j, is then post-processed by using double difference interferometric carrier phase measurements, leading to the baseline vectors between the fixed receiver i and all the survey points successively occupied by receiver j. An a priori knowledge of the coordinates of station A allows the determination of the coordinates of all the other unknown stations. A survey network of, say, 20 stations may be observed by GPS kinematic surveying in under an hour, with an accuracy comparable to that of static GPS.

2   PRINCIPLES OF KINEMATIC SURVEYING

It is well established that baselines, accurate to 10 mm $\pm$1 (or 2) ppm can be routinely determined by using static carrier phase observations. However, in order to obtain this level of accuracy a minimum observation time of about thirty minutes is required.

Consider the basic double difference equation

$$\Phi_{AB}^{ij}(\tau) = \frac{f}{c} \, \rho_{AB}^{ij}(t) + N_{AB}^{ij}(\tau o)$$

where

$\Phi_{AB}^{ij}(\tau)$ = double difference phase observable

$\rho_{AB}^{ij}(t)$ = range between stations A & B and satellites i & j

$N_{AB}^{ij}(\tau o)$ = initial double difference integer unknown.

The right-hand-side of the equation is made up of two terms, namely the range differences between the satellites and the receivers, and an initial double difference integer unknown. Using this technique, it is necessary to solve for the initial (at 'lock-on') double difference integer unknowns corresponding to pairs of satellites. These calculated 'integer' values are generally found to be close to an integer value (eg 187.9954 or 188.0023). These first estimates are then constrained to their true integer value (ie 188) in a subsequent solution. It is this 'constrained' solution which yields the most precise baseline vectors. The reason for requiring approximately 30 minutes of data is that it takes about this much time before the 'computed' double-difference integer unknowns settle to their theoretical values.

Once the integers have been recovered it only takes a few observations to obtain the precise baseline vector between any two points.

Similarly, if the integers are kept constrained, only a couple of observations are required to continuously determine the coordinate differences between successive baseline vectors sharing one end of the line.

This is the basic principle of kinematic surveying. Firstly, initial double difference integer unknowns are found. Then one of the two receivers is moved to an unknown station while maintaining phase lock and consequently holding on to the initial integer unknowns. On arrival at the new station, only a couple of observations are required in order to determine the new baseline vector to sufficient accuracy. The antenna is then moved on again to the next unknown point, and so on. Provided there is no loss of lock, which results in the observational data being contaminated by cycle slips, this process can be continued until all the unknown points have been surveyed.

One approach to finding the original baseline vector (and hence the initial integer unknown) is to leave the two antennas in static positions for about 30 minutes or so, at the start of each observation session. Once the initial double difference integer unknowns have beeen determined, kinematic surveying can begin by moving the roving antenna to the various points whose coordinates are to be determined.

A second technique for establishing the double difference integer unknonws at lock-on is simply to begin the kinematic GPS survey from two points with known 3-d coordinates. Consider again the two antennas (i and j) erected on static reference points (A and B) several metres apart. If the position of one of the two stations is known from, say, a pseudo-range solution in WGS 84 coordinates, and the baseline vector between antennas i and j has been previously established to sub-centimetre level, then it is possible to calculate the required integers. The method utilises a knowledge of the satellites' coordinates for that particular epoch, and range differences are calculated in terms of cycles, as in the equation at the beginning of section 2.

A third approach is described by Rémondi (1986). This avoids the need for an initial 30 minute position fix on the reference baseline, by determining the initial double difference integer unknowns rapidly. The two antennas of receivers i and j are initially situated on points A and B respectively. However, in this case, they are left in these positions only 1 or 2 minutes. Receiver i is then moved to point B, and receiver j to point A. Another 1 or 2 minutes of data is again collected. Assuming the two points are close to one another, the whole procedure will probably take less than 5 minutes. When post-processing the data, the apparent movement of the antennas will be twice the initial baseline vector. By averaging the pre-swap and post-swap baseline vectors, any systematic biases inherent in the few observations will be cancelled out, and the baseline vector is obtained to an accuracy sufficient for computing the initial double difference integers biases.

3 HARDWARE CONSIDERATIONS

Of fundamental importance to the concept of kinematic GPS surveying is the ability of the receiver to keep lock onto the satellite emitted carrier frequency phase reading while the antenna is in motion. Generally, if a receiver has a very narrow bandwidth, then more accurate carrier phase measurements can be taken. However, with a narrow bandwidth, if the antenna is moved the receiver cannot maintain lock onto the carrier frequency. With a widening of the bandwidth, the receiver can withstand a higher acceleration, but the resulting data may be more noisy. A solution to this problem would be changing from a narrow to a wide bandwidth, before moving the static antenna. This approach enabled Remondi (1985 and 1986) to perform his first experiments with kinematic GPS by using TI-4100 receivers.

The TI-4100 receiver has four possible user dynamics (UD) settings. These are 0, 1, 2 and 3 which correspond to a frequency of 0.7, 5, 8 and 16 Hz respectively. In turn, each of these UD's can withstand accelerations of 0, 6, 15 and 40 m/sec^2 respectively. Clearly, while the antenna is static, UD 0 should be utilised, but when the antenna is mobile a wider bandwidth setting has to be used. Tests (Remondi, 1986) have shown that sudden jerks in the motion of the antenna cannot be accommodated at UD levels 1 and 2. Consequently, it is appropriate to make use of UD 3 when the antanna is in motion. The process of collapsing the bandwidth when reaching the survey mark takes around fifteen seconds. Hence, it is necessary to increase the total time of observation at each station by this amount.

Some GPS receivers, such as the Trimble 4000 SX, do not need changing the bandwidth of the tracking loops. This reduces the time required at each survey point. Other receivers are not able to maintain lock on the carrier frequency when in motion, and cannot therefore be used for kinematic surveying.

4   INITIAL GPS KINEMATIC TRIALS

The first kinematic GPS surveying trials were conducted in Arkansas, during July 1985 (Remondi, 1985 and 1986). In the Arkansas Kinematic Survey, two TI-4100 receivers were used, with one receiver kept permanently on a fixed station with known coordinates, and the second 'roving' receiver moving to successively occupy the remaining four stations. During movement, a UD setting of 3, corresponding to a bandwidth of 16 Hz, was used to maintain lock on the carrier frequency. When the antenna reached a point for static measurements, the bandwidth was collapsed to a 0 UD setting to reduce the level of noise. It was concluded that the roving receiver had to be frequently brought back to the starting point, for the equivalent of a '0-position-update'.

The second experiment reported in the same paper (Remondi, 1986) was carried out on 12 and 13 August 1985, at the US National Bureau of Standards, again with TI-4100 receivers. An 'antenna swap', in order to determine the baseline vector between the two starting points, was incorporated in this second test. This was considered a great success, in that the initial baseline could be determined simply and accurately, and the double difference integer biases could then be easily calculated.

In the next phase of the trial, the position of one antenna was held fixed, while the other was moved between 6 other points, whose coordinates were precisely known. The discrepancies, between the GPS calculated coordinates and those found from terrestrial observations, ranged from a maximum of approximately 5 cm in Northings, 3 cm in Eastings, and 16 cm in the height component. However, it was noticed that these discrepancies increased almost linearly with time, from the start of the observation session. This was particularly evident, as the roving receiver was frequently returned to its initial base station, and the computed baseline vector compared with its known value. By removing this linear drift from the coordinates of all the other stations, the maximum discrepancies were reduced to 1.9, 1.6 and 4.7 cm respectively in each of the three directions. By improving the modelling techniques used, it was thought that this level of accuracy could be achieved routinely, if not bettered. However, the main conclusion was that, for the first time, centimetre accuracies could be achieved in seconds. This was appropriately chosen as the title of the resulting report (Remondi, 1986).

5 NOTTINGHAM KINEMATIC GPS TESTS

Three series of kinematic GPS tests have been carried out at Nottingham, by using TI-4100 and TRIMBLE 4000 series receivers. In each case, the test took place on the Electromagnetic Distance Measurement (EDM) Calibration Baseline at Bunny Park Farm. This testing facility consists of 7 pillars (labelled A to G) which are in a straight line orientated approximately East to West. Each pillar is constructed from reinforced concrete, with additional protection against accidental damage provided by an external plastic sleeve. Inter-pillar distances are regularly measured, and are known to the sub-millimeter level. However, the full 3-d vector of the baseline has never been obtained by using normal surveying methods. Of the 7 baseline pillars, only the middle 5 (B to F) were used in the kinematic GPS tests, because of the difficulties in quick access to the other points. The following table lists the inter-pillar distances.

From	To	Distance (m)
B	C	41.751
C	D	50.770
D	E	53.091
E	F	51.419
B	F	197.031

Inter-Pillar Distances

The first series of tests took place on the 21st and 22nd April 1987, using two Trimble 4000S GPS receivers. On the first day both receivers were kept static, with one situated on pillar B and the other on pillar F. This gave an opportunity for the precise baseline vector between the two points to be determined. This information was then used in the computation of the initial double-difference integer biases required for the tests on the following day. Using this information and the inter-pillar distances, it was possible to calculate the full 3-d coordinates of each of the intermediate points.

On the second day, the receivers were centred and left over the points B and F for approximately 25 minutes for checking purposes. Thereafter, the receiver on B remained static for the duration of the test, with the other receiver roving from point to point. Two sessions of measurements, each lasting about 45 minutes were carried out. This was done because the two corresponding sky plot alerts showed the four satellites of the first session to have a comparatively poor geometry with those of the second session a relatively good configuration (Appendix A and Appendix B).

The results of the first session, with one receiver permanently on B, and the second roving receiver moving from F to E, D and C in succession, demonstrated the linear drift experienced by Remondi (1986). After about 45 minutes, the drift amounted to 12 cm in X, 2 cm in Y and 8 cm in Z (see fig 1). It is worth noting that the best results (least drift) were achieved along the direction of the baseline and satellite configuration.

The results of the second session, with one receiver permanently at B, and the second roving receiver moving from E to D and C, were exceptionally good. This was probably due to correspondingly good satellite geometry (see Appendix B). Regrettably, a cycle slip occurred halfway through this session, which invalidated any of the following observations. However, during the first 20 minutes or so of this session, there was no perceptible linear drift, with the maximum drift after about 18 minutes amounting to 15 mm in X, 7 mm in Y and 5 mm in Z respectively (see fig 2). The corresponding double-difference phase residuals (shown in figure 3) are all less than 0.030 cycles or 6 mm. It is worth mentioning that, although the roving receiver was held over each pillar for about 5 minutes, only one or two (15-second) observations were sufficient to determine the corresponding coordinate differences to a very high accuracy.

6   CONCLUSIONS

Preliminary tests at Nottingham University confirm the validity of and the very high accuracies which can be achieved by Remondi's Kinematic GPS Surveying technique. The limited experiments show that the linear drift with time is very strongly correlated with the satellite-station geometrical configuration. Further tests are being carried out at Nottingham and elsewhere, to establish empirical guidelines to minimise the linear drift and the resulting necessity for frequent 0-position-updates.

ACKNOWLEDGEMENTS

The authors wish to acknowledge GPS Survey Services and, in particular,   Mr Barry Hogarth, for providing the TRIMBLE receivers which were used for the GPS Kinematic tests reported in this paper.

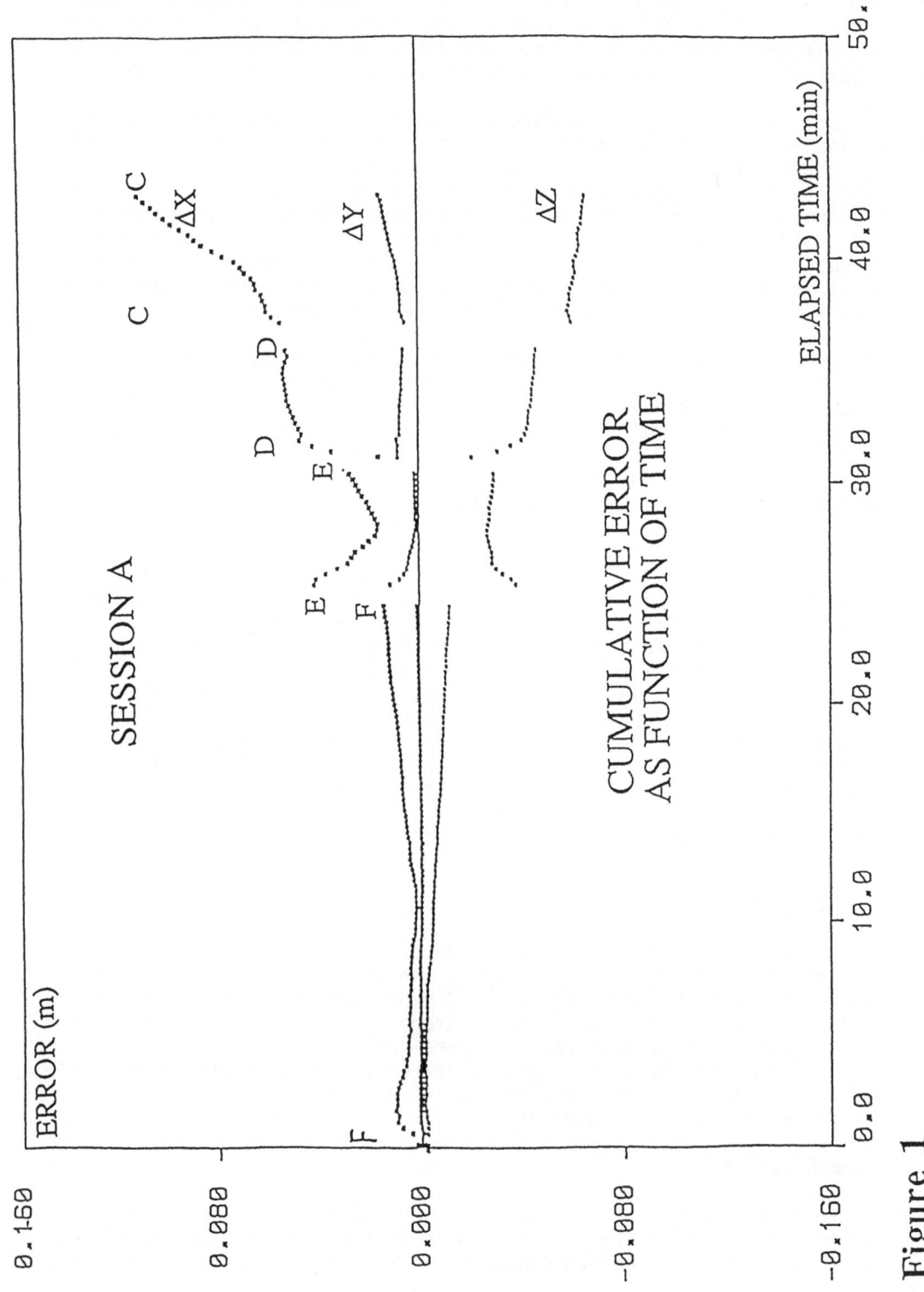

Figure 1

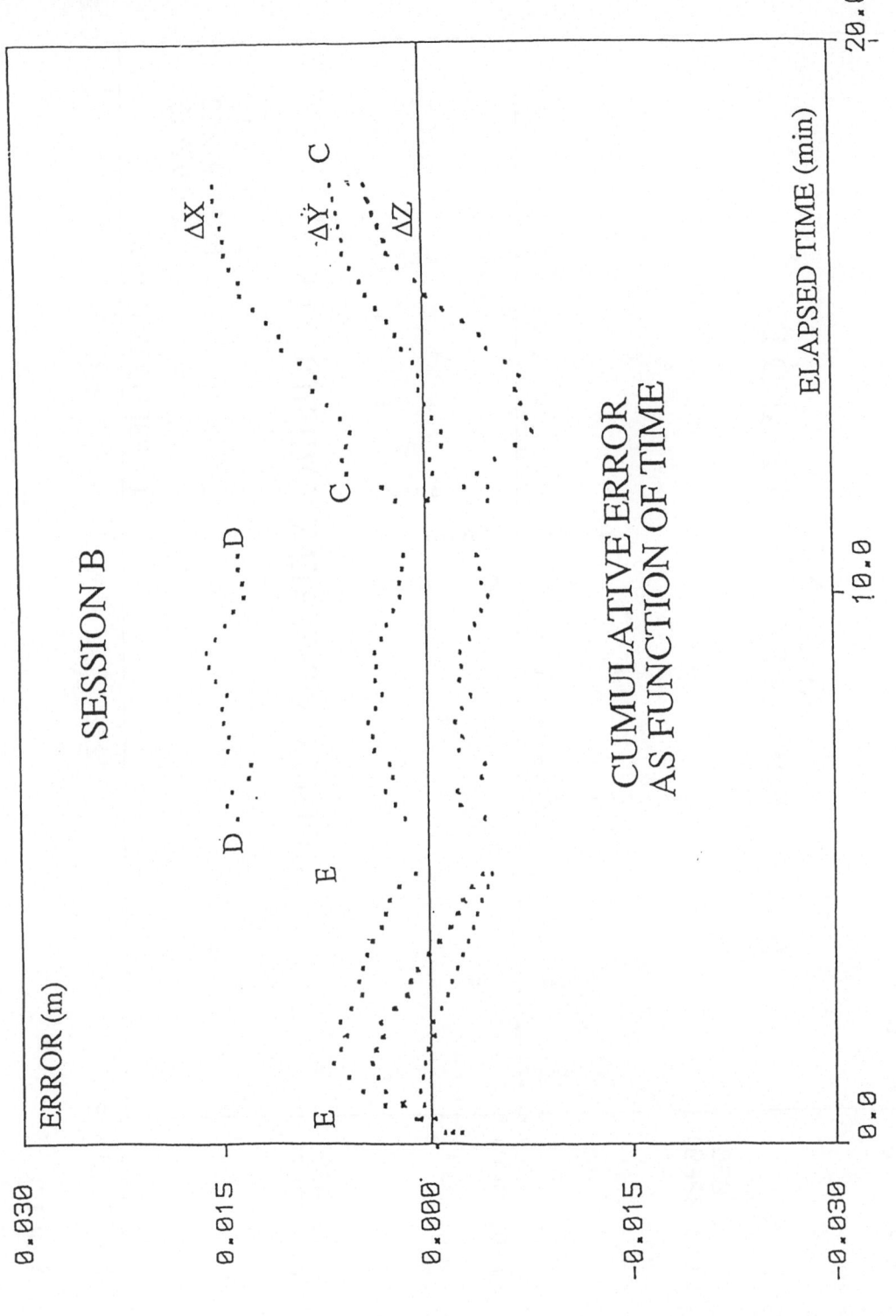

Figure 2

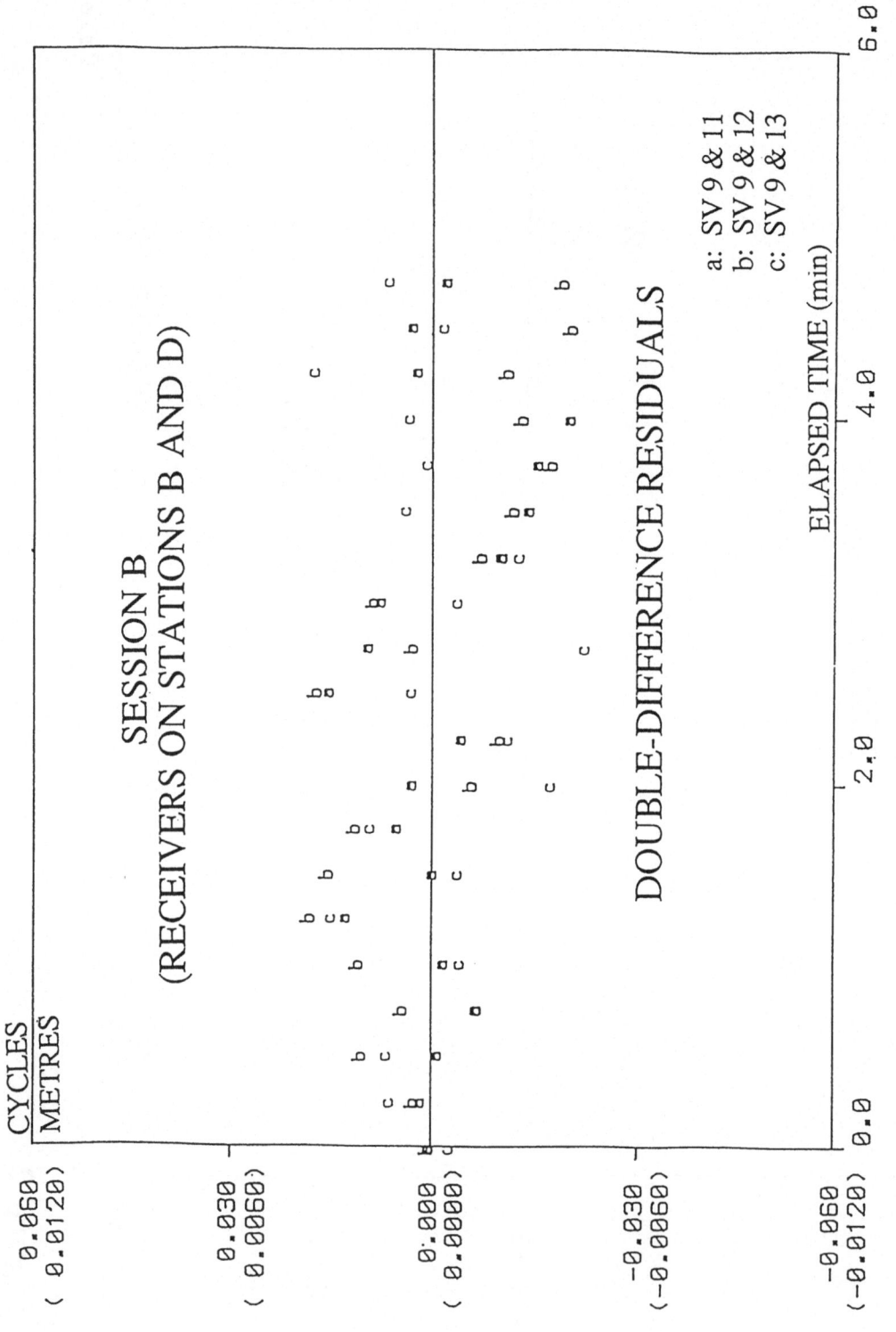

Figure 3

244

REFERENCES

Goad, C. C. (1985): Precise Relative Position Determination using Global Positioning System Carrier Phase Measurements in a Non-Difference Mode, Proc. First Int. Symp. on Precise Positioning with the Global Positioning System, Rockville, Maryland.

Hatch, R. R. (1986): Dynamic Differential GPS at the Centimeter Level, Proc. Fourth Int. Geodetic Symp. on Satellite Positioning, Austin, Texas.

Lachapelle, G., Hagglund, J., Falkenberg, W., Bellemare, P., Cassey, M. and Eaton, M. (1986): GPS Land Kinematic Positioning Experiments, Proc. of the Fourth Int. Geodetic Symp. on Satellite Positioning, Austin, Texas.

Remondi, B. W. (1986): Performing Centimeter-Level Surveys in Seconds with GPS Carrier Phase: Initial Results, Proc. Fourth Int. Geodetic Symp. on Satellite Positioning, Austin, Texas.

Remondi, B. W. (1985): Performing Centimeter Accuracy Relative Surveys in Seconds Using GPS Carrier Phase, Proc. First Int. Symp. on Precise Positioning, Rockville, Maryland.

SATELLITE GEOMETRY

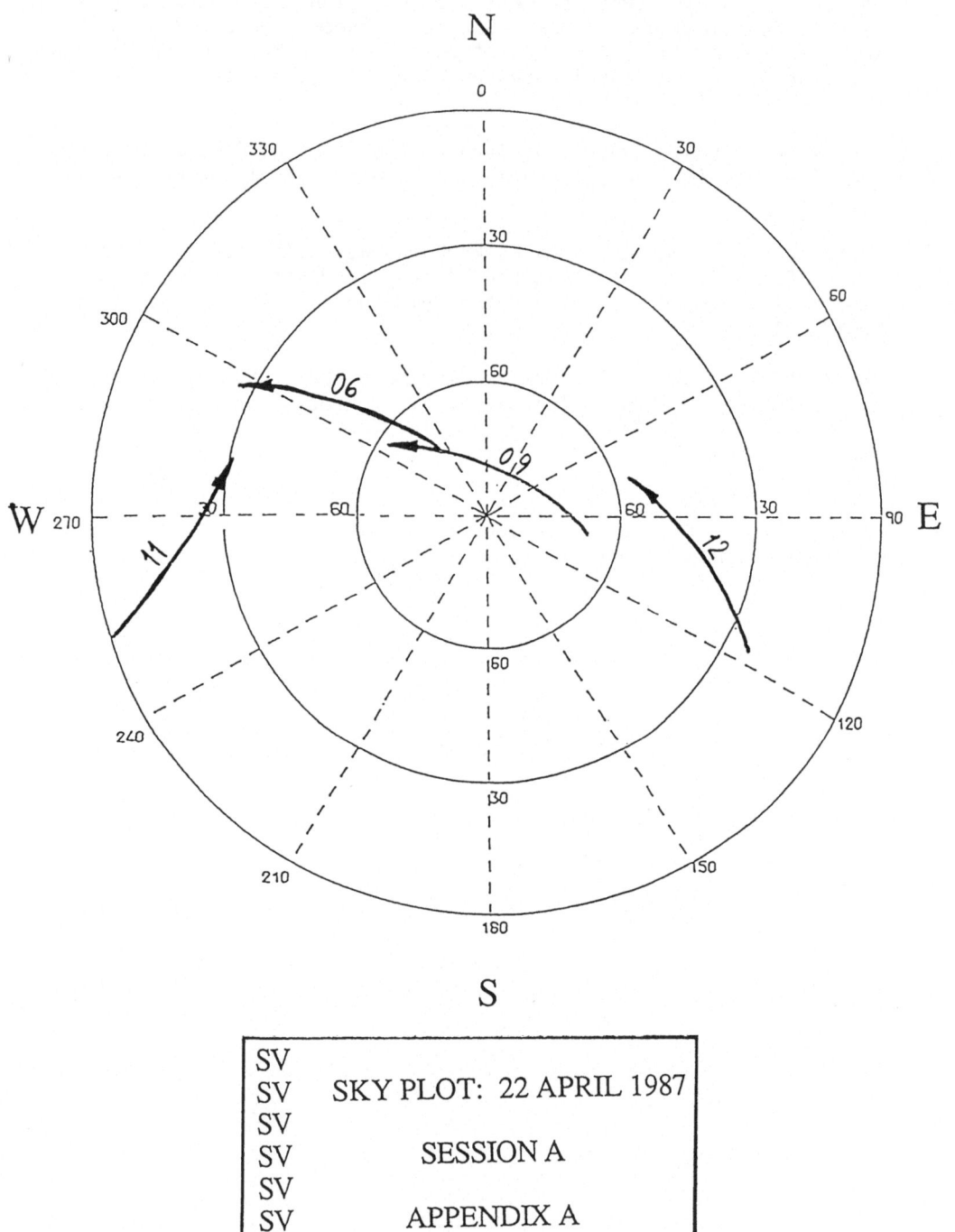

SV	
SV	SKY PLOT: 22 APRIL 1987
SV	
SV	SESSION A
SV	
SV	APPENDIX A

SATELLITE GEOMETRY

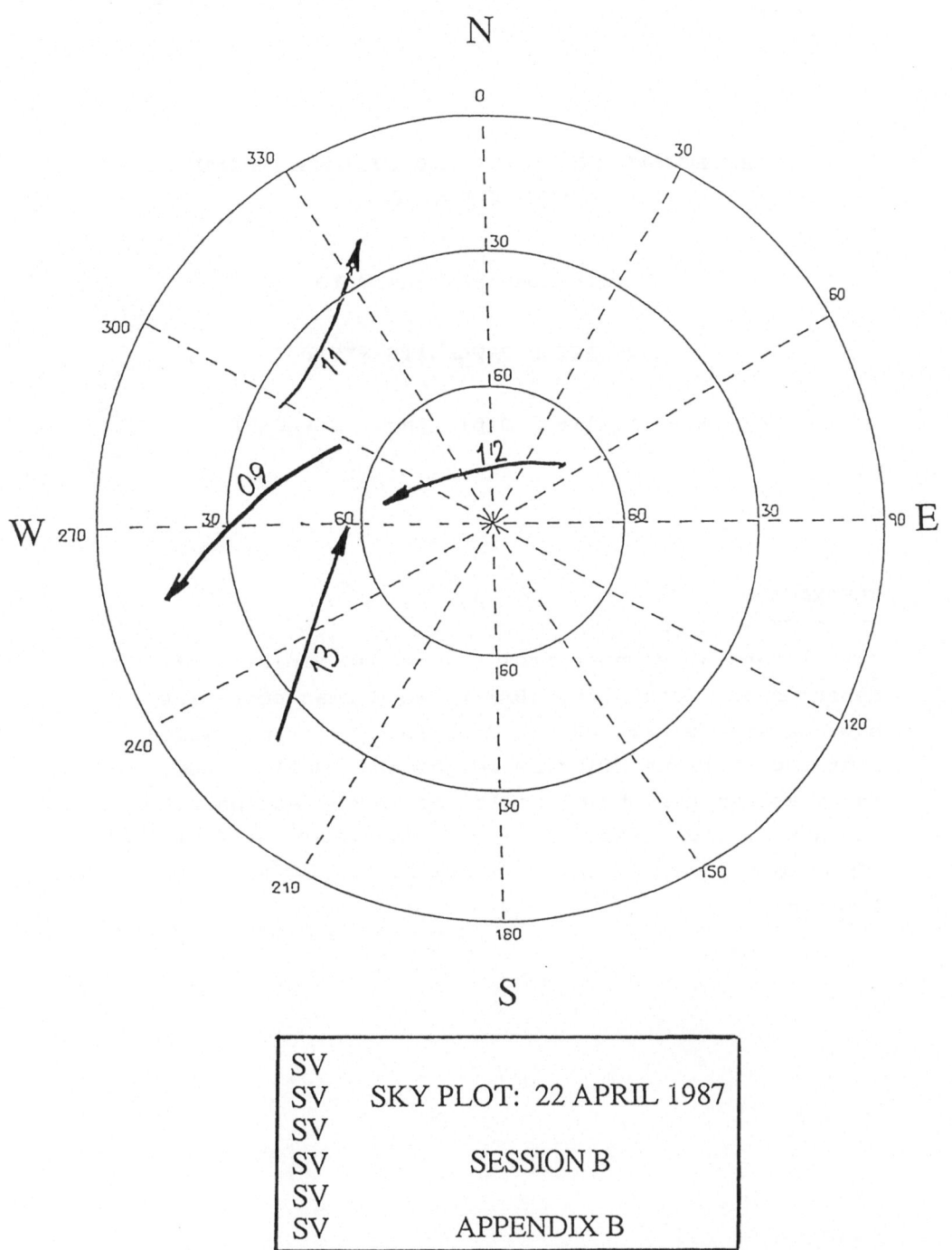

N

0

330

30

60

30

300

60

60

11

09

12

W 270

30

60

60

30

90 E

13

240

120

160

13

150

210

180

S

SV	
SV	SKY PLOT: 22 APRIL 1987
SV	
SV	SESSION B
SV	
SV	APPENDIX B

TECHNOLOGICAL UTILIZATION OF SPACE

with special regard to

NAVIGATION SATELLITE SYSTEMS

A.H. Stiller, Dipl.-Ing., DFVLR-PT

ABSTRACT

With financial support from the German Minister of
Research and Technology (BMFT) two German companies
have developed two GPS-C/A-Code-receivers for different applications with low weight and small volume.
The measured results of positions in connection with
the ABS of a car (Anti lock braking system) and in Diff.-
GPS-mode are very satisfying and in the range of 15 and/or
3 meters.

A   TECHNOLOGICAL UTILIZATION OF SPACE

A 1   What is the technologically usable space ?

The technically usable space begins just over the air
cover of the earth i.e. about 150 km above the surface of
the earth and ends at about 1/10 of the earth-moon
distance that is at about 38.000 km.
Below this distance satellites normally do not survive and
outside of the geostationary orbit (36.000 km) there are
only a few scientific satellites.

In addition, it is not recommendable to place highly
sophisticated satellites in the Van-Allen-Belt (about
1.100 - 11.000 km) because an adverse or even damaging
effect on electronic components.

A 2   What are artificial satellites ?

In our sense satellites are artificial moons of our earth
specially designed and built for their specific tasks.
Their weight (or mass) is in the range of some kilogrammes
to some tons, but normally their "empty weight" in the
orbit is in the range of 60 to 800 kg. In addition, each
satellite needs about the same amount of fuel in the form
of Hydrazine or partly in the form of a solid state motor
for the exact positioning and attitude control in its
precalculated orbit.
The in-orbit cost per kilogramme of such a satellite is in
the range of DM 0.1 to 1.0 million. The HF-output of each
of its transmitters is between 10 and 300 Watt. For this
HF-power and for housekeeping solar panels in the range of
1 to 50 m² are necessary.

A 3   What is the destination of artificial satellites ?

Besides research satellites and probes, the other,
artificial satellites can be classified into three
different technological destinations and three differnt
orbits:

a.) Remote sensing satellites (150-600 km)
b.) Navigation satellites (1.000 and 20.000 km)
c.) Communication satellites (36.000 km)

In the following only navigation satellite systems will be discussed.

B   **SATELLITE NAVIGATION SYSTEMS**

B 1   Terrestrial radio navigation aids

Since the beginning of the 20th century many radio navigation aids with different users, methods and frequencies (10 KHz till 14 GHz) have been designed and established.
Generally speaking it applies for terrestrial radio navigation aids that the higher the transmitting frequency the better the accuracy of the position but the shorter the operating range.

B 2   Advantages of satellite navigation systems

Only navigation systems with satellites allow a continuous worldwide navigation with the highest possible degree of accuracy of position.

B 3   Disadvantages of satellite navigation systems

The disadvantage of all satellite navigation systems is the relatively large distance between the transmitter and the user-receiver, because the transmitted HF-output power decreases with the square of the distance and continuous moving of the transmitter around the earth. To compensate this a high-precision time synchronization is necessary.

B 4   Existing satellite navigation systems

Since 1959 the following satellite navigation systems have
been established and are in operation

a.) TRANSIT (1959), managed by: US Navy, 150 and 400 MHz
b.) ZIKADE   (1967), managed by: Russ. Navy, 150 and 400 MHz
c.) GLONASS (1976), managed by: Russ. Air Force,
                                     1.25 and 1.6 GHz
d.) NAVSTAR-GPS (1974), managed by: US Air Force,
                                     1.22 and 1.57 GHz

In all four systems each satellite transmitts its Kepler's
orbit parameters (ephemerides). Additionally a highly
accurate time synchronization of the system is necessary.
It should also be mentioned that navigation satellites are
always turning in a certain angle to the plane of the
earth's equator (inclination).

In the TRANSIT- and Cicada-satellite navigation systems,
satellites are in a 1.000 km orbit and have a period of
108 minutes and each satellite covers 5-8 percent of the
earth's surface.
In the modern satellite navigation systems (GLONASS and
GPS), satellites are in a 20.000 km orbit with a period of
about 12 hours. Each satellite covers 30-35 percent of the
surface of the earth and has a three-dimensional attitude
control system to keep the correct predetermined in orbit.

B 4a   TRANSIT

TRANSIT (NAVY NAVIGATION SATELLITE SYSTEM = NNSS) or in
the geodetic applications shortly designated as DOPPLER-
System, was established by the US Navy in 1959. The free
use for civilian navigation purposed has been allowed
since 1967 by the owner of the system.

It is estimated that the amount of civilian TRANSIT-
receivers is in the range of 70.000. The operation will be
phased out five years after the full operation of GPS
(about 1996).

At the time being TRANSIT consists of six non-attitude-
stabilised satellites in a 1.075 km orbit which have a
period of 108 minutes. Therefore each TRANSIT-satellite is
only 5-18 minutes in the line of sight of the user and
transmitts always two coherent carrier-frequencies of 150
and 400 MHz respectively. The two frequencies are phase-
modulated and the dataframe contains the Kepler's parameters for the satellite concerned. The time of reference
will be taken of the Doppler-shift.
The TRANSIT-navigation message, its bitrate and dataframe
are published and available.

B 4b   CICADA (Zikade)

Since 1965 the Russian Marine (Navy) has built up a
satellite navigation system very similar to TRANSIT. It is
named CICADA and consists, at the time being, of 12 cosmos
satellites in a 1.000 km orbit with a period time of 105
minutes. Each Cicada-satellite continuously transmitts two
frequencies (150 and 400 MHz), but the higher frequency is
unmodulated and is only used for the correction of the
ionosheric influence. The number of user equipment is
unknown.

The Cicada navigation messages, their dataframe , bitrate
and scalefactors have already been decoded by British
scientists and published.

B 4c   GLONASS

GLONASS is the English abbreviation of GLObal NAvigation
Satellite System.
It is the Russian counterpart to the GPS-System. Similar
to the GPS-System the GLONASS-satellites are in the 20.000
km orbit, have a period time of 12 hours and an
inclination of 64.8°. Each satellite transmitts continuously two L-Band frequencies (1.6 and 1.25 GHz) but only
the higher frequency is modulated with the navigation
message. The second frequency (1.25 GHz) is unmodulated

and is only used for the correction of the ionospheric
time delay.

From 1982 till 1987 the Russians launched 27 GLONASS-
satellites into orbit, but some of them do not transmit.
The reason for that is unknown in Western countries. The
system is in a preoperational phase. The last six GLONASS-
satellites failed to reach their predicted orbit at
launch. The number of user equipment and its price is
unknown.

It could be assumed that every satellite of the GLONASS-
System has one or more atomic time references and a three
dimensional stabilised attitude control system to achieve
the highest possible accuracy of position.
The GLONASS-navigation messages, its dataframe, bitrate
and scalefactors are also already decoded by British
scientists and offered for sale.

B 4d   NAVSTAR-GPS

GPS-(NAVSTAR-GPS) is the English abbreviation of
NAVigation System with Time And Ranging - Global
Positioning System.
It has been designed, built up by the American Department
of Defense (DoD) since 1974. It is now in a preoperational
phase and ready for tests.

Since 1984 the owner of the system has allowed all civilian
users to use the C/A-code free of charge for worldwide
navigation applications in the range of accuracy of 100 m.
It could be assumed that 1991 the whole GPS-System will be
completed. Many experts are of the opinion that GPS then
will became the main navigation aid of the future.

The whole system will consist of the following three
parts:

- the spacesegment, consisting of 18 (later 24) satellites
  in six different but nearly circular orbits at an
  altitude above the earth of 20.000 km (12 hour orbit)
  with an inclination of 63°.

- the ground control segment consisting of a main control and up-load station and four monitorstations and

- the user segment consisting of an unlimited number of GPS-receivers for the three different kinds of applications :

  o GPS-receiver for highly accurate time synchronisation
  o GPS-receiver for navigation applications
  o GPS-receiver for geodetic purposes.

Each GPS-Satellite transmitts continuously two carrier-frequencies of 1.575 and 1.228 GHz, which are both modulated with PRN-signals. The navigation signals of the modulation are different for each satellite and contain mainly its Kepler's parameter, the highly accurate time and the truncated orbit parameters of all other GPS-satellites.

The GPS-dataframe duration is 30 s at a bitrate of 50 bits/s and has been published with all details for a lot of years. But the knowledge concerning the Y-code is confidential.

B 5   General-Satellites

General

According to the latest General Accounting Office (GAO) announcement the United States intend to start all the next 28 Block II-GPS-Satellites till 1991 either by ELV (Erpendable Launch Vehicle) or by the shuttle. Then a worldwide, highly precise and three dimensional navigation in the already prognosticated limits of accuracy will be possible.

B 5a   Structure of the GPS-Satellites

Weight
Each of the Block II-GPS-Satellites has

o   a launch weight of 1.715 kg and

o   an orbit weight of   815 kg.

It consists of two parts

o   the satellite bus system and

o   the navigation payload

B 5b   Bus system

The bus system itself consist of an integral box structure,
on which the two folding out solar panels are mainted with
rotating capability.
Additionally, the structure carries

- The Thermal Control System (TCS), which comprises mainly
  the 1.7 m² Louvers

- The Electrical Power and Distribution System (EPDS)
  For this the 8 m² solar panel surface generates 700 Watt
  at 27 ± 1 VDC, For the power consumption during the
  earth shadow phase three NiCd-batteries of 35 Ah are
  installed.

- The Telemetry and Telecommand System (TM/TCS)
  Each satellite informs the main control station about
  its own housekeeping data by a special S-band-connec-
  tion.
  The main control station itself can command each
  GPS-satellite by an other S-band-connection.
  The information about the dataframes of these TM/TC
  S-band-connections are not published. Their data rates
  are 0.5 and 4 Kbit/s.

- The Attitude and Velocity Control System (AVCS)
  Launched by the shuttle, 1 to 3 GPS-satellites are
  carried into a 300 km orbit with an inclination of
  26.5°. To achieve this for each satellite a payload
  assist module (PAM) is necessary, which brings each
  satellite to a certain rotation. To dampen the nutation

an active and passive nutation damper is nessecary. If
the nutation is small enough the STAR-37 XF-Solid Rocket
Motor (SRM) will be ignited and carry the satellite into
the 20.000 km orbit.
If launched by an Expendable Launch Vehicle (ELV) the
apogee phase of the launch is slightly different but
basically similar.

- The Attitude and Orbit Control System (AOCS)
  In the predetermined orbit (20.000 km) each GPS-
  Satellite is 3-axis-stabilized by
- two (each 22 N) and sixteen (each 0.4 N) hydrazine
  thrusters with
- two hydrazine tanks and
- four reaction wheels

B 5c Navigation payload

The navigation payload of each GPS-satellite mainly
consists of the following:

o the hardened navigation data unit
o the two navigation transmitter with the antennas
  L_1 = 1.575 GHz, PRN-Modul, C/A- and P-Code
  L_2 = 1.227 GHz, PRN-Modul, C/A- or P-Code
o the two Cesium-time standards
o the two Rubidium-time standards and
o the navigation data memory for 14 days

Lifetime

All Block II-GPS-satellites are designed for

a life time of six years with
a design goal of seven and a half years

C   Conclusions

C 1   Which accuracies of position are possible ?

Tab. 3

It is very difficult to predict achieveable accuracies of
position if you advance to the limits of what is physically
measureable.
In the terrestrial navigation aids the transmitters are on
a fixed position on the surface of the earth. But in all
satellite navigation systems the transmitters are always
revolving around the earth and are influenced of their
orbits by different attractions of the earth, the sun and
the moon. Therefore for each satellite position not only a
high accurate time information is necessary but Kepler's
parameters must be transferred as accurately as possible.

For a worldwide navigation normally an accuracy of the
position of 30 m is sufficient, but in some cases the
desired accuracy will be less. Then it is possible to
measure the position first with the known method and
additionally to this with Differential-GPS or measurement
of the phase of the HF-carrier.

C 2   Development of civilian C/A-Code GPS-receivers in Germany

In the last five years two German companies with the
support of the German Ministry of Research and Technology
(BMFT) have developed two different GPS-receivers for
navigation applications. Both companies have already not
only tested their GPS-receivers but have combined their
products with the Anti-lock-breaking-system of a car, or
the other has already tested the Diff-GPS for the possibility of aircraft blind landing Cat I.

Both receivers worked quite well and both companies have
demonstrated their capability to meet our high exspecta-
tions. Unfortunately the GPS-satellite to be launched are
behind schedule, therefore the two German companies cannot
sell their products and if the GPS-system will be completed
in the year 1991 other technologies with smaller and
cheaper receivers will be on the market.

For that reason we should do our best to encourage the two
companies to continue their work to develop a GPS-receiver-
system with the most advanced technologies at the cheapest
possible price for navigation and geodetic applications.

Sixth session: Kinematic Applications (continued)

Chairman: Prof. Ashkenazi, Nottingham

THE ANTENNA EXCHANGE: ONE ASPECT OF
HIGH-PRECISION GPS KINEMATIC SURVEY

by

Bernhard Hofmann-Wellenhof   and Benjamin W. Remondi

Abstract

This is the dream of many geodesists: millimeter-accuracy
for very short baselines in extremely short time of
observation. Is GPS a candidate to achieve this? In order
to reach this, we turn to GPS in the kinematic mode and
describe the method of the antenna swap which implies an
interchange of the antennas at the baseline stations while
continuing satellite tracking. The theory of the antenna
exchange is shown in detail with a systematic derivation of
all relevant equations.

1. Introduction

Is GPS capable of yielding millimeter-accuracy for very short baselines in extremely short observation times? Static GPS relative positioning might be used but even with dual frequency receivers 10–15 minutes occupation time is required plus separate data files and post-processing for each survey mark. Therefore we turn to GPS in kinematic survey mode where, as we have seen, only seconds of occupation time are necessary at each survey mark and one data file is sufficient for possibly hundreds of survey marks.

Kinematic survey processing requires a precisely-known starting vector between the reference mark and the initial location of the roving receiver. This starting vector will be available in the case where precise geodetic coordinates of the reference and the initial mark of the rover are known a priori. Often these will not be available. The antenna exchange is a technique to establish, within seconds, this precise starting vector in those circumstances where the precise geodetic coordinates are available for only one or even no survey marks. In the case of no survey mark one would use the code phase point-position solution for the reference site and the antenna exchange technique to establish the high-precision starting vector from the reference mark to the starting point of the roving receiver. In all cases one can compute the relevant integer ambiguity values once the coordinates of the reference mark and the initial location of the rover are known. These integer values can yield an improved starting vector and permit the subsequent kinematic survey (not discussed in this paper) to be directly processed in double-differenced mode. The triple-difference approach to the antenna swap yields nearly identical results to those obtained using double differences. The subsequent kinematic survey can be processed in either double difference mode (with the integer ambiguities) or in triple difference mode with the geodetic coordinates of the reference mark and the starting coordinates of the moving receiver. Both methods have advantages, the double difference approach is theoretically better.

In the presentation which follows the theoretical basis of the antenna exchange will be developed.

2. Millimeter or not millimeter — that is the question

This is the task: determine a very short baseline using relative positioning, with high precision and with a short observation period. Let us discuss a GPS solution step by step according to the task. Definitely, GPS allows one to measure, accurately, very short baselines. One can even go to the limit and measure a baseline of length zero, see Ashkenazi and Yau (1986). Relative positioning with high precision is probably the most favourite application of GPS for geodesists. As one example of a high precision engineering survey network we mention Ruland and Leick (1985). But now comes the most critical question: is GPS capable of delivering high precision results based on a few observations in a short observation time, e.g. 1 minute or less? If we speak of few observations then we mean 10 to 100. In this context we should not forget that soon one will be able to have 30 or more measurements in 30 seconds. Leafing through numerous papers, one finds decimeter precision for an aircraft in Mader et al. (1986), centimeter-level accuracy in Remondi (1985), Remondi (1986), discussions on millimeters in Hofmann-Wellenhof and Remondi (1985), Hofmann-Wellenhof (1985). All of the papers share one common point: they deal with kinematic surveys, meaning one of the two receivers is in motion. However, we are interested in baseline determinations between two fixed points with ten to one hundred epochs where one of the two stations is known. This is desirable for a subsequent kinematic survey where a precise starting vector is necessary. For a medium-lengthed (12.5 km) baseline, reasonable results with 30 minutes of data can be found, see Ladd et al. (1985), which is not yet sufficient for our goals. We are interested whether GPS can compete with conventional methods in determining a very short baseline in a very short time. A first hint how to achieve this is given in Remondi (1985) and in Remondi (1986). The latter paper mentions the antenna swap method to determine a very short baseline. This is the subject of the theory presented below. Results are already reported in Remondi (1986).

In practice the antenna swap may be used to determine the initial coordinates of the moving receiver. The remaining kinematic survey would not typically be accomplished via antenna swaps, rather, the receiver whose position was initialized by the antenna swap would move from mark to mark where the companion receiver would remain at rest after the antenna swap.

3. Double difference equations without antenna swap versus double difference equations with antenna swap

We start with modeling the phase by

$$\Phi_r^j(t) = -\frac{1}{\lambda} \cdot \rho_r^j(t) + N_r^j + f \cdot \Delta \tau_r^j(t) \qquad (3-1)$$

where $\Phi_r^j(t)$ is the measured phase (in cycles) at an epoch t and the right-hand side is the model for this measured phase. In detail, λ is the wavelength of the satellite emitted signal, $\rho_r^j(t)$ is the geometric distance between the ground station (receiver station) r and the satellite j at epoch t, moreover, N_r^j is the integer ambiguity, and $\Delta \tau_r^j(t)$ is the influence caused by clock errors. This error is multiplied by the frequency of the signal, in our case it is f, which is assumed to be equal for all satellites, therefore we did not use the superscript. To clarify, the subscript denotes something on the earth, the superscript denotes a satellite (the satellite is "higher" than the "lower" ground or receiver station). We did not model the atmosphere because it has a negligible influence for very small baselines and we also omitted other minor effects.

If we split up the clock error into two parts, namely $\delta_r(t)$ related to the receiver clock and $\delta^j(t)$ related to the satellite clock, we get

$$\Phi_r^{\,j}(t) = -\frac{1}{\lambda}\cdot\rho_r^{\,j}(t) + N_r^{\,j} + f(\delta_r(t) + \delta^j(t)) \quad . \tag{3-2}$$

We shall base subsequent formulas on this phase equation. We restrict ourselves to relative positioning of two stations. Therefore we form, between the two stations 1 and 2, the single difference

$$SD_{12}^{\,j}(t) = \Phi_2^{\,j}(t) - \Phi_1^{\,j}(t) \quad . \tag{3-3}$$

Inserting the result for the phase equation (3-2) twice into the equation above leads to

$$SD_{12}^{\,j}(t) = -\frac{1}{\lambda}\Big[\rho_2^{\,j}(t) - \rho_1^{\,j}(t)\Big] + N_2^{\,j} - N_1^{\,j} + f\cdot\delta_2(t) - f\cdot\delta_1(t)$$

$$\ldots \quad (3-4)$$

where we joyfully recognize the effect of cancelling out the satellite clock error. In order to eliminate the receiver clock errors, we form double differences by taking the single difference equation for two satellites j and k and by subtracting the corresponding equations:

$$DD_{12}^{\,jk}(t) = SD_{12}^{\,k}(t) - SD_{12}^{\,j}(t) \quad . \tag{3-5}$$

The result is

$$DD_{12}^{\,jk}(t) = -\frac{1}{\lambda}\Big[\rho_2^{\,k}(t) - \rho_2^{\,j}(t) - \rho_1^{\,k}(t) + \rho_1^{\,j}(t)\Big]$$

$$+ N_2^{\,k} - N_2^{\,j} - N_1^{\,k} + N_1^{\,j} \quad . \tag{3-6}$$

We can abbreviate this formula slightly by introducing

$$N_{12}^{\,j} = N_2^{\,j} - N_1^{\,j} \tag{3-7}$$

and

$$N_{12}^k = N_2{}^k - N_1{}^k \tag{3-8}$$

and finally

$$N_{12}^{jk} = N_{12}^k - N_{12}^j \tag{3-9}$$

thus, the double difference equation shows up as

$$DD_{12}^{jk}(t) = -\frac{1}{\lambda}\left[\rho_2{}^k(t) - \rho_2{}^j(t) - \rho_1{}^k(t) + \rho_1{}^j(t)\right] + N_{12}^{jk} \ . \tag{3-10}$$

The simplification achieved, namely replacing the four unknowns $N_1{}^j$, $N_1{}^k$, $N_2{}^j$ and $N_2{}^k$ by one new unknown N_{12}^{jk}, really makes sense because we are generally not interested in any of those quantities which must be dragged along as an inconvenient companion.

A little earlier we mentioned to deal only with relative positioning. This means that one of the two stations must be known. Let us assume station 1 to be known. In addition, the satellite coordinates are assumed to be available for any arbitrary epoch t. Putting all known terms on the left-hand side, we get

$$DD_{12}^{jk}(t) + \frac{1}{\lambda}\left[-\rho_1{}^k(t) + \rho_1{}^j(t)\right] = -\frac{1}{\lambda}\left[\rho_2{}^k(t) - \rho_2{}^j(t)\right] + N_{12}^{jk}$$
$$\ldots \quad (3-11)$$

where the right-hand side terms contain the unknowns. How many unknowns do we have in this double difference equation? There are four: the three station coordinates of station 2 contained in the ρ_2-terms and the unknown integer ambiguity N_{12}^{jk}.

Up to now we have always assumed the two receivers to be stationary at the two stations of the desired baseline vector. We now allow motion of the receivers as well. This means that we must slightly change the notation of the equations. In eq. (3-1) we used the subscript either for the ground station or for the receiver, the superscript for the satellite and the argument for the time. Now we also need a distinction between the receiver and the ground station. We take this into account by putting the receiver mark within parentheses. Then we may re-write eq. (3–2) as

$$\Phi^j(r,t) = -\frac{1}{\lambda}\rho_g^{\ j}(r,t) + N^j(r) + f\cdot\delta(r,t) + f\cdot\delta^j(t) \qquad (3\text{–}12)$$

where g denotes the location of the receiver r at time t, and j denotes the satellite.

Remark: We have introduced a redundancy in the notation with respect to the ρ-term on purpose. The receiver symbol r could be omitted since the location g of the receiver is sufficient and unique. Nevertheless, for at least some of the following equations it simplifies understanding.

Look at this equation carefully. Especially note that the ambiguity $N^j(r)$ depends on the starting receiver location and the satellites. It does not depend on time (as long as no interrupt of the signal occurs). We stress, once again, $N^j(r)$ remains the same as long as no loss of lock occurs. Consequently, if we start to move the receiver while tracking, the integer ambiguities will remain constant integral values.

For two receivers $R1$ and $R2$ we get the single difference

$$SD^j(R2\text{–}R1,t) = \Phi^j(R2,t) - \Phi^j(R1,t) \qquad (3\text{–}13)$$

and the double difference

$$DD^{jk}(R2\text{–}R1,t) = SD^k(R2\text{–}R1,t) - SD^j(R2\text{–}R1,t) \quad . \qquad (3\text{–}14)$$

Inserting (3–12) into (3–13) twice and subsequently (3–13) into (3–14) twice, we get the final result

$$DD^{jk}(R2-R1,t) = -\frac{1}{\lambda}\left[\rho_B{}^k(R2,t) - \rho_B{}^j(R2,t) - \rho_A{}^k(R1,t) + \rho_A{}^j(R1,t)\right]$$

$$+ N^{jk}(R2) - N^{jk}(R1) \qquad \cdots \quad (3–15)$$

where we assumed receiver R1 at the ground station A and R2 at B, see the left part of Fig. 3.1. For the integer ambiguities, we have used the short-hand notations $N^{jk}(R2) = N^k(R2) - N^j(R2)$ and $N^{jk}(R1) = N^k(R1) - N^j(R1)$. Now we consider two epochs t_1 and t_2, respectively, and the situation for the receivers as presented in Fig. 3.1. Explicitly:

epoch t_1:   receiver R1 at A   epoch t_2:   receiver R1 at B
              receiver R2 at B   receiver R2 at A.

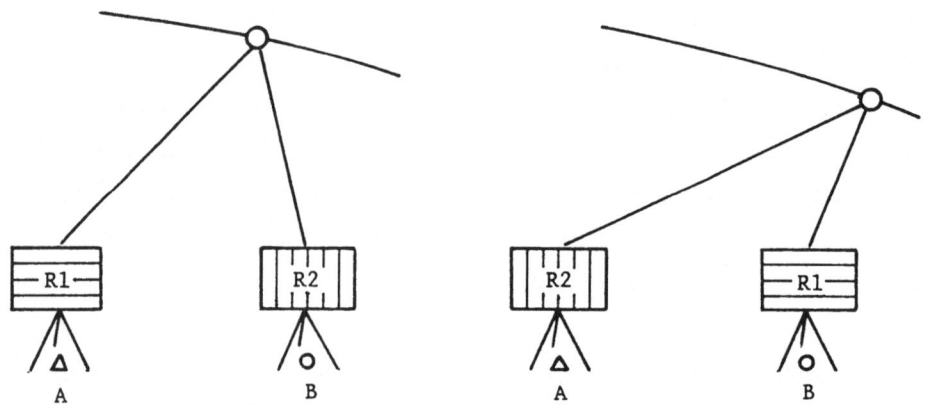

Figure 3.1. The antenna swap. The left part shows the original receiver situation, the right part the swapped situation.

What happened between the two epochs? We have performed a so-called antenna swap, see Remondi (1986). This means that we have taken the antenna of receiver R1 and moved it to station B and the antenna of receiver R2 and moved it to station A. Both receivers must continue to track while moving!

We set up the two double differences for the two epochs t_1 and t_2:

$$DD^{jk}(R2-R1,t_1) =$$

$$-\frac{1}{\lambda}\left[\rho_B^{\ k}(R2,t_1) - \rho_B^{\ j}(R2,t_1) - \rho_A^{\ k}(R1,t_1) + \rho_A^{\ j}(R1,t_1)\right]$$

$$+ N^{jk}(R2) - N^{jk}(R1) \qquad\qquad (3-16)$$

and analogously

$$DD^{jk}(R2-R1,t_2) =$$

$$-\frac{1}{\lambda}\left[\rho_A^{\ k}(R2,t_2) - \rho_A^{\ j}(R2,t_2) - \rho_B^{\ k}(R1,t_2) + \rho_B^{\ j}(R1,t_2)\right]$$

$$+ N^{jk}(R2) - N^{jk}(R1) \quad . \qquad\qquad (3-17)$$

Now we eliminate the integer ambiguities by subtracting (3-16) from (3-17). According to the remark after eq. (3-12) we may omit the receiver specifications in the ρ-terms. Thus we get

$$DD^{jk}(R2-R1,t_2) - DD^{jk}(R2-R1,t_1) =$$

$$-\frac{1}{\lambda}\left[\rho_A^{\ k}(t_2) - \rho_A^{\ j}(t_2) - \rho_B^{\ k}(t_2) + \rho_B^{\ j}(t_2)\right. \qquad\qquad (3-18)$$

$$\left. + \rho_A^{\ k}(t_1) - \rho_A^{\ j}(t_1) - \rho_B^{\ k}(t_1) + \rho_B^{\ j}(t_1)\right] \quad .$$

This equation is the point of the antenna swap. Do you see its strength? If not then consider two ordinary double differences without antenna swap meaning

epoch t_1:   receiver R1 at A   epoch t_2:   receiver R1 at A

receiver R2 at B   receiver R2 at B

then we obtain, by using (3–15) twice, the result

$$\text{DD}^{jk}(R2\text{–}R1, t_2) - \text{DD}^{jk}(R2\text{–}R1, t_1) =$$

$$- \frac{1}{\lambda} \Big[- \rho_A{}^k(t_2) + \rho_A{}^j(t_2) + \rho_B{}^k(t_2) - \rho_B{}^j(t_2) \tag{3–19}$$

$$+ \rho_A{}^k(t_1) - \rho_A{}^j(t_1) - \rho_B{}^k(t_1) + \rho_B{}^j(t_1) \Big] \quad .$$

Again we have omitted the receiver notation in the ρ-terms. Now compare (3–18) comprising an antenna swap and (3–19) without antenna swap. The signs of the terms depending on epoch t_2 are reversed! This is the huge advantage of the antenna swap procedure. Combining two equations in the form (3–18) delivers an excellent equation since the corresponding ρ-terms sum up whereas in (3–19) the ρ-terms almost cancel out for two nearby epochs. We can see this best if we consider an exaggerated example where $t_1 \approx t_2 \approx t_1 + \epsilon$ where $\epsilon = 1$ μsec (obviously, this is not realistic because you cannot perform an antenna swap in 1 microsecond; but it is a good demonstration example). We get

$$\text{DD}^{jk}(R2\text{–}R1, t_1+\epsilon) - \text{DD}^{jk}(R2\text{–}R1, t_1) \approx$$

$$\tag{3–20}$$

$$- \frac{1}{\lambda} \Big[2\rho_A{}^k(t_1) - 2\rho_A{}^j(t_1) - 2\rho_B{}^k(t_1) + 2\rho_B{}^j(t_1) \Big]$$

from the antenna swap eq. (3–18), but

$$0 \approx 0 \tag{3–21}$$

from eq. (3–19) where no antenna swap occurred. This example shows excellently why you cannot get reasonable results in a very short time by simply using double differences whereas by the antenna swap method you can. Speaking mathematically, double differences of two close epochs are linearly dependent in case of no antenna swap.

In geometric terms we see that in GPS static relative positioning time must elapse so that the geometry can change; whereas in GPS kinematic survey, in general, and the antenna exchange method in particular, no change of the earth-satellite geometry is required. For this reason, B.W. Remondi has observed that a geosynchronous GPS satellite (with essentially no Doppler shift) would substantially strengthen kinematic survey.

3.1 Relative positioning using the antenna swap method

In eq. (3–18) the left-hand side is the combination of phase measurement values and the right-hand side represents the corresponding model. For relative positioning the coordinates of one station must be known. In our case we assume station A to be known and station B to be unknown and to be determined. Assuming additionally known satellite positions at any arbitrary epoch, we may re-write eq. (3–18) by putting all known terms on the left-hand side:

$$
\begin{aligned}
&DD^{jk}(R2{-}R1,t_2) - DD^{jk}(R2{-}R1,t_1) \\
&\quad + \tfrac{1}{\lambda}\Big[\rho_A{}^k(t_2) - \rho_A{}^j(t_2) + \rho_A{}^k(t_1) - \rho_A{}^j(t_1)\Big] - \\
&\quad - \tfrac{1}{\lambda}\Big[-\rho_B{}^k(t_2) + \rho_B{}^j(t_2) - \rho_B{}^k(t_1) + \rho_B{}^j(t_1)\Big]
\end{aligned}
\qquad (3\text{--}22)
$$

The unknown station coordinates of B are non-linearly involved in the ρ-terms.

3.2 The linearization of the ρ-terms

Skip this section if you are familiar with the simple linearization. Nevertheless, we show the linearization in order to deliver a complete formula set without burdening the reader by leaving some steps of the antenna swap method to be solved. Consider from (3–22) e.g. the term $\rho_B{}^k(t_1)$ being the distance between station B and satellite k at epoch t_1, cf. Fig. 3.2.

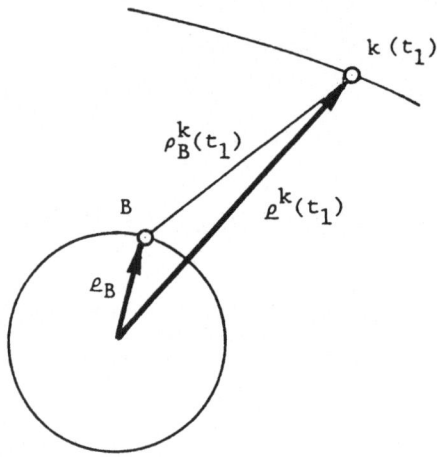

Figure 3.2. Geometric situation at epoch t_1. Here ϱ_B is the vector to the station, $\varrho^k(t_1)$ is the satellite vector.

The distance between station B and satellite k may be expressed by

$$\rho_B{}^k = |\varrho^k - \varrho_B| \tag{3–23}$$

where we have omitted the time argument for ρ and for the satellite position temporarily. This is the magnitude of the vector from the station to the satellite, see Fig. 3.2. To linearize this expression we set up a Taylor expansion

$$\rho_B{}^k = (\rho_B{}^k)_{appr} + \left[\frac{\partial(\rho_B{}^k)_{appr}}{\partial \varrho_B}\right]^T \cdot \Delta \varrho_B + \ldots \tag{3–24}$$

where $\Delta \varrho_B = \varrho_B - (\varrho_B)_{appr}$ and $(\rho_B{}^k)_{appr}$ is obtained by substituting the approximate coordinates $(\varrho_B)_{appr}$ into (3–23). The T denotes the transposition of the vector within the brackets. The partial derivatives follow immediately from

$$\frac{\partial \rho_B^k}{\partial \ell_B} = \frac{1}{2\rho_B^k} \cdot 2(\ell^k - \ell_B)(-1) = -\frac{\ell_B^k}{\rho_B^k} \tag{3-25}$$

where we have introduced the vector ℓ_B^k from station B to satellite k. As we can see, the result for the partial derivatives is the negative station to satellite unit vector (this is the same as the satellite to station unit vector). Therefore we obtain

$$\rho_B^k(t) = (\rho_B^k(t))_{appr} - \left[\frac{\ell_B^k(t)}{\rho_B^k(t)}\right]^T_{appr} \cdot \Delta\ell_B \tag{3-26}$$

which is the final result for the linearization where $\Delta\ell_B$ contains the unknown increments to known approximate coordinates of B and where we have added the time argument.

3.3 Linearized antenna swap equation

Applying (3–26) to (3–22) on the right-hand side 4 times and shifting the terms containing the approximate coordinates of B, which must be known, on the left-hand side results in the following lengthy formula

$$DD^{jk}(R2-R1,t_2) - DD^{jk}(R2-R1,t_1) + \frac{1}{\lambda}\Big[\rho_A^k(t_2) - \rho_A^j(t_2)$$

$$+ \rho_A^k(t_1) - \rho_A^j(t_1) - (\rho_B^k(t_2))_{appr} + (\rho_B^j(t_2))_{appr}$$

$$- (\rho_B^k(t_1))_{appr} + (\rho_B^j(t_1))_{appr}\Big] =$$

$$\tag{3-27}$$

$$= -\frac{1}{\lambda}\left\{\left[\frac{\ell_B^k(t_2)}{\rho_B^k(t_2)}\right]_{appr} - \left[\frac{\ell_B^j(t_2)}{\rho_B^j(t_2)}\right]_{appr} + \left[\frac{\ell_B^k(t_1)}{\rho_B^k(t_1)}\right]_{appr}\right.$$

$$\left. - \left[\frac{\ell_B^j(t_1)}{\rho_B^j(t_1)}\right]_{appr}\right\}^T \cdot \Delta\ell_B \quad .$$

Immediately we introduce auxiliary expressions by

$$\alpha^{jk}(t_1,t_2) = \text{left-hand side of } (3\text{--}27)$$

$$\beta^{jk}(t_1,t_2) = -\tfrac{1}{\lambda}\cdot\{\text{expression within braces of } (3\text{--}27)\}$$

(3–28)

and get

$$\alpha^{jk}(t_1,t_2) = \left\{\beta^{jk}(t_1,t_2)\right\}^T\cdot\Delta\rho_B \quad .$$

(3–29)

This is one equation with three unknowns $\Delta\rho_B$. Remember that t_1 is an epoch in the starting situation receiver R1 at A and receiver R2 at B whereas t_2 is an epoch of the swapped configuration, i.e. receiver R1 at B and receiver R2 at A.

Each additional satellite delivers one additional equation in the form of (3–29). One of the satellites is kept as a reference satellite. Having two more satellites available solves the problem. The minimum requirements for the antenna swap procedure is therefore 4 satellites and 2 epochs, one before the antenna swap and one afterwards. Naturally, 5 or 6 or more satellites are recommended to detect and correct cycle slips.

It should be mentioned that the use of a reference satellite results in a correlation of the equations for each epoch. This can be compensated with an appropriate weight matrix. Assume, as one example, a simple situation where we have six satellites at epoch t_1 before the swap and the same six at epoch t_2 after the swap and assume that those two epochs are the only ones. Then

$$\begin{bmatrix} 4 & 2 & 2 & 2 & 2 \\ 2 & 4 & 2 & 2 & 2 \\ 2 & 2 & 4 & 2 & 2 \\ 2 & 2 & 2 & 4 & 2 \\ 2 & 2 & 2 & 2 & 4 \end{bmatrix}\sigma^2$$

(3–30)

is the covariance matrix for the five double differences where σ^2 is the variance associated with the single difference observable. The weight matrix follows from the inverse of the covariance matrix. For more details see Remondi (1984). In general we measure more than one epoch at each receiver location. Then we can adjust in the usual way. Note that every epoch before the swap with every epoch after the swap (along with the corresponding weight matrix) forms one of the equations (3-27) and can be included in the normal equations.

The procedure mentioned above gives millimeter-level accuracy. One slight improvement should be mentioned here; once the solution of the baseline is obtained one can go back to the double difference equation at t_1 or the double difference equation at t_2 or both and compute the values of the ambiguities $N^k - N^j$ which are most likely slightly off with respect to integer values. Then one can fix them to exact integers and re-compute the desired station position again. This, theoretically, will give a better answer.

Speaking a little more precisely on the accuracy to be expected, we may say that the "millimeter level times the PDOP" is realistic. As a consequence, with 5 or 6 satellites PDOPs of about 1.0 will be common so that 1 mm results can generally be expected using this method.

4. Conclusion

In detail we have described a method to determine short baselines with high accuracy in very short time. The GPS relative positioning method is based on the so-called antenna swap method and allows for the elimination of the integer ambiguities. This can be achieved by an observation session in two steps. The first is the starting situation with receiver R1 at the known station A and receiver R2 at the unknown station B. Now we start to measure. In principle one epoch with 4 satellites measured simultaneously from the two stations is sufficient. In practice at least 5 satellites are required so that cycle slips can be detected. With 4 satellites, one would require repeated measurements and, nevertheless, achieve inferior results in sites of poor satellite geometry. The second step comprises the antenna swap. The antenna of receiver R1 is taken and moved to station B while the antenna of receiver R2 is moved to station A. While moving, both antennas must continue to track the satellite signals in order to stay with the same integer ambiguities. Having finished the antenna swap then again one measurement epoch from the two stations is necessary.

Once the resulting mark B coordinates are determined by the antenna exchange theory, herein presented, one can compute the integers and (theoretically) improve the solution of mark B using double differences.

The accuracy to be gained by this method is in the millimeter-level.

Acknowledgment: The main part of the research necessary for this publication was conducted while B. Hofmann-Wellenhof was a National Research Council Senior Scientist in Geodesy. The generous grant obtained from the National Research Council, National Academy of Sciences is gratefully acknowledged. The co-operation with many colleagues while in residence at the National Geodetic Survey at Rockville, Maryland, USA has kept alive the memory of a very pleasant time.

5. References

Ashkenazi, V.; J. Yau (1986): "Significance of discrepancies in the processing of GPS data with different algorithms." Proceedings of the Fourth International Geodetic Symposium on Satellite Positioning. Austin, Texas, April 28 – May 2, Vol. 1, pp. 689–703.

Hofmann-Wellenhof, B. (1985): "Millimeter accuracy in real time using GPS – dream or reality?" Proceedings of the Second SATRAPE Meeting. Saint-Mande, France, November 4–6, pp. 86–97.

Hofmann-Wellenhof, B; B. W. Remondi (1985): "Determination of the trajectory of a moving platform using GPS carrier phase." Proceedings on Inertial, Doppler and GPS Measurements for National and Engineering Surveys. Munich, Federal Republic of Germany, July 1–3, Vol. 20-2, pp. 443–461.

Ladd, J. W.; C. Counselman; S. A. Gourevitch (1985): "The Macrometer II^{TM} dual-band interferometric surveyor." Proceedings of the First International Symposium on Precise Positioning with the Global Positioning System. Rockville, Maryland, April 15–19, Vol. 1, pp. 175–180.

Mader, G.; W. Carter; B. Douglas; W. Krabill (1986): "Decimeter precision aircraft positioning using GPS carrier phase measurements." Technical Memorandum (unnumbered) of the National Geodetic Survey, Rockville, Maryland. Also published with the same title, but with G. Mader as the only author, in Proceedings of the Fourth International Geodetic Symposium on Satellite Positioning. Austin, Texas, April 28 – May 2, Vol. 2, pp. 1311–1325.

Remondi, B. W. (1984): "Using the Global Positioning System (GPS) phase observable for relative geodesy: modeling, processing, and results." The University of Texas at Austin, Center for Space Research, 360 pp.

Remondi, B. W. (1985): "Performing centimeter accuracy relative surveys in seconds using GPS carrier phase." Proceedings of the First International Symposium on Precise Positioning with the Global Positioning System. Rockville, Maryland, April 15–19, Vol. 2, pp. 789–797.

Remondi, B. W. (1986): "Performing centimeter-level surveys in seconds with GPS carrier phase: initial results." Navigation, Journal of The Institute of Navigation, Vol. 32, No. 4, 1985/86, pp. 386–400.

Ruland, R.; A. Leick (1985): "Application of GPS in a high precision engineering network." Proceedings of the First International Symposium on Precise Positioning with the Global Positioning System. Rockville, Maryland, April 15–19, Vol. 1, pp. 483–493.

GPS GEODESY AND KINEMATIC TOPOGRAPHY MEASUREMENTS AND DATA PROCESSING

by

Georges Pierre NARD

1 - ABSTRACT

For over 30 years SERCEL has been studying and manufacturing high accuracy trade electronical equipment used for maritime radiolocation and land-surveys. To answer the present and future needs in these fields, SERCEL developed 3 types of GPS receivers characterized by a very high accuracy, and respectively intended for : high accuracy real time dynamic positioning (TR5S), geodesy or surveys (NR52) and very accurate time measurement (NRT). Dedicated and original softwares have been developed, allowing the user, thanks to very accurate GPS data gathered by these receivers, to work out either fix position solutions with a decimetric precision within a few minutes, or with a centimetric precision within a few tens of minutes. These softwares are very flexible (multisite and multisession solutions), offer a very high computation speed using light processing means.
At last, a unique software, which allows accurate determination of the trajectory of any mobile, (land mobile, maritime mobile or a plane) has been set up and tested under severe conditions.
The accuracies obtained are about 10 to 50 centimetres, three-dimensional, with a fast data sampling rate, and a very low noise measurement. Original and effective answers have been brought to the difficult problem of temporary satellite masks.
These softwares and equipment have been proved as operational in various fields.
This paper describes the equipment and methods used and especially the results obtained.

2 - INTRODUCTION

For over 30 years, Sercel has been developing and manufacturing high accuracy equipment and systems for radionavigation and land-surveys.
In the accurate radiolocation field, historically, TORAN was the first one ; then SYLEDIS, for which a new generation of equipment has just come out, and then GEOLOC which has been commercially experimented for two years.
Numerous activities have been developed from the various SYLEDIS applications, mainly for offshore, but also for land, airborne and

helicopter-borne operations : 4000 Syledis Units, beacons or mobiles, have been put into service over more than 20 countries.

In the land-survey field, and for more than 20 years, Sercel has been studying developing and manufacturing more than 30 000 infrared EDM of the DISTOMAT family : DI10, DI3, DI4, DI5 and DI2000, which have been commercialized mainly by WILD.

Researches and developments conducted from years 1972 - 1975 for SYLEDIS, allowed Sercel technical teams to get used to the pseudo-random signals, to correlation and accurate phase measurement techniques, to the problems of geodesy, to the UHF techniques applied to accurate measurements of propagation time and so on .. any propitious factor making up a good experience to take up GPS studies and applications useful for high-accuracy positioning and localization.

A first prototype of GPS receiver has been manufactured between 1979 and 1981. It allowed us to study signals and behaviour of the GPS system.

A first generation of equipment has then been studied and developed. It has been operated commercially since 1986.

This first generation aims to meet with the first operational applications of GPS, to allow different scientific or service operators to become acquainted with the GPS system and to contribute to the coming out of new processing methods for GPS satellites in geodesy and static or kinematic surveys.

These receivers are essentially characterized by the very high accuracy they reach in the various types of application.

Simultaneously with the development of these pieces of equipment, original acquisition and data processing methods have been developed, making available high-performance softwares dedicated to geodesy and static or kinematic surveys.

3 - SERCEL GPS EQUIPMENT

This first generation of equipment includes 3 main units :
- The NRT which is especially intended for frequency-time measurements and synchronization.
- The NR52 which is a pure GPS raw data sensor (pseudoranges and carrier phases), used in geodesy and kinematic surveys.
- The TR5S which includes very powerful means of calculation for real time positioning. All these units are essentially characterized by their capability to make reliable and high-accuracy measurements at a very high speed.

3-1   THE NRT GPS RECEIVER IS USEFUL FOR FREQUENCY-TIME APPLICATION

Though the topics of this meeting are essentially geodesy and land-survey, it is worth pointing out the existence of this GPS receiver mainly used for very accurate time measurement and also to mention its good performance. Indeed, it has been recently proved that, processing in a special way the time measurements in common view over distances of about 1000 kilometres, and using GPS receivers and equipment able to guarantee instrumental accuracies of about 1 nanosecond, can lead to relative positioning accuracies equal to those obtained by the carrier phase "double difference" method (between 10^{-7} and 10^{-6} of the baseline length) (References 7 and 8).

The quality of these results, which are likely to surprise some experts in geodesy, is due to the fact that, thanks to the use of high-stability frequency sources (hydrogen maser or cesium), it is possible to use all the data from every GPS satellite, even if the observations are not made simultaneously.

One drawback of this method is that it becomes accurate only after at least 48 hours of observation.

However, one should not underestimate the increasing interest and importance of the presence of high accuracy GPS time receivers in numerous scientific laboratories. Those receivers regularly and systematically observe all the GPS satellites and can become points of permanent observation allowing for the knowledge of propagation errors and orbital errors of GPS satellites.

The Sercel GPS NRT receiver is capable of measuring with great precision the time difference between a precise and stable local source and the GPS time from 1 to 4 satellites simultaneously, by means of 4 channels, which observe the C/A code and the carrier phase of the L1 frequency.

The absolute instrumental accuracy is better than 1 nanosecond, antenna included, and the relative precision between channels reaches 0.2 nanosecond RMS (reference 7).

Figure 1 shows the photography of an NRT receiver and figure 2 gives the typical distribution function of the errors between GPS satellites observed in common view over a distance of about 700 kilometres by NRT receivers.

3-2 THE NR52 GPS RECEIVER IS USEFUL FOR GEODESY AND LAND-SURVEY

The NR52 receiver is a GPS raw data sensor. It can observe pseudoranges from C/A code with a 10-centimetre resolution, and the L1 carrier phase with a 1.5-millimetre resolution. The receiver is equipped with 5 independent channels. Each channel is entirely dedicated to the observation of pseudoranges and phases of one satellite, and to the data extraction from its numerical messages.

Figure 3 shows NR52 during land-survey observation.

Data collecting on the field is processed by a portable microcomputer of the PC type coupled to the receiver, the whole being supplied by a 12-volt battery.

The microcomputer controls the receiver, organizes satellite acquisition by the receiver and gathers data which can be "compacted" over a time multiple of 0.6 s (usually from 10 to 30 seconds). It also can state autocalibration sequences at the beginning and the end of an observation session (accuracy : 5 centimetres for pseudoranges, 1 millimetre for phase). Durations of observation periods, chosen satellites and observation mode can be selected and programmed.

The position of the antenna phase centre is known with an accuracy of 2 millimetres along the three axes. This accuracy is maintained for every antenna used, for every value of azimuth and for every site higher than 5°. This antenna, specially designed by SERCEL, is particularly well protected against close reflections, thanks to the use of absorbent materials and protection screens.

The NR52 receiver can be used indifferently for GPS raw data observation or collecting at fixed points as well as on board the mobiles even if those are moving at high speed. It can therefore be very useful for land-survey or geodesy, as well as for very accurate determination of the trajectory of mobiles, land vehicles, ships or planes.

Raw data from NR52 receiver are most often recorded to be post processed. However some users want an immediate processing of these raw data in real time, more particularly for applications on board the mobiles.

3-3 TR5S GPS RECEIVER IS USEFUL FOR LAND-SURVEY, GEODESY AND VERY-ACCURATE REAL-TIME NAVIGATION

TR5S receiver includes the same elements as an NR52. It has the same characteristics and performances. It can also collect raw data. However this receiver includes an additional high-performance processor allowing high-speed computation of accurate and independent position solutions every 0.6 second, for which both the C/A accurate pseudorange data and L1 carrier phase are used. It also exhibits very comfortable display and dialogue means, useful in very accurate navigation (refer to figure 4).
This receiver can therefore be used to record GPS raw data of pseudoranges and phase for further processing but also, and simultaneously, to compute three-dimensional positions in real time, the whole at a 0.6-second rate.
This receiver is also equipped with hardware and software which allow efficient operation in 3D or 2D mode, by using an internal frequency source or an external Cesium clock. The above mentioned modes can be operated either in direct GPS or DIFFERENTIAL mode in real time, using very accurate correction data from the differential correction receiver developed by Sercel (correction resolution equal to 1 centimetre).

4 - METHODS AND SOFTWARES USEFUL FOR GEODESY AND STATIC SURVEY

The usual land-survey or geodetic measurements operated by Sercel GPS receivers are most often performed with 2 or more NR52 units.
The receivers are used simultaneously at several points, the relative position of which must be determined, either in an ECEF frame associated with WGS84 or in a local frame.
Each reception unit records GPS raw data (pseudoranges and carrier phases) usually compacted over a 15-to 30-second period. This data collecting lasts from a few minutes to a few tens of minutes, depending on the aims of the survey and the method used.

4-1 DATA COMPACTING

The NR52 receivers are devised to process pseudorange and carrier phase measurements every 0.6 second with the accuracies and resolutions hereabove mentioned (para. 3).
When observations are made at a fixed point, and in order to reduce the amount of data to be recorded and processed, the NR52 receiver first proceeds with a compacting sequence which consists in smoothing the pseudoranges over a ΔT time multiple of 0.6 second, by comparing their evolution to the carrier phase's, the carrier phase being also sampled.
This operation is made simultaneously for the five channels in a receiver and the moments that characterize these observation events in the various receivers are synchronized according to an entirely automatic procedure which refers to the GPS time observed by the receivers.
The value of ΔT is arbitrarily chosen between 0.6 second and a few minutes when defining the mission.
This procedure allows the system to get eassly the whole data from at least 5 satellites. The data is received and recorded in a perfectly synchronous way in all the receivers. The most commonly used value of

ΔT is 15 seconds, which leads to a data recording density of about 1kbit/minute per receiver.

4-2   UTILIZED METHODS

The data from NR52 receiver are postprocessed according to the following two methods :

4-2-1 Method providing position solutions from simple differences on pseudorange innovations

Used alone, this method is interesting in land-survey, for which a precision of a few decimetres is generally sufficient.
The first step in this method consits in roughly determining the position of each receiver within 5 to 10 metres, using the raw data from GPS. This solution makes it possible then to linearize the calculations.
The second step consists in calculating the mean value, for the overall time T of the session, of the elements of the three-dimensional position vector. This vector is issued by the least-square solution, which is performed on the simple difference innovations of pseudoranges smoothed by the phases observed in the two receivers.

$$
\left| X_{(3D + T)} \right| = \frac{1}{n} \sum_{n=0}^{n = T/\Delta T} \left(A.A^T \right)^{-1}.A^T. \left| \left(PD_{i,l} - D_{i,l} \right)_{\Delta T} - \left(PD_{i,k} - D_{i,k} \right)_{\Delta T} \right|
$$

in which :   ΔT is both the compacting period and rank
A is the geometry-time matrix
PD is the pseudorange observed in the receivers
D is the distance calculated from the estimated position
i is the satellite rank
l, k... is the receiver rank.

This method exhibits the following advantages :
- Observation period reduced to less than 5 minutes in open areas, between 15 and 20 minutes in more encumbered areas.
- Precision of a few decimetres, (in the 1988 GPS constellation) almost independent of the distance up to 500 to 800 kilometres.
- Evolution toward a significantly better precision expected by the use of all the satellites in the final GPS constellation.

However, even in cases where it is possible to lengthen the observation period up to about one hour, the present method will never offer the precision reached by the phase carrier double-difference method, especially for short distances (less than 100 kilometres).
The method of simple differences operated on pseudorange innovations is generally used as a preliminary calculation prior to proceeding with the carrier phase double-difference method. This preliminary calculation allows the system to have a better knowledge of the initial position prior to taking the next step.

4-2-2 Method of carrier-phase double differences

This method is the most currently used when the best precision is needed, for either short or medium distances, and when it is possible to allow for sufficiently long observation periods.

This method consists in first, determining an approximate position, as accurate as possible, using the method previously described in para. 4-2-1.

Using this approximate position, the system is capable of calculating the distance to each satellite for a given time (ΔT) and throughout the observation period (T). Thoses distances make it possible to calculate the "theoretical" phases and their variations versus time.

The theoretical phases are subtracted from the carrier phase values, the latter being those actually observed for each satellite and each used receiver. The subtraction is performed at the beginning of the observation period, assuming an arbitrary integer value to the resulting phase residuals.

The position is refined through a least square operation, using the geometry matrix and determining the values of Δx, Δy, Δz with respect to the approximate value. The Δx, Δy, Δz values reduce the residuals of the double differences to a minimum :

$$\Delta\, ik, jl_{\Delta t} = \left(\Phi_{ik} - \Phi_{jk} \right)_{\Delta t} - \left(\Phi_{il} - \Phi_{jl} \right)_{\Delta t}$$

with i, j ...   satellite rank
and k, l ...   receiver rank

Depending on both the baseline between the two receivers and the quality of measurements and ephemeris parameters, it can be decided whether the phase integers for each difference can be determined or not. The system can then proceed to successive approaches, resuming the processing repeatedly, each time using the newly improved position and more advantageously, being acknowledged of the integers that could not be fixed in the previous calculations.

Thanks to this method and using a single GPS frequency received by well installed and well calibrated equipment, the phase integers can be determined successfully in a very wide proportion of cases, even for distances of a few hundreds of kilometres.

The method of phase double differences offers the following advantages :
- The accuracy for short distances (less than 100 kilometres) is currently the best : a few millimetres plus a few 10^{-7} of the distance value.
- A slight improvement can be expected when the final constellation of GPS satellites is available. However, this improvement is not so significant as afforded by the method of simple differences on the pseudoranges.

On the other hand, the method of phase double differences has two drawbacks :
- The observation time period cannot be less than 40 minutes or else, the interesting features of the method are somewhat degraded.
- There are numerous calculations ; therefore, the processing is somewhat long. However and thanks to both new methods and recent processing equipment, very short processing times are now obtained (about 1 minute).

4-3   POSTPROCESSING SOFTWARES FOR LAND-SURVEY AND GEODESY

SERCEL developed a unique postprocessing software permitting both methods described in para. 4-2 to be implemented.
This software, which is called "MISSION" can be installed and operated on microcomputers of the PC-compatible type. The main features of MISSION are :
- Processing and file managers, which considerably reduce the processing time. As an example, the time needed to find the solution of a 60-minute data base, using the double-difference processing and phase integer fixing, is less than 1 minute (Equipment used : Portable PC T3200). The time gain is very much appreciated, especially for the ultimate solution in a multisite network.
- Capability of processing multisite networks (up to 9 sites).
- Capability of accepting several satellite masks during a session (up to nine temporary signal losses allowed per satellite in a given session).
The GPS-MISSION software package is composed of five modules :

4-3-1 "MANAGER"

That program allows a user to define a mission and to assign a name to that mission. Then, the only thing required for all processings to be executed is that mission name (no file names to be handled).
MISSIONS and SITES are stored into two data bases that the user can modify.
A few terms have been adopted, with the following meanings :
MISSION :   a crew moving to several sites in order to record one or more sessions.
SITE   :   record location.
SESSION :   period of time with constant satellite configuration.
Defining a mission results in creating a data block that will be shared by all program modules, which avoids entering the same data repeatedly.
The final function of that program is to collect the data recorded at the various sites, updating its own almanac with the help of the almanacs acquired.

4-3-2   "FIELD"

The FIELD program - recorded on floppy disk by MANAGER - controls a receiver in accordance with the Mission definition file.
It decodes GPS data and stores them into specific files.
It makes it possible, at the end of a record campaign, to compute a pseudorange fix for each session(straight GPS point position).

4-3-3   "PROCESSOR"

The PROCESSOR program processes the data recorded by FIELD whose files have been collected by MANAGER (processing on a session-by-session basis).
First it selects an ephemeris and an Iono correction model - out of all those recorded - to be used for all sites.
Then, for all sites, it computes a pseudo-range solution (position and clock model), saving satellite positions, geometric parameters as

well as propagation corrections - Iono model : Klobuchar ; Tropo model : Hopfield-Black.

After selecting a holdfixed site, the program builds the phase double-differences for sites 1-2, 2-3, ... and for satellites 1-2, 2-3, ...

Then it computes a solution for non-holdfixed sites and presents the double-difference residuals.

On a request from the operator the program then computes a second solution with holdfixed phase integer cycles.

The program also provides a variance covariance matrix of the solution allowing the operator to state on the quality of each session and solution.

4-3-4 "TRANSFORMER"

Because the programs described above only accept site positions expressed with Latitude-Longitude coordinates in the WGS84 system, the TRANSFORMER program allows conversion from one datum to another, with or without projection (UTM, conic), and presents coordinates in the following formats : L-G-H, X-Y-Z, (ECEF), N-E-H (Proj.).

That program includes two data bases (Datum values and Projections) in which the operator can change, delete or add some data.

4-3-5 "PREDICTOR"

The PREDICTOR program is used to predict satellite availability periods. With the help of the updated almanac, for a given user position and date, that program calculates the positions of satellites and presents the following graphs : Elevation, Azimuth, GDOP (versus time).

That program may also be used to edit the almanac.

5 RESULTS OF STATIC LAND-SURVEYS

It is difficult, in a short paper as this one, to provide all the results we obtained in the campaigns we conducted.

As some reports about those results already exist (see ref. 1, 2, 4, 10), only the most recent results will be covered in this paper.

5-1 RESULTS OBTAINED WITH THE SIMPLE-DIFFERENCE METHOD ON PSEUDORANGES

A campaign was conducted in October 1987 in view to achieve accurate time transfers between the following three laboratories : OBSERVATOIRE DE PARIS, CNES in TOULOUSE (670 km apart) and SERCEL, NANTES (380 km apart). To accurately determine the relative positions of the three sites, we used the method of simple differences on pseudoranges, aided by simultaneous observations of only four satellites during sessions of 13 minutes each.

The satellites were tracked according to two configurations during three sessions over three consecutive days (October, 24, 25, 26th, 1987). The conditions were as indicated below.

SV PRN =	ALT DOP.	NORTH DOP.	EAST DOP.
6-9-11-12 at 2H30min UT	4	2.8	1.8
3-9-11-12 at 4H30min UT	3	1.6	2.6
3-9-11-13 at 5H15min UT	2.8	1.1	2.3

For the nine conducted sessions, the uncertainties upon the relative positions of the three sites were the following :

SITES	DISTANCES	σ ALT.	σ NORTH	σ EAST
NANTES-PARIS	340 km	18 cm	20 cm	25 cm
NANTES-TOULOUSE	630 km	24 cm	25 cm	34 cm

So, for each individual 13-min session, the mean standard deviation for the observed baselines is 0.4 metre, which corresponds, in average, to about 10^{-6} of the baseline length.
This method, which requires the use of several GPS receivers, is commonly used in the United States for fast land-surveys, using several NR52.

5-2   RESULTS OBTAINED WITH THE CARRIER PHASE DOUBLE DIFFERENCE METHOD

Two recent campaigns can be mentioned :

5-2-1 SPAIN campaign

The interest of this campaign was that it allowed the comparison on the same basis, of measurement results issued - at an interval of a few days - by GPS receivers from four different manufacturers.
This experiment was conducted at MADRIDEJOS with the SPANISH NATIONAL INSTITUTE OF GEOGRAPHY, 120 km south of MADRID.
The results are summarized in the table below :

MANUFACTURER	GPS MEASURED DISTANCES	DISCREPANCIES VERSUS AVERAGE	RELATIVE DISCREPANCIES/ AVERAGE
TRIMBLE	14 664.623	- 11.5 mm	- 8 × 10^{-7}
MACROMETER	14 664.629	- 5.5 mm	- 4 × 10^{-7}
SERCEL NR52	14 664.636	+ 1.5 mm	+ 1 × 10^{-7}
MAGNAVOX WM101	14 664.650	+ 15.5 mm	+ 11 × 10^{-7}
AVERAGE OF GPS MEASUREMENTS :	14 664.6345		
STANDARD DEVIATION :	11.62 mm (7.9 × 10^{-7})		

This campaign shows that first, the total uncertainty on the GPS measurements is slightly better than 10^{-6} whatever the equipment used, the configuration of satellites available and the day when the survey takes place, and second, the results obtained by at least two manufacturers (MACROMETER and SERCEL) correspond to each other and to the mean value of the GPS measurements within an uncertainty better that $\pm 2 \times 10^{-7}$.

The SPANISH NATIONAL INSTITUTE of GEOGRAPHY carried out a measurement with a RANGE MASTER 3 laser system which found a deviation of 8 millimetres with respect to the GPS ones, that is about 5×10^{-7}.

Except through uncountless measurements, it must be very tricky to state whether the optical measurement or the GPS measurement is closer to the exact distance (which is about 15 kilometres).

5-2-2 Campaign in WEST of FRANCE (february, 1988)

A much more thorough campaign, consisting of 66 sessions observed on 9 different baselines, was conducted in France between january and february, 1988. In this survey, the associated teams of SERCEL and of the INSTITUT GEOGRAPHIQUE NATIONAL implemented three SERCEL NR52 receivers and a DISTOMAT DI2000 manufactured by SERCEL-WILD.

Nine baselines, ranging from 100 metres to 300 kilometres were surveyed. Every day during the campaign, two 96-minute sessions corresponding to the following configurations were executed :

SESSION 1	PRN SV 6-8-9-11-12 from 20H30 min to about 22H
SESSION 2	PRN SV 3-9-11-12-13 from 22H30 min to about 0H

Only the three shortest baselines were calibrated through the optical equipment.

The table below allows a comparison between the results obtained by the GPS and the optical equipment's :

DATE	BASELINE	AVERAGE GPS DISTANCES (m)	GPS σ (mm)	OPTICAL DISTANCE (m)	OPT. σ (mm)	GPS - OPTIC. DISCREPANCY (mm)	NUMBER OF GPS SES-SIONS
21 to 24/01	SERCEL : M001 - M002	98.8876	3.7	98.8873	0.5	+ 0.3	8
28 to 31/01	SERCEL - AGFA	1180.438	4.4	1180.4385	2.8	- 0.5	6
28 to 31/01	SERCEL - MATRA	2728.9918	1.7	2728.991	0.8	+ 0.8	6

So, in this experiment, the results provided by both optic E.D.M. and GPS are similar to within one millimetre. This is partly due to the fact that the GPS result proceeds from the mean value calculated through 6 to 8 sessions over several successive days. This tends to prove that the random uncertainty (σ = 2 to 4 mm) for a session is larger than the effect of the offsets, which accounts for the performance gain due to the mean calculation. At this level of precision though, there is probably a part of chance too. That is why such experiments are worth renewing.

Other optical data for distances between 3 and 15 kilometres are still being postprocessed by the I.G.N.

As a complement, and for all the baselines, we can analyze the uncertainties between the GPS measurements themselves and the repeatability versus time using the table below.

DATE	BASELINE	AVERAGE GPS DISTANCES (m)	σ GPS (mm)	No. of sessions	Relative uncertainty per session	Baseline relative uncertainty
21 to 24/01	SERCEL : M001 M002	98.8876	3.7	8	3.7×10^{-5}	1.3×10^{-5}
21 to 24/01	SERCEL : M001 M003	167.3525	1.5	8	8.9×10^{-6}	3.1×10^{-6}
28 to 31/01	SERCEL - AGFA	1180.438	4.4	6	3.7×10^{-6}	1.5×10^{-6}
28 to 31/01	SERCEL - MATRA	2728.9918	1.7	6	6.2×10^{-7}	2.5×10^{-7}
02 to 05/02	SERCEL - CASSON	14149.9265	17	4	1.2×10^{-6}	6×10^{-7}
09 to 10/02	SERCEL - CASSON	14149.928	8.5	2	7×10^{-7}	4×10^{-7}
15 to 21/02	SERCEL - CASSON	14149.9216	6.1	7	4.3×10^{-7}	1.6×10^{-7}
02 to 05/02	SERCEL - LES TOUCHES	19666.4037	11	4	5.6×10^{-7}	2.8×10^{-7}
15 to 21/02	SERCEL - LES TOUCHES	19666.4108	11.5	13	5.8×10^{-7}	1.6×10^{-7}
09 to 12/02	CASSON - LA PLAINE	56840.4746	29.2	3	5.1×10^{-7}	2.9×10^{-7}
15 to 21/02	SERCEL - ST MANDE	340369.1886	158	5	4.6×10^{-7}	2.1×10^{-7}

In this table, we note that the uncertainties per session, which are a few millimetres for short distances, rise to 1 to 3 centimetres for distances between 20 and 50 kilometres, and up to 15 centimetres for a distance of 340 kilometres.

Concerning all the results in that campaign, the uncertainty for each 96-minute session of observation can be depicted by the following model :

$$\sigma_S = 3 \text{ mm} + (5 \times 10^{-7} \times D)$$

Taking into account the number of observations, we can appreciate the probability of failing to set the integers of the phase values in the double difference solutions, versus the distance. The result is given in the table below.

DISTANCE	0 to 5 km	15 to 20 km	60 km	350 km
Probability of not fixing integer	≪ 1 %	2 %	8%	25 %

This table shows the interest of having redundant data (number of tracked satellites, number of sessions).
As shown in figure 5 for a 236-km distance, the MISSION software package allows plotting of the residuals of the phase double-differences. In this figure, the mean quadratic value of the residuals varies between 3 and 6 centimetres.

6 - METHODS AND SOFTWARES USED FOR ACCURATE RECONSTITUTING OF MOBILE'S TRAJECTORIES BY POSTPROCESSING - RESULTS OF KINEMATIC SURVEYS

SERCEL developed unique softwares especially intended for kinematic surveys. These now operational softwares can be used to accurately reconstitute the trajectory of land vehicles as well as ships or planes (see ref.3).
For such surveys, the TR5S and NR52 receivers can be used indifferently to collect raw data of pseudoranges and carrier phase every 0.6 second, simultaneously from five satellites, in view to postprocess them.
The studies made by SERCEL showed that, in most applications, there is a possibility of taking advantage of a continuous observation of the carrier phase from the satellites : by refining the "pure phase" trajectories, free from any discontinuities and only slightly affected by noise, onto the "pseudorange" trajectories, which are unambiguous although very noisy, we can obtain elements of profiles, or partial trajectories with great accuracy.
For this purpose, redundant measurements in 3D + T mode are necessary. The receivers must then be fitted with five reception channels minimum ; they must feature a high sensitivity, a high precision in determining the phase (1 millimetre) and the pseudoranges, the capability of acquiring measurements at a rate of 0.6 second, the capability of tracking signals with very small S/N ratios, the capability of very fast re-locking (from a few tenths of a second to a few seconds) after possible yet temporary mask of a satellite. The TR5S and NR52 receivers exhibit all these capabilities.
The already known applications are : fast land-surveys of roads and railways, calibration of sea-borne radionavigation systems, calibration of aid systems to plane landing, calibration of flight trials and, in fact, all the applications for which the successive positions of a moving vehicle have to be accurately known.
Most of these applications need a precision from a few decimetres to 1 metre. To reach these precisions, differential postprocessing software have been devised and tested.
The theory of operation consists of collecting the pseudorange and carrier phase data, at a high sampling rate and in synchronous operation, in both the mobile receiver and the reference receiver,

which is located at a site whose position, expressed in the WGS84, is known with an accuracy of a few centimetres.

The collected GPS data are dated with great care. If needed, the data acquisition is synchronized - with an accuracy better than 1 millisecond - with other means also recording this data (video camera, magnetometer, etc.).

The synchronism between the GPS data collected by the different GPS receivers is thoroughly ensured ; which is particularly easy to achieve with the TR5S and NR52 since all the measurements made by these receivers are referenced to the transmission time of the satellites.

Data collected by both the mobile and the reference receiver are then postprocessed through a special software, which accurately reconstitutes the trajectory in differential mode. For this purpose, the software processes the pseudorange and carrier phase data in the best way. A special procedure allows the processing to detect and compensate for most of the possible cycle jumps of the phase. In some cases where the reception of one or more satellites is interrupted, (land vehicle under a bridge, screening by a wing of the plane during a turn), the processing delimits the continous sections of the trajectory and optimizes their connection.

6-1   POSTPROCESSING SOFTWARE FOR KINEMATIC SURVEYS

The software operating according to the above method fullfils a number of functions executed in succession. These functions are :
- Establishing differential corrections by means of pseudoranges and carrier phases observed at the fixed reference site. They consist of residuals from a mean solution of local time.
- Processing the pseudoranges and carrier phases observed at the mobile in two successive steps :
  . In the first step, the system determines - separately for each satellite - the sections in the trajectory where no reception discontinuities were detected. For each of these sections associated with a given satellite, the software refines the evolution of the carrier phase according to the corresponding evolution of the pseudoranges. This refining applies for the whole section concerned.
  . In the second step, the system determines the trajectory according to the series of the X, Y, Z values. Those values are calculated by means of the pure carrier phase values whose integers have been determined as accurately as possible thanks to the previous step.
- The software then proceeds with the analysis of the continuity of the carrier phases for each satellite, in order to detect then correct the possible cycle slips of the carrier phase.

  In a redundant solution providing a 3D + T position from the reception of at least five satellites, it is always possible to re-establish the phase continuity of a satellite for which one phase slip only was detected.

  If two or more discontinuities occur at the same time, it is difficult to reliably cancel the doubt about the phase if the periods of reception loss are greater than a few seconds.
- The software then verifies the measurement dating using the transmission time of the satellites as the reference.
- The software incorporates calculation particulars which ensure satisfactory processing of the data observed in 3D mode with a

reduced number of satellites thanks to the possible use of
additional high-stability frequency standards (cesium).
- Finally the program includes a module which optimizes the
trajectory by refining all the observed sections.

6.2 RESULTS OF KINEMATIC SURVEYS BY POSTPROCESSING

Experiments (both experimental and real) have been carried out to
determine the trajectory of land vehicles (trucks,trains), ships and
aircrafts . (Ref. 5, 6, 9).

In all the cases where it was possible to compare the trajectories
determined by GPS with those obtained by other surveying means, the
observed deviations were on the order of a few decimetres, whatever
the dynamic characteristics of the mobile moves.
- LAND VEHICLES
  Campaigns have been carried out on trucks and cars, notably in
  France ,in order to accurately verify the profile of some roads.
  Figure 6 shows a partial example of the results.
  In that type of survey, the coherence between the GPS measurements
  and the known land marks is always within 15 to 30 centimetres.
  The deviations on the vertical coordinates are more important,
  most often due to the difference between the map and the practical
  building of the road.
  A survey conducted in the United States,on the BURLINGTON NORTHERN
  company's behalf, provided the profile of the railway using NR52
  and TR5S receivers on board trains moving at more than 70 mph
  (Ref. 5). Compared with a few points calibrated by the NGS, the
  quality of the connection revealed an accuracy better than 50
  centimetres.
- SHIPS.
  Numerous measurements were gathered on sea-borne mobiles. Figure 7
  shows an example of altitude measurements using the GPS compared
  with high-accuracy optical measurements. The ship heave is also
  determined with an absolute mean accuracy of about 10 to 20
  centimetres. As shown in figure 7, the short-term uncertainty is
  excellent: standard deviation of about 2 centimetres.
- AIRCRAFTS.
  In May and September 1987, two important measurement campaigns
  were carried out in the in-flight trial centre (C.E.V.) of
  Bretigny near Paris (Ref. 6). The purpose of those campaigns was
  to demonstrate the quality of the trajectory determined by means
  of the differential GPS on board an aircraft flying at a speed of
  300 knots approximately.
  The measurements were made with two TR5S GPS receivers, one
  installed on board the plane, the other at a ground reference
  point. Measurements were carried out for numerous flying
  conditions, notably in view to simulate the approach of a "HERMES"
  space shuttle. Figure 9 shows an example of a passage during a
  level flight at about 1000 feet.
  The three lower curves show the X ,Y and Z deviations between the
  GPS-determined trajectory and the one obtained with the laser -
  based STRADA tracking system. The x-axis of the curves corresponds
  to the distance between the aircraft and the STRADA. Note the
  important noise of the horizontal tracking during the laser's
  locking step (distance between 6500 and 5800 metres). Outside this
  period, the observed mean and standard deviations are the
  following:

	Latitude	Longitude	Altitude
Mean offset	+13 cm	-43 cm	+52 cm
Standard deviation	28 cm	33 cm	9 cm

A detailed analysis of the quality of speeds Vx, Vy, Vz shows that a major part of the observed noise is due to the laser tracking. A spectral analysis of the noise observed on the speed resulting from the GPS measurements provides the following standard deviations :

$$\text{for } T=1 \text{ second} \qquad \sigma = 3 \text{ mm/s}$$
$$\text{for } T=10 \text{ seconds} \qquad \sigma = 1 \text{ mm/s}$$

All those results were obtained with unfiltered GPS measurements, the only filtering being the recording rate in the receivers (0.6 s).

7 - REAL-TIME KINEMATIC SURVEYS USING THE CARRIER PHASE

It is possible to take advantage of both the high sensitivity and the low noise of the GPS carrier phase measurements for a fast in-the-field determination of the positions of a set of points located close to a known position. This is the typical case of the laying out of marks close to an already calibrated reference point. Such an operation can be carried out with a centimetric accuracy provided that the following two conditions are fulfilled :

a) The surrounding of the moving receiver antenna should be clear from any obstacle that could deteriorate the phase continuity of the observed GPS signals. To avoid glaring errors, the phase continuity should be monitored with adequate means.

b) If this operation is carried out starting from a direct observation of the GPS ,then its duration should be as short as possible. As a matter of fact, the position variations associated either with changes in the propagation delays or with errors in relation to the satellite positions may reach a few centimetres per minute. Moreover, this method may be totally prohibited if the intentional degradation of the GPS comes into effect.

On the other hand, if a real-time differential method applied to the pure carrier phases is used, all the drawbacks described in b) are overcome and a centimetric accuracy can be ensured as long as the phase continuity of the measurements is also ensured. This remains true in a radius of several tens of kilometres around the reference point. In a specific configuration of some input parameters of the TR5S receiver,it can be contemplated performing high-accuracy dynamic layouts in real time as described above. However the future development of such methods is linked with the design of very small and easy-to-operate units. It is worth noting that the use of the GPS for land-surveys is just starting and the size reduction of the GPS receivers dedicated to that type of work will probably be comparable to the E.D.M.'s in the past. As an indication, within less than twelve years, SERCEL reduced the volume of DISTOMATs by a factor of 25 (between DI10 and DI4), this, on the basis of tens of thousands of units manufactured.

8 - REAL-TIME KINEMATIC SURVEYS USING PSEUDORANGES AND CARRIER PHASE IN DIFFERENTIAL MODE.

SERCEL has developed specific equipment for both transmitting and receiving the GPS corrections. In real-time operation, such equipment associated with TR5S-type GPS receivers, can offer an accuracy nearly equal to that obtained with the postprocessing methods described in para.6 (a few decimetres).
This equipment, mainly developed for real-time high-accuracy navigation can also be used for real-time kinematic surveys. Figure 10 shows a GPS differential correction transmitter currently under test on the west coast of FRANCE. This station can transmit the corrections relating to up to 8 GPS satellites, at the high rate of 2 to 6 seconds and over a maximum distance of about 1500 km.

9- CONCLUSIONS

Two methods of static land-survey based on the differential GPS principle have been tested using both accurate GPS equipment (TR5S and NR52) and dedicated softwares developed by SERCEL.
The so-called "simple-difference" method applied to pseudoranges offers an accuracy of about a few decimetres in a few minutes and for distances as long as several hundreds of kilometres.
When thoroughly implemented, especially as for the antennae, the so-called "double-difference" method applied on the L1 carrier phase demonstrates its capability to obtain millimetric accuracy for short distances and centimetric accuracy for distances up to several hundreds of kilometres.The SERCEL PC-installed software can provide a baseline solution in less than one minute of computation time.
The reconstituting of mobile's trajectories, even for fast moving units (300 knots on aircrafts) features an accuracy of about 10 to 30 centimetres when compared against the trajectory from an optical means . This was obtained thanks to the use of both suitable methods and software.
Developments and trials concerning real-time high-accuracy navigation and real-time topographic layout are under way.
In particular, SERCEL is now testing a high-rate (2 to 6 s) long-range (up to 1500 km) GPS correction transmitting station.
Both new methods and new (size-reduced) equipment used for land-survey and fast layout may be expected when the GPS system reaches its final configuration.

References :

1 - G. BEUTLER (1986) "Comparison between TERRAMETER and GPS
    results and how to get there". CERN accelerator school
    applied geodesy for particules accelerators. CERN GENEVA
    SWISS.

2 - C. BOUCHER - G. NARD (1985) "Capabilities of the TR5S SERCEL GPS
    receiver for precise positioning" 1st international
    symposium on precise positioning. ROCKVILLE MD.

3 - R. BROSSIER al. (1986) "Use of the TR5S GPS receiver in air borne
    photographic surveys" Internat. Society for photogrammetry
    and remote sensing". STUTTGART F.R.G.

4 - R. GOUNON (1986) "GPS differentiel résultats d'enregistrements
    simultanés à des distances de 400 à 1000 km" SERCEL
    RN 80-86.

5 - R. GOUNON (1987) "BURLINGTON NORTHERN RAILROAD. GPS
    trajectography results". SERCEL Report RN30-87.

6 - R. GOUNON (1987) "GPS trajectography on board a plane - Results
    from trials at Bretigny C.E.V. centre". SERCEL report
    RN 60-87.

7 - R. GOUNON - J.N. LELONG (1988) "GPS high precision time transfer
    methods and related equipment". 2nd EFTF. NEUCHATEL SWISS.

8 - B. GUINOT - W. LEWANDOWSKI (1988) : "Nanosecond time comparisons
    in Europe using GPS" 2nd EFTF. NEUCHATEL SWISS.

9 - G. NARD (1987) "Matériels et applications du GPS différentiel de
    haute précision, hybridation avec d'autres moyens".
    3ème Colloque national sur la localisation en mer.
    PARIS - RUEIL.

10 - P. WILLIS (1986) "The IGN Geodetic Software System for GPS data
    analysis". 4th symp on Geodesy - AUSTIN - TEX.

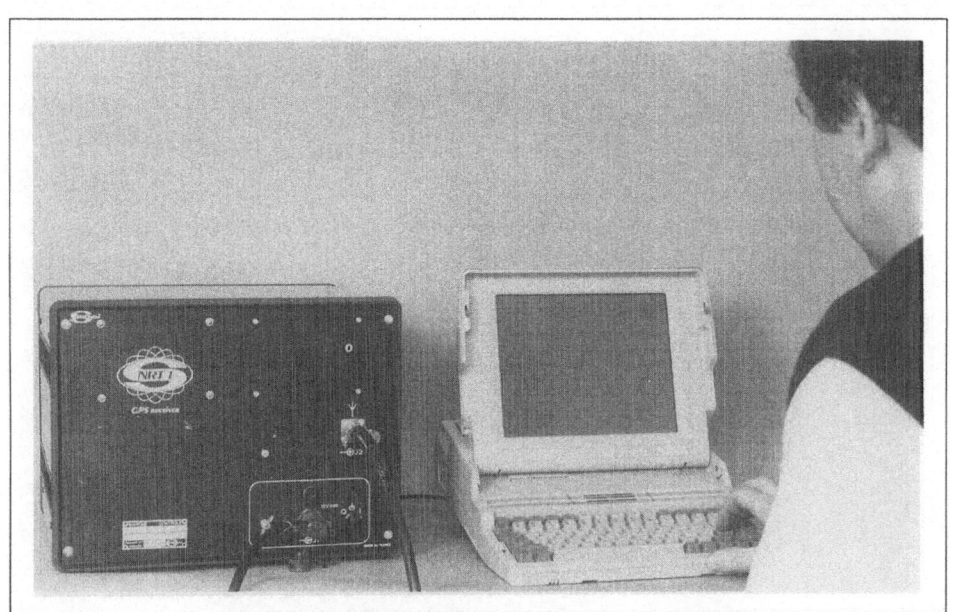

FIGURE 1 - NRT SERCEL GPS TIME RECEIVER
used for nanosecond level time measurements

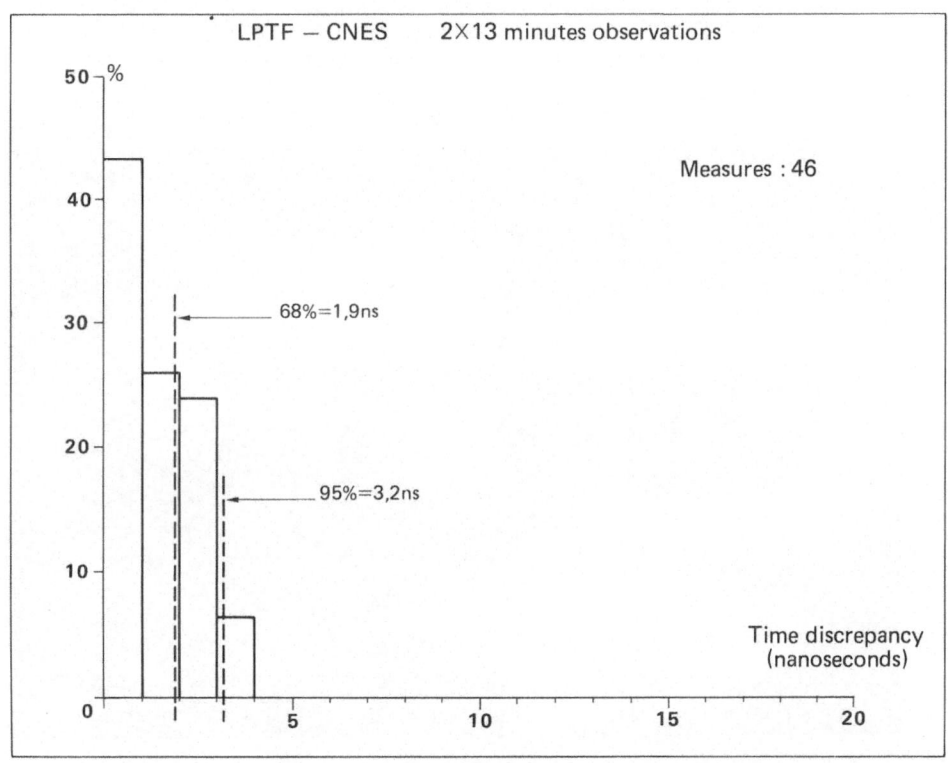

FIGURE 2 - Example of GPS time measurements discrepancies
(between simultaneous different satellites observations)
as measured with NRT SERCEL GPS RECEIVER

FIGURE 3 - NR 52 SERCEL GPS RECEIVER
used for geodesy and static or kinematic surveys

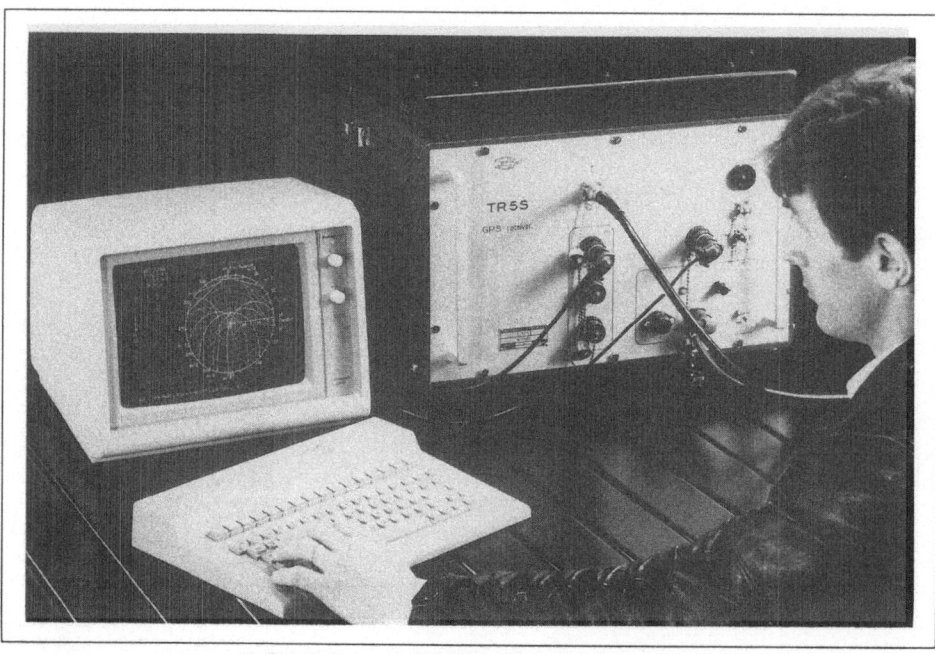

FIGURE 4 - TR5S SERCEL GPS RECEIVER
used for accurate real time navigation or surveys on board vehicles,
boats or airplanes

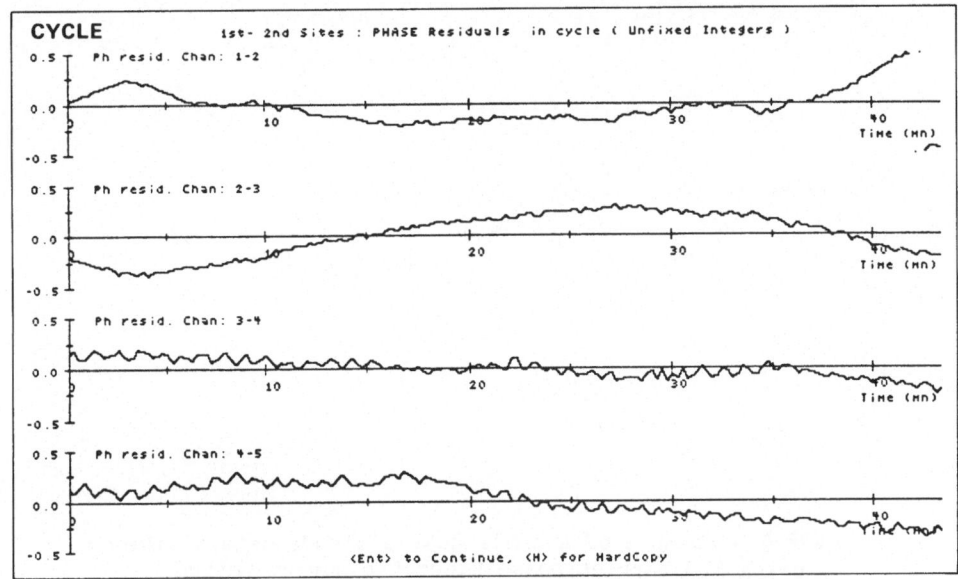

FIGURE 5 - Example of carrier phase double difference
residuals of a 5 satellites solution for a distance
between receivers of 236 kilometres
The RMS residual ranges from 3 to 6 centimetres

Pt Topo	Lat	Long	Alt	Distance
46	-0.61	-0.21	-2.05	2.15
47	-0.55	-0.20	-1.35	1.47
48	-0.54	-0.21	-1.10	1.24
49	-0.57	-0.21	-0.78	0.99
50	-0.47	-0.13	-0.55	0.73
51	-0.47	-0.10	-0.37	0.61
52	-0.39	-0.06	-0.27	0.48
53	-0.49	-0.05	-0.29	0.57
54	-0.47	-0.03	-0.29	0.55
55	-0.56	-0.02	-0.45	0.71
56	-0.33	-0.01	-0.55	0.65
57	-0.46	-0.01	-0.76	0.89
58	-0.71	-0.04	-1.05	1.26
59	-0.52	-0.03	-1.26	1.36
60	-0.64	-0.03	-1.51	1.64
61	-0.89	-0.01	-1.67	1.89
62	-0.52	0.02	-1.79	1.86
63	-0.38	0.03	-1.76	1.80
64	-0.49	0.07	-1.77	1.83
65	-0.16	0.03	-1.73	1.74
66	-0.29	0.08	-1.72	1.74
67	-0.20	0.06	-1.72	1.73
68	-0.41	0.13	-1.74	1.79
69	-0.76	0.15	-1.66	1.83
70	0.25	-0.00	-2.15	2.16
71	0.09	0.02	-2.34	2.34
72	0.04	0.02	-2.42	2.42
73	0.02	0.03	-2.18	2.18
74	0.21	0.09	-2.14	2.15
75	0.09	0.05	-2.11	2.11

MOY/ETY Lat Long Alt

Lat : -0.37 0.29
Long : -0.02 0.10
Alt..: -1.38 0.69
Mes : 30
        ↑   ↑
       Bias   σ

FIGURE 6 - Example of DIFFERENTIAL GPS KINEMATIC SURVEY
on board a truck running at 80 km/hour.
The comparison with land marks shows bias of 1 metre or so
(probably due to land marks altitude uncertainty).
The standard deviation is 0.3 to 0.6 metre. AVERAGE DISTANCE of GPS
reference = 45 kilometres - numerous obstructions

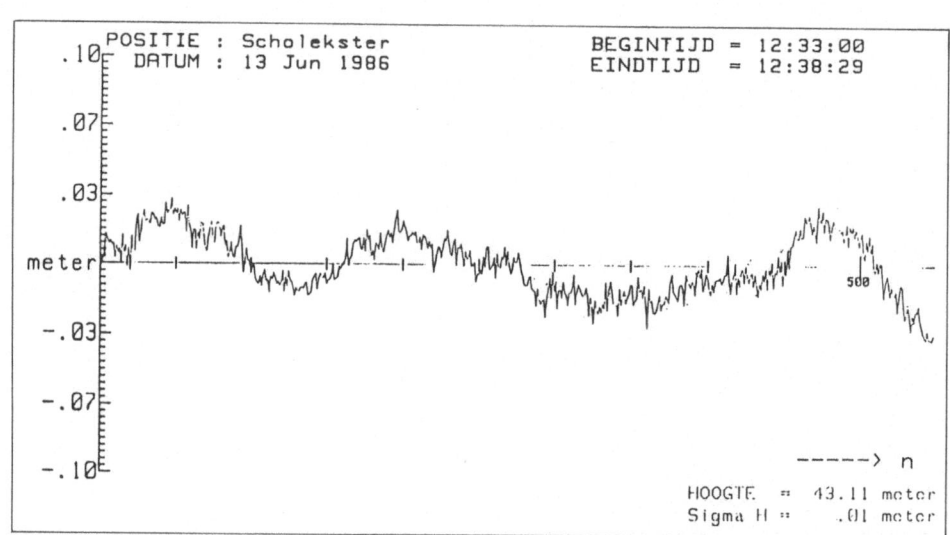

FIGURE 7 - Example of a small boat altitude determination
using GPS carrier phase kinematic survey method.
The comparison is made with EDM and angular infrared tracking.
Measurements made with 2 TR5S SERCEL GPS receivers by Dutch R.W.S.
The standard deviation is less than 2 centimetres.

FIGURE 8 - Example of GPS kinematic survey of a Caravelle plane flying
at 450 km/hour. The survey is made with 2 TR5S SERCEL RECEIVERS.
Atop the mast is the antenna of the TR5S reference receiver.
Comparison of positions is made against data from a laser tracking
system of C.E.V. BRETIGNY

298

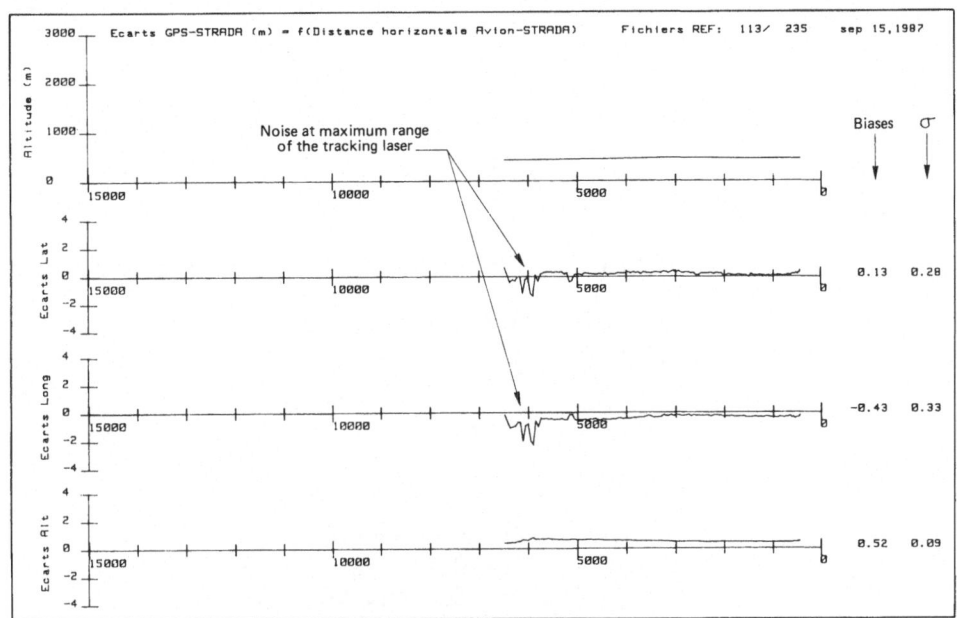

FIGURE 9 - Example of results of GPS kinematic survey
on board CARAVELLE (V = 450 kilometres/hour). T = 0.6 second
Bias comparison with laser tracking is less than 0.5 metre X-Y-Z
and standard deviations (without the noisy part coming from the laser
jitter at long range) is less than 10 centimetres

FIGURE 10 - DIFFERENTIAL GPS PSEUDORANGES AND CARRIER PHASE
CORRECTIONS TRANSMITTING STATION FROM SERCEL
CURRENTLY UNDER TRIALS IN FRANCE
This light weight station can be set up within 48 hours.
It allows for very accurate navigation or real time differential
kinematic surveys with reliable corrections transmissions
over more than 1500 kilometres.

DEVELOPMENTS FOR THE OPERATIONAL USE OF GPS

IN KINEMATIC MODES

by

Günter Seeber
Frank Heimberg
Andreas Schuchardt
Gerhard Wübbena

Abstract

This is a short version of the paper giving the essential remarks and findings. The use of carrier smoothed pseudoranges in relative mode is demonstrated with the operational software package GNAV/GNAVC.

Short Version

The potential of GPS for use in kinematic positioning has been discussed and demonstrated by various authors. With respect to marine applications some requirements and limitations are valid, such as

- generally it is not possible to start with static positioning on a known point

- antenna change techniques are not applicable

- relative informations have to be available in real time

- final data processing using all available information has to be performed in real time.

The accuracy requirements for marine applications of GPS vary between several tengths of meters and subdecimeter level. Main fields of applications are

- precise navigation

- hydrography near shore and in estuaries

- sea bottom mapping using swath sonar systems like SEABEAM or HYDROSWEEP

- marine geophysics using seismic and magnetic techniques

- marine geodynamics with respect to sea floor positioning and sea floor spreading.

GPS may be used in different modes. The easiest way is the stand alone receiver onboard providing $\pm$ 10 m on-line position accuracy with P-code availability and $\pm$ 10 - 30 $\overline{m}$ with C/A-code only. These numbers may change, once the system will be definitely available.

In order to model or to eliminate systematic effects such as orbital errors or ionospheric propagation influences, a relative mode may be used. Pseudorange corrections from one or more land-based reference stations have to be communicated to the marine user in order to improve the onboard position solution. With this concept, a relative position accuracy of 2-3 m can be achieved at sea, such being sufficiently accurate for most marine scientific purposes.

A further improvement is possible by the use of carrier phase measurements. Once the ambiguity problem is solved, this method can provide sub-decimeter accuracy in real-time. Provisions have to be made in order to avoid cycle slips or to recover the ambiguity solution within a short time, when cycle slips occur.

In all cases when highest accuracy is necessary for scientific purposes, the use of carrier phase observations at sea is essential. Phase observations are also useful in order to determine orientation and attitude of a marine vehicle by use of two or three antennas simultaneously.

With respect to carrier phase measurements the following procedures are possible in kinematic mode

(1) pure carrier phase integration

(2) combination of code and carrier phases

(3) carrier smoothed pseudoranges.

The first concept requires the solution of ambiguites before starting and/or the start on a known point. The solution can be improved by antenna changing techniques. Cycle slips however may introduce serious problems and force the user to go back to the starting point. The use of more than 4 satellites may help in recovering the lost cycles. In general, at the time being, this technique cannot be used in an operational application environment.

The second concept works well with a P-Code receiver. Through combination of code and carrier measurements the ambiguities can be resolved after less than 1 minute, thus making cycle slips a minor problem. With a forthcoming generation of low-noise receivers, the technique may be applicable also with C/A-Code. Furtheron the use of more than 4 satellites will help in the recovery of cycle slips.

For the time being, the third concept is the only one to be used in an operational and most general way.

At the Institut für Erdmessung, University Hannover, a software package for the operational positioning of moving platforms has been developed. The basic concept is shown in Fig.1. On a known reference station pseudorange corrections are computed with the correction program GNAVC. On the moving platform the "Geodetic Navigation Program" GNAV is used to provide best position estimates to the user.

The main features of GNAV and GNAVC are

- receiver independent (accepts any data in any sequence)

- accepts code measurement (P- and C/A-Code)
  carrier phases
  Doppler frequency

- relative mode possible with range corrections

- range corrections from reference station with parabolic representation

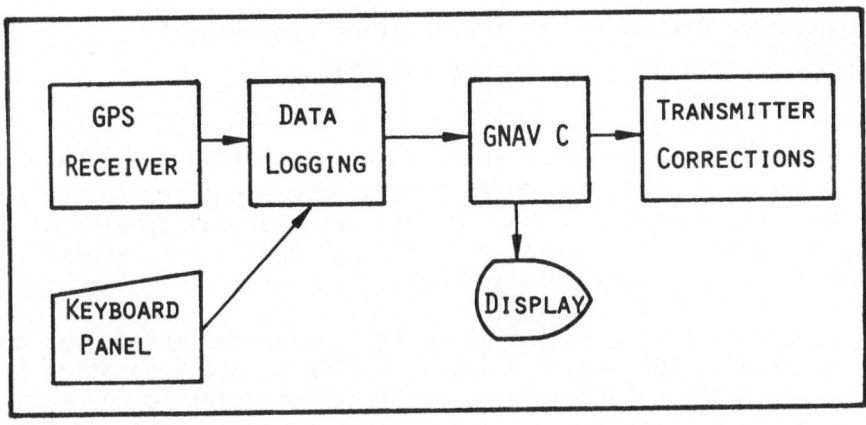

REFERENCE STATION

MOVING STATION

Fig. 1: Block concept of GNAV-GNAVC, Geodetic Navigation Software

- standard ionosphere and troposphere
- carrier smoothed pseudoranges
- two 4-parameter KALMAN filter for position and velocity
- fixed height possible, also with 4 satellites.

Figure 2 shows the potential of this concept, verified in a "Zero Baseline Experiment". Two TI4100 receivers were connected to the same antenna, thus the baseline being 0 m. One receiver was treated as a reference receiver, the other as the "moving" navigation receiver. The "real time positions" vary within 1 - 2 m.

Figure 3 demonstrates the potential of the navigation model. Two antennas were mounted on the research vessel METEOR with a separation of 36 m. Two independent GNAV solutions were run (without relative corrections). The distance coming from the two independent solutions differs less than 2 meters during a 4 hours period.

The full version of this contribution including further developments will be published in "Marine Geodesy Journal".

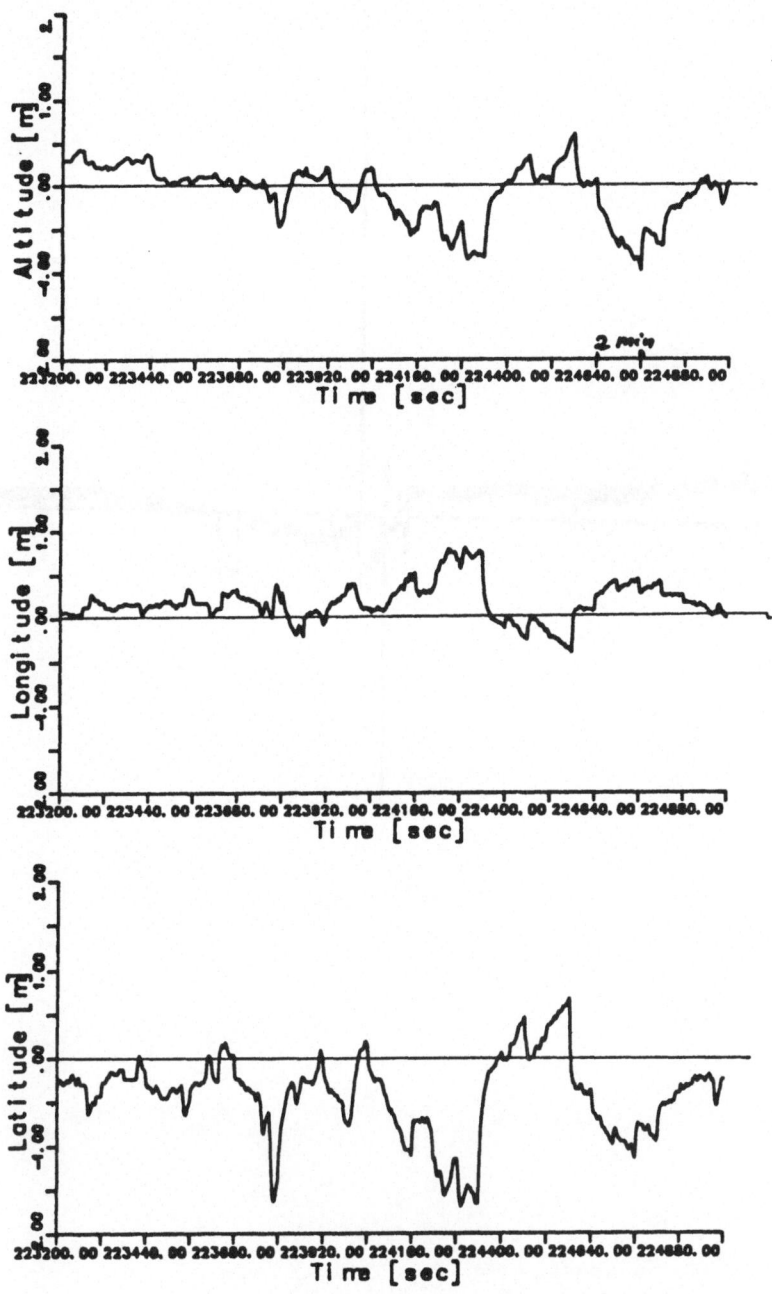

Fig. 2: Zero baseline experiment using GNAV and GNAVC

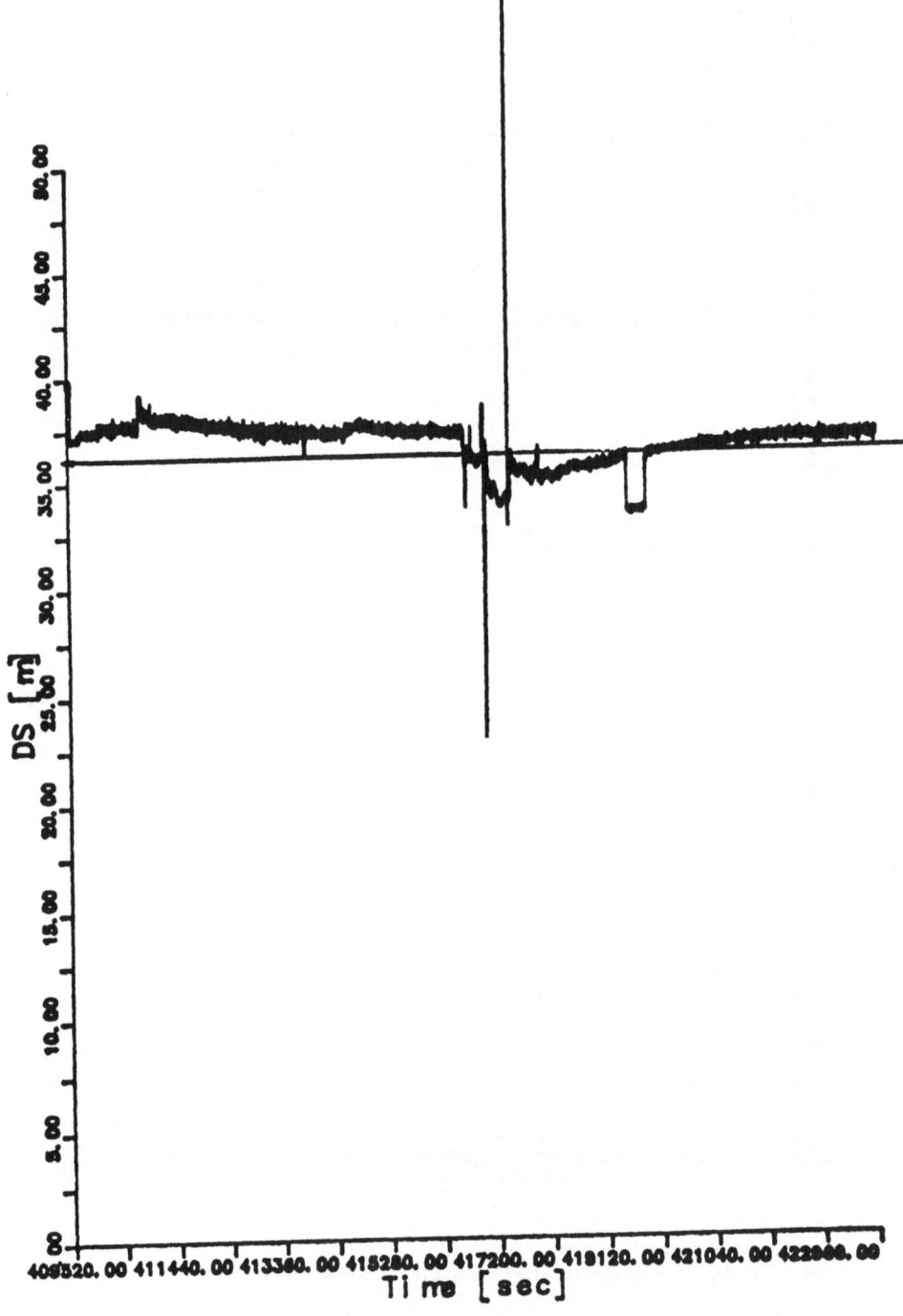

Fig. 3: Distance between two antennas on research vessel METEOR, from two independent GNAV solutions (correct value 36 m)

TERRESTRIAL AND AIRCRAFT DIFFERENTIAL KINEMATIC GPS POSITIONING

by

Günter W. Hein
Herbert Landau
Gerald Baustert

Abstract

During 1986-87 IAPG has carried out several differential kinematic GPS positioning tests on land using GPS carrier phases and pseudo-ranges, TI 4100 dual-frequency receivers and commercial TI 4100 Navigator as well as GESAR 32.1 software. In August 1987 the concept was used to determine the coordinates of a photogrammetric camera in a flying air-craft. The paper presents the results of the land and flight tests and analyzes it in various ways. The models and requirements for kinematic GPS surveying are discussed.

It is shown that accuracies in relative position of ± 1 - 2 cm in the kinematic mode (over a range of 10 km) are possible when using carrier phases and pseudo-ranges in two frequencies.

1. INTRODUCTION

Although the NAVSTAR/GLOBAL POSITIONING SYSTEM (GPS) satellite configuration is not yet fully completed, it seems that - from the research point of view - most problems in precise relative static 3d-positioning using carrier phase data are solved. Present efforts are carried out towards shorter observation times, more operational and user-friendly software, improved antenna design, combination with other data types to solve for cycle slips, orbit improvement techniques and the modelling of the ionospheric effects in single frequency measurements. Achievable accuracies for static relative positioning are in the range of 0.1 ... 1 ppm (part per million), corresponding to 1 ... 10 cm over 100 km.

There are, however, many applications where the (relative) position of a moving vehicle, ship or aircraft is needed with much higher accuracy than the navigation mode of a GPS receiver using only pseudo-ranges can offer (10 ... 15 m absolute). Examples can be found in hydrographic surveying, 3d-seismic surveys for offshore exploration, airborne gravimetry and photogrammetric aerotriangulation. On the other hand, standard surveying would appreciate an enormous saving in time if GPS positioning could be done in seconds instead of staying one or more hours at stations.

Remondi (*1985a, b; 1986*) reported initial results of such a kinematic test survey over a small network (30 ... 400 m) using carrier phase observations of TI 4100 dual frequency receivers where it was demonstrated for the first time that also accuracies equivalent to the static observation scenario are in principle possible. In marine kinematic positioning (and other fields), methods using a combination of pseudo-range and phase measurements were presented (*Cannon, 1987; Kleusberg, 1986; Kleusberg et al, 1986; Lachapelle et al, 1986, 1987a, b, c*) suitable for single frequency receivers and real-time applications and resulting in accuracies in the 1 m level. Thereby the phase data were not corrected for cycle slips. For the determination of camera station positions for aerotriangulation a feasibility study conducted by NASA and NOAA's National Geodetic Survey (NGS) verified accuracies in the decimeter range (*Krabill and Martin, 1987; Lucas and Mader, 1987; Mader et al, 1986; Mader, 1986a, b*). Thereby a discrete pointwise analysis was applied. However, the cycle slip problem still remained as a serious complication. A road test of *Cannon et al* (*1986*) concluded that the implementation of corresponding algorithms is difficult and results are unreliable. Nevertheless, kinematical geodesy and system integration is a challenge for the near future (*Schwarz, 1987*). It is worthwhile to mention that twenty years ago *Moritz* (*1967, 71*) published the first reports on that subject.

This paper summarizes some experiences of the Institute of Astronomical and Physical Geodesy of the University FAF Munich in *high precision* kinematic GPS positioning of the last two years. Three land tests and one kinematic aircraft positioning test are discussed. The last was conducted within a series of trials to determine the position of a photogrammetric camera in space for aerotriangulation applications.

In chapter 2 the term kinematic positioning is defined and requirements for it are summarized. Chapter 3 describes the land and flight tests. Afterwards the applied strategy in analyzing the data is presented. Finally, results and conclusions are pointed out.

2. STATIC, KINEMATIC AND DYNAMIC MODELLING OF GPS OBSERVATIONS

2.1 Definitions

For the reader not so familiar with all published papers in this field it might be helpful to define static, kinematic and dynamic modelling since some recent publications have brought some confusion between the different terms.

In *static* relative GPS positioning we assume that stations are occupied with receivers for a time ranging from a few minutes to several hours. The subsequent data reduction does not consider the parameter time and any movement of the stations or receivers (except those of the satellites) and thus, refers to *one* observation epoch.

If a vehicle, ship or aircraft with a receiver on board is in motion and we have to model it, we require either knowledge of the forces driving it or measurements of the motion. The first type is called *dynamic* modelling, the second one *kinematic*.

There is only one case with respect to the analysis of GPS observations where we can speak of dynamic modelling, namely the determination of the orbits of the GPS satellites (so far a force model was used).

Thus, all geodetic efforts to determine the position of a moving receiver can be called kinematic positioning, e.g., analysis of observations referring to different time epochs t_j by use of a kinematic position, velocity or acceleration vector. Although the problem is of kinematic nature, the analysis is split up in two main categories. Most frequent in use is a determination of discrete positions at the corresponding measurement epochs without taking into account the vehicle motion from observations beforehand (see e.g., *Cannon et al, 1986; Kleusberg et al, 1986; Mader, 1986a, b*). To some extent the model itself can be considered as a static one.

With respect to land kinematic GPS surveying applications the procedure is as follows. We start from a station with known coordinates where the antenna is first put on a tripod centered over it. After a short while the antenna is carried to a moving car. Usually, it is mounted during the move on the roof of the car in order to keep continuous tracking. After arriving at the next station to be determined, the antenna will be taken from the car and mounted on the tripod which was put over the station beforehand. Again, for a while the antenna is in rest before travelling with it continues to the next station. Theoretically, the period of the rest should guarantee only that, at least, one GPS measurement is recorded. That means something like 5 seconds when assuming a time interval of 3 seconds between subsequent measurements. However, in practice the period will be a little bit extended, say to 20 seconds or even one minute in order to check for gross errors or failures, or simply to have convenient time to make a notice to the field books. Those measurements (in thirty seconds we would get, for example, already ten observations) can yield the basis for a determination of the (relative) coordinates of the corresponding station. Although the measurement process is of *kinematic* nature, the applied algorithms for the positioning are essential the same as in the static observation scenario (forming differences to a non-moving monitor station). *Schwarz et al (1987)* have entitled this approach as *"no dynamics"* model.

In contrast to that kind of analysis in GPS kinematic positioning, we can define a Kalman filter algorithm taking into account *all* observations along the trajectory. In general, such a modelling might be superior to a so-called "no dynamics" approach (*Schwarz et al, 1987*). Especially for aircraft and ship positioning where it is impossible to have the carrier of the receiver in rest for some seconds, and the path of the move is also needed, Kalman filtering is the appropriate tool.

There is no doubt that for the removal of inevitable cycle slips in GPS carrier phase observations, also in the "no dynamics" model observations along the trajectory are needed. The trajectory itself, however, is only of secondary interest there.

2.2 Requirements for relative kinematic GPS positioning

In the following we summarize some of the requirements needed for the kinematic use of GPS observations. In particular, we like to stress out those allowing us to come down to an accuracy better than 1 m.

Receiver characteristics. With respect to the hardware the *number of tracking channels* plays an important role for the detection and removal of cycle slips in the carrier phase observations. Every channel more than four offers the possibility to solve problems through redundancy.

For the precise applications having in mind we assume that both *pseudorange and continuous phase* measurements are available. For the geodetic user it may be of secondary interest whether the receiver is of multichannel, multiplexing or sequential type, although multichannel receivers may have a signal-to-noise ratio superior over the other ones.

The use of *dual-frequency* receivers is highly desirable for two reasons, the necessity of correcting the observations for the effect of the ionosphere and for detecting and correcting cycle slips. Although the ionospheric effects can partly be eliminated in differential positioning the residual effects may range - depending on the length of the baseline - in the decimeter level or even approach the one meter. In order to determine relative 3d-coordinates within a few centimeters in the kinematic mode, dual-frequency receivers are absolutely necessary. There is at present no alternative to that as long as no system integration, e.g. with inertial surveying instruments, can be realized.

The ultimate accuracy ratio between *P and C/A code* is founded by the corresponding chip code ratio of 10. However, the finally achieved measurement accuracy is still a function of other parameters, like carrier-to-noise ratio, tracking bandwidth, etc. The P code is less sensitive to multipath errors. Since for *high precision* applications *over long distances* always carrier phases of two frequencies are needed, there is only the possibility of using a code-free squaring receiver if the U.S. Department of Defense's policy of selective availability is once realized and the P code is restricted for civilian use. Real-time kinematic applications are difficult then because of the need of satellite ephemerides from an external source, and the higher computation load in using *only* phase data. In addition, the receivers should have an appropriate interface to establish the necessary *communication link* between those used for differential positioning.

Up till now there is only the Texas Instruments' TI 4100, a P code multiplex receiver that can measure both pseudorange and continuous phase in *two* frequencies, which is suited for differential kinematic positioning in the cm-level. No experiences are available for the recently offered dual-frequency code free Macrometer II system, Mini-Mac 2816, of Aero Service Division of Western Atlas International. Other firms, no doubt, are also planning dual-frequency instruments (*McDonald et al, 1987*).

Receiver software, tracking bandwidth, sampling interval. The tracking bandwidth is the most important parameter in kinematic applications. On the one hand, the bandwidth should be wide enough to maintain phase lock in a dynamic environment. The more wide the bandwidth the more unfavourable is the achievable accuracy. Detection and removal of cycle slips may become from a certain point even impossible. Thus, based on the type and magnitude of movement, e.g. in particular the appearing velocities and accelerations, an optimal tracking bandwidth has to be chosen minimizing phase lock losses and maximizing the signal-to-noise ratio of the code and phase measurements. With respect to terrestrial surveying applications the receiver software has an advantage that can switch during the measurement campaign between stationary and dynamic modes. When the receiver is in short rest (a few seconds) at a station to be determined one would be able to change the bandwidth to the most narrow setting which may result in a higher accuracy. In some operating software the change in recording time interval is fixed with the user dynamics class.

Antenna. A dynamic environment requires with respect to physical size, shape and ruggedness higher requirements to a GPS antenna. Whereas in terrestrial kinematic surveying applications the standard antenna at least from some of the equipments for static positioning may be used, is a use on an aircraft, for example, impossible. The microstrip-type antenna appears to have the potential to fulfill most of the geodetic kinematic positioning requirements in the near future. For more details see, e.g., *Lachapelle et al (1987a)*.

Satellite coverage limitations. A more serious problem is posed by the fact that - at least based on the present state-of-the-art in high precision kinematic positioning algorithms - a *permanent* observation of the *same four* satellites during the mission is necessary *for cm-accuracy*. Considering in addition a good geometry of the user location and the four satellite orbits, let's say with a PDOP (Position Dilution of Precision) < 6 and an elevation mask angle of 10°, *only a daily period of approximately ~ 1.5 h* (Fig. 1) *is available* for differential kinematic positioning in the cm-range. Receivers having more than four channels may help in developing data reduction algorithms capable in switching over to different four-satellite-configurations during a measurement mission. There is no doubt that methods resulting in accuracies in the meter range do not require such a constraint.

2.3 A brief survey of available data reduction algorithms for differential kinematic GPS positioning

Tab. 1 summarizes the characteristics of the main possibilities in using GPS for differential kinematic positioning. There are only three main trends in modelling.

Tab. 1: Characteristics of main differential kinematic positioning methods

DIFFERENTIAL POSITIONING MODEL / CHARACTERISTICS		DIFFERENTIAL POSITION CORRECTIONS	PHASE-SMOOTHED PSEUDORANGE ——————— CARRIER-SMOOTHED CODE	PHASE DIFFERENCING (1) "NO DYNAMICS" APPROACH (2) KALMAN FILTERING
Relative positioning accuracy	C/A code	5 ... 15 m	1 ... 3 m	1 ... 3 cm
	P code	3 ... 5 m	0.5 ... 1 m	
Pseudorange observations		needed	needed	(needed)
Carrier phase observations		not needed	yes	yes
Cycle slip detection		not needed	yes	yes
Cycle slip removal		not needed	no	yes
Dual frequency use		only for higher accuracy (iono-spheric effects)	only for higher accuracy (iono-spheric effects)	needed
P code		only for higher accuracy needed	only for higher accuracy needed	not needed using codeless receivers
Real-time application		easily possible	possible	possible with external sensors
References		Kalafus et al. (1986)	Cannon (1987) Cannon et al.(1986) Hatch (1982,1986) Kleusberg (1986) Kleusberg et al. (1986) Lachapelle et al. (1986,1987a,b,c) Seeber et al.(1986)	(1) Krabill, Martin (1987) (1) Lucas, Mader (1987) (1) Mader (1986a,b) (1) Remondi (1985a,b) (1986) (1),(2) Schwarz et al. (1987) (2) Bastos, Landau (1988) (2) Landau (1987)

DILUTION OF PRECISION FOR NAVSTAR-GPS

STATION : SYDNEY AIRPORT   DATE : 18-10-1987

Latitude : -33°56'40"
Longitude : 151°10'34"
Height   : 60 meter

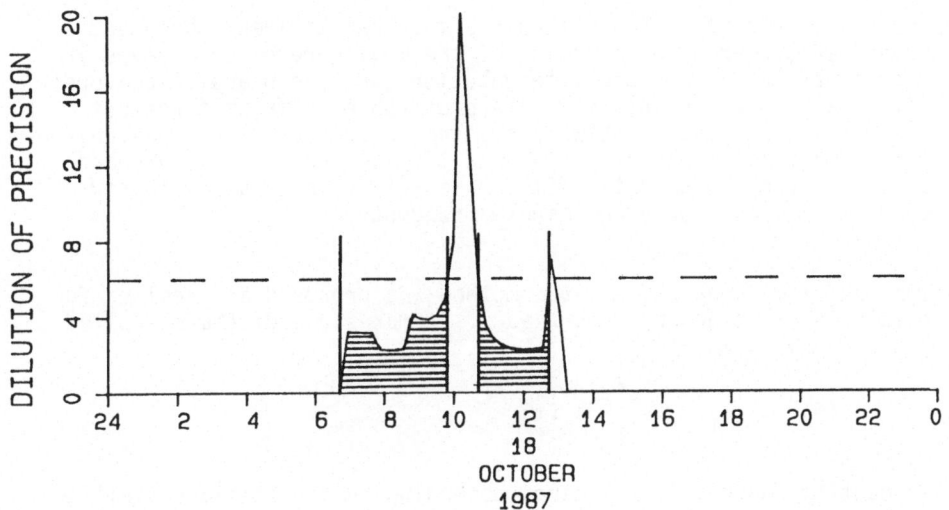

<u>Fig. 1</u> Example for the daily observation time suited for high-precision
kinematic positioning with PDOP < 6

The *differential position correction* method compares a known location with that predicted by GPS pseudorange observations and broadcasts the so-determined corrections to nearby users to improve their position solution. The *phase-smoothed pseudorange* (or *carrier-smoothed code*) method uses carrier phases (or differences in time and/or combinations of it, eventually in two frequencies) to filter the pseudorange observation errors. Different variations of such algorithms are developed.

The *phase differencing* method is essentially the same applied in stationary differential geodetic positioning. Discrete position determination at a specific time epoch ("no dynamics" approach) as well as use of sequential algorithms like Kalman filtering are possible. The ultimate highest accuracy achieved with that type of data reduction, however, has to be paid by the many restrictions and requirements outlined in chapter 2.2. In particular, the detection and *necessary removal of cycle slips* in the carrier phases poses the main difficult problem. Two methods are in use:

(1) The *phase velocity trend* method assumes that between subsequent phase measurements the dynamics of the moving receiver behaves in such a way that a certain interpolation can take place. Receivers with small sampling intervals are preferable. With that respect the TI 4100 Navigator software has the advantage in the availability of the two measurements over a 160 msec interval before and after the code Fundamental Time Frame (FTF). The phase velocity trend method uses only one single frequency.

(2) The *dual-frequency phase ratio* method was proposed and applied for static observations by *Goad (1986)*. It makes use of the so-called *ionospheric residual*,

$$\delta\Phi(t) \;=\; \Phi_1(t) \;-\; \frac{f_1}{f_2}\,\Phi_2(t) \tag{1}$$

a quantity, where all the terms depending on the station-satellite range and the clock errors have dropped out. f_1, f_2 are the frequencies of L_1 and L_2, and Φ_1, Φ_2 are the corresponding carrier phases.

Approximate quantities for the cycle slip values can be found from the phase-range combination

$$\Delta\rho_i(t) \;=\; \lambda_i[\Phi_i(t) - \Phi_i(t_1)] - [\rho(t) - \rho(t_1)] \tag{2}$$

where i indicates L_1 or L_2. λ_i, Φ_i is the corresponding wavelength and carrier phase, respectively. ρ is the observed range derived from code measurements. t is the actual, and t_1 the initial epoch.

Whereas *Mader (1986a, b)* uses a linear regression through the quantities (1) and (2) to fix the cycle slips, have *Landau (1987)*, *Bastos and Landau (1988)* a Kalman filter applied to improve the reliability of the method.

Due to unpredictable dynamics and possible unfavourable signal-to-noise ratio (for example, because of a too wide opening of the bandwidth) both methods may fail or produce unsatisfactory results under certain circumstances. Thus, a careful planning of the kinematic survey has to take place.

A further improvement in detecting and removing cycle slips in high-precision kinematic missions can be achieved by use of *external sensors, integration with other instruments* or *redundant measurements*, e.g. with a second GPS receiver, and/or with more than four channels. As external sensors we can use an atomic clock, laser altimeters (may reduce the height component, e.g., over sea), or inclination monitors. The integration of a GPS receiver with a low-cost strapdown system is the most promising combination of different data types, see e.g., *Wong et al (1985)*. A further method, suited however only for post-mission analysis, is the use of photogrammetric images taken during a flight. The relative orientation parameters of overlapping images provide the basis for an interpolation along the path of flight. This method is currently investigated at our institute.

3. DESCRIPTION OF LAND AND FLIGHT TESTS

In 1986 and 1987 several tests on kinematic GPS positioning were carried out. Three tests on land and one aircraft positioning test are analyzed in this paper. Tab. 2 presents an overview of the tracking software, bandwidth and the recorded measurement rate of the used Texas Instruments TI 4100 dual-frequency receivers.

The *TI 4100 Navigator* operating software (*Texas Instruments, 1984*) as well as the special version 32.1 of the *GESAR* (Geodetic Satellite Receiver) software (*Darnell and Hawkins, 1986*) allow four different level of user dynamics (Tab. 3). The main difference between the two software is the capability of GESAR 32.1 to select the measurement rate and to change the user dynamics class anytime during the mission. This is an advantage in terrestrial surveying applications where it is in most cases possible to bring the antenna in short rest and to switch the dynamics class to the stationary case. Lowering the user dynamics (class) improves the system accuracy. On the other hand it was mentioned already in 2.3 that the TI 4100 Navigator software provides the two phase measurements over a 160 msec time interval before and after the code FTF which may be helpful in using the phase velocity trend method for detecting cycle slips.

Two land tests were carried out in an area called "Werdenfelser Land" approximately 50 km south of Munich along an east-west traverse. Seven stations with separations between 400 m up to 1000 m were observed over 3 km (Fig. 2). A preliminary analysis of the test "Werdenfelser Land 1" was presented by *Landau (1987)*.

The test at the Sydney airport was conducted during the sabbatical of the first author with A. Stolz from the University of New South Wales (*Hein and Stolz, 1988*) and the help of the Australian army. The observed traverse around the run way (Fig. 3) has a length of about 5 km with 6 stations belonging to the precise airport geodetic 3d-network with an estimated station accuracy of approx. ± 5 mm.

The scenario for all the three land tests was identical. The control stations to be observed in kinematic mode were equipped with survey tripods carefully centered over the survey marks. Before and after the

Tab. 2: Analyzed relative kinematic positioning tools

PROJECT	TYPE OF TEST	DATE	TIME (UTC)	TI 4100 OPER- ATING SOFTWARE	TRACKING BANDWIDTH		MEAS- URE- MENT RATE
					STATIC RECEIVER	MOVING RECEIVER	
WERDENFELSER LAND 1	Land Vehicle (Van)	17 Sept 1986	4:30 - 6:00	TI Navi- gator	0.7 Hz	5 Hz	3 sec
WERDENFELSER LAND 2	Land Vehicle (Van)	11 Dec 1986	22:50 - 0:40	GESAR 32.1	0.7 Hz	16 Hz	1 sec
SYDNEY AIRPORT	Land Vehicle (Jeep)	18 Oct 1987	8:30 -10:00	TI Navi- gator	5 Hz	5 Hz	3 sec
MÜNCHEN- FREIMANN 1	Flight Photogrammetric Aircraft	17 Aug 1987	8:50 - 9:40	GESAR 32:1	0.7 Hz	16 Hz	1 sec

Tab. 3: Used operating software for the Texas Instruments TI 4100

USER CODE		DESCRIPTION	TRACKING BANDWIDTH	MAXIMUM ACCELERATIONS	MEASUREMENT INCORPORATION RATE	
TI 4100 Navigator	GESAR 32.1			[m/sec^2]	TI 4100 Navigator	GESAR 32.1
0	1	Stationary (lowest), default in GESAR	0.7 Hz	0	3.0 sec	variable
1	2	Low dynamics, default in TI Navigator	5.0 Hz	6	3.0 sec	variable
2	3	Medium dyna- mics (light aircraft or rough seas)	8.0 Hz	15	1.2 sec	variable
3	4	High dynamics (helicopter)	16.0 Hz	40	1.2 sec	variable

PATH OF KINEMATIC POSITIONING

Project : Werdenfelser Land

Date : 11-12-1986   GPS Week : 361   Day of year : 345

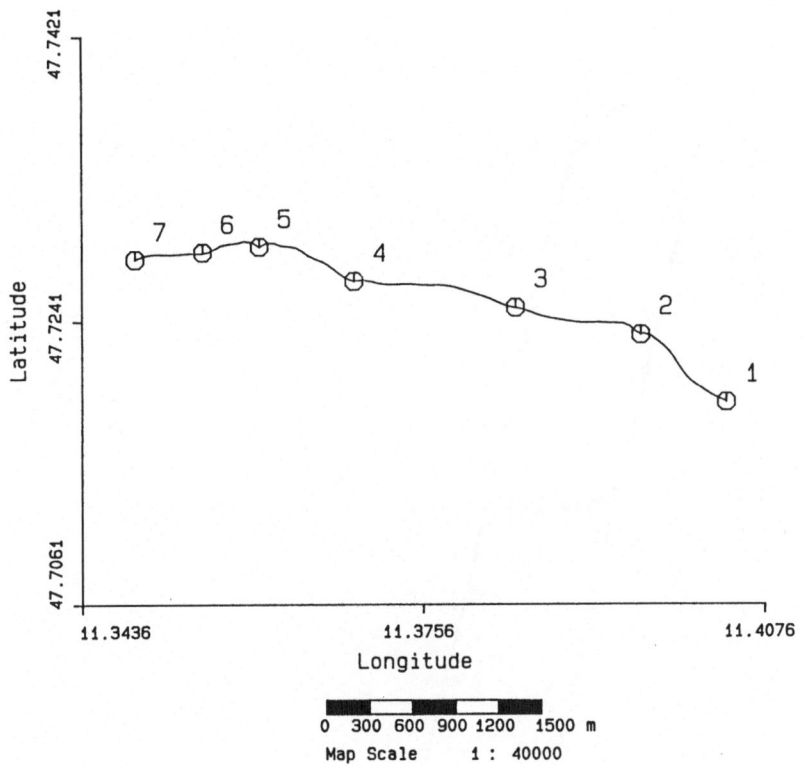

Fig. 2 Location of the traverse used in tests "Werdenfelser Land 1 + 2"

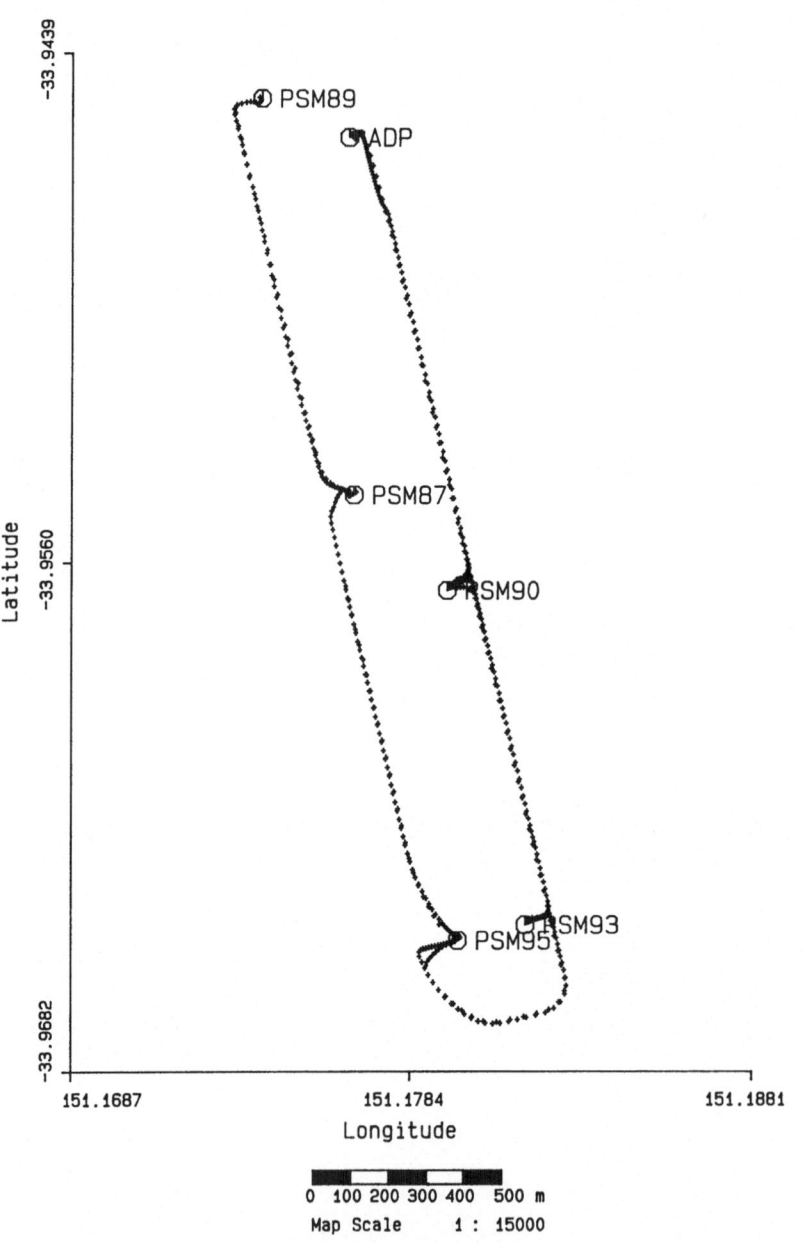

PATH OF KINEMATIC POSITIONING

Project : Sydney Airport

Date : 18-10-1987   GPS Week : 406   Day of year : 291

Fig.3 Location of the traverse used in test "Sydney airport"

318

move (double run) a 10 - 15 minutes static positioning part took place
where the "moving receiver observed at the first station to get initial positioning values with respect to the static receiver located a
few kilometers far from that. During travelling the GPS antenna (TI 4100
standard antenna) was mounted on the roof of a car. It was carried carefully by hand then from the car to the considered survey station and
put for a period of 10 ... 40 sec on the tripod. The speed of travelling in the test "Werdenfelser Land 1 + 2" was up to 75 km/h whereas at
the Sydney airport we drove with a velocity less than 25 km/h. In the
test "Werdenfelser Land 2" the tracking bandwidth of the receiver was
set to the stationary mode (0.7 Hz) during the rest at the stations and
changed again before travelling. For the satellite geometry see Figs. 7 to 9.

The documented flight test "München-Freimann 1" out of a series of trials
within a joint project with the Institute of Photogrammetry and Carto-
graphy of the University FAF Munich on aerotriangulation without ground
control was carried out on 17 Aug. 1987 with a DORNIER DO 28 aircraft.
A dual-frequency GPS flight antenna DM C146-2-1 from Dorne & Margolin
Inc., N.Y., was mounted on top of the roof mid of the location between
the two wings (Fig. 5). The photogrammetric camera RMK 15/23 from Zeiss,
Oberkochen was time-tagged with the one-pulse-per-second (1 pps) of the
TI 4100 via a frequency generator (accuracy 0.25 msec). The time offset
between the 1 pps impulse and the GPS time was recorded from the J2 port
of the TI 4100 using a RS-232C interface whereas the offset between the
electronic impulse of the camera shutter and the frequency counter was
kept via a HP-IB interface. Both are incorporated in a recording HP-IL
loop monitored by a simple HP 71B pocket computer (Fig. 6).

Due to a long bad weather period and the failure of a GPS satellite
PRN 9 just before the test flights, only three flights could be carried
out. Fig. 4 shows the path of the "München-Freimann 1" flight. The aircraft started in Neubiberg airport and was flying over a photogrammetric
test area located in the north of Munich city. The DO 28 climbed up to
approx. 920 m altitude above ground (corresponding to an orthometric
height of 1420 m) and reached a velocity of 54 m/sec (~ 190 km/h). The
before-mentioned satellite PRN 9 which could not be used restricted the
observation window and the geometry of the only four remaining satellites
was not very favourable, see the polar plot of the satellite passes
(Fig. 10). In addition, after the flight experiments we found out that
the used flight antenna from Dorne & Margolin Inc. shows a significant
lower signal-to-noise ratio in L2 than the standard TI 4100 antenna.
Whether this obviously hardware-caused fact is common for all antennas
of this type or only specific for the used one is not known to the authors.

Tabs. 4 to 7 summarize the observation statistics of the four kinematic
GPS positioning tests. "Bad data" which were not used for deriving the
positioning results are those where loss of lock and/or cycle slips
occured or where the signal-to-noise ratio dropped below 40 dB. (in stationary applications a value of 36 dB is sufficient.) Due to the low
signal-to-noise ratio in L2 during the flights as mentioned before the
limiting value for rejecting observations was lowered.

Whereas the cycle slips found in the data of "Werdenfelser Land 1 + 2"
are mainly due to shadowing by trees near the street, are those ones in
test "Sydney airport" occured in the short connection between the stations and the concrete drive way when the jeep drove through a bumpy

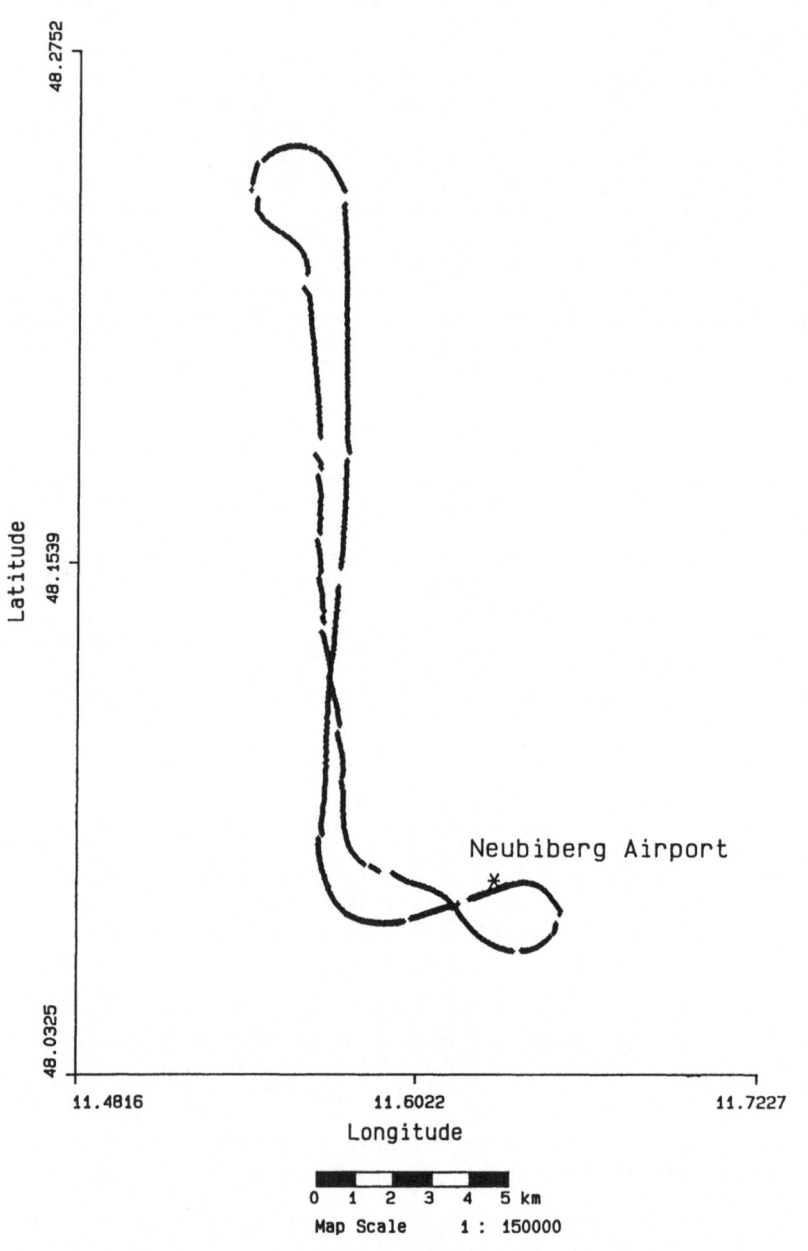

PATH OF GPS-FLIGHT

Flight : DO 28   Flying Remote

Date : 17-8-1987   GPS Week : 397   Day of year : 229

Neubiberg Airport

Longitude

Latitude

0 1 2 3 4 5 km

Map Scale   1 : 150000

Fig. 4   Path of GPS flight in test "München-Freimann 1"

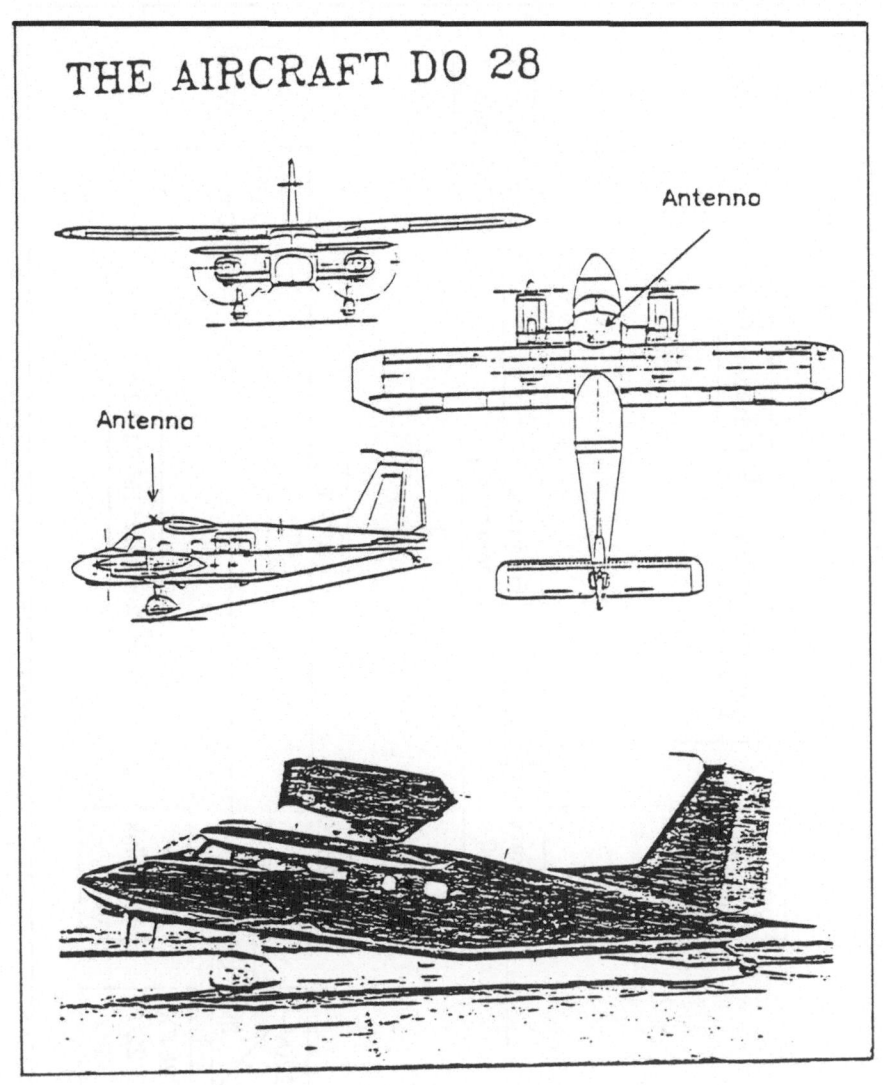

Fig. 5 Aircraft used for photogrammetric flight test

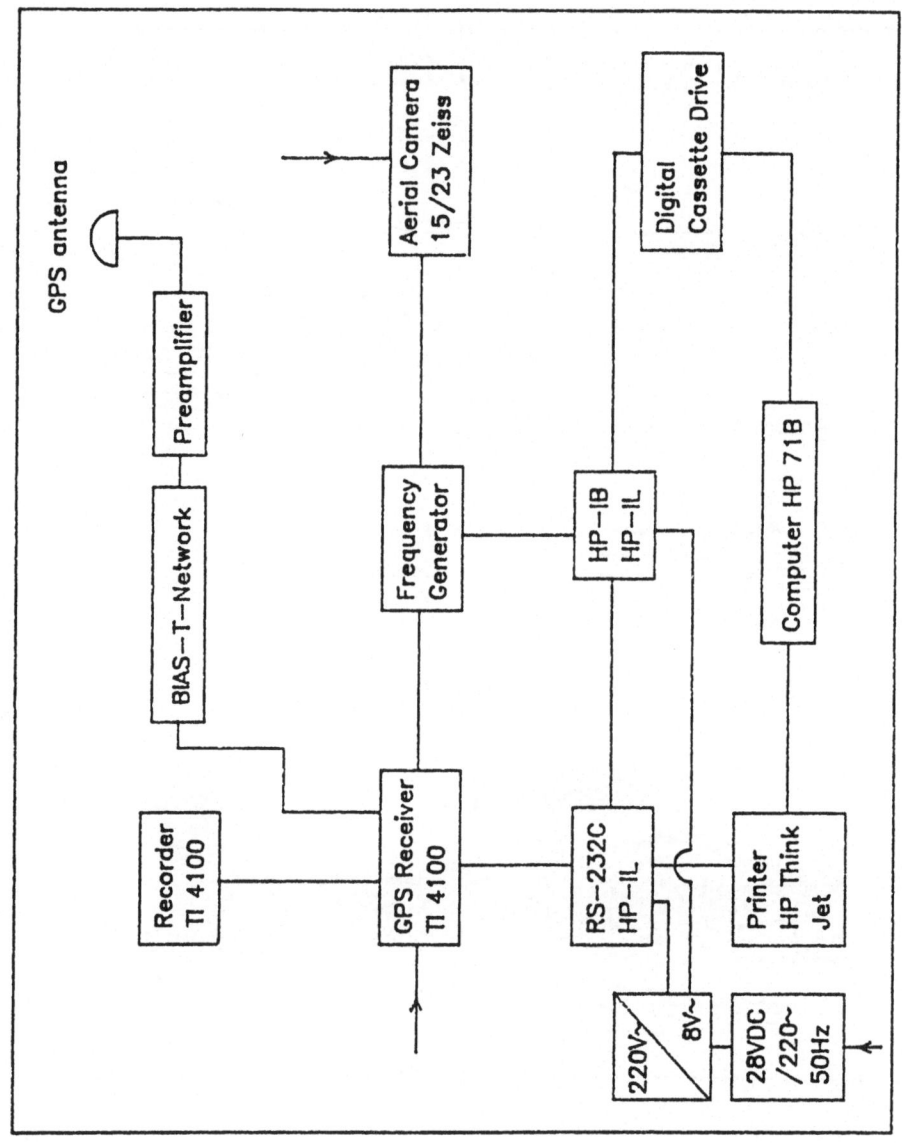

Fig. 6 Synchronisation of the TI 4100 GPS receiver and the photogrammetric aerial camera

SATELLITE PASSES OF THE NAVSTAR-GPS SYSTEM

STATION : WERDENFELSER LAND 1   DATE : 17-9-1986

Latitude   : 47°43'0"
Longitude : 11°24'23"
Height   : 620 meter

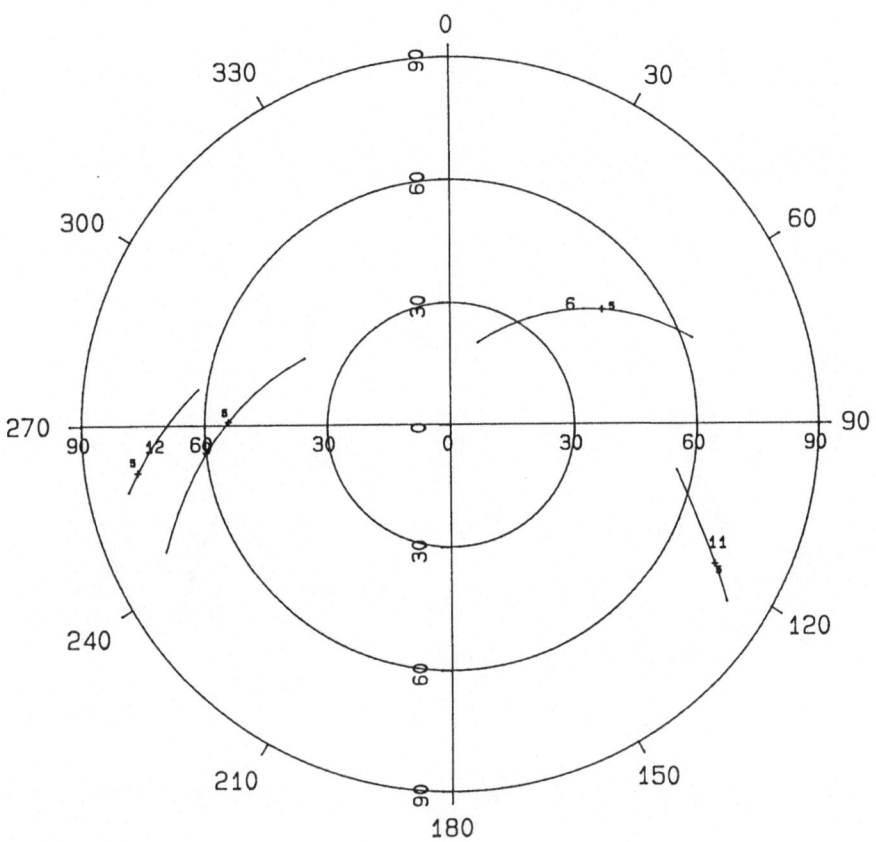

Fig. 7   Polar plot of satellite passes in land test "Werdenfelser Land 1"

SATELLITE PASSES OF THE NAVSTAR-GPS SYSTEM

STATION : WERDENFELSER LAND 2   DATE : 11-12-1986

Latitude   : 47°43'0"
Longitude : 11°24'23"
Height   : 620 meter

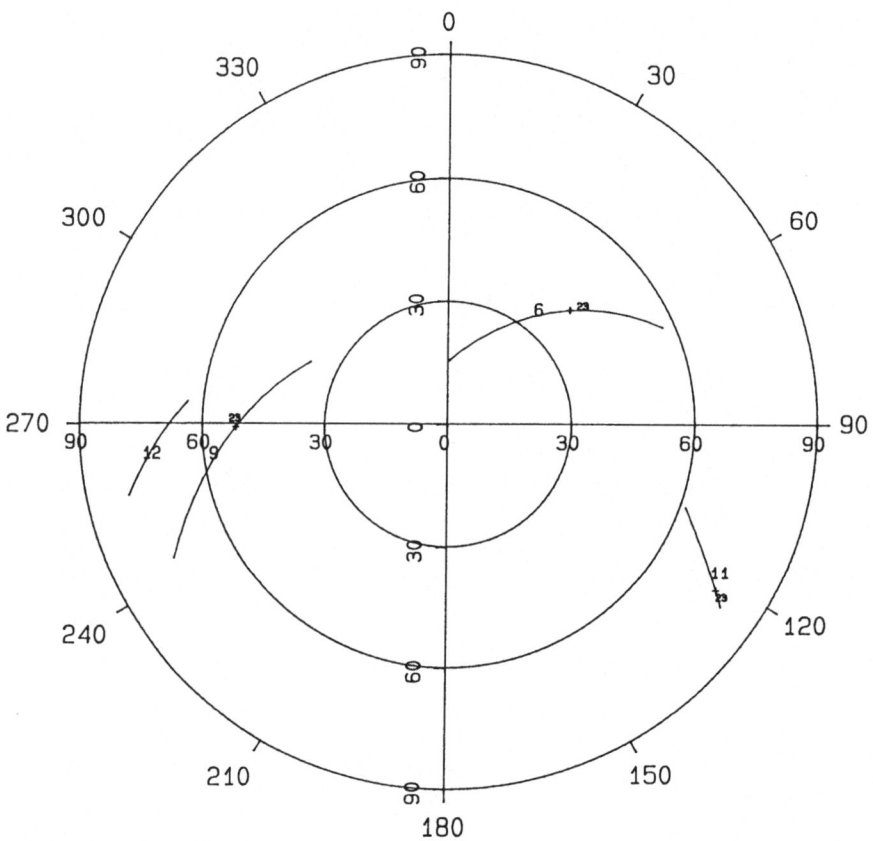

Fig. 8 Polar plot of satellite passes in land test "Werdenfelser Land 2"

SATELLITE PASSES OF THE NAVSTAR-GPS SYSTEM

STATION : SYDNEY AIRPORT   DATE : 18-10-1987

Latitude : -33°56'40"
Longitude : 151°10'34"
Height   : 60 meter

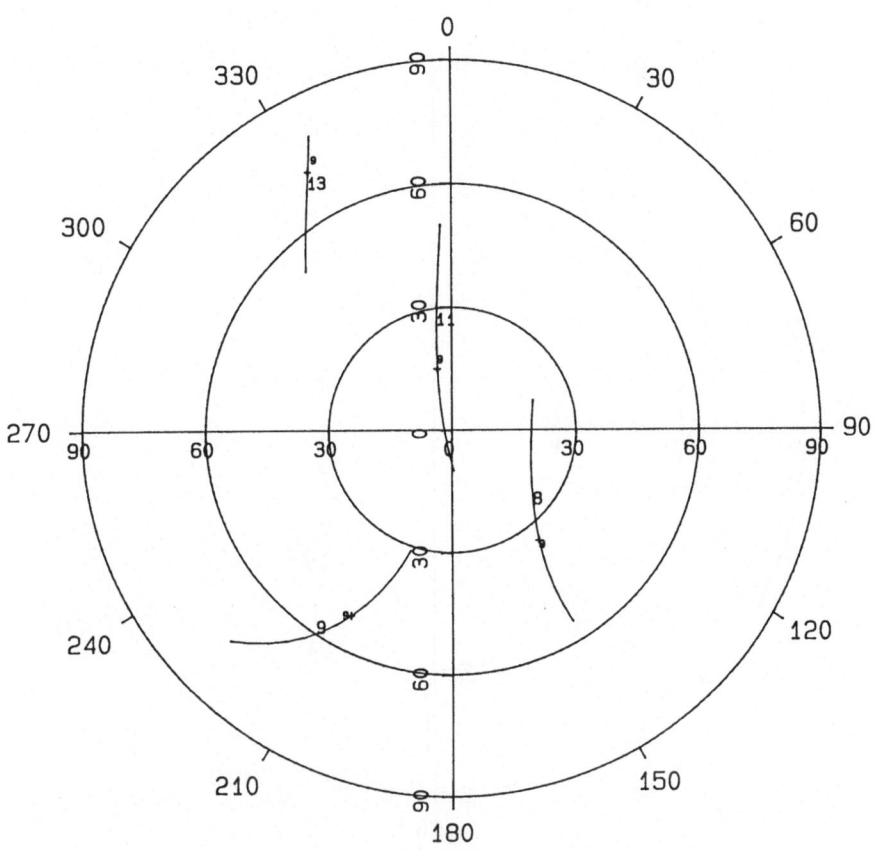

Fig. 9 Polar plot of satellite passes in land test "Sydney airport"

SATELLITE PASSES OF THE NAVSTAR-GPS SYSTEM

STATION : NEUBIBERG AIRPORT   DATE : 17-8-1987

Latitude   : 48°5'0"
Longitude : 11°38'0"
Height   : 570 meter

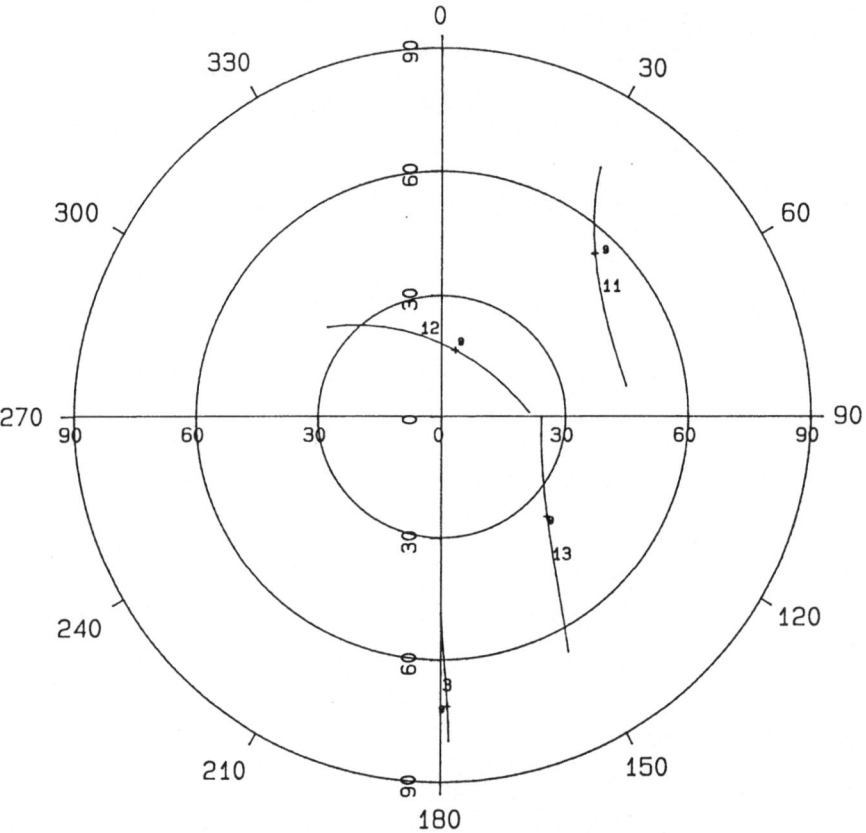

Fig. 10 Polar plot of satellite passes in flight test "München-Freimann 1"

Tab. 4: Observation statistics for the moving receiver in land test "Werdenfelser Land 1"

Project: Werdenfelser Land 1 Date: 17-09-86		Satellites			
		6	9	11	12
Elevation range		23^0-18^0	66^0-67^0	15^0-35^0	15^0-37^0
Average signal-to-noise ratio	L_1	42.7 dB	41.7 dB	42.3 dB	42.9 dB
	L_2	41.0 dB	40.0 dB	41.3 dB	41.4 dB
Bad data	L_1	6.9%	6.3%	5.4%	8.8%
	L_2	6.9%	6.4%	5.6%	8.8%
Number of cycle slips		10	7	9	10

Tab. 5: Observation statistics for the moving receiver in land test "Werdenfelser Land 2"

Project: Werdenfelser Land 2 Date: 11-12-86		Satellites			
		6	9	11	12
Elevation range		55^0-23^0	34^0-61^0	10^0-38^0	10^0-36^0
Average signal-to-noise ratio	L_1	42.3 dB	41.4 dB	42.1 dB	42.2 dB
	L_2	42.4 dB	41.4 dB	42.2 dB	41.9 dB
Bad data	L_1	4.6%	3.4%	2.9%	4.8%
	L_2	4.6%	3.4%	1.7%	4.7%
Number of cycle slips		17	10	5	25

Tab. 6: Observation statistics for the moving receiver in land test "Sydney Airport"

Project: Sydney Airport Date: 18-10-87		Satellites			
		8	9	11	13
Elevation range		69^0-23^0	25^0-70^0	56^0-88^0	12^0-46^0
Average signal-to-noise ratio	L_1	41.8 dB	41.5 dB	41.4 dB	41.9 dB
	L_2	40.7 dB	40.4 dB	40.4 dB	40.6 dB
Bad data	L_1	3.2%	4.5%	4.0%	3.6%
	L_2	3.2%	4.5%	4.0%	4.8%
Number of cycle slips		4	7	7	5

Tab. 7: Observation statistics for the moving receiver in flight test "München-Freimann 1"

Project: München- Freimann 1 Date: 17-08-87		Satellites			
		3	11	12	13
Elevation range		11°-36°	39°-21°	71°-74°	48°-66°
Average signal-to-noise ratio	L_1	40.3 dB	42.6 dB	42.3 dB	43.4 dB
	L_2	35.8 dB	36.8 dB	39.7 dB	39.4 dB
Bad data	L_1	38.4%	24.4%	4.1%	1.7%
	L_2	38.4%	24.4%	4.1%	1.7%
Number of cycle slips		39	58	16	7

grass area. The large number of cycle slips in the observations of satellites 3 and 11 are caused by the tilted aircraft (roll angle) when due to the low elevation angle (see Fig.10) of those satellites the wings shadowed the antenna. This explains also the high percentage of bad data there.

Assuming that all tests posed more or less an environment of practical surveying we may conclude that an average of 3 - 5 % bad data and a number of 5 to 10 cycle slips per channel and hour seem to be a realistic case for the TI 4100.

4. DATA REDUCTION ALGORITHM

For the *cycle slip fixing* the dual-frequency phase ratio method was applied. Approximate values were found using the phase-range combination of eq. (2) and final values by the ionospheric residual, eq. (1). In contrast to the linear regression algorithm used by *Mader (1986a, b)*, a Kalman filter was applied (*Bastos and Landau, 1988; Landau, 1987*). Thereby, the dynamic system matrix was approximated by use of a simple quadratic polynomial over time (*Goad, 1986*). The *system driving noise matrix* Q is a 3×3 matrix defined empirically for the phase-range combination (2) by

$$Q_1 = \text{diag} (10^{-8}, 10^{-10}, 10^{-12}) \ [\text{m}^2, -, \text{m}^{-2}] \tag{3}$$

and for the *ionospheric residual* (1) by

$$Q_2 = \text{diag} (10^{-6}, 10^{-8}, 10^{-10}) \ [\text{cycles}^2, -, \text{cycles}^{-2}] \tag{4}$$

The *measurement noise matrix*, in our case, defined as a simple scalar (only one measurement per epoch) is given for the phase-range combination by

$$r_1 = 2.25 \ [m^2] \tag{5}$$

and for the ionospheric residual by

$$r_2 = 2.5 \cdot 10^{-3} \ [cycles^2] \ . \tag{6}$$

After computing the approximate cycle slip values by (2) the final corrections are found by (1) through varying the approximations within a small search window, e.g. [-4, 4] and looking for the minimum in $\delta\Phi$. *Bastos and Landau (1988)* have stated that in order to guarantee a unique result the approximate values have to be better than 3 cycles. This is a very high requirement which may limit the reliability of the method in presence of a low signal-to-noise ratio.

Again, in the relative positioning model a Kalman filter was applied for the combination of single-difference carrier phases and pseudo-ranges (*Landau, 1987*). For further modelling of clock errors and of ionospheric effects the reader is referred to (*Landau, 1988*).

The briefly described data reduction algorithm is implemented in the software "DYNAMITE" (*Dynamic Trajectory Estimation*) written by H. Landau in FORTRAN 77 for the DEC VAX family under VMS.

5. ANALYSIS OF RESULTS

In Tab. 8 to 13 a detailed comparison between the L_1- and L_2-derived relative positions as well as between forward and backward runs for the three land tests are presented. In addition the coordinate differences between the results of the two tests over the same traverse in the "Werdenfelser Land" using different operating software and tracking bandwidths are shown in Tab. 14. For one baseline a stationary GPS measurement of one hour was available which could be compared to the kinematic results, see Tab. 15.

With respect to the test "Sydney airport" precise station coordinates of the airport's 3d-geodetic network were available. The accuracy of those values in the local A.M.G. system is estimated to be better than ± 5 mm, so that true errors could be computed for the kinematic survey (Tab. 16). These were derived after transforming the determined WGS 84 coordinates via a three-dimensional similarity transformation into the A.M.G. system. Although the transformation parameters (translation, rotation, scale) are here not of interest, they agree very well with preliminary ones for whole Australia.

The results of the kinematic differential positioning of the photogrammetric aircraft cannot fully be assessed since not all camera positions are already determined by photogrammetric block adjustment. However, the differences analyzed so far show rms-errors of 11 cm. A summary statistics (rms-errors) of the three land tests are presented in Tab. 17.

From the many tables we can see that excellent results in 1 - 2 centimeter level were achieved by the combination of carrier phase and pseudo-range observations in the kinematic mode. No significant differences could be noticed between the horizontal position and height accuracy, nor between the results of L_1 or L_2. The slightly larger errors for the $\Delta\varphi$ - values (in comparison to $\Delta\lambda$) of the tests in the Werdenfelser Land are due to the present GPS satellite constellation in Europe. Thus, the results of the test "Sydney airport" confirm this statement.

Tab. 8: Comparison of positions derived from L_2 frequency with positions derived from L_1 in test "Werdenfelser Land 1"

Project: Werdenfelser Land 1						
Station	L_2-L_1			L_2-L_1-mean		
	Δx (cm)	Δy (cm)	Δz (cm)	Δx (cm)	Δy (cm)	Δz (cm)
1	-0.1	-0.1	-0.1	-0.5	-0.3	0.6
2	0.6	-0.4	0.2	0.2	-0.6	0.9
3	6.8	0.8	0.6	6.4	0.6	1.3
4	-1.2	0.0	-1.4	-1.6	-0.2	-0.7
5	-2.7	-0.7	2.3	-3.1	-0.9	3.0
6	4.7	1.0	-1.7	4.3	0.8	-1.0
7	-2.5	-0.6	-1.3	-2.9	-0.8	-0.6
6	1.1	0.7	1.1	0.7	0.5	1.8
5	-1.2	0.2	-1.7	-1.6	0.0	-1.0
4	0.6	0.0	-2.4	0.2	-0.2	-1.7
3	0.5	0.5	0.1	0.1	0.3	0.8
2	1.6	1.5	-3.4	1.2	1.3	-2.7
1	-3.4	-0.9	-0.9	-3.8	-1.1	-0.2
Mean	+0.4	+0.2	-0.7	0.0	0.0	0.0
RMS	±1.9	±0.5	±1.1			

Tab. 9: Differences between forward and backward run in test "Werdenfelser Land 1"

Project: Werdenfelser Land 1												
	Carrier phases						Pseudo-ranges					
Station	L_1			L_2			L_1			L_2		
	Δx (cm)	Δy (cm)	Δz (cm)	Δx (cm)	Δy (cm)	Δz (cm)	Δx (m)	Δy (m)	Δz (m)	Δx (m)	Δy (m)	Δz (m)
1	-0.7	-0.9	0.6	-4.0	-1.7	-0.2	-1.18	-1.71	-1.44	-0.54	-0.45	-0.18
2	-4.3	-2.3	0.3	-3.3	-0.4	-3.3	0.61	1.50	0.60	-7.32	-1.98	2.49
3	-0.6	-0.7	-3.7	-6.9	-1.0	-4.2	-5.44	-0.04	2.78	8.07	2.23	1.62
4	-0.2	0.2	-1.5	1.6	0.2	-2.5	4.55	0.86	2.15	2.72	0.62	2.13
5	4.8	0.8	1.5	6.3	1.7	-2.5	4.90	1.18	-4.55	-1.86	-1.24	-0.09
6	1.8	0.8	0.4	-1.8	0.5	3.2	2.84	-0.18	3.92	-1.75	-0.22	-2.15

Tab. 10: Comparison of positions derived from L_2 frequency
with positions derived from L_1 in test
"Werdenfelser Land 2"

Project: Werdenfelser Land 2						
Station	L_2-L_1			L_2-L_1-mean		
	Δx (cm)	Δy (cm)	Δz (cm)	Δx (cm)	Δy (cm)	Δz (cm)
1	-0.6	-0.2	0.4	-0.8	-0.1	0.7
2	-0.7	0.2	1.1	-0.9	0.3	1.4
3	3.5	0.9	0.2	3.3	1.0	0.5
4	-1.3	-1.3	-3.4	-1.5	-1.2	-3.1
5	-2.8	0.0	2.3	-3.0	0.1	2.6
6	-2.7	-0.3	0.0	-2.9	-0.2	0.3
7	7.3	-1.5	-0.5	7.1	-1.4	-0.2
6	-2.6	-0.1	-0.5	-2.8	0.0	-0.2
5	-3.8	-0.7	0.6	-3.6	-0.6	0.9
4	3.6	0.8	-0.6	3.4	0.9	-0.3
3	1.3	1.3	-0.6	1.1	1.4	-0.3
2	3.2	0.6	-2.7	3.0	0.7	-2.4
1	-2.2	-0.4	0.4	-2.4	-0.3	0.7
Mean	+0.2	-0.1	-0.3	-0.1	0.0	0.0
RMS	±2.2	±0.6	±1.0			

Tab. 11: Differences between forward and backward run in test "Werdenfelser Land 2

Project: Werdenfelser Land 2												
Station	Carrier phases						Pseudo-ranges					
	L_1			L_2			L_1			L_2		
	Δx (cm)	Δy (cm)	Δz (cm)	Δx (cm)	Δy (cm)	Δz (cm)	Δx (m)	Δy (m)	Δz (m)	Δx (m)	Δy (m)	Δz (m)
1	1.3	0.4	-2.3	-0.3	0.2	-2.3	-0.84	0.08	2.27	-5.66	-0.93	4.28
2	-2.0	-1.0	-1.1	1.9	-0.6	-4.9	-1.98	-0.34	2.96	-5.32	-0.99	3.62
3	0.3	-0.5	-1.6	-1.9	-0.1	-2.4	0.45	-0.09	-0.12	1.68	-0.16	1.27
4	-3.2	-1.3	-2.4	1.7	0.8	0.4	0.90	0.15	0.02	-2.14	-1.05	-3.93
5	0.5	0.6	0.4	-0.5	-0.1	-1.3	2.49	0.84	-0.44	-1.76	0.40	2.10
6	-0.9	-0.4	1.3	-0.8	-0.2	0.8	-5.98	-0.81	1.79	4.99	0.91	0.13

Tab. 12: Comparison of positions derived from L_2 frequency
with positions derived from L_1 in test
"Sydney Airport"

Project: Sydney Airport						
Station	L_2-L_1			L_2-L_1-mean		
	Δx (cm)	Δy (cm)	Δz (cm)	Δx (cm)	Δy (cm)	Δz (cm)
ADP	0.1	-1.8	0.6	-1.1	-2.1	1.0
PSM 90	-0.6	-0.9	0.5	-1.8	-1.2	0.9
PSM 95	1.1	-0.2	0.8	-0.1	-0.5	1.2
PSM 87	3.4	-0.6	1.6	2.2	-0.9	2.0
PSM 89	0.9	1.1	-1.1	-0.3	0.8	-0.7
PSM 87	0.2	1.4	0.5	-1.0	1.1	0.9
PSM 95	-0.4	0.2	-1.7	-1.6	-0.1	-1.3
PSM 90	2.4	1.7	-0.3	1.2	1.4	0.1
ADP	4.1	1.9	-4.2	2.9	1.6	-3.8
Mean	+1.2	+0.3	-0.4	0.0	0.0	0.0
RMS	±1.5	±0.9	±1.2			

Tab. 13: Differences between forward and backward run in test "Sydney Airport"

Project: Sydney Airport												
Station	Carrier phases						Pseudo-ranges					
	L_1			L_2			L_1			L_2		
	Δx (cm)	Δy (cm)	Δz (cm)	Δx (cm)	Δy (cm)	Δz (cm)	Δx (m)	Δy (m)	Δz (m)	Δx (m)	Δy (m)	Δz (m)
ADP	2.9	0.4	3.1	6.9	4.1	-1.7	-1.89	0.86	-1.68	-2.98	0.48	-0.91
PSM 90	-1.6	4.9	1.8	1.4	7.5	2.6	1.40	0.42	-1.96	-2.46	0.48	0.56
PSM 95	1.7	-1.2	2.4	0.2	-0.8	-0.1	-2.34	0.03	0.40	-5.52	1.00	0.83
PSM 87	0.9	2.6	-4.2	-2.3	4.6	-5.3	-0.61	-0.17	-0.91	9.25	-1.47	2.18

Tab. 14: Comparison of coordinate differences from "Werdenfelser Land 1" and
"Werdenfelser Land 2" (see also Tab. 2)

$\Delta\varphi$	Werdenfelser Land 1		Werdenfelser Land 2		Difference in mean
	L_1^* [m]	L_2^* [m]	L_1 [m]	L_2 [m]	L_1^*/L_2^* - L_1/L_2 [cm]
1 - 2	-493.863	.833	.808	.779	-5.5
1 - 3	-708.880	.845	.835	.801	-4.5
1 - 4	-928.440	.419	.421	.387	-2.6
1 - 5	-1188.340	.347	.354	.375	+2.1
1 - 6	-1161.363	.328	.330	.334	-1.4
1 - 7	-1125.481	.482	.521	.457	+0.8
$\Delta\lambda$					
1 - 2	590.917	.912	.909	.907	+0.7
1 - 3	1464.977	.976	.976	.970	+0.4
1 - 4	2585.347	.345	.342	.346	+0.2
1 - 5	3242.477	.474	.484	.481	-0.7
1 - 6	3645.057	.053	.056	.053	+0.1
1 - 7	4124.890	.889	.840	.888	+2.6
ΔH					
1 - 2	6.489	.477	.505	.495	-1.7
1 - 3	11.960	.917	.944	.921	+0.6
1 - 4	16.770	.770	.770	.771	-0.1
1 - 5	14.510	.505	.551	.556	-4.6
1 - 6	11.483	.449	.454	.467	+0.6
1 - 7	9.542	.553	.582	.536	-1.2

Tab. 15: Comparison of relative kinematic positioning results
with those derived in stationary models

Baseline Components	Werdenfelser Land 1 Static - Kinematic			Werdenfelser Land 2 Static - Kinematic		
	L_1 [cm]	L_2 [cm]	Mean [cm]	L_1 [cm]	L_2 [cm]	Mean [cm]
$\Delta\varphi$(5-6)	4.7	-1.3	1.7	0.1	-3.5	-1.7
$\Delta\lambda$(5-6)	-1.8	-0.7	-1.3	-0.9	0.0	-0.5
Δh(5-6)	-3.4	-1.0	-2.2	3.5	2.3	2.9

Tab. 16: Discrepancies of relative kinematic positioning results of "Sydney Airport" to precise terrestrial airport 3d-network (true errors)

| STATION | RESIDUALS WITH RESPECT TO THE LOCAL A.M.G. REFERENCE SYSTEM (East, North, Height) | | | | | | | | |
| | L_1 | | | L_2 | | | MEAN L_1/L_2 | | |
	E	N	H	E	N	H	E	N	H
PSM 90	1.0	-1.7	-2.1	2.9	-0.9	-1.7	1.9	-1.3	-1.9
PSM 95	0.4	-0.4	1.1	0.3	0.0	0.3	0.4	-0.1	0.7
PSM 87	1.4	3.1	-0.4	1.3	3.0	-3.3	1.4	3.0	-1.9
PSM 89	0.3	-0.3	-1.0	0.5	-0.9	0.1	0.4	-0.6	-0.5
PSM 87	-1.2	-0.2	2.3	-1.6	0.9	3.1	-1.4	0.3	2.7
PSM 95	0.6	0.5	-2.0	0.9	-0.4	-0.2	0.8	0.0	-1.1
PSM 90	-2.5	-1.2	2.1	-4.4	-1.9	1.7	-3.5	-1.5	1.9

Tab. 17: Summary of statistics for the relative kinematic GPS positioning tests on land. RMS errors (1) and (2) computed from differences between forward and backward run. RMS errors (3) and (4) computed from residuals to precise terrestrial coordinate system.

| RMS ERRORS IN CM | WERDENFELSER LAND 1 | | WERDENFELSER LAND 2 | | SYDNEY AIRPORT | |
	HORIZONTAL	HEIGHT	HORIZONTAL	HEIGHT	HORIZONTAL	HEIGHT
(1) L_1	1.3	1.9	0.9	1.5	2.3	2.4
(2) L_2	1.8	2.9	1.1	1.4	4.6	2.7
(3) True errors L_1/L_2 single run	——	——	——	——	2.3	1.7
(4) True errors L_1/L_2 double run	——	——	——	——	0.6	0.4

6. CONCLUSIONS

It is shown that accuracies of 1 - 2 cm (over distances up to 10 km) can be achieved in terrestrial kinematic differential GPS positioning using dual-frequency receivers. We also believe that with this accuracy the potential of GPS is fully employed (at least over this short range). The various error sources of GPS (the noise of a phase measurement is already of the order of 3 mm) in combination with errors in centering exactly tripod and antenna over a survey mark will always result in a realistic minimum error budget of a baseline of about 1 cm (independent of the length).

The various tests with respect to different tracking bandwidth between 0.7 Hz and 16 Hz did not show differences in the achieved results of the land tests. However, it could be seen, especially in the data of the flight tests where we had a tracking bandwidth of 16 Hz, that the reliable detection and removal of cycle slips is more complicated or even questioned. This means, that for practical surveys the tracking bandwidth has to be chosen very carefully. On the data reduction side more efforts have to be concentrated on an improved and reliable cycle slip correction algorithm. Even in case that future receiver hardware will minimize or avoid cycle slips caused by the own electronics, the transfer of the kinematic method into practical surveying may be still limited to areas without high buildings or many trees where again a shadowing of the GPS antenna and subsequent loss of lock may appear. Thus, efforts are already carried out towards the integration with external sensors, e.g. low-cost strapdown systems to solve the cycle slip problem in a unique way.

It can be concluded that the application of GPS in the kinematic mode meets nearly all requirements of practical surveying and results in considerable saving of time in comparison to the stationary mode.

ACKNOWLEDGEMENTS. The support of the German Research Foundation (Deutsche Forschungsgemeinschaft) is gratefully achnowledged. The flights as well as the installation of the equipment in the aircraft was done by Wehrtechnische Dienststelle für Luftfahrzeuge, Manching. We further thank the Bundesamt für Wehrtechnik und Beschaffung, Koblenz, the U.S. National Geodetic Survey, Rockville, Md. for borrowing the flight antenna from Dorne & Margolin, Inc., N.Y., Zeiss, Oberkochen for lending the prototype of a photogrammetric camera RMK 15/23 with electronic shutter impulse as well as Prof. Dr. Lange and his coworker, Institut für Grundgebiete der Elektrotechnik und Mechanik, University FAF Munich, for building the frequency generator for the synchronization of GPS receiver and photogrammetric camera. The test at Sydney airport was carried out with the help of the Australian Army Survey Squadron. We are grateful to Lieutenant Colonel H.E. Hansen and Capt. J.D. Mobbs, Paddington, N.S.W., Australia. The support of the University of New South Wales, Kensington, in particular Prof. Dr. Stolz, during the sabbatical of the first author at that University is gratefully acknowledged.

REFERENCES

BASTOS,L. H.LANDAU (1988): *Fixing Cycle Slips in Kinematic GPS-Applications Using Kalman Filtering.* Submitted to Manuscripta Geodaetica.

CANNON,M.E., K.P.SCHWARZ, R.V.C.WONG (1986): *Kinematic Positioning with GPS. An Analysis of Road Tests.* In: Proceedings of the Fourth International Geodetic Symposium on Satellite Positioning, April 28 - May 2, 1986, Austin, Texas, Vol. 2, pp. 1251-1267.

DARNELL,A.R., L.L.HAWKINS (1986): *User's Guide to GESAR Version 1.1 (With Amendments from B.G.Jones of 19 June 1986).* NSWC, Dahlgren, VA.

GOAD,C. (1986): *Precise Positioning with the Global Positioning System.* In: Proceedings of the Third International Symposium on Inertial Technology for Surveying and Geodesy, Sept. 16-20, 1985, Banff, Canada, pp. 745-756.

HATCH,R. (1986): *Dynamic Differential GPS at the Centimeter Level.* In: Proceedings of the Fourth International Geodetic Symposium on Satellite Positioning, April 28 - May 2, 1986, Austin, Texas, Vol. 2, pp. 1287-1298.

HEIN,G.W., A.STOLZ (1988): *Kinematic GPS for Local Crustal Deformation Monitoring.* Paper to be presented at the Spring Meeting of the American Geophysical Union, Baltimore.

KALAFUS,R.M., A.J.VAN DIERENDONCK, N.A.PEALER (1986): *Special Committee 104 Recommendations for Differential GPS Service.* In: Navigation. Journal of the Institute of Navigation, Vol. 33, No. 1, pp. 26-41 and Vol. 33, No. 2, pp. 160-161.

KLEUSBERG,A., S.H.QUEK, D.E.WELLS, G.LACHAPELLE, J.HAGGLUND (1986): *GPS Relative Positioning Techniques for Moving Platforms.* In: Proceedings of the Fourth International Geodetic Symposium on Satellite Positioning, April 28 - May 2, 1986, Austin, Texas, Vol.2, pp. 1299-1310.

KLEUSBERG,A. (1986): *Kinematic Relative Positioning Using GPS Code and Carrier Beat Phase Observations.* In: Marine Geodesy, Vol. 10, No. 3/4, pp. 257-274.

KRABILL,W.B., C.F.MARTIN (1987): *Aircraft Positioning Using Global Positioning System Carrier Phase Data.* In: Navigation, Journal of the Institute of Navigation, Vol. 34, No. 1, pp. 1-21.

LACHAPELLE,G., M.CASEY, R.M.EATON, A.KLEUSBERG, J.TRANQUILLA, D.WELLS (1987a): *GPS Marine Kinematic Positioning Accuracy and Reliability.* In: The Canadian Surveyor, Vol. 41, No. 2, pp. 143-172.

LACHAPELLE,G., W.FALKENBERG, J.HAGGLUND, D.KINLYSIDE, M.CASEY, P.KIELLAND, H.BOUDREAU (1987b): *Shipborne GPS Kinematic Positioning with Emphasis on Survey Launch Applications.* Presented at: General Assembly of the International Association of Geodesy, Vancouver, Aug. 87, 16 pp.

LACHAPELLE,G., W.FALKENBERG, M.CASEY, (1987c): *Use of Phase Data for Accurate Differential GPS Kinematic Positioning*. In: Bulletin Geodesique, Vol. 61, pp. 367-377.

LACHAPELLE,G., J.HAGGLUND, H.JONES, M.EATON (1984): *Differential GPS Marine Navigation*. In: PLANS 84, San Diego, Nov.28-30, 11 pp.

LACHAPELLE,G., J.HAGGLUND, W.FALKENBERG, P.BELLAMEARE, M.CASEY, M.EATON (1986): *GPS Land Kinematic Positioning Experiments*. In: Proceedings of the Fourth International Geodetic Symposium on Satellite Positioning, April 28 - May 2, 1986, Austin, Texas, Vol. 2, pp. 1327-1344.

LACHAPELLE,G., J.LETHABY, M.CASEY (1984): *Airborne Single Point and Differential GPS Navigation for Hydrographic Bathymetry*. In: The Hydrographic Journal, Vol. 34, October 1984, pp. 11-18.

LANDAU,H. (1987): *Precise Kinematic GPS Positioning. Part 1: Experiences on a Land Vehicle Using TI 4100 Receivers and Software*. Paper presented at the IUGG Meeting, Vancouver 1987, 16p.

LANDAU,H. (1988): *Zur Nutzung des Global Positioning Systems in Geodäsie und Geodynamik: Modellbildung, Softwareentwicklung und Analyse*. Ph.D. Thesis (in preparation). Institute of Astronomical and Physical Geodesy, University FAF Munich

LUCAS,J.R., G.MADER (1987): *Successful Demonstration of Aerotriangulation without Ground Control*. Manuscript, NOAA/NGS, 11 pp.

MADER,G. (1986a): *Decimeter Level Aircraft Positioning Using GPS Carrier Phase Measurements*. In: Proceedings of the Fourth International Geodetic Symposium on Satellite Positioning, April 28 - May 2, 1986, Austin, Texas, Vol. 2, pp. 1311-1325.

MADER,G., W.CARTER, B.DOUGLAS (1986a): *Decimeter Precision Aircraft Positioning Using GPS Carrier Phase Measurements*. Submitted to Navigation, Journal of the Institute of Navigation, 28 pp.

MADER,G. (1986b): *Dynamic Positioning Using GPS Carrier Phase Measurements*. Manuscripta Geodaetica, Vol. 11/4, pp. 272-277.

McDONALD,K.D., B.W.PARKINSON, C.P.McDONALD (1987): *A Survey of GPS User Equipment, Applications and Receiver Technology Trends*. Presented at The Institute of Navigation Satellite Division First Technical Meeting, Colorado Springs, CO, Sept. 21-25, 1987.

REMONDI,B.W. (1985a): *Performing Centimeter Accuracy Relative Surveys in Seconds Using GPS Carrier Phase*. In: Proceedings of the First International Symposium on Precise Positioning with the Global Positioning System, Rockville, MD, pp. 789-798.

REMONDI,B.W. (1985b): *Performing Centimeter-Level Surveys in Seconds with GPS Carrier Phase: Initial Results*. NOAA Technical Memorandum NOS NGS-43.

REMONDI,B.W. (1986): *Performing Centimeter-Level Surveys in Seconds with GPS Carrier Phase: Initial Results*. In: Proceedings of the Fourth International Geodetic Symposium on Satellite Positioning, April 28 - May 2, 1986, Austin, Texas, Vol. 2, pp. 1229-1250.

SCHWARZ,K.P. (1987): *Approaches to Kinematic Geodesy*. In: IAG, Section IV (Eds.): Contributions to Theory and Methodology, XIX General Assembly of the IUGG, August 9-22, Vancouver, Canada, pp. 29-47, Publication 60006, Calgary.

SCHWARZ,K.P., M.CANNON, R.V.C.WONG (1987): *The Use of GPS in Exploration Geophysics - A Comparison of Kinematic Models*. Paper presented at the IUGG Meeting, Vancouver 1987, 20 pp.

SEEBER,G., A.SCHUCHARDT, G.WÜBBENA (1986): *Precise Positioning Results with TI 4100 GPS Receivers on Moving Platforms*. In: Proceedings of the Fourth International Geodetic Symposium on Satellite Positioning, April 28 - May 2, 1986, Austin, Texas, Vol. 2, pp. 1269-1286.

TEXAS INSTRUMENTS TI 4100 NAVSTAR Navigator (1984): *Owner's Manual*. Texas Instruments, GPS Navigation Systems, Lewisville, TX.

WONG,R.V.C., K.P.SCHWARZ, J.HAGGLUND, G.LACHAPELLE (1985): *Integration of Inertial and GPS-Satellite Techniques for Precise Marine Positioning*. In: Marine Geodesy, Vol. 9, No. 2, pp. 213-226.

Comparison of Horizontal Positions (L2–L1)

Project: Werdenfelser Land 1

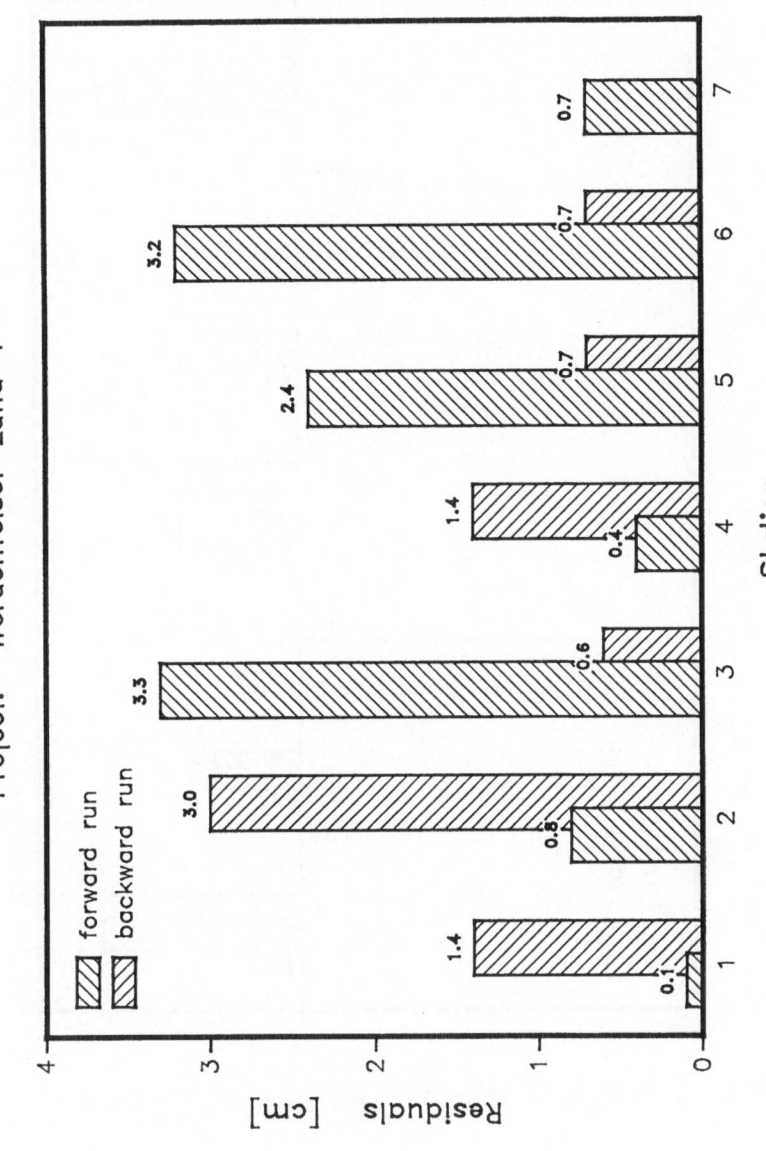

Figure 11

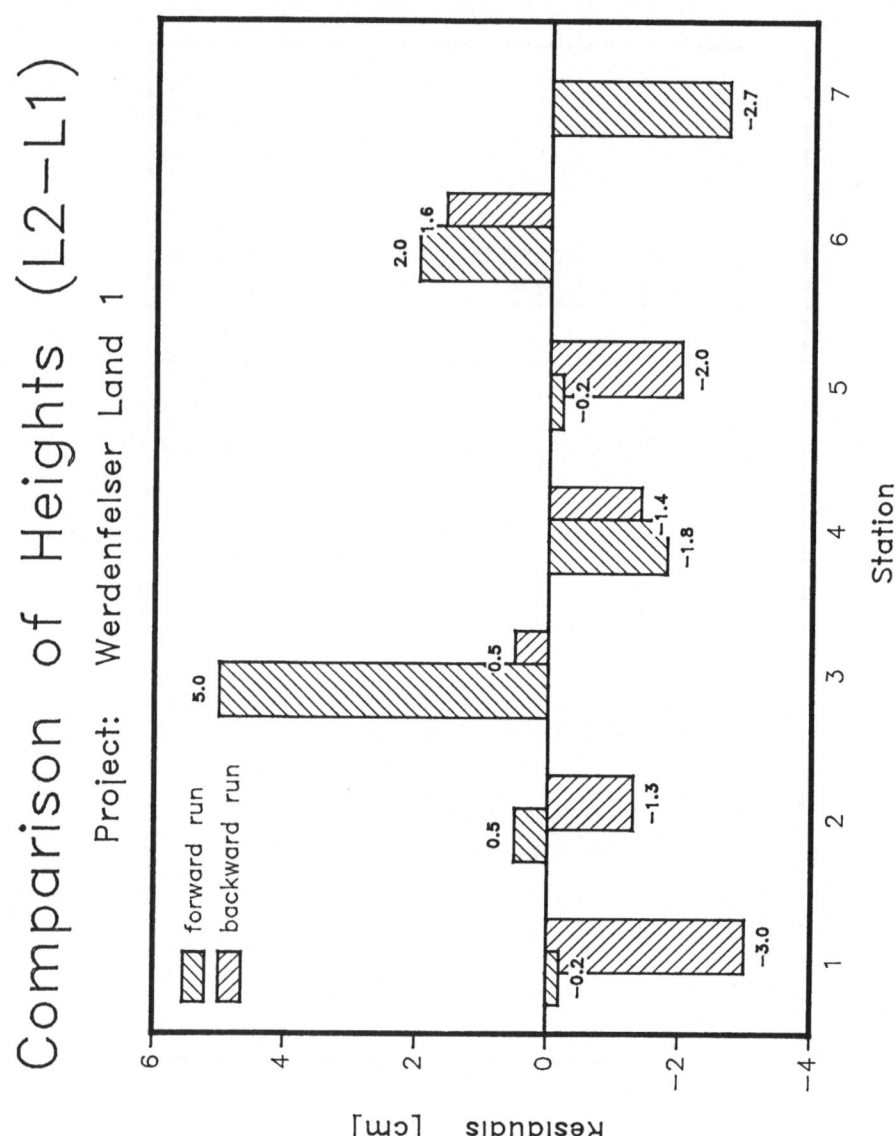

Figure 12

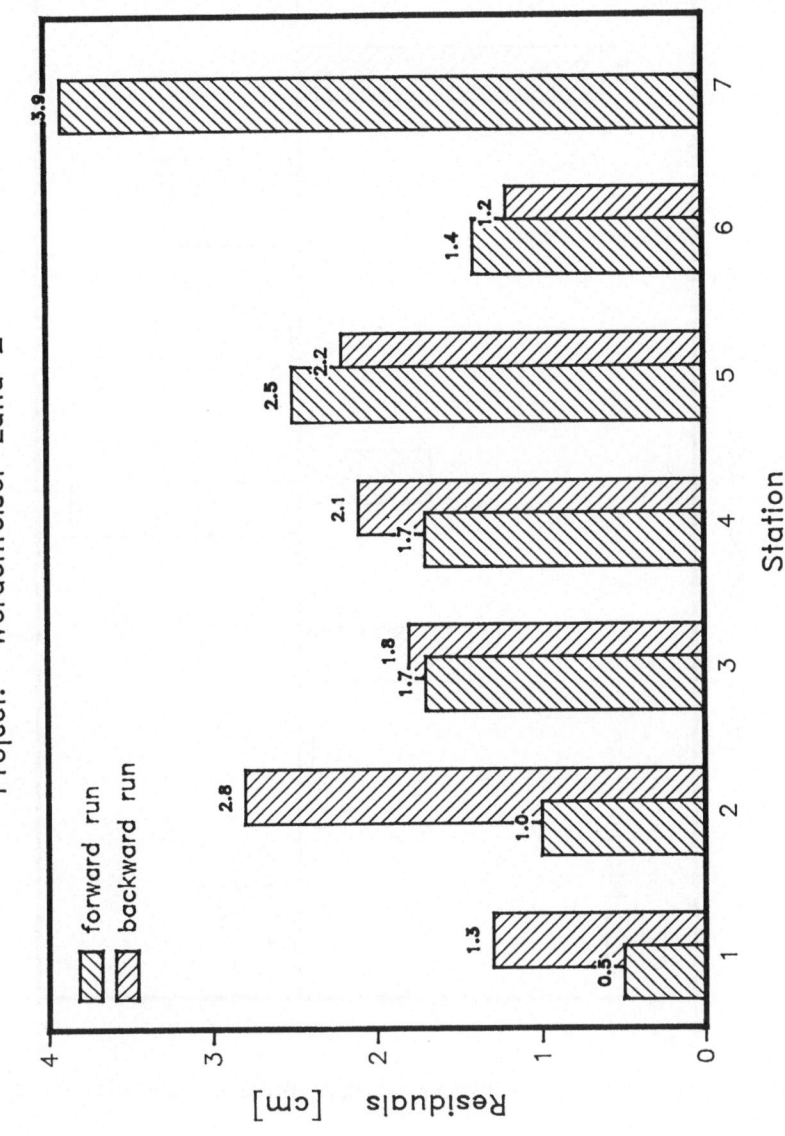

Comparison of Horizontal Positions (L2−L1)

Project: Werdenfelser Land 2

Figure 13

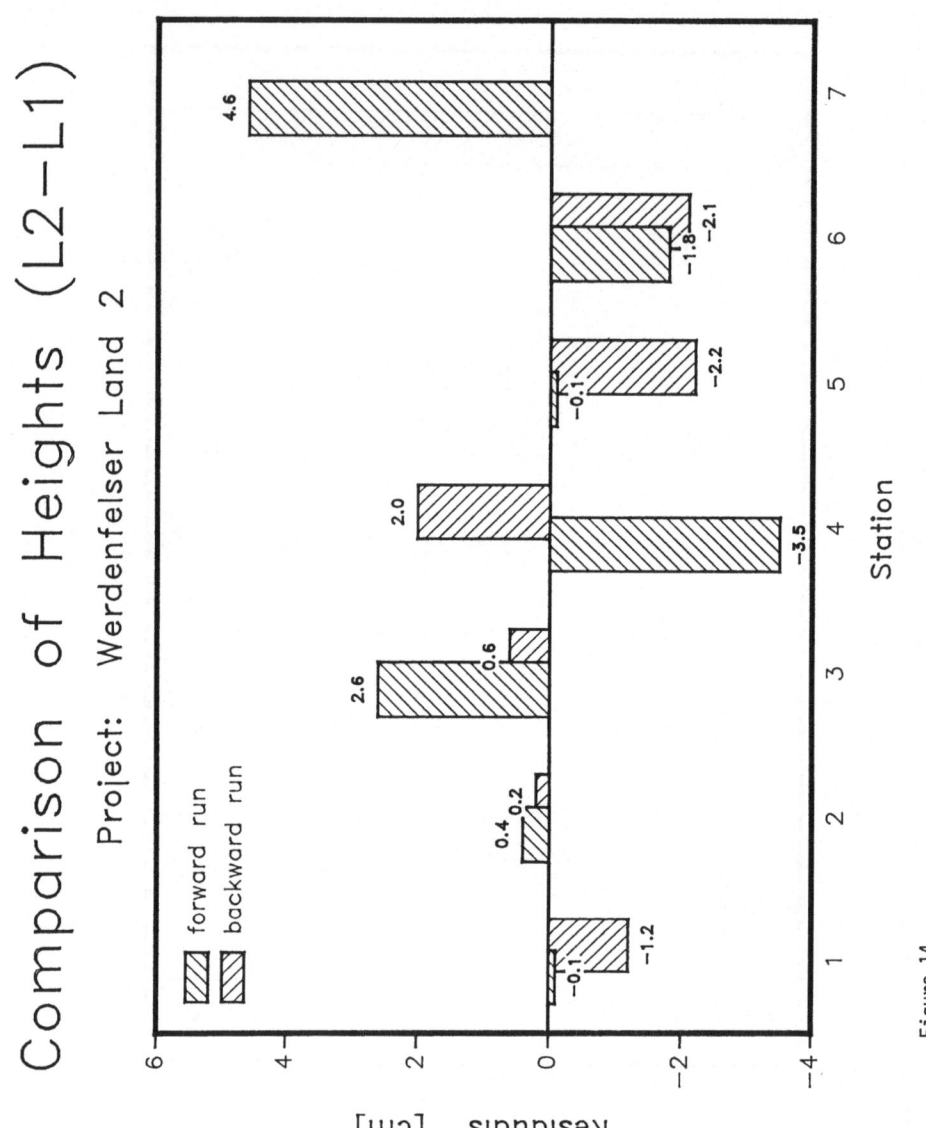

Figure 14

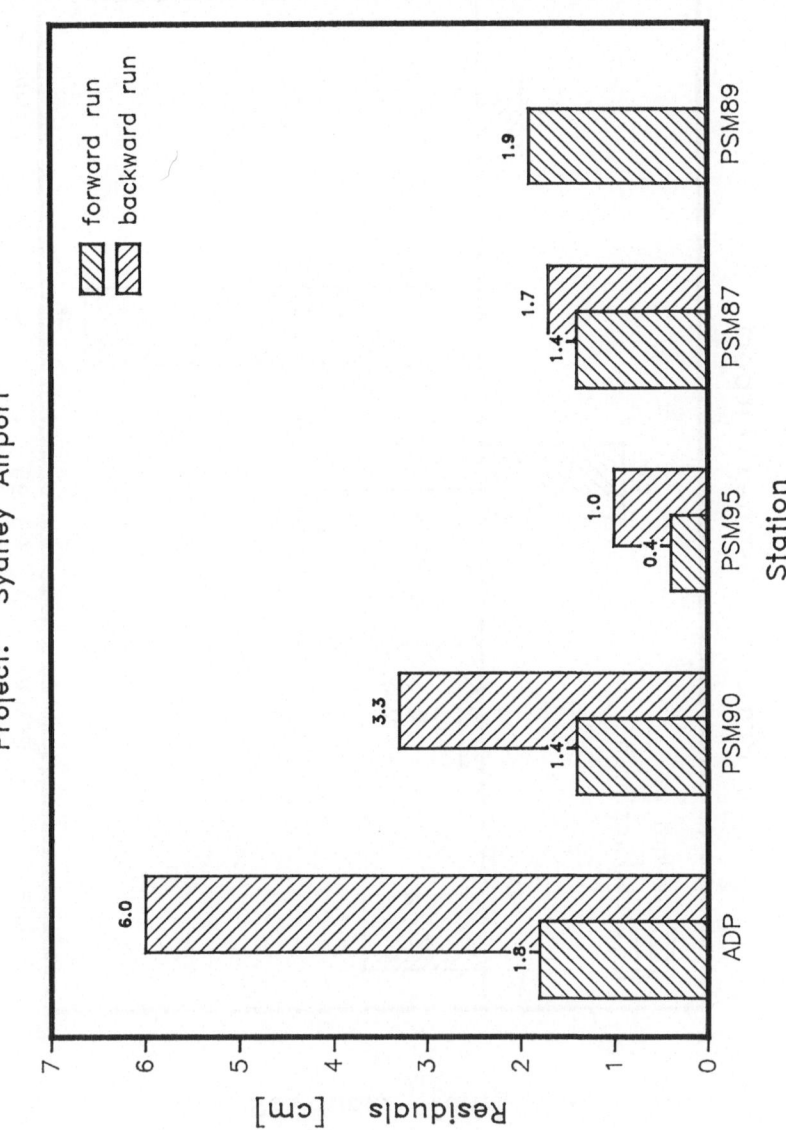

Figure 15

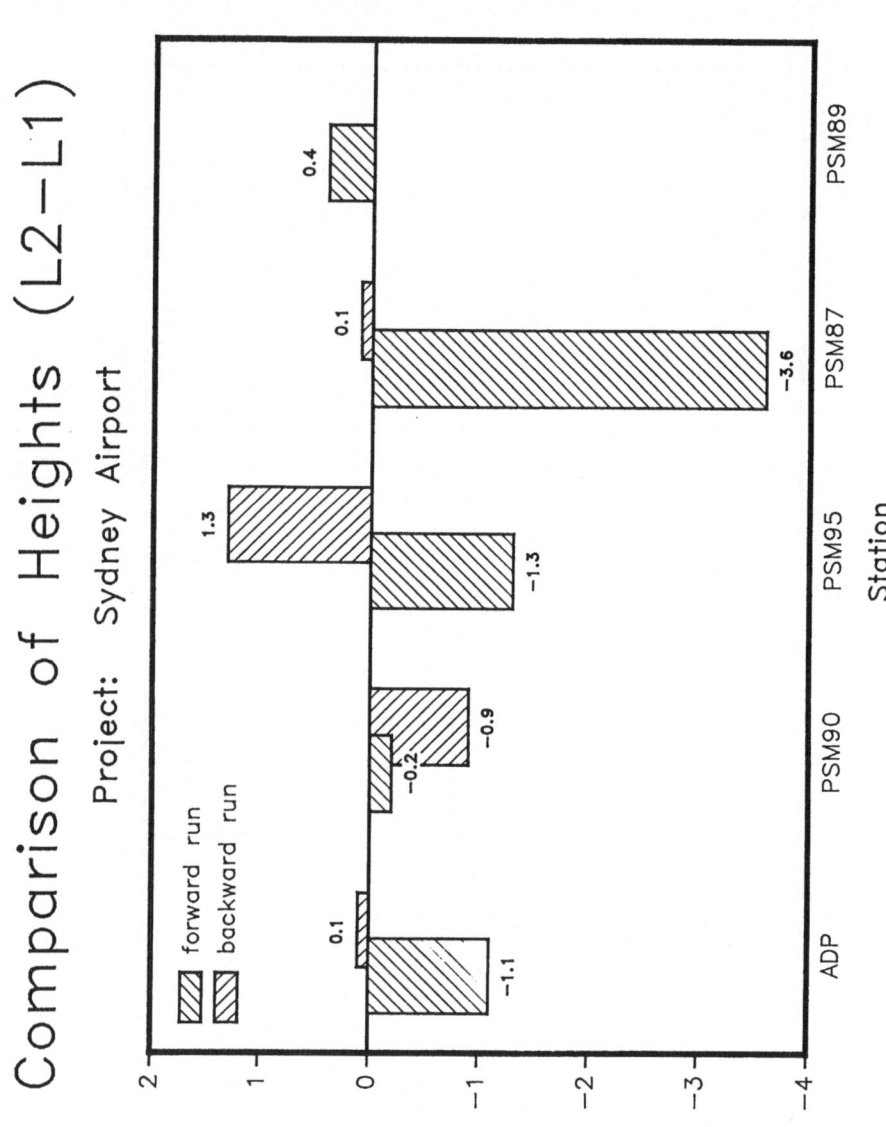

Comparison of Heights (L2–L1)

Project: Sydney Airport

Figure 16

True Errors of Horizontal Positions (L2/L1)

Project: Sydney Airport

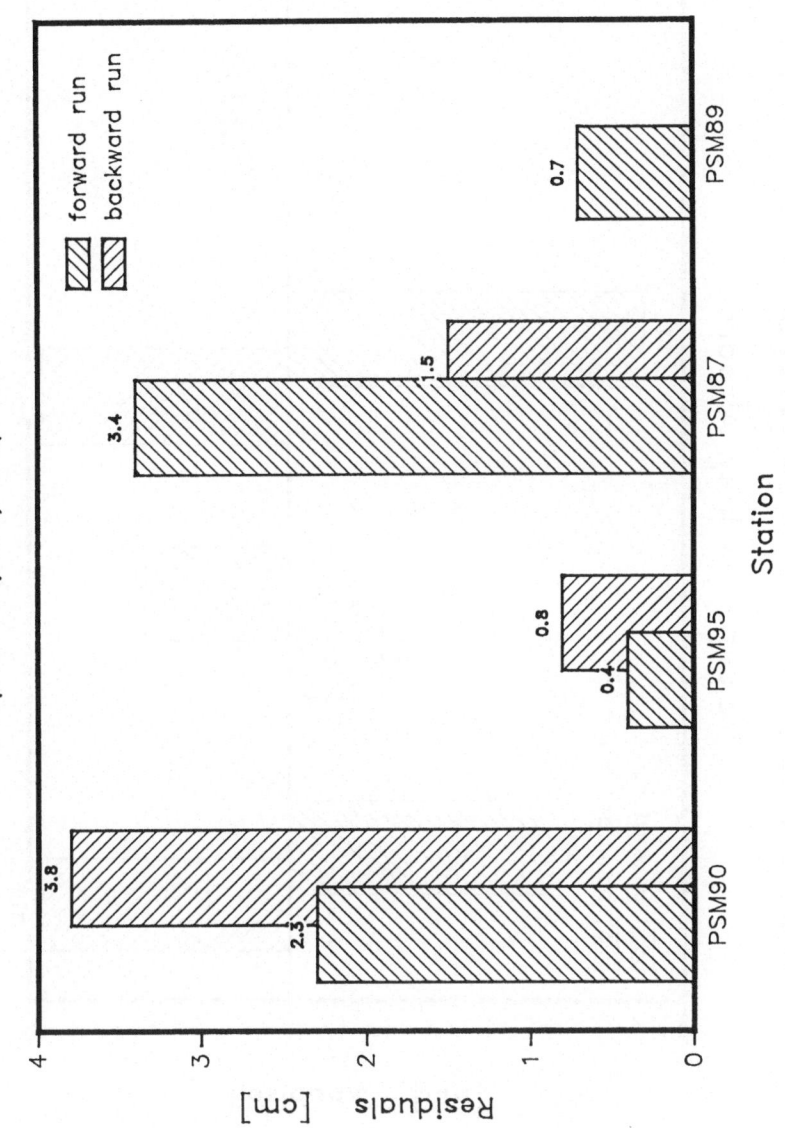

Figure 17

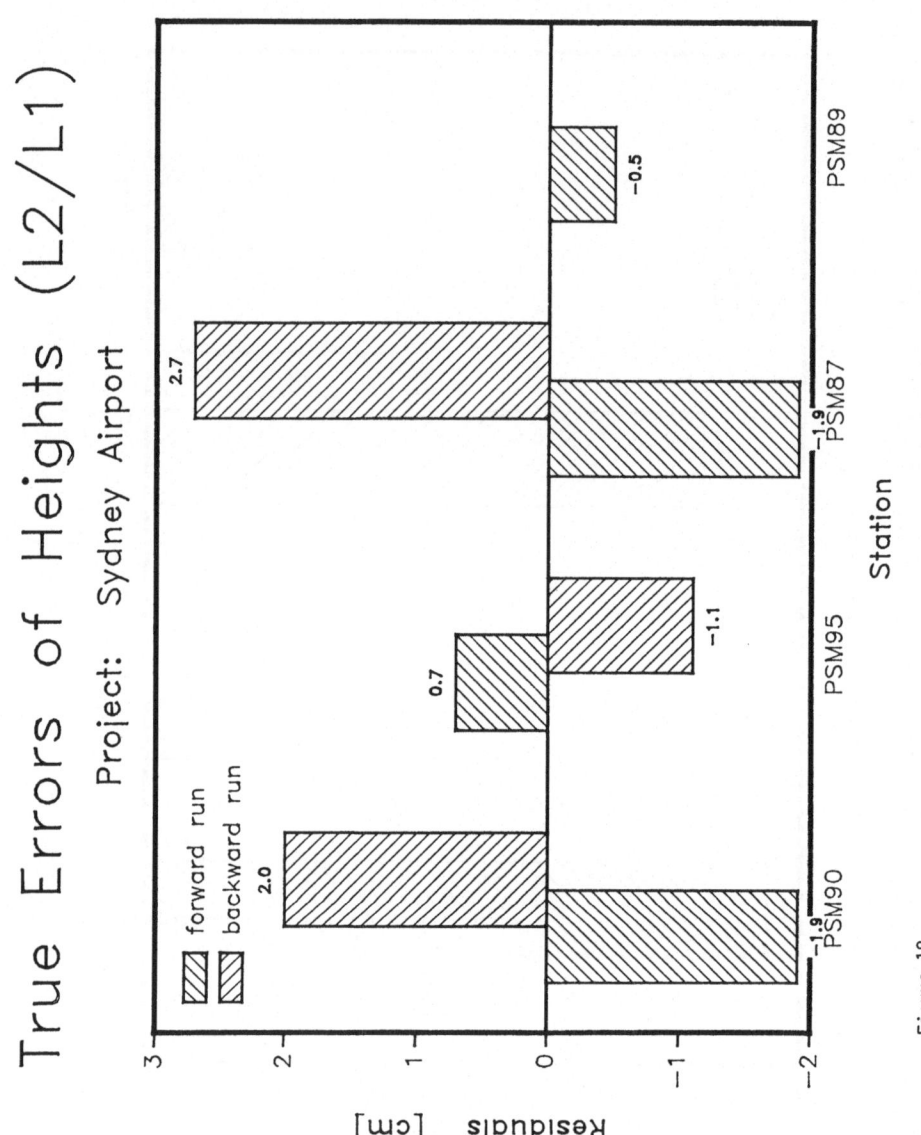

Figure 18

IONOSPHERIC RESIDUAL

STATION : Sydney Airport Remote   Channel : 1

Date : 18-10-1987   GPS Week : 406   Day of year : 291

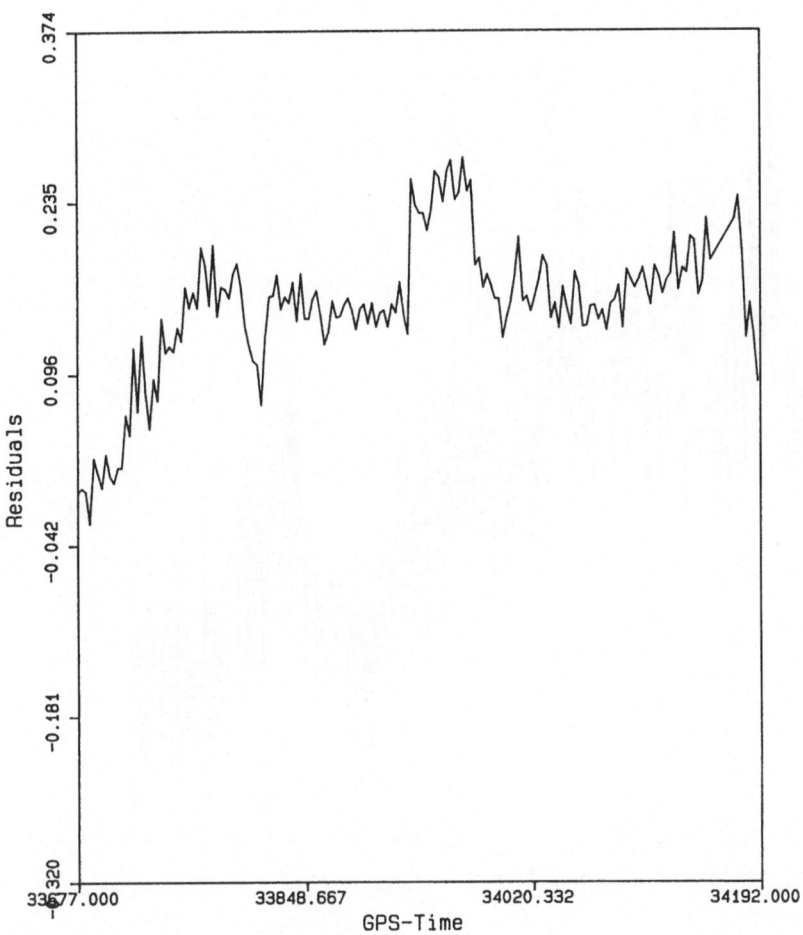

Fig. 19 Example of the noise in the ionospheric residual in the land test
         "Sydney airport"

IONOSPHERIC RESIDUAL

STATION : DO 28 – Flying Remote   Channel : 4

Date : 17-8-1987   GPS Week : 397   Day of year : 229

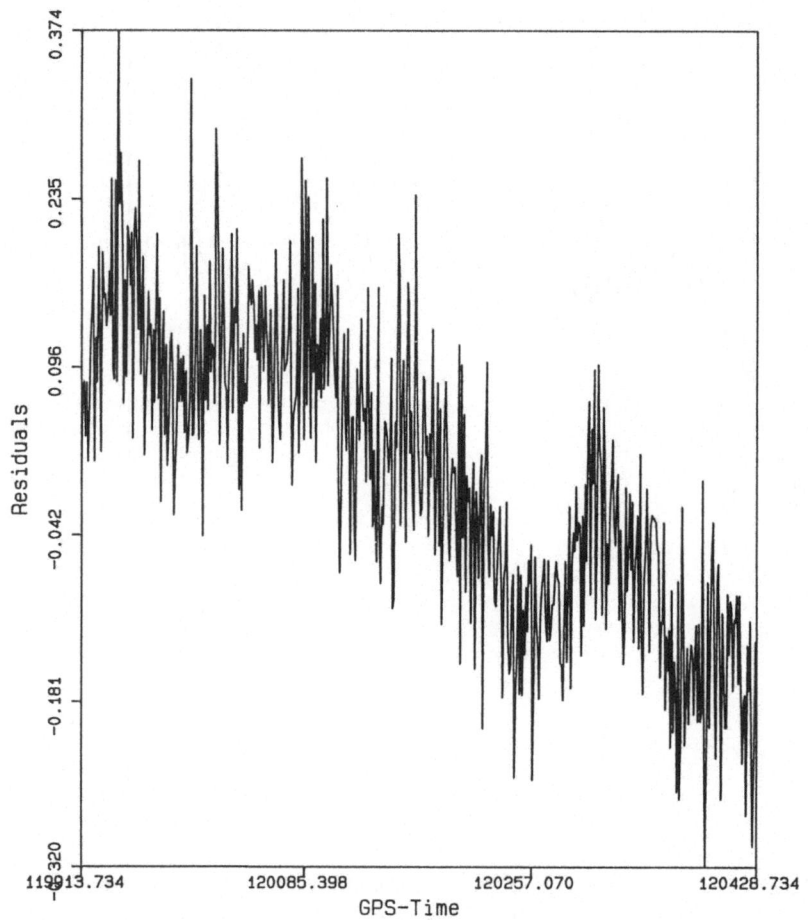

Fig. 20   Example of the noise in the ionospheric residual in the flight
test "München-Freimann"

FIELD VALIDATION OF GPS PHASE MEASUREMENTS

by

Paul A Cross and Noordin Ahmad

Abstract

The GPS phase observable is well known to be subject to a number of errors, the most common of which is the so-called *cycle slip*. A cycle slip occurs when the receiver momentarily loses lock on a satellite, resulting in an error in the integer part of the phase observable of one or more cycles. It is essential to locate such errors and, if possible, repair them by adding back the appropriate number of cycles. A number of strategies are available. Here they are reviewed and a new strategy based on data collected at a single station only is presented. The new strategy has been implemented for the Trimble 4000S system and a computer program that can run on a microcomputer produced. This program can be used in the field to check the quality of GPS phase data before leaving the observing site.

1. INTRODUCTION

The GPS signal is rather complex and can be used in a number of different ways for the determination of absolute and relative positions, depending on the purpose of the survey and on the equipment available, King et al (1987). For relative positioning the most commonly used observable is the so-called *accumulated phase observable* sometimes referred to as the *continuous phase observable*.

Essentially the pseudo-random codes and data message are removed from the broadcast signal (either by signal squaring or by use of prior knowledge of them) to yield signals of frequencies equal to or (in the case of signal squaring) double that of the L1 carrier and, in the case of a dual frequency receiver, also the L2 carrier. Then, as explained in Remondi (1985), the phase difference between the received carrier and a signal generated inside the receiver is measured at prescribed epochs of time (as kept by the receiver oscillator). This phase difference is known as the *fractional phase*. Also the phase difference is constantly monitored and each time it changes from 360 degrees to zero a counter is incremented (note that this phase difference is not constant due to the varying Doppler shift of the received signal due to the satellite motion). The counter then yields the so-called *integer count* and the sum of this with the fractional phase yields the accumulated phase observable.

If the receiver momentarily (or for a more significant period of time) loses 'lock' on the satellite (e.g. due to radio interference or the satellite passing behind an obstruction) and then 'locks-on' again the fractional phase will be unaffected (i.e. it will be identical to that which it would have been without a loss of lock) but the integer count will be in error (a number of cycles will have been lost). The result will be that the accumulated phase observable will be in error by an integer number of cycles and the phenomenon known as a *cycle slip* is said to have occurred.

Essentially there are two problems associated with cycle slips: *detection* and *repair* (i.e. determining the number of missing cycles and correcting the observations). This paper addresses these problems. It begins with a brief review of a number of published techniques for cycle slip detection and repair and then introduces a new method that, although not as effective as some of the existing methods at dealing with very small slips, has the very special advantage of needing data from only a single site. Hence it can be used for the field validation of the data. Details of some test results with the new method are included.

350

2. REVIEW OF CURRENT METHODS

Various methods have been developed to detect and correct cycle slips and to edit other types of gross error (such as clock jumps). Some of the methods are dependent only on the data collected by the receiver while others can only be applied if external information, such as good quality approximate station coordinates, is made available. This section will review some of the 'well-known' methods that have been, and are being, used successfully.

One of the earliest approaches was to edit manually the data by examining residuals on the computer screen after the single or double difference solutions had been carried out. This method works if the station positions are known quite accurately. The manual editing is, however, rather time consuming and takes most of the processing time if there are many cycle slips in the data. The cycle slip can be approximately repaired by use of a knowledge of the sizes of the residuals.

Alternatively the cycle slip can be repaired by introducing another integer ambiguity as an additional parameter (Wei, 1985). Such parameters become part of the unknown vector which is solved for in the estimation process. This method has a number of disadvantages although it does have the obvious advantage of yielding a direct least squares estimate of the size of the cycle slip. The disadvantages include the following.

(a) The presence of a large number of additional parameters will dilute the final solution.

(b) If there are a large number of cycle slips the normal equation matrix involved in final computation may become rather large and special methods may be required to handle it.

Fitting a piecewise continuous polynomial to the single difference or double difference observations is a method that has been successfully implemented by the University of New Brunswick in their GPS software DIPOP (Beutler et al, 1984, Vanicek et al, 1985) and by the Geodetic Survey of Canada in their software VECA (Delikaraoglou, 1984).

From an initial (either manual or automatic) analysis of the residuals the data is divided into slip free subintervals. A low degree polynomial is first used to fit the residuals of the differences between the observed single differences and their computed counterpart. The polynomials between adjacent subintervals are compared and if there has been a cycle slip the first terms of the polynomials should be different by an amount equal to the size of the cycle slip. This amount is added to the all future equations. The process is then repeated with the double difference observations.

Remondi (1984) has proposed an automatic approach to cycle slips editing. Using the triple difference algorithm, which is insensitive to satellite and receiver clock errors and to the unknown integer ambiguity, the residuals are examined automatically to identify any large values and hence any large cycle slips. Note that the effect of a cycle slip remains local to individual triple differences (because triple differences involve differencing with respect to time) and hence cycle slips are usually easily spotted in their respective residuals (although it is unlikely to be seen in only one residual). The corresponding discontinuities in the double differences are then searched for and repaired. The process is usually repeated in a subsequent double difference solution where the final editing of any small cycle slips takes place.

All the methods mentioned in the foregoing involve data from at least two stations. Goad (1985), however, has proposed a simple method of fixing cycle slips at a single station when a dual frequency receiver has been used. The technique is based on the following relationship.

$$\phi^i_j(t)_{L2} = \phi^i_j(t)_{L1} * fL_2 / fL_1 \qquad ..(1)$$

The linear combination d where d is given by

$$d = \phi^i_j(t)_{L1} - \phi^i_j(t)_{L2} * fL_1/fL_2 \qquad ..(2)$$

will then give

$$d = N^i_j(L_1) - fL_1/fL_2 * N^i_j(L_2) + A/fL_1 * (1-fL_1^2/fL_2^2) \qquad ..(3)$$

where

$\phi^i_j(t)_{L1}$ — the accumulated phase observable for L_1 from satellite i at receiver j at time (t)

$\phi^i_j(t)_{L2}$ — the accumulated phase observable for L_2 from satellite i at receiver j at time (t)

fL_1, fL_2 — frequencies of L_1 and L_2 respectively

$N^i_j(L_1)$ — the integer ambiguity for satellite receiver combination i,j on L_1

$N^i_j(L_2)$ — the integer ambiguity for satellite-receiver combination i,j on L_2

A — the ionospheric effect

From (3) the coefficient of $N^i{}_j(L_1)$ is 1.0 and the coefficient of $N^i{}_j(L_2)$ is -1.28. A single slip would therefore cause d to change by either 1.0 or -1.28 depending on whether the slip was in L1 or L2. If both frequencies are affected the change in d would be 0.28.

3. THE NEWCASTLE METHOD

At present the design of procedures for the collection and processing of GPS phase data is simpler than it will be in the future. This is because there is limited satellite coverage, typically phase observations are possible on four or five satellites for about four hours per day, and it is usual to occupy sites for the complete period available (unless the stations are very close, there is generally not time to change site during the observing window). The data from the various simultaneously occupied sites can then usually be brought together for processing before the following day when field work begins again. If there are problems with the data then sites can be reoccupied. When the full constellation is available the situation will be more complicated and groups of receivers will be able to be deployed continuously. This will mean that communication between observing teams will be required and plans for moving receivers made when sufficient good quality data has been collected at the various sites. It will then be essential to have techniques for the field verification of the data.

Realizing this fact, the Newcastle approach is to produce a piece of software that can be used to detect cycle slips at a single station. Moreover, bearing in mind the fact that not all future receivers will be dual frequency, the software needs to be suitable for both single frequency (L1 only) and dual frequency receivers. The ability to be independent of other sources of information, especially station coordinates, is also considered important, i.e. the method should be such that only the phase data itself needs to be taken into consideration.

3.1 The approach

As a satellite approaches a receiver, the received frequency will increase due to the Doppler effect, and vice versa as the satellite moves away. A plot of accumulated phase against time will look similar to Fig. 1.

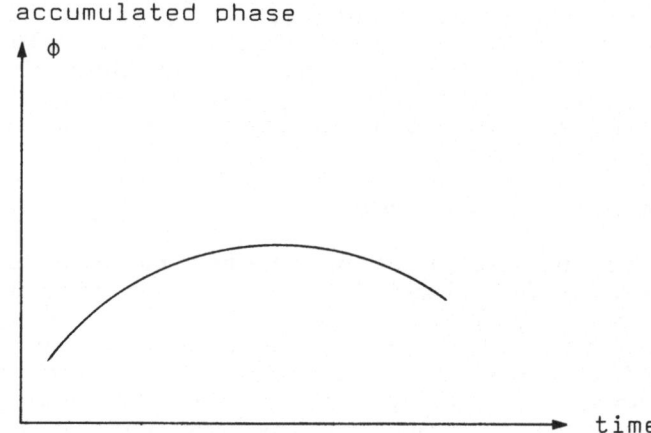

Fig. 1:   Plot of accumulated phase against time

The curve essentially represents the changing geometry due to the satellite motion.   Note the smoothness of the curve.   This smoothness is the basic factor in the development of the Newcastle approach.   As the accumulation of the beat phase increases or decreases steadily with time, the rate of accumulation will behave similarly.

Let the curve of accumulated phase (φ) against time be represented by a function f(t).   The tangent at any point on the curve f(t) represent the rate of accumulation (df(t)/dt) at that point, as in Fig. 2.

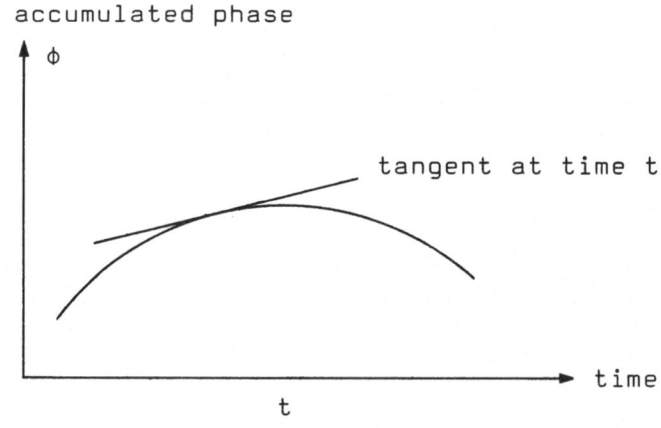

Fig. 2: Slope of accumulated phase curve

Now if a cycle slip of magnitude X occurs at time t, the graph of accumulated phase against time will be as Fig. 3 and the corresponding graph of rate against time is as Fig. 4. Both graphs show the 'jump' when a cycle slip is encountered. The curves are no longer smooth. In Fig. 4 the jump can be completely ignored without affecting the smoothness of the curve. This is not the case with Fig. 3. This suggests that by interpolating the data before and after the jump on a rate-curve will give the expected rate at the time when the slip occurs. The interpolated rate is compared with the computed rate (from the real data) and the difference is multiplied by the period of the jump to give the value of the cycle slip.

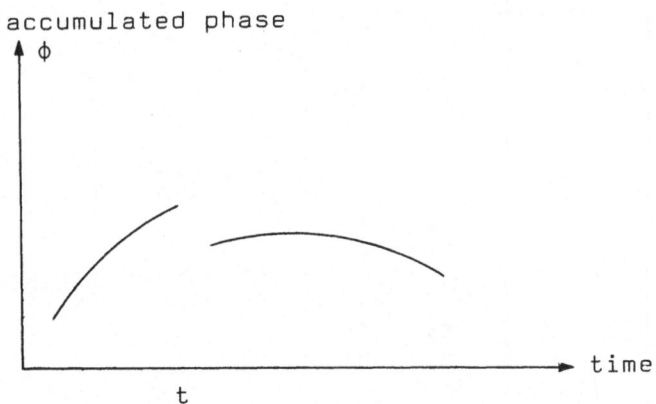

Fig. 3: Cycle slip in accumulated phase plot

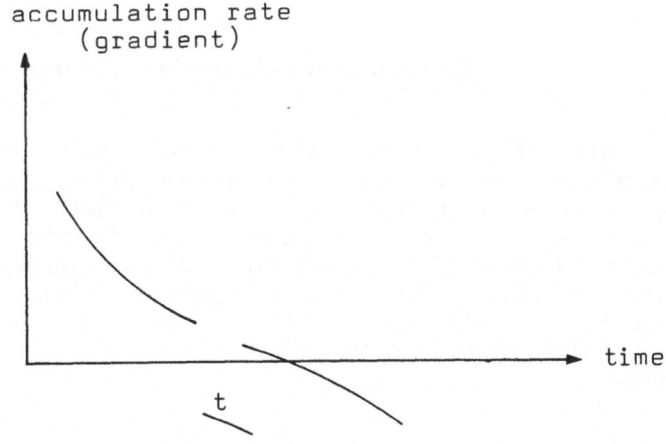

Fig. 4: Cycle slip in rate of accumulation curve

The jump is detected using Newton's divided-difference method and the interpolation is carried out using least squares fitting. Table 1 below shows the Newton's divided-difference technique. This method has been adopted because it, along with the more complicated Lagrangian approach, is one of the very few that is able to handle unequally spaced data (note that the Trimble 4000S system does not record the accumulated phase at exactly even intervals of receiver clock time, although they are all approximately 15 seconds).

x_i	$f[x]$	$\Delta f[x]$	$\Delta^2 f[x]$	$\Delta^3 f[x]$
x_0	f_0			
		$\dfrac{f_1 - f_0}{x_1 - x_0}$		
x_1	f_1		$\dfrac{(f_2-f_1)}{(x_2-x_1)} - \dfrac{(f_1-f_0)}{(x_1-x_0)}$	
		$\dfrac{f_2 - f_1}{x_2 - x_1}$		. . . . .
x_2	f_2		$\dfrac{(f_3-f_2)}{(x_3-x_2)} - \dfrac{(f_2-f_1)}{(x_2-x_1)}$	
		$\dfrac{f_3 - f_2}{x_3 - x_2}$		
x_3	f_3		. . . .	
		$\dfrac{f_4 - f_3}{x_4 - x_3}$		. . . . .
x_4	f_4			

Table 1 : Newton's divided-difference method

After cycle slip detection and the corresponding repair, the data can now be used to generate single, double or triple difference data. Bear in mind that the data is not truly "clean". At Newcastle a double differencing processing strategy is used and, after convergence of the solution, a first difference approach with the residuals is used to isolate any small cycle slips that are still left in the data (note that the sizes of the first differences of the residuals are the estimates of the cycle slips).

3.2 Implementation and status

FORTRAN software using the single station algorithm was initially developed and implemented on the University of Newcastle's mainframe Amdahl computer. The software detects and repairs cycle slips and prepares double difference files for a second FORTRAN program which carries out baseline estimation and searches for further cycle slips in the double difference residuals as described earlier.

Also, a version for the first stage (i.e. error detection using the single station approach) has been written in PASCAL for microcomputers. This latter program has been named TRIMSLIP and is now in use with a number of government agencies and a commercial company who are able to run the program on the same field computer that is used for data logging. Hence they can check the quality of their data on site. TRIMSLIP also gives a summary of the data that has been collected and identifies other data problems besides cycle slips. In particular it can locate problems with initial epochs (poor data is often recorded immediately on locking-on to a satellite, especially when it is at a low altitude) and receiver clock jumps. At the moment it can only accommodate Trimble data.

The next step is to automate the complete the process: initial detection of the larger cycle slips and other data problems, generating double difference data files, estimating baselines from double differences, searching for remaining cycle slips and editing the original double differences ready for a further baseline estimation. Parallel work at Newcastle is leading to the development of a rigorous network double difference processing package.

4. RESULTS AND DISCUSSION

The foregoing method has been used to assess the data quality from a 1986/87 GPS campaign on the Newcastle test network using four Trimble 4000S GPS receivers. This network has four stations with baseline lengths varying between about 3km to about 35km. Table 2 contains a sample of the results that have been achieved from one of the observing sessions using the single station approach.

The phases observed at the four stations were corrected for the cycle slips shown in Table 2 and double difference baseline estimation was carried out. Table 3 then shows the cycle slips that were subsequently found by this method (i.e. those that were missed by the single station method).

It can be seen from Table 3 that, except for measurements
at time 394575 (measured in GPS seconds since the
beginning of the week), the single station method has
repaired all cycle slips within one cycle. In fact the
only place at which one cycle slip still remained was at
station CL at time 396465. Note that the reason why
complete rows of single cycle slips appear in Table 3 is
that the remaining cycle slip is in the measurements to
SV9 which, by chance, has been chosen as the reference
satellite in the double differencing.

Station	SV No	Time (GPS secs)	cycle slip
PP	9	394575	21
		396360	-17985092
		396405	-196819
		396435	-71487
	11	394575	-34
	12	394575	43
	13	391020	458633
		391080	229039
		394575	-61
GC		no cycle slips detected	
MM	9	396750	-18275906
		396780	-72084
CL	9	396465	-18726637

Stations

PP - Physic Penthouse   GC - Gosforth Church
MM - Mickley Moore   CL - Collier Law

Table 2: Cycle slips detected by single station method

It can be seen by examining Table 2 that at time 394575 a
loss of lock occurred on all satellites at station PP. In
fact, although the recovered number of missing cycles is
small, the receiver at PP did not record for a period of
10 minutes 45 seconds. In such circumstances a rather long
curve has to be fitted to the accumulated phase rate curve
and cycle slips cannot be expected to be determined very
accurately.

Baseline	Time (GPS secs)	Final correction in double differencing (SV combination) 9:11 9:12 9:13		
PP - MC	394575	23	-13	15
GC - CL	396465	1	1	1
MC - CL	396465	1	1	1
PP - CL	394575	23	-13	15
	396465	1	1	1
GC - MM		none detected		
PP - GC	394575	22	-13	15

Table 3: cycle slips detected from double differences

Tests have also been carried out with simulated cycle slips in the same data set, i.e. known cycle slips of varying sizes have been injected into the data and an attempt made to recover them. It has been found that, so long as there is no loss of data, i.e. the period of loss of lock occurs during the interval between the recording of the accumulated phase at two adjacent epochs, the single station method can repair cycle slips perfectly. As data loss increases so the performance of the method deteriorates due to the need to fit longer curves. The rate of this deterioration is a function of the rate of accumulation of phase, the higher the rate the better the cycle slip detection, i.e. the method is least effective at the point of closest approach of the satellite.

This point does not in any way affect the usefulness of the method as a field validation procedure as losses of lock longer than the recording interval will be seen by the fact that data is missing.

5. CONCLUSIONS

Tests with real and simulated data show that the Newcastle single station cycle slip detection method can, in the case of very short losses of lock, repair cycle slips perfectly. The method performs less well in the case of longer losses of lock with errors of around twenty cycles still remaining after about ten minutes.

The method has hence been shown to be an extremely useful field data validation tool.

REFERENCES

BEUTLER G, DAVIDSON D A, LANGLEY R B, SANTERRE R, VANICEK P A and WELLS D E (1984) Some theoretical and practical aspects of geodetic positioning using carrier phase difference observations of GPS satellites. University of New Brunswick, Department of Surveying Engineering, Technical Report No 109.

DELIKARAOGLOU D (1984) 'VECA-1' A Vector Adjustment Program for Differential GPS Observations - A Reference Guide, Technical Report No 2, Geodetic Survey of Canada.

GOAD C C (1985) Precise positioning with the Global Positioning System. Proc. 3rd Int. Symp. on Inertial Technology for Surveying and Geodesy, Banff, Canada.

KING R W, MASTERS E G, RIZOS C, STOLZ A AND COLLINS J (1987) Surveying with GPS. Dummler, Bonn, 128pp.

REMONDI B W (1984) Using the Global Positioning System (GPS) phase observable for relative geodesy: modelling, processing and results. PhD dissertation, University of Texas at Austin.

REMONDI B W (1985) Global Positioning System carrier phase: description and use. Bulletin Geodesique, vol 59, p361-377.

VANICEK P, KLEUSBERG A, LANGLEY R B, SANTERRE R and WELLS D E (1985) On the elimination of biases in processing differential GPS observations. Proc. 1st Int. Symp. on Precise Positioning with the Global Positioning System, Rockville, USA.

WEI Z (1985) GPS positioning software at the Ohio State University: Franklin County results. Proc. 1st Int. Symp. on Precise Positioning with the Global Positioning System, Rockville, USA.

ACKNOWLEDGEMENT

The authors are grateful to GPS Survey Services Ltd, especially Mr B Hogarth, for the loan of GPS equipment and for financial support of GPS research at the University of Newcastle upon Tyne.

Seventh session: Software

Chairman: Prof. Goad, Columbus

STATIC POSITIONING WITH THE GLOBAL POSITIONING SYSTEM (GPS): STATE OF THE ART

by

G. Beutler
I. Bauersima
W. Gurtner
M. Rothacher
T. Schildknecht

Abstract

We are now looking back at five years of practical experience with the test configuration of the GPS. This time span allowed to gain good insight into the capabilities and the problem areas of this system. Processing techniques evolved from pure baseline, session, and L_1 carrier oriented to network, project, and dual frequency oriented modes. Several institutions developed software capable of orbit improvement and demonstrated that the ultimate accuracy to be obtained by GPS is close to that of VLBI for regional surveys,and, close to that of the most advanced terrestrial techniques in small size networks.
In this paper we review the history of static positioning with GPS, we discuss processing techniques necessary to obtain highest accuracies, and we give an estimation of GPS accuracies obtainable today.

1.   Remarks Concerning History

1.1   Receivers for Geodetic Applications

The first receiver producing observations relevant for geodesy certainly was the Macrometer V-1000. This receiver passed the FGCC test (FGCC = Federal Geodetic Control Committee) in January 1983. But already in 1982 the V-1000 proved the amazing potential of GPS for geodesy. The Macrometer V-1000 demonstrated its capability of measuring baselines of up to a few thousand kilometers with an accuracy of a few parts per million (using very basic processing techniques), and, more important, it showed that the accuracy limit obtainable with GPS over short baselines of a few kilometers was of the order of one millimeter (Bock et al., 1984, Goad et al., 1984).
The Macrometer V-1000 is a single frequency instrument measuring only the L_1 carrier phase. It does not decode the navigation message, nor does it use the C/A- (or P-) code to reconstruct the carrier phase. This concept gave the receiver a high degree of independence from the signal structure, but it had two severe disadvantages: (1) the orbit information had to be supplied from an external source, (2) the synchronization between receiver clocks had to be done by pre- and post-calibration before and after the session. The M.I.T. team writing the post-processing software for the Macrometer receiver also developed the essential ideas for processing GPS carrier phases: Use of carrier phase differences, ambiguity resolution over short baselines. In our opinion it was this latter concept which ultimately led to the breakthrough of GPS in geodesy.

The second receiver to be of interest for geodesy was the TI-4100. It is a dual frequency receiver measuring code (C/A or P) and phase. Moreover the receiver allows to decode the broadcast orbit- and satellite clock-information. A nice aspect of this and all other code receivers is the possibility to process observations without using information from external sources: The orbits may be extracted from the satellite messages, the receiver clocks may be synchronized with respect to GPS time by very simple code processing algorithms. Many pioneer exploits are due to the TI-4100. In our context (static positioning) certainly the accuracy demonstration of the order of 10^{-7} for regional surveys is outstanding (see e.g. (Beutler et al., 1987/2)). It should also be kept in mind that today all permanent tracking sites used for civilian orbit determination are occupied with TI-4100's. We also remind the reader of the fact that the first kinematic surveys (Remondi, 1986) were performed with this instrument.

Although there were other prototype receivers available in this pioneer era from 1982 to 1985 (e.g. the AFGL dual band receiver, a Macrometer-type instrument, and the SERIES-X receiver developed by JPL) it is fair to state that the Macrometer V-1000 and the TI-4100 were the working horses of these early days. Both receivers are still in service today, but they have been outnumbered by C/A code receivers which measure the L_1 carrier phase, and, in future, will also reconstruct the L_2-carrier phase without using P-code information. According to our knowledge so far two receiver types, the Trimble 4000 SX and the WM-101, were extensively used for geodetic purposes. Both receivers will be updated to measure also L_2. It should be mentioned however that the Mini-Mac instrument offers this capability already now. Certainly there will be more C/A code receivers with L_2 capability available in future.

1.2 Processing Techniques

The software developed for the Macrometer V-1000 instrument by the M.I.T. team (C.C. Counselman, Y. Bock, S. Gourevitch, R. King, and others) was the pioneer software setting the standard not only for the V-1000 but for all receivers used for geodesy.

The original software for the V-1000 is
- baseline oriented: The program accepts only the measurements of two receivers at the same time.
- session oriented: Only the observations of one session (60 consecutive observation epochs 1 to 5 minutes apart) can be processed in the same program run.
- The observable used is the double difference observable.
- The software is capable of resolving the phase ambiguities on short baselines. This capability brought millimeter precision/accuracy over short (1 km or less) baselines.

A first milestone showing different software-developements and concepts was the First International Symposium on Precise Positioning with the Global Positioning System 1985 in Rockville (Goad, 1985/1). It became clear there that most major centers developing software for static GPS were leaving the baseline- in favour of the session-mode or even the campaign mode (where all observations recorded simultaneously or even all observations of a campaign were also processed in the same program run). One problem which had to be solved in this context was the correlations problem: Each double difference observation is formed from four so-called one-way phases recorded simultaneously. This means that all double difference observations pertaining to one epoch are mathematically correlated. Now everybody agrees that it is best to take these correlations into account correctly. There are, however, several ways in which this may be done. Goad proposed to use the undifferenced carrier phases (see Goad, 1985/2 or Bauersima, 1984). The paper had a strong echo, and at times it seemed that a religious war on the only correct way of processing GPS carrier phase observations might break out. Later on (Beutler et al., 1986/2, 1987/1) it became clear that the same problem could be handled in a satisfactory way using the classical double difference observable. It also became obvious that it was possible to estimate orbits using the double difference phase observable (Abbot et al., 1985, Beutler et al., 1985). In the latter paper it was moreover demonstrated that orbit improvement techniques can be used successfully even in small size networks, if only very bad a priori orbits are available -- a fact which did not seem to be of interest to many people.

Only one year after Rockville the Fourth International Geodetic Symposium on Satellite Positioning 1986 in Austin, Texas offered the next possibility to compare GPS software developments. Due to the small time interval between the two symposia more than one "déjà vu" was encountered in Austin. From our point of view important contributions could be found in the branch "orbit determinations" -- most of the investigators worked with the observations from the "March 1985 High Precision Baseline Test". The "Fiducial Point Concept", mentioned already at the Rockville symposium, became a standard term after the Austin symposium. Several groups (the M.I.T. team, JPL, and the Bernese team) came essentially to the same conclusion: Combining GPS carrier phase observations made from (known) VLBI sites with observations from other unknown sites, and using orbit improvement techniques leads to regional surveys accurate to one part in 10^7 or better. Moreover it was shown that comparable accuracies could be obtained when the positions of the fiducial points were also estimated and not assumed as known (Beutler et al., 1986/1). This fact is of importance for regional surveys in remote areas where no VLBI positions are available. Also the presentation of JPL's GIPSY software showed that there is a more general approach to processing GPS data than using the double difference observable in connection with conventional least squares techniques (Davidson et al., 1986). An aspect important for dual frequency P-code/phase receivers which seems to be worth mentioning is the method to resolve the so-called wide lane ambiguity a special linear combination of P-code phase observations (Wübbena et al., 1986). (The method was already sketched at the Rockville symposium by the same author).

So far the latest major event important for static GPS positioning was the 19th General Assembly of IUGG held in summer 1987 in Vancouver. Since the proceedings from the GPS session are still missing we confine ourselves to a few remarks: What was true for the 1986 Austin symposium was even more pronounced in Vancouver. Many contributions were not really new -- the same things as one year earlier were done in a more professional way. Interesting contributions dealt with unmodeled biases or with plain applications. Our general impression in Vancouver was that it is now more or less clear what will be achievable in static positioning with GPS. We will tell our opinion in chapter 3 of this paper.

We do not think, however, that all problems are solved: On the contrary, we believe that we are now entering the phase where the tools have to be developed to solve these known problems without (blood), sweat, and tears. This statement is certainly true if we keep in mind that the GPS constellation will have 18 or more satellites after 1991. Global applications, civilian orbit determination services (regional or global), the use of GPS for the IERS (International Earth Rotation Service) still offer most interesting questions waiting to be solved.

2.   Software Tools

In this chapter we would like to present some considerations one has to go through when developing a software system for high precision static applications. We will illustrate these considerations by the approach(es) chosen in the Bernese GPS Software Version 3.0, which is the successor of the Bernese Second Generation GPS Software (Gurtner et al., 1985).

2.1 Main Objectives

If you develop a software system, you first have to define who is going to use it, second, and most important, the problems that have to be covered, and third, you have to define guidelines how to structure the system.

For the Bernese GPS Software Version 3.0 we answered these questions in the following sense:

A. It is a system for
    - Scientific use
    - Professional use in agencies concerned with high accuracy surveys

B. The following problem areas have to be covered:
    - Rapid processing of small-size single frequency surveys
    - Pure orbit determination (fiducial point concept)
    - Free network approach for regional surveys
    - Ambiguity resolution on long baselines
    - Ionosphere and troposphere modeling capabilities
    - Full simulation capability

    In order to reach all these objectives we found the following general features to be mandatory:
    - All GPS observables may be used
    - Different linear combinations of L_1 and L_2 may be used
    - Data from all receiver types may be processed
    - Combination of single and dual frequency receivers is possible
    - Baseline/session/campaign/multiple campaign processing is allowed
    - Simultaneous estimation of a large number of different parameter types is possible.

C. Guidelines for the Program Structure

    Programming Rules
    - Highly modular on the program and subprogram level
    - Computer-system independence by
       - using standard Fortran 77 (where possible)
       - accessing all data files via the OPEN statement using translation tables for internal and external file names
       - consequent use of the PARAMETER statement: limitations are only due to computer system
    - No constants in code: all general constants are in a user accessible file
    - Documentation available for each program unit

    For the parameter estimation program we observed the following principles:
    - Consequent use of the double difference observable
    - Least squares estimation with the possibility to model correlations correctly
    - Fully automatic ambiguity book-keeping
    - No numerical integration in parameter estimation program: orbits are updated in a separate program
    - New parameter types may be introduced very easily
    - Estimated parameters may be used as a priori values for subsequent program runs

Program Stucture
We defined five program groups
- Transformation part (generating Bernese formats from individual
               receiver formats)
- Orbit part   (generate source independent orbit representa-
                   tion, update orbits, generate orbits in precise
                   orbit format, ...)
- Processing part (code processing (single station), dual fre-
                   quency code and phase pre-processing, parameter
                   estimation)
- Simulation part (generate simulated GPS sessions: code and/or
               phase, L_1 or L_1/L_2)
- Service part   (compare coordinate sets, orbits, edit/browse
                   binary data files)

The system contains 22 different program units.

Data Structure
All data relevant for the system belong to the following three
groups:
- General information   (e.g. constants, geodetic da-
                             tum(s), earth potential, pole and
                             UT1-UTC, ...)
- Campaign specific information (e.g. observation files, coordi-
                             nates, orbits, ...)
- Program option files

2.2   Processing Techniques

2.2.1 Functional Flow Diagram of Processing

In Figure 1 the main components of the Bernese GPS Software Version
3.0 are laid out.

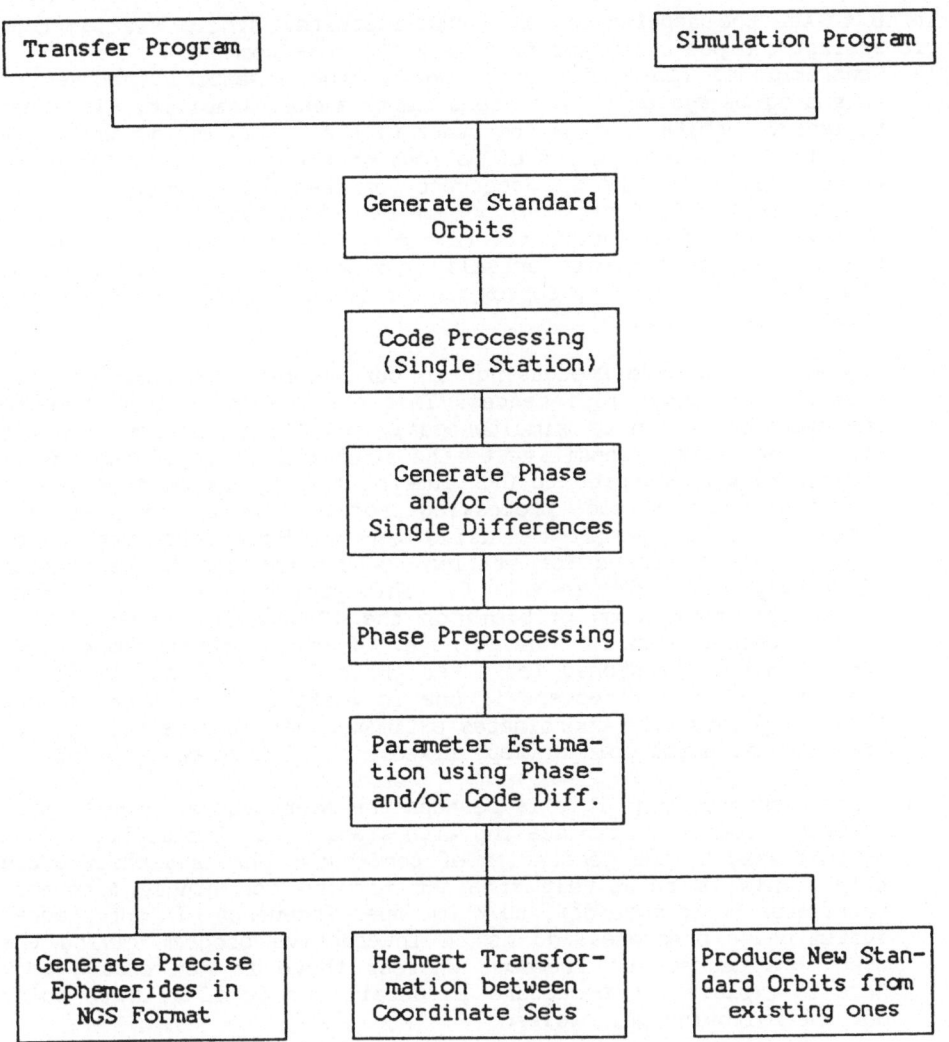

Figure 1: Functional Flow Diagram of Static GPS Processing
(Bernese GPS Software Version 3.0)

Three tasks, the generation of standard orbits, single station code processing, and phase preprocessing ask for some comments:

(1) Generation of Standard Orbits: One of the essential features of this program system is its orbit improvement capability. In order to improve orbits, one has to start with a priori orbits which are solutions of the equations of motions of the satellites. Therefore we never use directly the Broadcast (or Precise) Ephemerides (different orbit determinations) but we use these positions as pseudo observations in an orbit determination process. This guarantees that our standard orbits actually are solutions of the equations of motion and that they approximate the a priori positions in the least squares sense.

(2) Single Station Code Processing: In our parameter estimation program we use double differences. This observable is sensitive to the synchronization of simultaneously operating receivers on the microsecond level, sensitive to the synchronization of the receiver clocks with respect to GPS time on the millisecond level. In the single station code processing procedure we invoke a special parameter estimation program using C/A- or P-code observations of one receiver to solve for station coordinates and for a station clock polynomial (degree 4 to 6). This procedure solves all mentioned synchronization problems on the 10-nanosecond level. This is more than enough for our purpose. Of course these clock terms are automatically stored for later phase (or code) double difference processing. Moreover, if one is working in an area without fiducial points, the coordinates estimated here may be used to define one "fiducial point" (kept fixed) in relative positioning.

(3) Phase Preprocessing: This is perhaps the most crucial program part in every program system dealing with static GPS. The ultimate goal of this step is the generation of continuous phases without cycle slips. This is relatively straight forward for single frequency receivers, it is more difficult for dual frequency. In our program system phase preprocessing is an interactive program giving the user some degrees of freedom. Whereas there is not enough room here to explain our procedure in detail, we would like to point out the following principles:

- Screening is done separately for each baseline.
- All linear combinations (see Table 1) are checked in the same (interactive) program run.
- In a first step large slips (typically 100 cycles or more) are detected and removed automatically (using a polynomial fit).
- In a second step all remaining slips are detected and removed using either.
  - a triple difference solution.
  - a polynomial fit of the triple differences.
- The program displays a list of all cycle slips detected in the second program part. The program user may accept them or not. This strategy stands for "safety first".
- General Comment: Preprocessing is absolutely harmless for short baselines (typically 3-5 minutes per baseline). It may become more labour-intensive (typically 5-10 minutes) for long baselines (500 or more kilometers) in regions where the ionosphere shows strong and rapid variations (in time and space).

2.2.2 The Observable

In the Bernese GPS Software Version 3.0 we decided to use the double difference as basic observable. This means that in the estimation process we are (more or less) independent of the satellite clock performance and that receiver clock synchronization only matters on the microsecond-level.

Since we want to be able to process phase and code observations -- even in the same program run -- weighting of the observations is necessary. The weight ratio between code and phase observations is an input variable. We use the ratio 1:10'000 for P-code:phase, 1:100'000 for C/A-code:phase.

Let us specify the double difference observable and the linear combinations we consider useful in our parameter estimation program in a more specific way:

Let: $\phi_{ki}^{j}(t)$ one-way observation of satellite j from receiver k at its time t in the carrier L_i, i=1,2.

A "one-way observation" is a raw phase or code observation as recorded by one receiver. If two receivers k_1 and k_2 are observing two different satellites j_1 and j_2 at time t, the double difference observable is formed as

$$d_{k_1,k_2,i}^{j_1,j_2}(t) = (\phi_{k_1,i}^{j_1} - \phi_{k_2,i}^{j_1}) - (\phi_{k_1,i}^{j_2} - \phi_{k_2,i}^{j_2}) , \quad i=1,2 \quad (1)$$

Our parameter estimation program GPSEST actually reads the single differences stored in files (the quantities (...) in eqn. (1)) and forms a maximal set of linearly independent double differences for each epoch: If n satellites are observed from m stations at time t, at maximum (m-1)*(n-1) linearly independent double differences may be formed at time t. From eqn. (1) one may easily see that all double difference observations at time t are mathematically correlated. The method to take into account these correlations in GPSEST is described in (Beutler et al., 1986/2, 1987/1). If observations from dual frequency instruments are processed, (m-1)*(n-1) equations of type (1) may be formed for each epoch and for each carrier (L_1 and L_2).

Moreover it is possible to form many different linear combinations of L_1 and L_2. Omitting station subscript and satellite superscript, and using cycles as units, we may write for a special linear combination ℓ

$$\phi_\ell = \alpha_{\ell_1} \cdot \phi_1 + \alpha_{\ell_2} \cdot \phi_2 , \quad \ell=[1,2,]3,... \quad (2)$$

Each of these linear combinations may in principle be used for processing. In order to do that we have to know the (artificial) frequency ν_ℓ and the corresponding (artificial) wavelength λ_ℓ.

Since
$$\phi_1 = \nu_1 \cdot t$$
$$\phi_2 = \nu_2 \cdot t \quad (3)$$
and
$$\phi_\ell := \nu_\ell \cdot t$$

(the time origin is not of interest) we have

$$\nu_\ell \cdot t \;=\; \alpha_{\ell_1} \cdot \nu_1 \cdot t + \alpha_{\ell_2} \cdot \nu_2 \cdot t \tag{4}$$

or

$$\nu_\ell \;=\; \alpha_{\ell_1} \cdot \nu_1 + \alpha_{\ell_2} \cdot \nu_2 \tag{5}$$
$$\lambda_\ell \;=\; c/\nu_\ell \tag{6}$$

where   c is the speed of light
ν_1, ν_2 are the frequencies of carriers L_1, L_2.

Denoting the observable expressed in length units by

$$\phi'_\ell \;:=\; \lambda_\ell \cdot \phi_\ell \qquad , \qquad \ell = 1, 2, \ldots \tag{7}$$

we may write instead of eqn. (2):

$$\phi'_\ell \;=\; \beta_{\ell_1} \cdot \phi'_1 + \beta_{\ell_2} \cdot \phi'_2 \tag{8}$$

where

$$\beta_{\ell i} = \alpha_{\ell i} \frac{\lambda_\ell}{\lambda_i} \qquad\qquad i = 1, 2 \tag{9}$$

The linear combinations actually used in GPSEST are summarized in Table 1.

Table 1

Carriers and their Linear Combinations

Phase	Name	λ	β_{ℓ_1}	β_{ℓ_2}
ϕ'_1	L_1 carrier	19 cm	1	0
ϕ'_2	L_2 carrier	24 cm	0	1
ϕ'_3	Ionosphere free Linear Combination (LC)	0 cm	$\dfrac{\nu_1^2}{\nu_1^2 - \nu_2^2}$	$-\dfrac{\nu_2^2}{\nu_1^2 - \nu_2^2}$
ϕ'_4	Geometry free LC	∞ cm	1	-1
ϕ'_5	Wide Lane	86 cm	$\dfrac{\nu_1}{\nu_1 - \nu_2}$	$-\dfrac{\nu_2}{\nu_1 - \nu_2}$

Different linear combinations have different advantages:
- The advantage of ϕ'_1 and ϕ'_2 is their small measurement noise (the noise of an ϕ'_3 observation is roughly three times the noise of ϕ'_1 or ϕ'_2, the measurement noise in ϕ'_4 is roughly 1.4 that of ϕ'_1 (or ϕ'_2), the one of ϕ'_5 is roughly 5 times that one of ϕ'_1 (or ϕ'_2)).
- The advantage of ϕ'_3 is the elimination (or essential reduction) of ionospheric path delay.
- The advantage of ϕ'_4 is that it is independent of receiver clocks and of geometry. It only contains the ionosphere and (for phase observations) the ambiguity parameters.

372

- The advantage of ϕ'_5 is its long wavelength: Whereas systematic, unmodeled errors (like e.g. orbit errors) influence ϕ'_1 and ϕ'_5 in the same way, its wavelength is roughly 4 times longer. This means that ambiguity resolution in this linear combination is much less affected by systematic errors.

These properties essentially dictate the processing techniques for dual frequency instruments.

2.2.3 Parameter Space

Although the geodesist is primarily interested in site coordinates, other parameter types (may) occur when processing GPS observations. Some of them have to be estimated, if highest accuracy is required (e.g. orbit parameters for large networks), some of them (e.g. ionosphere parameters) are relevant only under certain conditions. The double difference phase observable contains an unknown but integer number of cycles which usually is called ambiguity parameter. In small networks it is possible to "resolve" these ambiguities by assigning the proper integer values to them. Ambiguity resolution brings a substantial improvement of the quality (factors of 5-10 are not uncommon) of the relevant parameters. In Table 2 we list the parameter types that may be estimated in the Bernese GPS Software Version 3.0. If single difference observations were to be processed, more receiver clock parameters would have to be estimated, if zero differences were to be processed, satellite clock parameters would have to be added to this Table (Bauersima, 1984).

Table 2
Parameter Types

Name	Comments
Ambiguity	For phase observations only. Initially one per satellite and baseline exclusive one reference satellite per baseline. If ambiguity resolution was successful, these parameters must no longer be estimated.
Coordinates	X, Y, Z coordinates in WGS-72 or WGS-84. Transformations into local systems (e.g. NAD-27 etc.) are available (including statistical information).
Orbit	Each satellite arc is characterized by at most 8 parameters. GPSEST may estimate any selection. Six parameters are the osculating elements at the start time of the arc. Two others model radiation pressure. In a program run more than one arc per satellite is allowed.

List of Orbit Parameters:

a	semimajor axis
e	numerical eccentricity
i	inclination of osculating orbit plane with respect to equator of the earth (system 1950.0)
node	right ascension of ascending node
per	argument of perigee
u0	argument of latitude at osculation time
p0	direct radiation pressure parameter
p2	y-bias
	For explanation of p0, p2 see (Beutler et al., 1986/1)

Name	Comments
Clock offset	Receiver clock offsets for individual stations. It is possible to estimate more than one offset per station. If the code processing was o.k., it should not be necessary to estimate clock offsets.
Troposphere parameters for indiv. stations	It is possible to solve for one or more tropospheric zenith delays for each station. The user specifies the time interval for which each parameter is valid.
Local troposphere model	The tropospheric zenith correction is defined to be a polynomial of degree n < 5 of height. If you work without a troposphere model and you use this option with n=1, you may use your GPS receivers as very good barometers. (Nobody thinks that this is fun).
Ionosphere models	You may estimate the coefficients of a Taylor series development of the electron density of a single layer model.
Receiver height bias	You may solve for receiver dependent height biases.

A priori weights may be assigned to most of the parameters in Table 2.
(For more information see (Beutler et al., 1987/2)).

2.2.4 Processing Strategies

In this chapter we discuss how to reach some of the objectives stated in section 2.1. Especially we will deal with the following problems:
- Rapid processing of small size single frequency surveys
- Pure orbit determination (fiducial point concept)
- Free network approach for regional surveys
- Ambiguity resolution on long baselines

We will understand by "good orbit information" orbits that have been generated in an a posteriori analysis using the phase observable.

Working with good orbit information in small networks

If you are working in a network of, let us say, 10 km * 10 km, the strategy is simple: You process the data session by session or even baseline by baseline, solve for the ambiguity parameters and store them for subsequent use. If you process data from dual frequency and single frequency receivers in the same program run (or if you process data from dual frequency receivers only) you may introduce one ionosphere model parameter per session (mean electron density) for the final solution using all files of the campaign. This should eliminate the ionosphere induced scale in your results.

Working with good orbit information in medium size networks

If you are working with dual frequency instruments in a network of, let us say, 300 km * 300 km, the safest strategy to resolve a maximum of ambiguity parameters is the following:

(1) Start with an L_3-solution without solving for ambiguity parameters. This gives you excellent a priori coordinates for the following steps.

(2) Use the coordinates previously estimated as fixed coordinates for an L_5 solution where you may possibly solve for a simple ionosphere model (one parameter). Under normal conditions you should have no problems to resolve the L_5 ambiguities.

(3) Make an L_1 solution and do not forget to specify that the L_5 ambiguities are used as known for this program run. Estimate a simple ionosphere model (1 to 10 parameters) depending on the size of your net. In networks smaller than 100 km * 100 km you should be able to resolve all L_1 ambiguities now. If some ambiguities remain unresolved, this is not a catastrophe either. It may be a good idea to work with a relatively high cut off angle (25 to 30 degrees).

(4) Make an L_3 ambiguity fixed solution. You may try to resolve the few ambiguity parameters which were not resolved previously.

The above recipe certainly is not the "one and only" way. If your net is not too big, you may skip step 1 and solve for the coordinates of all but one receivers in step 2. Usually this should work. Under the same assumptions you may also try to skip step 3 and resolve the L_1 ambiguities in the fourth of the steps given above. You should keep in mind however that the L_4 observable has the very nice characteristic not to depend on your a priori orbit or your a priori coordinates. If the irregular part of the ionosphere is not too big, step 3 is therefore the safest way.

Working with good orbit information in large size networks

If the distances between the stations of your network are of the order of 1000 km or more you will probably have no chance to resolve the ambiguities. The recipe for such cases is simple:
(1) Make an L_3-solution without solving for ambiguity parameters. This gives you the best coordinates under these circumstances.

Working without good orbit information in large size networks

(1) Make an L_3-solution without solving for ambiguity parameters. Solve for 6 to 8 orbit parameters using a priori variances for the orbit elements. An example for such an analysis may be found in (Beutler et al., 1987/2).

Produce good orbit information if fiducial points are known

Let us assume that within a continent we have four or more fiducial points equipped with dual frequency GPS receivers. The observations of these receivers are used to solve a pure orbit determination problem. We suggest that you combine data from 3 to 5 days. You impose no or only very small a priori constraints on your orbit elements. From a purely theoretical point of view it would be preferable to combine the observations from fiducial and non-fiducial points and solve simultaneously for all parameters that are acutally unknown (namely orbit elements and coordinates of the non-fiducial stations).

3. GPS Accuracy for Static Applications

It is an interesting question what precisions and what accuracies are obtainable today if you have a good software tool at your disposal. In (Beutler et al., 1987/3) we tried to give a reliable answer by using the entire GPS material available to our group. The meaning is of course that each survey is treated with adequate methods. If you process data of a continent wide survey without solving for orbit parameters, you would end up with different accuracy considerations. In Table 3 and in Figure 2 we reproduce the result published in (Beutler et al., 1987/3), and we include the rule which in our opinion gives a realistic accuracy estimate for static GPS applications (for more information see (Beutler et al., 1987/3):

Table 3

GPS Campaigns Processed in Berne and Related Accuracy Information

Campaign	Year Organized by	Sites	Size (km)	Accuracy (m) formal actual	
Ottawa	83 Canad. Energy, Mines, Resources	4	10 x 60	.004	---
Quebec	84 Energie et Resources, Quebec	16	2 x 3	.001	.0006*
California	84 JPL	2	140	.015	---
CERN	84 CERN	7	12 x 12	.002	.004
Alaska	84 U.S. NGS	8	800 x 1600	.050	.070
HPBL	85 JPL	9 2	2000 x 4200 240	.050 .020	.065 .030+
November 85	85 JPL	4	2000 x 4200 240	--- .020	---x
Turtmann	85 Swiss Geodetic Commission	7	4 x 6	.002	.003
Iceland	86 UNAVCO	52	250 x 450	.020	---o
Turtmann	86 Swiss Geodetic Commission	8	4 x 6	.002	.003
Europe	86 Univ. of Bonn	10	1200 x 1400	---	---
Alaska	86 U.S. NGS	8	800 x 1600	.030	.030
Switzerland	87 Swiss Geodetic Commission	12	180 x 180	.010	---

*   Special experiment on a 500 m baseline, which was measured with invar wire

+   Series-X baseline processed as independent baseline using orbits estimated with all other observations of HPBL test

x   Observations used to estimate orbits only

o   Not yet ready

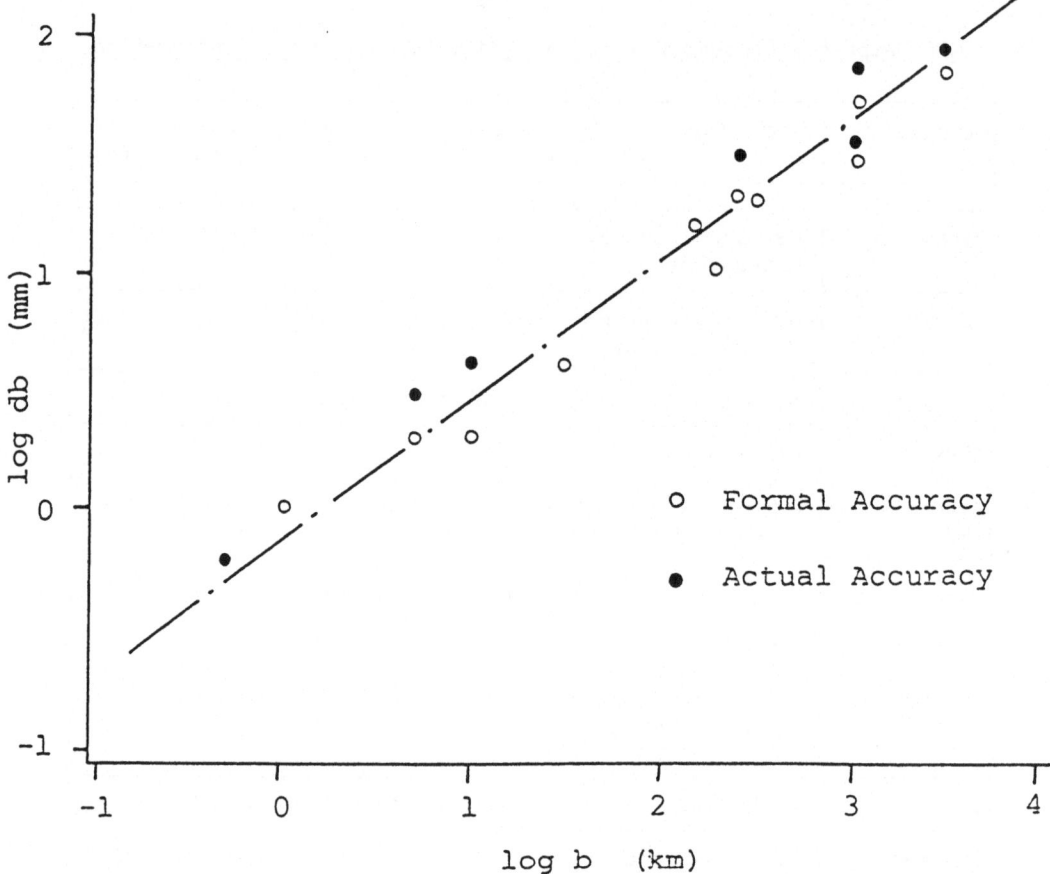

Figure 2: GPS accuracies as a function of baseline length

$$db/b = (2b)^{-1/2}$$

==================

Where: db is the baseline accuracy in mm
      b   is the baseline length in km

References

Abbot, R.I., Y. Bock, C.C. Counselman III, R.W. King, S.A. Gourevitch, B.J. Rosen (1985). "Interferometric determination of GPS satellite orbits." Proceedings of the First International Symposium on Precise Positioning with the Global Positioning System. Rockville, May 1985, pp. 63-72.

Bauersima, I. (1984). "Navstar/Global Positioning System (GPS)(V). Satellite Radiointerferometry." Mitteilungen der Satelliten-Beobachtungsstation Zimmerwald.

Beutler, G., W. Gurtner, I. Bauersima, R. Langley (1985). "Modelling and Estimating the Orbits of GPS Satellites." Proceedings of the First International Symposium on Precise Positioning with the Global Positioning System, Rockville, May 1985, pp. 99-112.

Beutler, G., W. Gurtner, M. Rothacher, T. Schildknecht, I. Bauersima (1986/1). "Determination of GPS Orbits using Double Difference Phase Observations from Regional Networks." Proceedings of the Fourth International Geodetic Symposium on Satellite Positioning, Austin, Texas, April 1986, pp. 319-355.

Beutler, G., W. Gurtner, I. Bauersima, M. Rothacher (1986/2). "Efficient Computation of the Inverse of the Covariance Matrix of Simultaneous GPS Carrier Phase Difference Observations." Manuscripta geodaetica, Vol. 11, pp. 249-255.

Beutler, G., I. Bauersima, W. Gurtner, M. Rothacher (1987/1). "Correlations between simultaneous GPS double difference carrier phase observations in the multistation mode: Implementation considerations and first experiences." Manuscripta geodaetica, Vol. 12, pp. 40-44.

Beutler, G., I. Bauersima, W. Gurtner, M. Rothacher, T. Schildknecht, G.L. Mader, M.D. Abell (1987/2). "Evaluation of the 1984 Alaska Global Positioning System Campaign with the Bernese GPS Software." Journal of Geophysical Research, Vol. 92, No. B2, pp. 1295-1303.

Beutler, G., I. Bauersima, S. Botton, W. Gurtner, M. Rothacher, T. Schildknecht (1987/3). "Accuracy and biases in the geodetic application of the Global Positioning System." Paper presented at the 19th IUGG General Assembly, Vancouver, Canada, 1987.

Bock, Y., R.I. Abbot, C.C. Counselman, S.A. Gourevitch, R.W. King, A.R. Paradis (1984). "Geodetic accuracy of the Macrometer V-1000." Bulletin Géodésique, Vol. 58, No. 2, pp. 211-221.

Davidson, J.M., C.L. Thornton, S.A. Stephens, S.C. Wu, S.M. Lichten, J.S. Border, O.J. Sovers (1986). "Demonstration of the Fiducial Point Concept using Data from the March 1985 GPS Field Test." Proceedings of the Fourth International Geodetic Symposium on Satellite Positioning, pp. 1019-1028.

Goad, C.C. (1985/1). "Proceedings of the First International Symposium on Precise Positioning with the Global Positioning System." Rockville, May 1985.

Goad, C.C. (1985/2). "Precise relative position determination using Global Positioning System carrier phase measurements in a nondifference mode." Proceedings of the First International Symposium on Precise Positioning with the Global Positioning System." Rockville, May 1985, pp. 347-356.

Gurtner, W., G. Beutler, I. Bauersima, T. Schildknecht (1985). "Evaluation of GPS Carrier Difference Observations: The Bernese Second Generation Software Package." Proceedings of the First International Symposium on Precise Positioning with the Global Positioning System, Rockville, 1985, pp. 363-372.

Remondi, B.W. (1986). "Performing centimeter level surveys in seconds with GPS carrier phase: Initial results." Proceedings of the Fourth International Geodetic Symposium on Satellite Positioning, pp. 1229-1249.

Wübbena, G., A. Schuchardt, G. Seeber (1986). "Multistation positioning results with TI-4100 GPS receivers in geodetic control networks." Proceedings of the Fourth International Geodetic Symposium on Satellite Positioning, Austin, Texas, April 1986, pp. 963-978.

GPS CARRIER PHASES AND CLOCK MODELING

by

Gerhard Wübbena

Abstract

The first part of the paper describes some basic relations concerning GPS carrier phases and linear combinations of the L1 and L2 phases. The so called wide lane and narrow lane signals are described with respect to the original L1-L2 signals and the code phases. A new 'signal', the 'extra wide lane' is introduced.

The second part of the paper deals with the problem of the simultaneous adjustment of carrier phases from different receiver types. Since different receivers do not generally observe carrier phases at the same epochs, the satellite clock errors cannot be eliminated through single differences. Hence, the satellite clocks have to be modeled in some way. A clock model is described, and the results of some tests are presented.

1.0 GPS Carrier Phases

This section describes the basic relations between carrier phases and satellite clocks as well as receiver clocks. Some linear combinations of the L1 and L2 carrier phases are examined with respect to the original phases and the PRN-code phases.

1.1 The Carrier Phase Observable

All signals transmitted by a GPS satellite are directly derived from the satellite clock. Thus, the phase of a transmitted carrier can be described by

$$\Phi_t^i(t_t) = t_t^i(t_t) \cdot f + \epsilon_\Phi(t_t) \tag{1}$$

where

t_t - is the transmission time,

$\Phi_t^i(t_t)$ - is the transmitted phase of the satellite i at the time t_t in [cycles],

$t_t^i(t_t)$ - is the satellite clock time at the time t_t,

f - is the nominal frequency of the carrier signal and

$\epsilon_\Phi(t_t)$ - is a time dependent phase delay due to satellite hardware.

This situation is shown in figure 1, where the satellite phase delay is neglected.

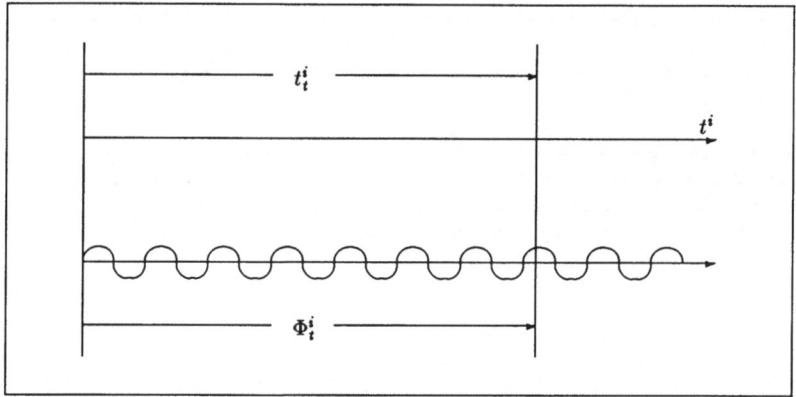

Figure 1: Transmitted Carrier Phase

The transmitted phase approximately propagates with the speed of light through the space and reaches the receiver j at the time of reception t_r:

$$\Phi_{r,j}^i(t_r) = \Phi_t^i(t_t). \tag{2}$$

The receiver measures the so called carrier beat phase, which can be represented as

$$\left(\varphi_{m,j}^i(t_r) + N_j^i\right) = \Phi_{r,j}^i(t_r) - \Phi_{0,j}(t_r), \tag{3}$$

with

$\varphi_{m,j}^i(t_r)$ - the measured carrier beat phase,

N_j^i - the carrier phase ambiguity and

$\Phi_{0,j}(t_r)$ - the phase of a receiver generated reference signal.

The phase of the reference signal is directly related to the receiver clock through

$$\Phi_{0,j}(t_r) = t_{r,j}(t_r) \cdot f_0 + \epsilon_{\Phi_j}(t_r),\tag{4}$$

where

$t_{r,j}(t_r)$ - is the receiver clock time at t_r,

f_0 - is the nominal frequency of the reference signal and

$\epsilon_{\Phi_j}(t_r)$ - is a time dependent phase delay due to receiver hardware.

The combination of the above equations leads to

$$t_t^i(t_t) = \frac{\varphi_{m,j}^i(t_r) + N_j^i + t_{r,j}(t_r) \cdot f_0}{f} + \frac{\epsilon_{\Phi_j}(t_r)}{f} - \frac{\epsilon_\Phi(t_r)}{f}.\tag{5}$$

This equation shows, that the carrier phase measurement is essentially an ambiguous reading of the satellite clock at the time of signal transmission. The two phase delay terms behave exactly as receiver respectively satellite clock errors.

The observation equation for phase measurements can be written as the code phase pseudorange equation with one additional ambiguity term

$$PR = c\left(t_{r,j}(t_r) - t_t^i(t_t)\right),\tag{6}$$

$$PR = \left|\vec{X}^i - \vec{X}_j\right| + c\,\delta t_j - c\,\delta t^i + c\,\delta t_A - \frac{c}{f}N_j^i.\tag{7}$$

Here is

c - the velocity of light,

$\vec{X}^i$ - the geocentric cartesian coordinate vector of the satellite i,

$\vec{X}_j$ - the geocentric cartesian coordinate vector of the receiver j,

δt_j - the receiver clock error,

δt^i - the satellite clock error,

δt_A - the residual atmospheric delay error and

$\frac{c}{f}N_j^i = \lambda N_j^i$ - the ambiguity term, where λ is the carrier wavelength.

1.2 Linear combinations of carrier phases

With dual frequency phase measurements the first order ionospheric phase delay can be corrected. The corrected phase consists of a linear combination of the L1 and L2 phases. Unfortunately, the coefficients of this linear combination are non-integers, which has the effect, that the corrected phase does not have an integer ambiguity. From this the problem raises, that no ambiguity fixing can be done, if only the ionospheric corrected phase is used.

There is an infinite number of linear combinations that can be obtained from the two carrier phases, but only a few of them are useful for the GPS adjustment. If ambiguities shall be fixed to integers, the coefficients of the linear combinations have to be integers. Further the ionospheric effect on the derived phase has to be analyzed. Only those combinations which do not increase the ionospheric effect should be used. An important criterion is the wavelength of the resultant signal and the noise of ranges derived from the signal with respect to the noise of the ranges derived from the original signals.

The wavelength of a linear combination can be determined by the following formulas. If the original signals are derived from the same clock and hardware delays are neglected, the relation between the two phases is given by

$$t^i(t) = \frac{\Phi^i_I(t)}{f_I} = \frac{\Phi^i_{II}(t)}{f_{II}} \ . \tag{8}$$

Here the notations $(\cdot)_I$ and $(\cdot)_{II}$ are for the L1 and L2 signals respectively. The phase of the linear combination is

$$\Phi^i(t) = n_I \cdot \Phi^i_I(t) + n_{II} \cdot \Phi^i_{II}(t) \ , \tag{9}$$

which gives in combination with equation (8)

$$t^i(t) = \frac{\Phi^i(t)}{n_I \cdot f_I + n_{II} \cdot f_{II}} \ , \tag{10}$$

where the denominator is the frequency of the derived signal. With the speed of light this can be scaled to the wavelength by

$$\lambda = \frac{c}{f} = \frac{c}{n_I \cdot f_I + n_{II} \cdot f_{II}} \ . \tag{11}$$

1.2.1 The wide and narrow lanes

There are two linear combinations which are of special interest for the adjustment of GPS carrier phases.

The first is the difference of the L1-L2 phases

$$\Delta\varphi = \varphi_I - \varphi_{II} \ , \tag{12}$$

the second is the sum of the two phases

$$\Sigma\varphi = \varphi_I + \varphi_{II} \ . \tag{13}$$

Using the equations from section 1.1 the corresponding signal transmission times can be computed by

$$t_\Delta = \frac{\Delta\varphi + \Delta N + t_r \cdot \Delta f_0}{\Delta f} = \frac{\varphi_I - \varphi_{II} + N_I - N_{II} + t_r \cdot f_{0,I} - t_r \cdot f_{0,II}}{f_I - f_{II}} \tag{14}$$

and

$$t_\Sigma = \frac{\Sigma\varphi + \Sigma N + t_r \cdot \Sigma f_0}{\Sigma f} = \frac{\varphi_I + \varphi_{II} + N_I + N_{II} + t_r \cdot f_{0,I} + t_r \cdot f_{0,II}}{f_I + f_{II}} \ . \tag{15}$$

In this equations the time dependency notation and the phase delay terms are neglected for simplicity.

The wavelength for the phase difference signal is 86.2 cm and 10.7 cm for the phase sum signal. The difference signal is also called 'wide lane' and the sum signal 'narrow lane'.

The first order effect of the ionosphere on the L1 and L2 phases can be written as

$$\delta\Phi_{iono,I} = \frac{C_{iono}}{f_I} \tag{16}$$

$$\delta\Phi_{iono,II} = \frac{C_{iono}}{f_{II}} \ , \tag{17}$$

or in terms of transmission times as

$$\delta t_{iono,I} = \frac{C_{iono}}{f_I^2} \tag{18}$$

$$\delta t_{iono,II} = \frac{C_{iono}}{f_{II}^2} \ . \tag{19}$$

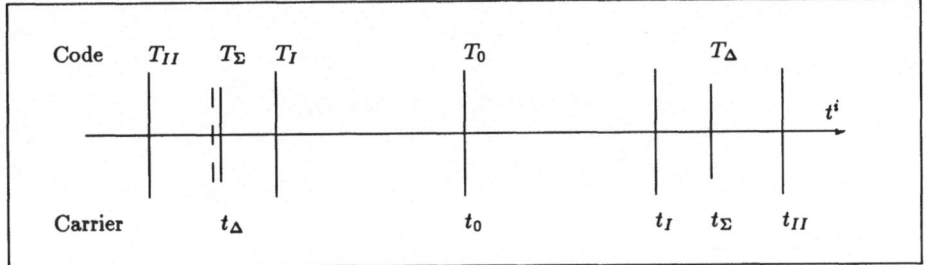

Figure 2: Signal Transmission Epochs

The factor C_{iono} is a function of the total electron content along the propagation path of the signal.

The insertion of equations (16) and (17) into (12) and (13) gives the ionospheric delay for the wide and narrow lanes:

$$\delta\,\Delta\Phi_{iono} = C_{iono}\left(\frac{1}{f_I} - \frac{1}{f_{II}}\right) = C_{iono}\frac{f_{II} - f_I}{f_I\,f_{II}} = -C_{iono}\frac{\Delta f}{\Pi f}\,, \tag{20}$$

$$\delta\,\Sigma\Phi_{iono} = C_{iono}\left(\frac{1}{f_I} + \frac{1}{f_{II}}\right) = C_{iono}\frac{f_{II} + f_I}{f_I\,f_{II}} = +C_{iono}\frac{\Sigma f}{\Pi f}\,, \tag{21}$$

and the corresponding transmission time delays to

$$\delta t_{iono,\Delta} = -\frac{C_{iono}}{\Pi f} \tag{22}$$

$$\delta t_{iono,\Sigma} = +\frac{C_{iono}}{\Pi f}\,. \tag{23}$$

Through the first order ionospheric effect the narrow lane is advanced by the same amount as the wide lane is delayed, i.e. the ionospheric corrected transmission time epoch computes to

$$t_0 = \frac{t_\Delta + t_\Sigma}{2}\,. \tag{24}$$

The PRN-code phases propagate with the group velocity, the corresponding ionospheric delays are

$$\delta T_{iono,I} = -\frac{C_{iono}}{f_I^2} \tag{25}$$

$$\delta T_{iono,II} = -\frac{C_{iono}}{f_{II}^2}\,. \tag{26}$$

If the signals of different carriers are received at the same epoch, they were transmitted at different epochs. This situation is illustrated in figure 2. Below the satellite time scale the carrier phase transmission epochs and the corresponding epochs of the wide and narrow lanes as well as the ionospheric corrected epoch are drawn. The respective PRN-code transmission epochs are drawn above the time scale and denoted with capital T's.

The sum and difference epochs for the PRN-codes are computed by

$$T_\Sigma = \frac{T_I \cdot f_I + T_{II} \cdot f_{II}}{\Sigma f} \tag{27}$$

and

$$T_\Delta = \frac{T_I \cdot f_I - T_{II} \cdot f_{II}}{\Delta f}\,. \tag{28}$$

Taking the first order ionospheric effect into account, it can be shown that

$$t_\Delta = T_\Sigma \tag{29}$$

and

$$t_\Sigma = T_\Delta \,. \tag{30}$$

It can further be shown that

$$T_\Sigma \;=\; \frac{T_I + T_{II}}{2} + \frac{\Delta f}{\Sigma f}\,\frac{T_I - T_{II}}{2} \tag{31}$$

$$T_\Sigma \;=\; \bar{T} + 0.0625\,\Delta T \,. \tag{32}$$

1.3 Ambiguity resolution

The ambiguity resolution is required for high precise positionings, especially in small and medium sized networks. If ambiguity parameters are estimated in the adjustment, they will be affected by umodelled systematic effects like orbital biases, tropospheric errors etc. The influence of these errors in terms of fractional cycles decreases with increasing wavelengths, i.e. the longer the wavelength the better the ability to fix ambiguities to integers. There is no way to work with different wavelengths if only single frequency observations are used, but with dual frequency measurements the linear combinations can be used to improve the ambiguity estimation.

1.3.1 Ambiguity resolution for the wide and narrow lanes

The strategy of ambiguity resolution depends on the influence of the ionosphere in single differenced observations. Like other systematic effects, this is a function of the interstation distances.

Since the wide lane has the longest wavelength the first step will be the resolution of the wide lane ambiguity. This can be done through the geometric approach, i.e. the estimation of ambiguity parameters in the adjustment, if the ionospheric error is below some fractional wavelength. For instance, if an ionospheric error of 1 ppm is assumed, the fractional ambiguity error will be less than 0.25 cycles for distances up to 200 km.

An alternative method to solve the wide lane ambiguity is the combination with dual frequency P-code observations. The combination of equations (14) and (29) gives a direct expression for the ambiguity

$$\Delta N = \Delta f \cdot T_\Sigma - (\Delta\varphi + t_r \cdot \Delta f_0) \,. \tag{33}$$

Under normal conditions, i.e. no strong multipath signals, the ambiguity can be found within a few minutes of observations (Wübbena, et.al. 1986). Since only a combination of observables is used here, the method is independent on interstation distances and geometry and works even in a kinematic mode. All systematic errors, except multipath, drop out in equation (33).

For longer distances the ionospheric errors cannot be negelected, therefore the narrow lane ambiguity has to be estimated in order to compute the ionospheric corrected observation with equation (24). Once the wide lane ambiguity is known, the narrow lane wavelength increases to about 21.4 cm. The reason for this are the conditions that

if ΔN is even - ΣN has to be even and

if ΔN is odd   - ΣN has to be odd,

because both are linear combinations of the L1-L2 ambiguities. Another point to be mentioned is that the ionospheric corrected signal has an integer ambiguity (wavelength 10.7 cm) if the wide lane ambiguity is known.

There are two ways to resolve the narrow lane ambiguity. The first is the estimation of $\Sigma N/2$ through the geometric adjustment, the second is the estimation of ΣN through ionospheric modeling.

Since the difference in the signal transmission epochs $t_\Sigma - t_\Delta$ is only a function of the ionosphere the narrow lane ambiguity can be estimated together with a regional ionosphere. This method works even over long distances (1000 km) if ionospheric irregularities are not too big (Wübbena and Seeber, 1987).

1.3.2 Extra wide laning

For short interstation distances the ionospheric effect will be small, this allows the direct estimation of the narrow lane ambiguity through the condition

$$t_\Sigma - t_\Delta \overset{!}{=} 0 \qquad (34)$$

for single or double differences. This condition can only be fulfilled if the wide lane ambiguity is correctly resolved, because the even – odd condition of the previous section has also to be fulfilled. For instance, assume that the initial ambiguities were correct and the estimated wide lane ambiguity is 0.6 cycles, which rounds to 1 cycle if it is fixed. Through the condition in equation (34) the narrow lane ambiguity will then be estimated to 8 cycles since

$$\lambda_\Delta \approx 8\lambda_\Sigma, \qquad (35)$$

in contradiction to the even – odd condition which allows values of 7 or 9, so the error in the wide lane ambiguity is detected and the correct value of 0 will be chosen.

Using this technique, which i would like to call 'extra wide laning', essentially means that the ambiguity for a 1.72 m wavelength has to be resolved. This can lead to dramatic savings in observation time for small sized networks.

2. Clock modeling

Although the GPS satellites are equipped with highly precise oscillators the observation errors due to the satellite clock errors cannot be neglected in geodetic adjustments. The modeling of clock errors is done in different ways. The most often used approach is to eliminate clock errors through observation differences. This technique is valid as long as the assumption is valid, that the differenced observations contain the same clock error, i.e. the observation epochs have to be the same within a certain amount of time.

Receivers of the same type normally observe at the same epochs, but receivers of different types generally take observations at different epochs. Figure 3 illustrates a situation where receivers of type X and Y observe at the same measurement rate but at different epochs. The time difference between the observation epochs is denoted with Δt. In general also the measurement rate will be

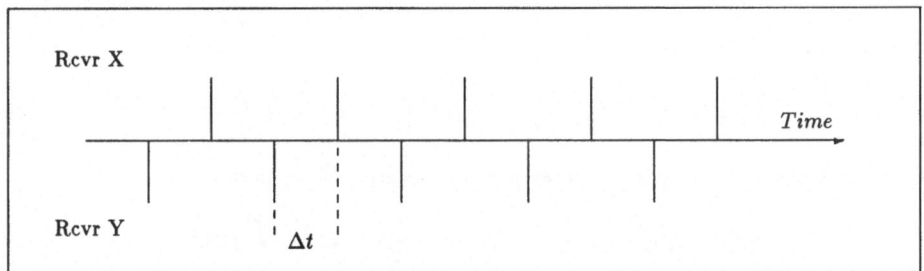

Figure 3: Observation Epochs of Different Receiver Types

different. Because satellite clock errors are varying with time, it is obvious that they cannot be eliminated through single or double differences. The same situation exists for receiver clock errors if sequencing receivers are used.

One solution of the problem could be the generation of normal points. In this case the observations and consequently the clock errors are fitted through a polynomial over a certain time interval. From the polynomial a derived observation at the normal point is computed. These values can be

387

treated as observations taken at the same epochs, so they can be adjusted with standard models. Since different measurement epochs are used for the normal point evaluation, the normal point observations will have different errors. The statistics of these errors is dependent on the clock type, the measurement rate and the length of the normal point intervals. In general the normal point errors are correlated and cannot be treated as white noise. If the error statistics are not introduced in the geodetic adjustment, the variance and covariance estimation of the results will be incorrect.

Since a reliable variance and covariance estimation is often a requirement of geodetic GPS users a more sophisticated clock model should be introduced.

2.1 Clock Models

The reading of a clock i can be represented as

$$t_i(t) - t_{0,i} = \int_{t_0}^{t} f_i(t)dt ,$$ (36)

with

t_0 - an arbitrary reference epoch,

$t_{0,i}$ - the clock reading at the reference epoch and

$f_i(t)$ - the frequency of the oscillator.

A standard model for the frequency of atomic and precise crystal oscillators is

$$f_i(t) = f + \Delta f + \dot{f}(t - t_0) + \tilde{f}(t) ,$$ (37)

where

f - is the nominal clock frequency,

Δf - is a frequency bias,

$\dot{f}$ - is a frequency drift and

$\tilde{f}(t)$ - are random frequency errors.

Inserting equation (37) in (36) gives

$$t_i(t) = t_{0,i} + (t - t_0) + \frac{\Delta f}{f}(t - t_0) + \frac{\dot{f}}{2f}(t - t_0)^2 + \frac{1}{f}\int_{t_0}^{t} \tilde{f}(t)dt .$$ (38)

The rearranging of terms and a new notation yields the error of the clock i

$$\delta t_i(t) = t_i(t) - t = a_0 + a_1(t - t_0) + \frac{a_2}{2}(t - t_0)^2 + \int_{t_0}^{t} y(t)dt .$$ (39)

It is

a_0 - the clock bias,

a_1 - the clock drift,

a_2 - the ageing and

$\int_{t_0}^{t} y(t)dt$ - the integrated random fractional frequency error.

The random fractional frequency error is usually characterized by the Allan variance, which is defined as an infinite sum

$$\sigma_y^2(\tau) = \frac{\sum_{k=0}^{\infty}(\bar{y}(t_k + \tau) - \bar{y}(t_k))^2}{2}, \tag{40}$$

where

$$\bar{y}(t) = \frac{\int_t^{t+\tau} y(t)dt}{\tau}, \tag{41}$$

is the average fractional frequency error over the time interval τ. Figure 4 shows the Allan variance

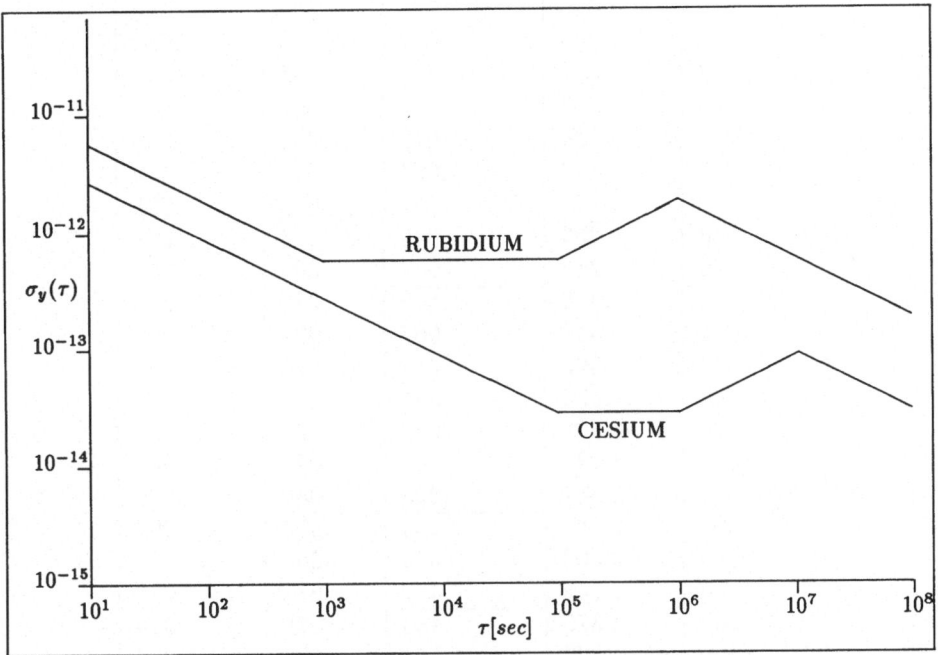

Figure 4: Allan Variance of Typical Oscillators

of typical rubidium and cesium clocks. Similar plots can be obtained for precise crystal oscillators and hydrogen masers.

The stability of an oscillator can also be characterized in the frequency domain through the spectral density of the fractional frequency errors. Both characterizations can be obtained from each other through simple transformation formulas. The spectral density of frequency errors can be used to derive covariance functions of clock errors and clock error differences.

The clock error changes in time can be described by a stationary process. The process parameters can be used in a geodetic GPS adjustment program to filter the clock errors.

The variances of the difference in clock errors $\sigma\left(\int_t^{t+\tau} y(t)\right)$, scaled to range errors, is shown in table 1 for some typical clocks.

In this table those rows are marked, where the clock error exceeds a quarter of a wavelength for the L1-L2 signals and the wide and extra wide lanes. This may give some information about the maximum time interval which can be used to predict carrier phases for cycle slip detection and resolution.

τ (seconds)	Crystal (TI4100)	Rubidium (Satellites)	Cesium (CS)	Hydrogen Maser
1	.007	.006	.003	.000
2	.013	.008	.004	.000
3	.020	.010	.005	.000
4	.026	.012	.006	.000
5	.032	.013	.006	.000
6	.039	.015	.007	.000
7	.045	.016	.008	.000
8	.051	.017	.008	.000
9	.058	.018	.009	.000
10	.064	.019	.009	.000
20	.127	.029	.013	.000
30	.190	.037	.016	.001
40	.253	.045	.018	.001
50	.316	.052	.020	.001
60	.378	.060	.022	.001
70	.441	.067	.024	.001
80	.503	.074	.026	.002
90	.566	.081	.027	.002
100	.628	.088	.029	.002
200	1.248	.156	.041	.004
300	1.867	.223	.050	.006
400	2.483	.290	.058	.008
500	3.098	.356	.066	.010
600	3.712	.422	.072	.012
700	4.325	.488	.079	.014
800	4.937	.554	.085	.016
900	5.548	.620	.090	.018
1000	6.158	.685	.096	.020
2000	12.236	1.331	.143	.040
3000	18.284	1.966	.184	.060
4000	24.314	2.593	.222	.080
5000	30.330	3.213	.258	.100
6000	36.333	3.827	.293	.120
7000	42.324	4.437	.328	.139
8000	48.305	5.044	.362	.159
9000	54.276	5.646	.396	.179
10000	60.237	6.246	.429	.199
20000	119.436	12.114	.753	.396
30000	178.065	17.820	1.068	.593
40000	236.255	23.414	1.378	.789
50000	294.064	28.920	1.685	.986
60000	351.517	34.353	1.989	1.181
70000	408.629	39.722	2.290	1.377
80000	465.409	45.034	2.589	1.572

Table 1: Variance of Clock Error Changes [meters]

2.2 Clock Modeling Tests

A testsoftware was written in order to the verify the suitability of such clock models in the geodetic adjustment of GPS carrier phases. A Kalman filter algorithm was chosen to estimate a state vector with the following unknowns

- 3 coordinates per station
- 3 clock parameters per station
- 3 clock parameters per satellite
- 1 ambiguity per station and satellite
- 1 new ambiguity per unrecovered cycle slip.

All parameters are constant, except the clock parameters, where a dynamical model is introduced with a stochastical process for the clock errors in form of white noise or random walk.

The program automatically fixes ambiguities to integers if their estimated variances reach a certain limit and the fractional ambiguity errors are below 0.2 cycles.

A one hour test data set from two TI4100 receivers was used to simulate receivers with different observation epochs. The length of the baseline was approximately 5 km. For the first run all measurements at the 3 second sampling rate were used to verify the software. The results are within 1-3 mm in all coordinate components compared to a GPS network solution which agrees with the terrestrial reference in the level of 5 mm.

Sampling Rate [sec]	Sampling Epoch Offset [sec]	Time to fix Ambiguities [min]	Error			Standard deviation		
			ΔX [mm]	ΔY [mm]	ΔZ [mm]	σ_x [mm]	σ_y [mm]	σ_z [mm]
6	3	15	0	0	1	1	1	1
12	6	17	-1	-1	1	2	1	2
18	9	23	-1	-1	1	2	1	3
24	12	28	1	-2	3	3	1	4
30	15	38	3	2	6	4	2	5

Table 2: Influence of Clock Errors

The clock modeling was tested through five runs with different sampling rates and sampling epoch offsets (Δt if figure 3) between the two receivers of half the sampling rate. The baseline components are compared with those of the first run. The results are summarized in table 2, which shows the coordinate errors and the estimated standard deviations for the five runs as well as the observation time used to fix the ambiguities.

The results lead to the conclusion that with an adequate clock model geodetic relative positioning with subcentimeter accuracy can be done, even if different receivers with observation epoch offsets of up to 15 seconds are used. The clock model allows a realistic variance estimation for the unknown parameters. It should be mentioned that the results were obtained with some satellites operating rubidium clocks. The operational satellites will all be equipped with precise cesium frequency standards, so the errors should decrease to some better level.

3. References

Fell, P.J. (1980): Geodetic Positioning Using a Global Positioning System of Satellites. Ohio State University, Report #OSU-DGS 299.

Wells,D. et.al.(1986): Guide to GPS Positioning. Canadian GPS Associates. 1986.

Wübbena, G., A. Schuchardt, G. Seeber (1986): Multistation Positioning Results with TI 4100 GPS Receivers in Geodetic Control Networks. Proceedings of the Fourth International Geodetic Symposium on Satellite Positioning, Austin, Texas, April 1986, pp. 963-978.

Wübbena, G., G. Seeber (1987): Analysis of Multiple TI 4100 GPS Measurements over a 920 km VLBI Baseline. Paper presented at the 19-th IUGG General Assembly, Vancouver 1987.

WORLD: A MULTIPURPOSE GPS-NETWORK COMPUTER PACKAGE

by

Erik W. Grafarend and Wolfgang Lindlohr

Abstract

WORLD is a multipurpose package to compute geodetic positions in geometry and gravity space. Here undifferenced GPS carrier beat phase
observations are processed in the free network mode, namely by the
prototype program called PUMA. Within two alternative model formulations, the classical *Gauß-Markov Model* and the so-called *Mixed Model*,
simultaneously estimated / predicted parameters are those of type
(i) Cartesian ground station coordinates *(geodetic positioning)*,
(ii) Cartesian satellite coordinates *(orbit determination)*,
(iii) receiver- and satellite-specific bias terms, *(iv)* initial
epoch ambiguities and *(v)* proportional tropospheric corrections. The
Mixed Model parameters appear from linearization as a point of *stochastic prior information*. Namely the weight matrices of stochastic
prior information, e.g. for orbit parameters, is assumed to be known.
Estimators of type *BLUE* and predictors of type *inhom BLIP* and *hom BLUP*
are used. Chapter four discusses in all detail the real analysis of
GPS *satellite networks of free type*. Most notable are the estimated
bias terms α, β, γ in a *twofold classification model*. The operability
of PUMA is demonstrated by the use of multistation phase observations
(Wild-Magnavox WM 101-receivers) in a local Berlin network (six
station network). It is documented that in spite of the advanced phase
observation modelling an internal relative baseline accuracy (utmost
length 30 km) of the order of *3 to 5 ppm* is achievable. In addition,
the influence of *orbital prior information* on ground station measures,
point position as well as accuracy, is demonstrated.

1. The components of the "WORLD"

WORLD is a computer package for processing large sets of data within satellite and terrestrial geodetic networks, namely for very accurate positioning in geometry and gravity space. Some components of WORLD are described in *Figure 1.1*.

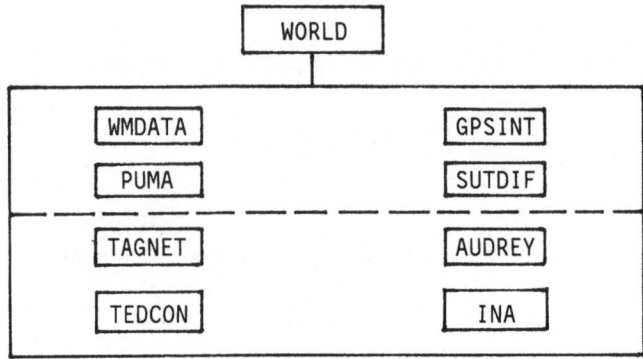

Figure 1.1: The overall structure of WORLD

The four programs in the upper dashed box are related to GPS carrier beat phase processing. WMDATA and GPSINT are preprocessing programs for the core programs PUMA and SUTDIF. The program WMDATA decodes the tape-recorded original measurements of a WM101-receiver, for instance. The 1st main processor PUMA contains all adjustment procedures for massive undifferenced carrier beat phase observations as being described in all detail by *W. Lindlohr* (1988b). In contrast, SUTDIF is the 2nd main processor which supplies triply differenced phase observations: it has been developed for WORLD by *K. Eren* (1987). Finally the four programs in the lower dashed box are processors of terrestrial observations, e.g. in threedimensional terrestrial networks or in inertial surveying networks.

The program AUDREY is a processor for an adjustment computation of heterogeneous terrestrial observations within threedimensional "integrated" geodesy. It is a new version of the wellknown TAGNET software being described by *J. Zaiser* (1984). The main new feature is the implementation of sparse matrix techniques, namely by the HEART-OF-GOLD package developed by *D.G. Milbert* (1984). In addition, variance-covariance component estimation for any type of observation will be incorporated. The applicability of the operational software for "integrated" geodesy has been efficiently demonstrated by *K. Budde* (1988). Finally the TEDCON program produces sets of Cartesian station coordinates for various epochs including their full variance-covariance matrices for *local deformation analysis*. For the development of the program INA extensive simulation and comparative studies of geodetic inertial measurement systems have been carried out, cf. *D. Schröder et al* (1988).

2. GPS phase observational equations

2.1 Undifferences versus fully differenced (triple differenced) carrier phases

The GPS carrier beat phase observable is defined by the difference of the satellite oscillator phase $\phi^S(t^S)$ at the time t^S of transmission and the ground receiver reference phase $\phi_r(t_r)$ at the actual time instant t_r of reception plus the integer ambiguity N_r^S ("natural number")

$$\Phi_r^S(t) := \phi^S(t^S) - \phi_r(t_r) + N_r^S . \tag{2.1}$$

Of course, all terms within (2.1) have to be related to the nominal time instants of reception of the GPS signal. A detailed series expansion of *B. Taylor* type of (2.1) following *D.E. Wells et al* (1986) and *B. Schaffrin and E. Grafarend* (1986) leads to the undifferenced carrier beat phase observational equation (2.2)

$$\Phi_r^S(t) = -\frac{f}{c} [1 - c^{-1} \dot{\rho}_r^S(t)] \rho_r^S(t) +$$
$$+ [1 + c^{-1} \dot{\rho}_r^S(t)] \alpha_r(t) + \beta^S(t) + \gamma_r^S + \tag{2.2}$$
$$+ (1 + \delta_r)d_{trop} - d_{ion} + u_r^S(t)$$

$$\rho_r^S(t) := \|\underset{\sim}{x}^S(t) - \underset{\sim}{x}_r\| \tag{2.3}$$

$$\dot{\rho}_r^S(t) := \|\underset{\sim}{x}^S(t) - \underset{\sim}{x}_r\|^{-1} \langle \underset{\sim}{x}^S(t) - \underset{\sim}{x}_r | \dot{\underset{\sim}{x}}^S(t) \rangle \tag{2.4}$$

$$\gamma_r^S := \phi^S(t_o) - \phi_r(t_o) + N_r^S . \tag{2.5}$$

f denotes the GPS-L1 carrier frequency, c the speed of light in vacuo, d_{trop} the tropospheric model correction term, d_{ion} the single frequency ionospheric model correction term and u the stochastic measurement error. $\alpha_r(t)$ is a collection of the receiver-specific bias, $\beta^S(t)$ of the satellite-specific bias, γ_r^S of the pair-specific bias according to a *twofold classification model*. Finally δ_r contains the tropospheric bias term. It should be mentioned that the pair-specific bias parameters γ_r^S are non-integer quantities. Only after differencing with respect to receiver and satellite in reference or sequential manner, the *initial epoch ambiguities* are of *integer type*.

Once we apply the mechanism of *triple differences* between ground receivers ("Δ"), between satellites ("∇") and between epochs ("δ") due to the notation of *W. Lindlohr and D.E. Wells* (1985), we come up with the observational equation for *triple differences* of carrier beat phases of type (2.6) which explicitly has been derived by *K. Eren* (1987)

$$\delta\nabla\Delta\Phi_r^S(t) = -c^{-1}f[1-c^{-1}\dot{\rho}_r^S(t)]\delta\nabla\Delta\rho_r^S(t) + \delta\nabla\Delta u_r^S(t) \tag{2.6}$$

2.2 The Gauß-Markov Model versus the Mixed Model

We give only a short review of the basic features of two alternative adjustment models, namely the classical *Gauß-Markov Model* of full rank and without restrictions and the *Mixed Model*. The starting point is the nonlinear functional model, the expectation of the original undifferenced phase observation $Y := \Phi_r^S(t)$ as being described by (2.7),

$$E\{\underset{\sim}{Y}\} = A_1 \underset{\sim}{\Xi} + A_2 \underset{\sim}{\eta} \tag{2.7}$$

where the *desired parameters* are collected in the vector $\underset{\sim}{\Xi} := [\underset{\sim}{x}_r, \underset{\sim}{x}^S, \dot{\underset{\sim}{x}}^S]$ containing the Cartesian coordinates of the ground station, of the satellite centre and of its velocity vector. Actually we have merged these unknowns in order to be able to perform *simultaneous relative positioning* and *orbit determination*. The vector $\underset{\sim}{\eta}$ is filled by *nuisance parameters* of type receiver-, satellite-, pair-specific and tropospheric bias. The estimation postulate of type BLUE leads after linearization at a *fixed position* $\underset{\sim}{\Xi}_0$, namely by introducing

$$\underset{\sim}{y} = A_1 \underset{\sim}{\xi} + A_2 \underset{\sim}{\eta} + \underset{\sim}{u} \; , \quad \begin{bmatrix} \underset{\sim}{\Psi}_0 := A_1 \underset{\sim}{\Xi}_0 \\ \underset{\sim}{y} := \underset{\sim}{Y} - \underset{\sim}{\Psi}_0 \\ \underset{\sim}{\xi} := \underset{\sim}{\Xi} - \underset{\sim}{\Xi}_0 \end{bmatrix} \; , \tag{2.8}$$

to the solution

$$\begin{bmatrix} \hat{\underset{\sim}{\xi}} \\ \hat{\underset{\sim}{\eta}} \end{bmatrix} = \begin{bmatrix} A_1^T P_y A_1 & A_1^T P_y A_2 \\ A_2^T P_y A_1 & A_2^T P_y A_2 \end{bmatrix}^{-1} \begin{bmatrix} A_1^T P_y \underset{\sim}{y} \\ A_2^T P_y \underset{\sim}{y} \end{bmatrix} \; . \tag{2.9}$$

An alternative approach which results in the so-called *Mixed Model* takes advantage of the fact the approximate values being needed for linearization are quantities of *stochastic prior information*: a second linearization at a *stochastic position* $\underset{\sim}{X}_0 := \underset{\sim}{\Xi}_0 + \underset{\sim}{o}$ ($\underset{\sim}{o}$ denotes the so-called stochastic null vector) defines the *Mixed Model* (2.10)

$$\underset{\sim}{y} = A_1 \underset{\sim}{x} + A_2 \underset{\sim}{\eta} + \underset{\sim}{u} \; , \quad \begin{bmatrix} \underset{\sim}{Y}_0 := A_1 \underset{\sim}{X}_0 \\ \underset{\sim}{y} := \underset{\sim}{Y} - \underset{\sim}{Y}_0 \\ \underset{\sim}{x} := \underset{\sim}{\Xi} - \underset{\sim}{X}_0 = \underset{\sim}{\xi} + \underset{\sim}{o} \end{bmatrix} \; . \tag{2.10}$$

A further decomposition of the now *stochastic unknown vector* $\underset{\sim}{x}$ is into

$$\underset{\sim}{x} = \underset{\sim}{\kappa} + \underset{\sim}{e} \; , \tag{2.11}$$

where $\underset{\sim}{\kappa} := \underset{\sim}{\kappa}_0 - \underset{\sim}{X}_0$ contains the differential *bias vector of prior information* and $\underset{\sim}{e}$ its error which constitutes the basis of the *variance-covariance* matrix of prior information.

The *stochastic unknowns* $\underset{\sim}{x}$ are *predicted* by inhom BLIP (best inhomogeneously linear prediction) or hom BLUP (best homogeneously linear unbiased prediction) while the *fixed unknowns* $\underset{\sim}{\eta}$ are *estimated* by BLUE (best linear unbiased estimation) according to (2.12) and (2.13)

$$
\begin{bmatrix} \tilde{\underset{\sim}{x}} \\ \hat{\underset{\sim}{\eta}} \end{bmatrix} = \begin{bmatrix} N_{11}+P_\kappa & N_{12} \\ N_{21} & N_{22} \end{bmatrix}^{-1} \begin{bmatrix} A_1^T P_y \underset{\sim}{y} + P_\kappa \underset{\sim}{\kappa} \\ A_2^T P_y \underset{\sim}{y} \end{bmatrix} \qquad \text{inhomBLIP} \qquad (2.12)
$$

$$
\begin{bmatrix} \tilde{\underset{\sim}{x}} \\ \hat{\underset{\sim}{\eta}} \end{bmatrix} = \begin{bmatrix} N_{11}+P_\kappa & N_{12} \\ N_{21} & N_{22} \end{bmatrix}^{-1} \begin{bmatrix} A_1^T P_y \underset{\sim}{y} + aP_\kappa \underset{\sim}{\kappa} \\ A_2^T P_y \underset{\sim}{y} \end{bmatrix} \qquad \text{homBLUP} \qquad (2.13)
$$

$$
N_{ij} := A_i^T P_y A_j \ , \qquad a := \frac{\underset{\sim}{\kappa}^T P_\kappa (N+P_\kappa)^{-1} A^T P_y \underset{\sim}{y}}{\underset{\sim}{\kappa}^T P_\kappa (N+P_\kappa)^{-1} N \underset{\sim}{\kappa}} \ . \qquad (2.14)
$$

For more details of the *Mixed Model* we refer to *B. Schaffrin* (1985). Numerical examples referring to WORLD are given by *E. Grafarend and B. Schaffrin* (1988) and *W. Lindlohr* (1988a).

3. GPS networks, examples

From 2 November to 6 November, 1987, a GPS survey took place in the city area of Berlin. During the first two days *five receivers* have been sim- ultaneously operated while during the last three days even *seven re- ceivers* had been available. (WM 101 single frequency receivers of Wild Magnavox Satellite Survey Company). The Institute of Geodesy and Photo- grammetry, Technical University of Berlin, the Institute of Geodesy, University FAF München, the Institute of Surveying Engineering, Tech- nical University of Braunschweig, and the Department of Geodetic Science, Stuttgart University, have taken part in this project.

A total of 19 stations formed the GPS network. *Figure 3.1* illustrates the local area network in a local horizontal plane, here only consisting of 6 stations which were involved in the PUMA computation. The baseline lengths vary from about 6.2 km at minimum to about 31.3 km at maximum. Due to the fact that we have used differencing techniques with respect to station 5, the length of 5 independent baselines is about 20.4 km.

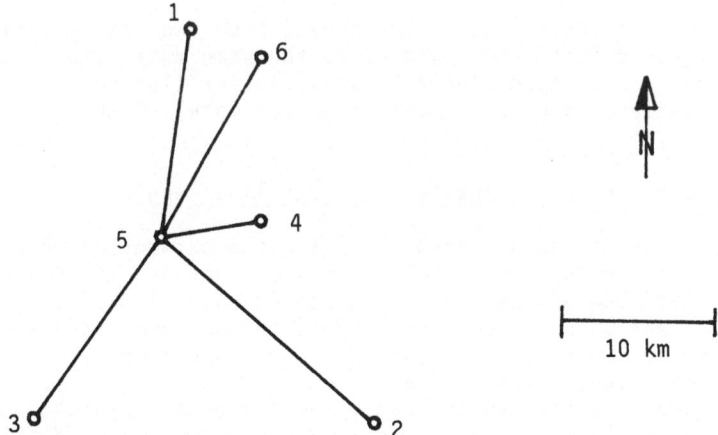

Figure 3.1: Part of the GPS Berlin network

In order to create regularly shaped data sets the original WM 101 measurements have been preprocessed by the so-called PoPS-Software, e.g. *E. Frei et al* (1986), namely in order to detect cycle slips from binary cassette data. The decoding procedure was performed on the basis of the WMDATA program kindly supplied by the Institute of Geodesy and Photogrammetry of the Technical University of Berlin. For further details we have to take reference to *W. Lindlohr* (1988b).

For demonstrating the capabilities and the performance of PUMA and SUTDIF we created two limited data sets. The first one consists of simultaneous observations from 6 ground stations to 5 GPS-satellites for the time span 4.01 to 4.33 UTC on the third observation day, i.e. November 4th, 1987. Considering that the compressed measurements are recorded on tape in full minute intervals we arrive at

$$n_1 = R \cdot S \cdot T = 6 \cdot 5 \cdot 33 = 990 \tag{3.1}$$

undifferenced phase observations which are used for all results computed by PUMA.

Due to central memory limitations the second data start from the same configuration, but uses two minute intervals between observation epochs and creates

$$n_2 = (R-1) \cdot (S-1) \cdot (T-1) = 5 \cdot 4 \cdot 16 = 320 \tag{3.2}$$

triple difference observations which are the basis for the computations by SUTDIF.

4. Review of results

4.1 Residuals from undifferenced carrier phase observations

In order to demonstrate that the observation equation for undifferenced phase observations in form of (2.2) are appropriate to describe the behaviour of GPS phase measurements, their estimated residuals are investigated. All unmodeled effects are reflected in the residuals being determined from the BLUE, or one of the prediction methods.

In *Figure 4.1* the residual time series from two rather arbitrarily chosen ground stations, 3 and 6, to the same satellite, PRN 11, are graphically displayed. Their total variation lies in the range of 0.2 to 0.3 cycles being equivalent to approximately 4 to 6 cm.

4.2 Geodetic relative positioning: baseline length

Although the Cartesian coordinates are the parameters which are estimated in the adjustment process, they are rather weakly determined quantities. The GPS carrier phases being some kind of biased range observable are insensitive with respect to the translational part of the *datum defect*. Due to very high correlations between the coordinates along their axes, derived quantities as inter-station coordinate differences and baseline length, can be estimated or predicted with sufficient internal accuracies. It can be concluded that undifferenced carrier phase observations are a useful tool for geodetic relative positioning.

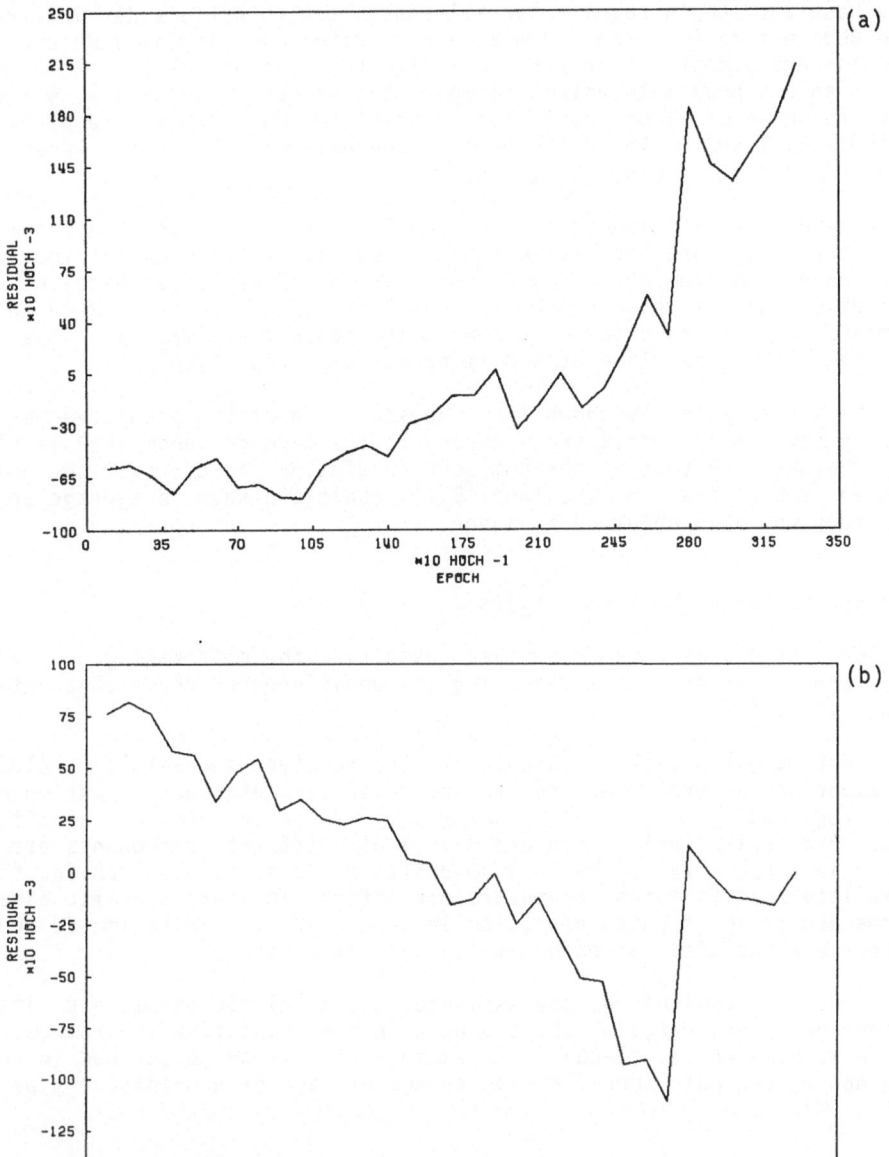

Figure 4.1: Estimated undifferenced carrier phase residuals:
station 3 - PRN 11 (a), station 6 - PRN 11 (b).

The baseline length results for all possible combinations derived from the BLUE estimation, the inhom BLIP prediction and the hom BLUP prediction are summarized in *Table 4.1*. The third column of the table contains some approximate values to which the values in columns 4, 6 and 8 have to be added. Their estimated or predicted mean errors are given in columns 5, 7 and 9. The difference of the determined baseline lengths can be found in columns 10, 11 and 12.

In general the differences between the BLUE estimation and the inhom BLIP prediction are the smallest ones, usually in the order of some mm and reach a maximum of 1.2 cm. Especially for three longer baselines the discrepancies between BLUE and hom BLUP as well as inhom BLIP and hom BLUP are of the order of a few cm and reach their maximal values for the 25.5 km baseline at 4.0 cm or 3.7 cm, respectively.

As it is theoretically known from the structure of the predicted variance-covariance matrix the accuracy in the case of inhom BLIP is always better than that of the hom BLUP prediction. In our case the predicted mean errors from the inhom BLIP prediction show in average an accuracy which is about 30 % higher.

4.3 Estimation of bias parameters

Besides the coordinates or derived quantities the different groups of *bias parameters* are estimated using the undifferenced phase observation equation (2.2).

In order to get a better insight into the receiver and satellite clock behavior the general model for an epochwise parameter determination has been replaced by *polynomial representations*. For the receiver-specific bias terms drift, drift rate and change of drift rate components are given in *Table 4.2a*. In *Table 4.2b* drift and drift rate values for the satellite-specific bias parameters are listed. The pair-specific bias terms are given in units of cycles in *Table 4.2c*. Their estimated mean errors are included for each case in the same units.

The order of magnitude of the estimated clock related parameters fits rather good into range which is quoted in the respective literature, cf. *R.W. King et al.* (1985), for example. As already described in paragraph 2, the pair-specific bias parameters are of non-integer type.

4.4 The general case of simultaneous orbit determination

In the classical navigation solution using the GPS-system a moving receiver is simultaneously measuring pseudo-ranges to four satellites. Using phase observations one might think at a first glance that the reversal, i.e. an Cartesian orbit determination for one satellite from four co-observing ground stations on the earth should be possible. In this sense "Cartesian" means that 3 parameters per satellite and epoch have to be determined. But the redundancy design investigations, cf. *E.W. Grafarend et al. (1986)* show that with the inclusion of the three double indexed bias parameters, at least two satellites have to commonly tracked from eight or more receivers. The results of the redundancy design are summarized in *Table 4.3* with respect to the minimal number

from	to	Approximate values [m]	BLUE (a) [m]	[cm]	inhom BLIP (b) [m]	[cm]	hom BLUP (c) [m]	[cm]	(a)-(b) [cm]	(b)-(c) [cm]	(c)-(a) [cm]
1	2	31320 +	1.582	± 3.1	1.591	± 3.2	1.583	± 4.2	-0.9	+0.8	+0.1
1	3	28520 +	1.878	± 5.5	1.894	± 6.1	1.865	± 7.1	-1.6	+2.9	-1.3
1	4	15520 +	7.281	± 2.7	7.290	± 2.7	7.288	± 3.7	-0.9	+0.2	+0.7
1	5	15590 +	7.263	± 4.0	7.275	± 4.3	7.262	± 5.2	-1.2	+1.3	-0.1
1	6	7040 +	6.723	± 5.9	6.720	± 6.5	6.728	± 7.6	+0.3	-0.8	+0.5
2	3	25530 +	3.749	± 8.5	3.746	± 8.6	3.709	±11.3	+0.3	+3.7	-4.0
2	4	16380 +	3.548	± 4.2	3.549	± 4.3	3.541	± 5.7	-0.1	+0.8	-0.7
2	5	20460 +	1.672	± 5.8	1.671	± 5.9	1.671	± 7.8	+0.1	0.0	-0.1
2	6	25560 +	1.792	± 2.8	1.792	± 2.8	1.785	± 3.8	0.0	+0.7	-0.7
3	4	19110 +	6.071	± 8.4	6.072	± 8.6	6.041	±11.2	-0.1	+3.1	-3.0
3	5	14160 +	4.003	± 7.5	4.004	± 7.7	3.981	±10.1	-0.1	+2.3	-2.2
3	6	28350 +	2.606	± 7.3	2.599	± 7.4	2.578	± 9.7	+0.7	+2.1	-2.8
4	5	6230 +	9.048	± 8.8	9.047	± 7.4	9.050	±11.7	+0.1	-0.3	+0.2
4	6	11380 +	7.743	± 4.5	7.737	± 4.5	7.739	± 6.0	+0.6	-0.2	-0.4
5	6	14320 +	2.851	± 7.0	2.843	± 7.2	2.843	± 9.4	+0.8	0.0	-0.8

Table 4.1: Estimated / predicted baseline lengths: BLUE (a), inhom BLIP (b), hom BLUP (c)

	$\hat{\alpha}_{1r}$ $\hat{m}(\hat{\alpha}_{1r})$	$\hat{\alpha}_{2r}$ $\hat{m}(\hat{\alpha}_{2r})$	$\hat{\alpha}_{3r}$ $\hat{m}(\hat{\alpha}_{3r})$
r	$[10^{-7}$ sec/sec]	$[10^{-14}$ sec/sec^2]	$[10^{-18}$ sec/sec^3]
1	-1.801 ± 0.0000078	-0.473 ± 0.085	-4.081 ± 0.27
2	-0.331 ± 0.0000079	$+0.406 \pm 0.085$	$+1.648 \pm 0.27$
3	-1.770 ± 0.0000079	$+2.060 \pm 0.085$	$+1.699 \pm 0.27$
4	-1.557 ± 0.0000079	$+1.591 \pm 0.085$	-0.081 ± 0.27
5	-1.776 ± 0.0000079	$+6.897 \pm 0.085$	-5.020 ± 0.27
6	-0.227 ± 0.0000079	$+6.477 \pm 0.085$	-7.890 ± 0.27

(b)

s	$\hat{\beta}_1^s$ $\hat{m}(\hat{\beta}_1^s)$	$\hat{\beta}_2^s$ $\hat{m}(\hat{\beta}_2^s)$
	$[10^{-11}$ sec/sec]	$[10^{-16}$ sec/sec^2]
PRN 03	$+0.718 \pm 0.057$	$+4.630 \pm 1.79$
PRN 11	-1.710 ± 0.894	-1.361 ± 1.83
PRN 12	-1.989 ± 0.552	$+1.740 \pm 1.79$
PRN 13	-0.448 ± 0.860	$+4.715 \pm 1.82$

(c)

r	s	$\hat{\gamma}_r^s$ $\hat{m}(\hat{\gamma}_r^s)$	r	s	$\hat{\gamma}_r^s$ $\hat{m}(\hat{\gamma}_r^s)$
1	1	128719510.393 ± 0.401	2	1	128230622.562 ± 0.468
1	2	112840420.291 ± 0.401	2	2	108383402.362 ± 0.536
1	3	131348791.096 ± 0.401	2	3	129127294.085 ± 0.537
1	4	130660579.384 ± 0.401	2	4	126425655.209 ± 0.458
1	5	128038617.125 ± 0.401	2	5	126362681.569 ± 0.465
3	1	127268044.840 ± 0.468	4	1	128566546.032 ± 0.468
3	2	112641924.094 ± 0.537	4	2	110161475.566 ± 0.536
3	3	131244123.291 ± 0.537	4	3	116512977.292 ± 0.537
3	4	130360955.942 ± 0.458	4	4	111133954.616 ± 0.458
3	5	128687250.627 ± 0.466	4	5	123947880.600 ± 0.465
5	1	127563944.148 ± 0.468	6	1	127775620.333 ± 0.468
5	2	112445044.750 ± 0.536	6	2	108832919.719 ± 0.536
5	3	131485821.748 ± 0.537	6	3	128374822.227 ± 0.537
5	4	118901229.696 ± 0.458	6	4	126179948.261 ± 0.458
5	5	128205620.441 ± 0.465	6	5	127157955.389 ± 0.465

Table 4.2: Estimated receiver-specific (a), satellite-specific (b) and pair-specific (c) bias parameters.

of receivers, satellites and observations epochs as well as the actual
number of phase observations and the redundancy itself.

R	S	T	m	ST	R+ST	m-n
8	2	25	400	50	58	0
9	2	15	269	30	39	1
10	2	11	220	22	32	0
11	2	10	217	20	31	3
12	2	9	212	18	30	4
13	2	8	205	16	29	3
14	2	7	196	14	28	0
18	2	6	215	12	30	1
28	2	5	280	10	38	0
Minimum values						
8	2	5	196	10	28	0

Table 4.3: Redundancy design for simultaneous ground station and
orbit determination using undifferenced phase
observations

4.5 Simultaneous orbital improvement from triple difference observations

Using the triple difference phase observations defined in eq. (2.2) the
influence of stochastic prior orbital information on different ground
station position measures has been investigated. These consist of ground
station coordinates and coordinate differences as well as of baseline
lengths and baseline length ratios. For the investigations a cycle slip
free part of the original phase observations of a GPS-survey in Berlin
was used. Triple difference observations have been created by reference
differencing between receivers and satellites and by sequential differencing between epochs, cf. *W. Lindlohr and D.E. Wells* (1985). The program SUTDIF uses these regularly shaped triple difference observations
as main input quantities as well as the prior information variance-covariance matrix both for cartesian station and satellite coordinates.
The variance-covariance matrix of cartesian satellite coordinates have
been derived from the respective matrix in the local moving triad by
the general error propagation law. For the analysis different variations
of the stochastic a priori orbital constraints between 0 and 100 m for
the along, across and out-of-plane components have been used. The temporal correlation between the consecutive epochs of the stochastic orbital information has a rather insignificant influence on all four
types of station position measures. Although the datum defect of the
satellite geodetic network is removed by introducing the constraints
as a kind of pseudo-observation, the origin of the network is very weakly defined from triple difference observations as well. Therefore ground
station coordinates are by far the most weakly estimable parametes, just
in contradiction to the relative position measures. The influence of the
orbital prior information on the derived position measure values and
mean square errors is rather similar. Depending upon increasing orbit
relaxation the derived quantities themselves show a systematic and de-
creasing drift. Their mean errors exhibit a rather stable behaviour
with respect to the variation of the orbital constraints. If the stochastic prior information is too weak, i.e. the a priori standard deviations are too large, one gets into extremely ill-conditioned adjustment problems where the inversion of the extended normal equation system

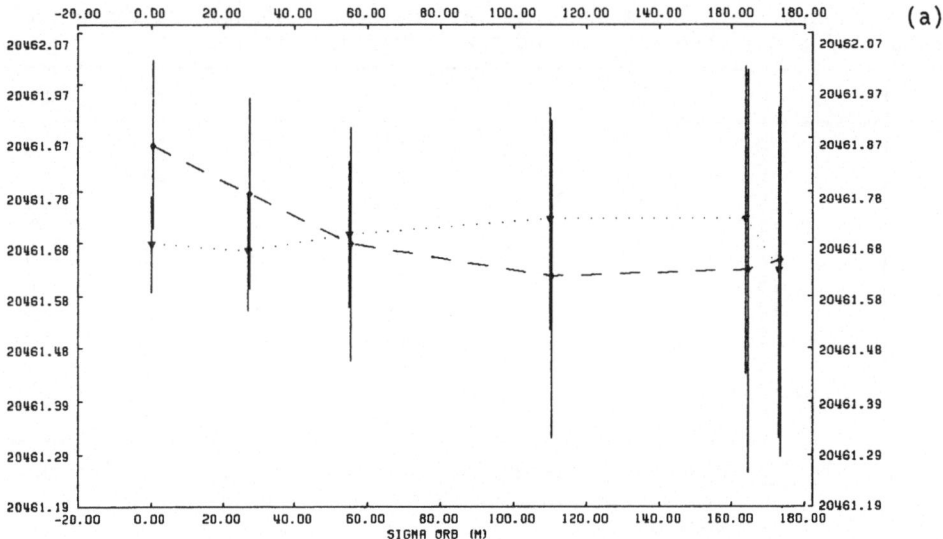

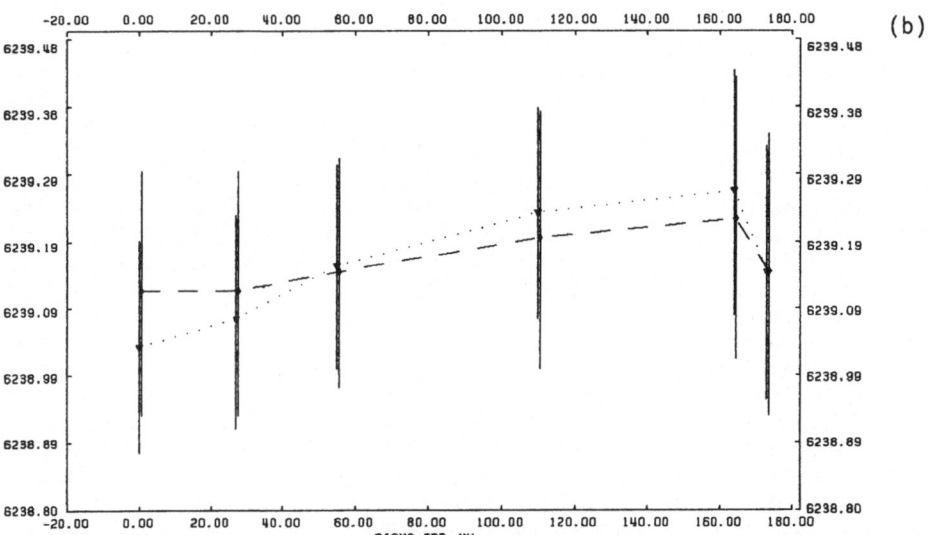

Figure 4.2: Influence of orbital prior information on ground station
position measures: baseline 2-5 (a), baseline 4-5 (b).
A priori mean ground station coordinate error of 50 m
(dotted line) or 1000 m (dashed line).

fails, especially with respect to a small size local network and the short total observation time span. A very detailed description of the complete investigation of the special triple difference observation data set is given by *W. Pachelski et al (1988)*.

In *Figure 4.2* the influence of orbital prior information is exemplarily demonstrated for the longest baseline 2-5 and the shortest baseline 4-5. In dependence of the point related a priori mean errors the variation of the baseline length and its predicted mean error is shown. The dotted line represents an a priori mean station coordinate error of 50 m and the dashed line a corresponding error of 1000 m for each of the Cartesian ground station coordinates.

5. Conclusions

Besides all types of differenced carrier beat phases, the undifferenced phase observations are as well a useful tool for geodetic relative positioning. In contradiction to the classical Gauß-Markov model, the more general mixed model allows the incorporation of stochastic prior information which consists of the triple of the approximate coordinates X_0, the differential vector κ and the variance-covariance matrix C_κ of prior information. Using the special limited phase data set with roughly thousand observations, a relative positioning accuracy in the range of 3 to 5 ppm can be obtained for baselines of up to 30 km in length. In addition to the Cartesian ground station parameters, Cartesian coordinates per satellite and epoch for orbit determination purposes and four types of nuisance parameters can simultaneously be estimated or predicted.

6. References

Budde, K. (1988): Adjusting large threedimensional networks: strategies and computations. Submitted Paper, Third SIAM Conference on Applied Linear Algebra, Madison, May 23-26.

Eren, K. (1987): Geodetic network adjustment using GPS triple difference observations and a priori stochastic information. Technical Report No. 1, Institute of Geodesy, University of Stuttgart.

Frei, E., R. Gough and F.K. Brunner (1986): PoPS: a new generation of GPS post-processing software. Proceedings, Fourth International Geodetic Symposium on Satellite Positioning, Austin, Vol. 1, pp. 455-473.

Grafarend, E.W., W. Lindlohr and D.E. Wells (1985): GPS redundancy design using the undifferenced phase observation approach. Proceedings, Second Meeting of the European Working Group on Satellite Radio Positioning, Saint-Mandé, pp. 100-107.

Grafarend, E.W. and B. Schaffrin (1988): Von der statischen zur dynamischen Auffassung geodätischer Netze. Zeitschrift für Vermessungswesen, Vol. 112, No. 2, pp. 79-103.

King, R.W., E.G. Masters, C. Rizos, A. Stolz and J. Collins (1985): Surveying with GPS. Monograph No. 9, School of Surveying, University of New South Wales, Kensington.

Lindlohr, W. and D.E. Wells (1985): GPS design using undifferenced carrier beat phase observations. Manuscripta Geodaetica, Vol. 10, No. 4, pp. 255-295.

Lindlohr, W. (1988a): Alternative modeling of GPS carrier phases for geodetic network analysis. Submitted Paper, International Workshop High Precision Navigation, Stuttgart-Altensteig/Wart, May 17-20.

Lindlohr, W. (1988b): PUMA: processing of undifferenced GPS carrier beat phase measurements and adjustment computations. Technical Report No. 5, Institute of Geodesy, University of Stuttgart, in preparation.

Milbert, D.G. (1984): Heart of gold: computer routines for large, sparse, least squares computations. NOAA Technical Memorandum, NOS NGS-39, Rockville.

Pachelski, W., D. Lapucha and K. Budde (1988): GPS network analysis: the influence of stochastic prior information of orbital elements on ground station position measures. Technical Report No. 4, Institute of Geodesy, University of Stuttgart, in print.

Remondi, B.W. (1984): Using the Global Positioning System (GPS) phase observable for relative geodesy: modeling, processing, and results. CSR-84-2, Center for Space Research, The University of Texas at Austin.

Schaffrin, B. (1985): Das geodätische Datum mit stochastischer Vorinformation. Deutsche Geodätische Kommission, C-313, München.

Schaffrin, B. and E.W. Grafarend (1986): Generating classes of equivalent linear models by nuisance parameter elimination: applications to GPS observations. Manuscripta Geodaetica, Vol. 11, No. 4, pp. 262-271.

Schröder, D., N. Chi Thong, S. Wiegner, E.W. Grafarend and B. Schaffrin (1988): A comparative study of local level and strapdown inertial systems. Manuscripta Geodaetica, Vol. 13, No. 4, in print.

Wells, D.E., K. Doucet and W. Lindlohr (1986): First order geodetic network design: some considerations. Proceedings, Fourth International Symposium on Satellite Positioning, Austin, Vol. 1, pp. 801-819.

Wells, D.E., W. Lindlohr, B. Schaffrin and E.W. Grafarend (1987): GPS design: undifferenced carrier beat phase observations and the fundamental differencing theorem. Department of Surveying Engineering Technical Report No. 116, University of New Brunswick, Fredericton, N.B.

Zaiser, J. (1984): Ein dreidimensionales geometrisch-physikalisches Modell für konventionelle geodätische Beobachtungen: Beobachtungsfunktionale, Parameterschätzung und Deformationsanalyse. Deutsche Geodätische Kommission, C-298, München.

PC PREPROCESSING OF GPS - TI 4100 NAVIGATOR DATA

by

Roman Galas

Abstract

A program-package for the data transfer from TI 4100 cassettes to hard
and floppy disks, using the M5450XL and CDS 1.30 data terminals, is
described. The data are separated onto 4 different files, containing only
valid records, and then saved on floppy disks for archive. These binary
files can be transformed into an editable form if desired. The problem of
cycle slip detection, separate for each station, is discussed. A method
enabling a direct data transfer from the TI 4100 receiver, without the
tape recorder, is in preparation.

1.   General overview.

The preprocessing software for the data collected by the TI 4100 Navstar navigator. for a personal computer. is currently under development at the Institut for the astronomical and physical geodesy at the Technical University Munich. It is designed to create one coherent system in order to transform the raw field observations into clear data files. The workhorse of this program — package is GPSTIPS (GPS TI 4100 Navigator Data Preprocessing Software).

The software presented below has been designed to treat GPS data independently for each space vehicle constellation at a specific site . Its structure is shown in Fig. 1. There are three major stages:

The first one is the data transfer. It requires one of two alternative inputs:
- TI 4100 field data cassettes. or
- on—line data acquisition from the TI 4100 receiver into the preprocessing personal computer via serial port.

The results are:
- B — Files. binary files containing only valid TI 4100 reports .
- INFO file . describing the data and containing all necessary information for later use.

The second stage edits the collected data. It is composed of three programs which shape four files. The first two files contain the doppler accumulated carrier phase for both L1 and L2 frequencies. the third one contains pseudorange measurements and in the last one there are the brodcast ephemeris. The user can edit and inspect all Receiver—Ranging Measurement reports . User State Solutions and Space Vehicle Navigation Data blocks.

The third one is still under development and is designed to fix and repair cycle slips.

The major modules of the software are described in more detail in the following sections.

2.   Data transfer.

The main task of the stage 1 is to transfer the GPS field data into a personal computer. The preprocessor accept TI 4100 binary data from cassettes or archival data from floppy disks. These archival floppy disks must be created previously with the program — package. presented here. of course. The additional possibility, the collection of the GPS TI 4100 reports in on — line mode with interfaced Atari 1040ST computer is also proposed. Another extension of the system is the program creating a full set of files in the format required by GEOMARK processing software.

2.1   Transfer of the data from TI 4100 data cassettes.

The field data cassette is the standard interface between TI 4100 receiver and our preprocessing software. The data can be transferred from a cassette by utilizing either the CDS 1.30 data terminal from Mess + System and Technik GmbH or the Memtec Corporation´s cassette reader model 5450XL. There are. certainly two different programs to load the data into

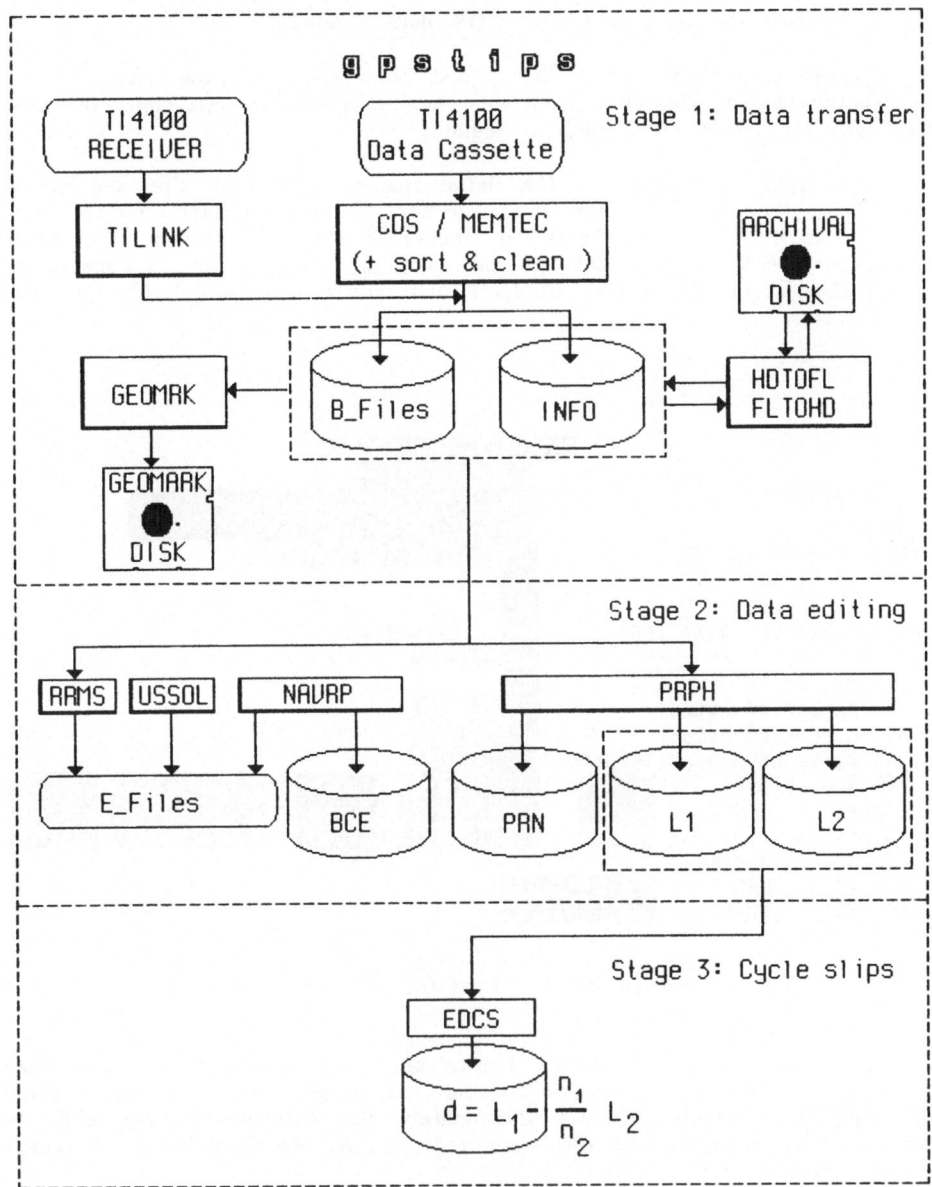

Fig. 1. Preprocessing flow chart.

the microcomputer. One of them cooperates with CDS and the other one with MEMTEC device. On the output of this stage there are four binary files, containing:

- receiver - ranging measurements,
- navigation messages,
- user solutions.
- error log data.

(B − Files) and one information file, all on the hard disk.

2.1.1 Loading the data with the CDS data terminal.

The program is written for a 9600 baud communication rate. That is why the transmission of the data from one full standard cassette into the host microcomputer needs only about 6 minutes.

The data transfer starts with the definitions of a current site and customer names, then asks for the total number of data cassettes in the current SV constellation and finally it checks if there is enough free space on the hard disk for the data to be read in (Fig. 2) . It should be noticed that about 270 kbytes of hard disk space is required for one full data cassette.

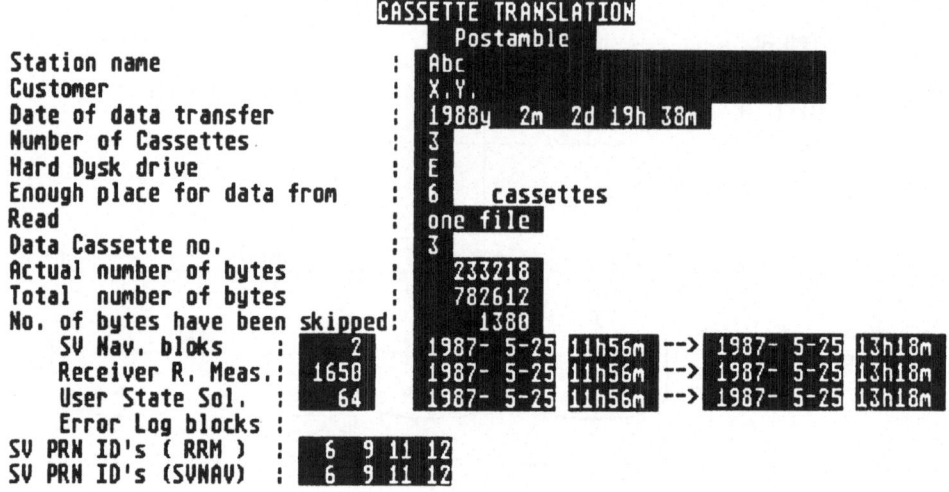

Fig. 2. Screen of the TI 4100 cassette translation.

If the observations from one SV constellation are collected on more then one cassette the user is responsible for the proper order of the inserted tapes into the cassette drive. Unfortunately the software is not able to sort the data according to the time tags in the Receiver—Ranging Measurement reports.

In order to allow the transmission of only one file from a data cassette, the user must decide during the initialization step if a file mark should be skipped and afterwards if all the following data (i. e. until clear leader) or only the data up to the next file mark should be loaded. This tools allow the user to save observations from more then one SV constellation on a cassette and afterwards to elaborate each of them separately by successive extraction of the proper files.

Another advantage of this approach is, that data cassettes must not be erased prior to the field acquisition . But in this case the file marks must be set twice, at the begining and at the end of data collection for the specific constellation.

Transmission stops only when a file mark is reached or when there are no more data on the current tape, respectively to user's require.

When the loading from a tape is finished, the data are searched for invalid blocks and are separated into four different files according to the TI 4100 reports. During the verification of the Receiver—Ranging Measurement Data blocks the software reports also the loss of signal and of phase — lock indicators (gaps) . In Space Vehicle Navigation Data blocks only the NPRMP map array is checked.

During this data transmission the helpful information is screened (Fig. 2). The number of bytes which have been skipped, indicates that some parts of the data are invalid i.e. an incomplete or invalid report or an CRC error has appeared, on the tape. The SV PRN IDs from receiver measurements and from navigation message and also the maximum and the minimum PDOP are reported too. The number of TI 4100 reports is the number of only valid ones in every case. Start and stop times of each type of report can indicate that the proper data were loaded and the cassettes were inserted in the proper order. The time history of created files is also available. On the output there are four binary not printable files and an information file. This last one contains description of the data which has been loaded and the quantities required later when GEO-MARK files are created.

2.1.2   Loading the data with the MEMTEC terminal.

This modul performs the task of the GPS data transfer from TI 4100 receiver data cassettes into the host computer very similar to the above described program, but there is one very important restriction . The cassette must be erased prior to the field data registration . Another disadvantage is the transfer time of a full data cassette of about 28 minutes. It is due to the fact that the MEMTEC buffer data terminal can work with a baud rate not greater than 2400.

2.2   Data collection by a user's computer system.

An alternative approach of the data acquisition has been investigated in order to omit the time consuming process of the GPS data transfer from expensive field data cassettes into a preprocessor. A serial Atari 1040ST personal computer has been interfaced to a TI 4100 receiver for the data acquisition. The external computer provides the data storage and control.

Unfortunately there is no possibility to command the receiver through the Atari, because this computer has only one serial port . Therefore the following procedure can be applied. On start—up the program TILINK goes at once into a "wait state". It means: the program only tests if RS—232C computer buffer is empty. Now the user is obliged to perform appropriate operating functions with the hand—held control display unit of the receiver. When the software program detects the data coming from the receiver it begins to store them in computer memory and next reverts to the "wait state " again. The operator must command with the CDU which kind of report should be loaded into PC. If the link buffer is empty longer then 3 minutes, the program asks if the user wishes to finish the data acquisition. The received data is converted from the transparent binary text format to the recorder format and transferred onto the hard disk. Finally the collected observations are examined for validity and separated into

B—Files. Up to date, due to the handle convenience, TILINK cooperates with a hard disk only. It can be expected that the procedure described above guarantees dumping of the data direct onto a floppy diskette. In this case of the data collection a floppy disk is the field recording medium instead of a cassette and therefore the data should be preprocessed afterwards.

2.3   Archiving of the data.

This section is done to preserve the field data unchanged through the processing and to restore them if necessary. After the B—Files are created they can be archived on floppy disks either one sided 5 1/4 inch or one— and double sided 3.5 inch. The software program estimates the required number of floppy disks. When the rewrite process is done this number is checked and in case of wrong estimation the INFO files are corrected on each floppy disk.

2.4   GEOMARK link.

This module produces a full set of files for GEOMARK processing software in formats as it is described in the GEOMARK manual. The four not printable internal GEOMARK files . i.e. Receiver—Ranging Measurements . User State Solutions . Navigation Message Data and Error Log, are created on the base of B—Files by stripping the original ones of all unnecessary information. If more then four satellites have been found in the receiver measurement data, the software does not allow to create these exchange files. The bytes are reversed then according to the way in which the data is interpreted by the Intel 8088 processor. The first two editable files : Station Data Control file and Site ID file, are easy created because all necessary information is already recorded in the INFO file. Remaining editable GEOMARK files : meteorological data, GEORES and precise ephemeris, may have only standard values, but they must resident on an exchange floppy disk.
The transfer part of this section rewrites all created files onto one sided floppy disks , under MS—DOS, for future use.

3.   Data editing.

The main purpose of the stage 2 is to produce editable phase, pseudorange and broadcast ephemeris files . It has just two parts.

The first one creates the receiver measurement, ´the user solution and the navigation report files (E — Files) and then allows the user to edit and to inspect all data items from the binary data base. The first two files are decoded and scaled under the formats and units given in the TI 4100 Owner´s Manual.

In order to inspect the Receiver—Ranging Measurement Data blocks simply with any editor (usually this file is to big to be edited as a hole) , the decoded data is separated into a number of files. Each of them contains hundred blocks , and the last letter of the file name indicates which hundreds it is.

In order to decode the navigation message , first the bits are reconstructed, then the relevant values are extracted and scaled. The necessary additional information to perform it has been found in Interface Control Document ICD—GPS—200 and in Van Dierendoc at al. (1978).

In the second part of this stage doppler influenced carrier phase information for both L1 and L2 frequencies (Ph—Files) and pseudoranges (PRN — File) are extracted . Because the phase measurement is the biased continuous accumulated cycle count, the observables are corrected for the value of the observable at the first epoch determined modulo one cycle. The fixed frequency bias is removed with the elapsed time from the first phase observable . These values are given at the start and at the end of FTF in units of cycles. Pseudorange measurements are scaled from the unit of time to the distance unit.

4. Cycle slip detection.

It is well known that a careful detection and repairing of the cycle slips is required because their frequent occurence in TI 4100 data affects significantly geodetic utilizing of the carrier phase data. This section should perform introductory rectification of the integer cycles in mode " one station — one satellite ". before the data is used for final computations. to obtain on the output of this preprocessing software the data as "clear" as possible. The accuracy of the level of few cycles is expected. First the well - known method proposed by Goad (1985) was implemented. The files containing the quantity $\delta_i = L_1 - (\nu_1 / \nu_2) L_2$ are created and cycle slips are fixed. The operator can visually follow on the screen the occurence of the large cycle slips in the carrier phase data.
Also other algorithms are in preparation.

5. Conclusion.

GPS TIPS has been designed with the modular structure to enable possible future modifications and improvements. Its implementation on the Atari 1040ST computer guarantees realy a low price of the hardware. However it can be expected that its implementation on another personal computer. e.g. IBM PC or compatible . can be easy performed. The table below shows the time necessary for some operations for one full transfer tape.

Tab.1. Time budget of the data preprocessing (in minutes).

	CDS	MEMTEC
1. Reading of the tape	~ 6	~ 28
2. Sorting, checking & saving on H.D.	~ 1.1	
3. Creating of an archival disk.	~ 4	
4. Creating of GEOMARK disk.	~ 4	
5. Ph and PRN files (for $\Delta T = 30$ sec.)	~ 4	

All programs. including device interface subroutines . are written in the C language. The pull - down menu technique and the interactive technique have been used to allow the user an easy handling with the system.

6. Acknowledgments.

This work, performed at the Institute for astronomical and physical geodesy TU Munich, was a part of the research program of an Alexander v. Humboldt fellowship. I am deep indepted to the Foundation for making my researches possible.

I am heartily grateful to Prof. R. Sigl for all discussions, his indispensable support and patience. The research purposes would be also impossible to perform without fruitfull, close cooperation with Prof. K. Deichl and Dr. A. Bauch, and their continuing support. The assistance of Dr. D. Egger in solving computer problems was essential. I would also like to thank Ing. G. Dichtl and Ing. E. Kistler for discussions and remarks. Let me use this occasion to express my truly deep gratitude for my pleasant and effective scientific stay in Munich to all of the members of IAPG TUM.

7. References.

Goad C.C. (1985) : Precise Positioning with the Global Positioning system. Proceedings of the Third International Symposium: Inertial Technology for Surveying and Geodesy. Banff, Canada, pp. 745-756.

Van Dierendock A.J., Russel S.S., Kopitzke E.R., Birnbaum M. (1978): The GPS Navigation Message, Navigation, Vol. 25, No.2, pp. 55-73.

Eighth session: Geodynamics

Chairman: Prof. Beutler, Bern

GEODYNAMICS OF ICELAND STUDIED WITH THE AID OF TERRESTRIAL GEODETIC AND GPS
EXPERIMENTS

by

Wolfgang R. Jacoby

Abstract

In extended abstract form a very brief review of the geodynamics of Iceland
is presented. It is attempted to show that Iceland is an excellent field to
do geodynamic GPS experiments on seafloor spreading, hotspots, and mantle
plumes as well as the "plumbing system" feeding the rift volcanoes. The recent
rifting episode offers the opportunity to test geodynamic hypotheses by GPS
in a way not possible by terrestrial methods.

1. Introduction

A number of topical geodynamic questions can be tackled with the aid of GPS in conjunction with other geosciences. The essential components of our present geodynamic picture of Iceland are very briefly reviewed without going into details or background information. One of the purposes of this review is to guide interested geodesists to this information where the arguments for the present picture are given. It is admittedly still very speculative and needs to be proven or disproven by solid observations.

I shall develop the picture from depth to surface, although it did develop in the opposite direction. We shall begin with the hypothetical mantle plume from which melt rises and flows through the "plumbing system along the deep plate boundary. From there it gets into upper crustal levels, particularly the Krafla magma chamber and the Krafla fissure swarm to form new Icelandic crust by dyke formation and volcanism. The Krafla rifting episode since 1975 is used as a model for these processes.

2. Iceland as a hotspot on the Mid-Atlantic Ridge

Iceland is an anomalous part of the Mid-Atlantic Ridge caused by excessive volcanism the products of which build up land to more than 2 000 m elevation within an ocean basin which itself is anomalous in many respects (Fig. 1; Jacoby, 1979). Thus Iceland is called a hotspot; there is nothing hypothetical about hotspots as defined by volcanism. For physical reasons we must also postulate a strong heat source in the mantle capable of producing the excess melt, but its form is not directly known. The current assumption is that the heat is advected by a mantle plume, i.e. a localized upwelling of mantle convection rising from unknown depths. There is indeed indirect

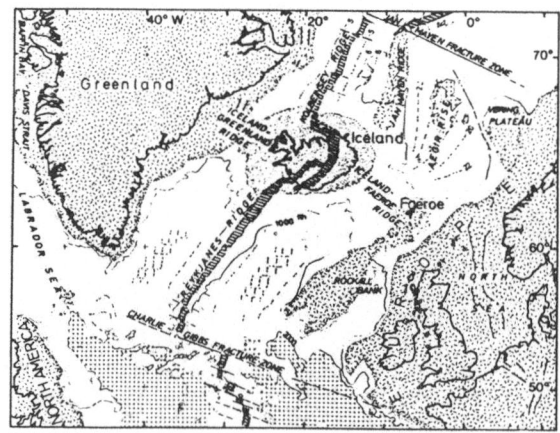

Fig. 1 Schematic map of North Atlantic and Iceland. Iceland and shelf: dotted; magnetic anomalies (partly numbered): dashed lines; depth contours 1000 and 2000 fathoms; greater than 2000 fth depth: regular dots; continental crust: irregular line pattern; major grabens indicated.

evidence for Iceland to be underlain by anomalous hot, partially molten mantle material in the state of internal convection with upwelling below about the center (Fig. 2) from seismic (RRISP Working Group, 1980; Gebrande et al., 1980), magneto-telluric (Arnason, 1981; Beblo & Björnsson, 1980; Beblo et al., 1983), and gravity data (Jacoby et al., 1983). The argument for internal convection in the plume is mainly the discrepancy between the observed Bouguer anomaly and the gravity effect calculated on the basis of the seismic model and the assumption of the usual velocity-density relationship.

Fig. 2 A Bouguer anomaly profile across Iceland along the RRISP seismic profile, compared with the gravity effect calculated for the seismic model on the basis of the usual velocity-density relationship. BA: Bouguer anomaly; FA: Free Air anomaly; vertical hatching: crust; oblique hatching: normal lithospheric upper mantle of the Reykjanes Ridge flank; below Iceland: anomalous upper mantle (asthenosphere, mantle plume)

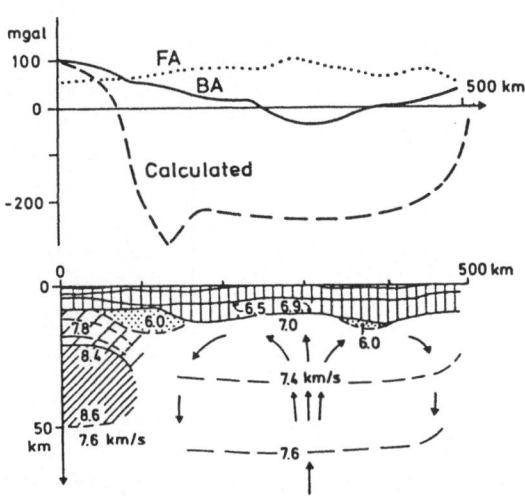

From volcanological observations it is evident that the feeding of melt into the crust is an episodic process. The latest episode which we use as a model for the activity occurred (or is still occurring) in the Krafla volcanic system (Fig. 3) consisting of a central volcanic complex (with a caldera) and a fissure swarm. Prior to the Krafla activity unusually strong earthquakes of thrust type began to occur below Bârdarbunga volcanic complex at the NW edge of Vatnajökull (P. Einarsson, pers. comm., 1986), more than 100 km south of Krafla (Fig. 3). This was interpreted by P. Einarsson as evidence for a deeper melt reservoir in the mantle to have begun to empty itself into the crust-mantle transition zone that makes up the "plumbing system" following the plate boundary. The correlation between the magmatic and rifting activity of Krafla, as described below, and the seismic activity of Bârdarbunga supports Einarsson's interpretation. A discussion of the process has been recently given by Jacoby et al. (1988).

The Krafla rifting episode started with an eruption of a fissure near the caldera center in December 1975, accompanied and followed by rapid subsidence of the caldera floor, caused by magma chamber deflation, and by rifting,

Fig. 3
Tectonic sketch map of
NE Iceland (after
Björnsson, 1985).
Tertiary plateau basalt:
hatched; see also
legend on figure.

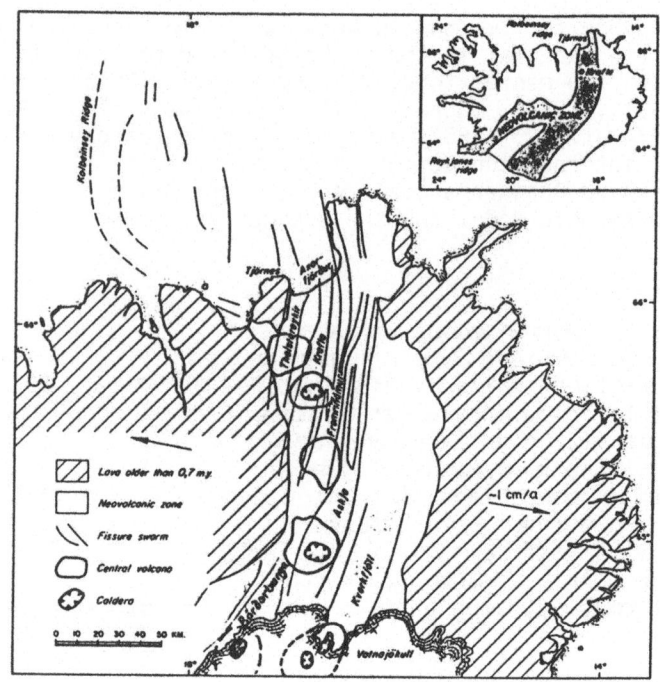

graben subsidence and shoulder uplift with intense seismicity along the
northern section of the Krafla fissure swarm.Subsequently the caldera floor
began to rise again by magma chamber inflation; it continued to rise until
another deflation-rifting event occurred. This kind of activity has continued to at least September 1984 and is depicted by the elevation change
of a benchmark at the caldera center (Fig. 4). The inflation-deflation plus
rifting, seismicity and sometimes eruption activity had some correlation

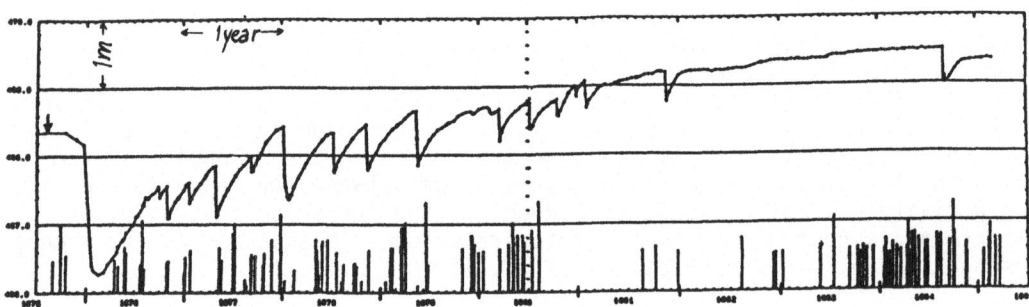

Fig. 4 The inflation-deflation activity of the Krafla magma chamber shown by the elevation
of a benchmark (from Björnsson et. al., 1985). Vertical lines represent earthquakes in the
Bárdarbunga region; line length: magnitude of quakes (after P. Einarsson, pers. comm. and
data from Skjálftabréf, Science Institute, University of Iceland, Reykjavík, 1977-1984).

with the unusual seismicity of the Bárdarbunga volcanic complex (see Fig.5). Thus it appears that a connection exists between Bárdarbunga and Krafla; it could be hydraulic through the melt "plumbing system" along the plate boundary in the there elevated crust-mantle transition zone or/and through elastic stress in the thin lithosphere. A cartoon (Fig. 5) illustrates this hypothetical model.

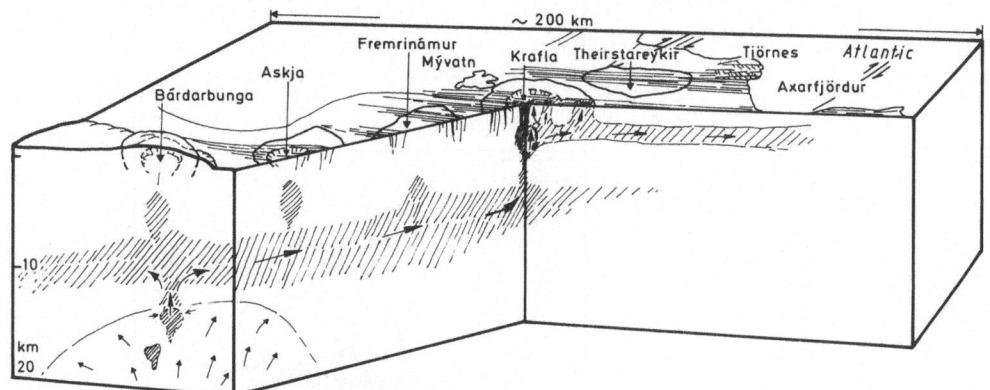

Fig. 5 Block diagram of the NE Iceland plate margin as a cartoon of the melt "plumbing system" connecting the upwelling plume centered below Bárdarbunga with the Krafla system which has been active since 1975.

So far there is merely spotty evidence for the above model. There are terrestrial geodetic observations of horizontal and vertical displacements as well as of gravity changes, cumulative during certain time intervals before and during the rifting episode (Kanngieser, 1982; Möller & Ritter, 1980; Ritter, 1982; Torge & Kanngieser, 1980; Tryggvason, 1986; 1987; Wendt et al., 1985). The observations have been taken in a network spanning about 50 km to either side of the Krafla fissure swarm and 20 km north-south at the latitude of the Krafla caldera or somewhat to the south (horizontal displacements) and along a profile from Akureyri to Vopnafjördur (elevation and gravity changes). Fig. 6 shows the horizontal displacements for four time intervals between 1971 and 1982, and Fig. 7 shows the elevation and gravity changes as well the changes in Bouguer anomaly and in Free Air anomaly from 1975 to 1980.

We have modelled the horizontal displacements by assuming rectangular vertical dykes with normal stress acting on the walls embedded in an elastic half space (Zdarsky, 1987; Jacoby et al., 1988). The best fitting dykes were found for each interval by least-squares minimizing the residuals of the computed displacements; the non-linear problem of adjusting dyke geometry and location was solved by systematic search. For the four intervals Fig. 8 shows the results, i.e. the dyke locations and depth extent. The results are supported by the seismicity that occurred during the time intervals in question.

The elevation and gravity changes were modelled (Altmann, 1987; Jacoby et al., 1988) with a "two-dimensional" horizontal line source of expansion (e.g. caused by intruding hot melt) in an elastic half space; a mass line at the same depth would give an equivalent effect. By optimizing the least-squares

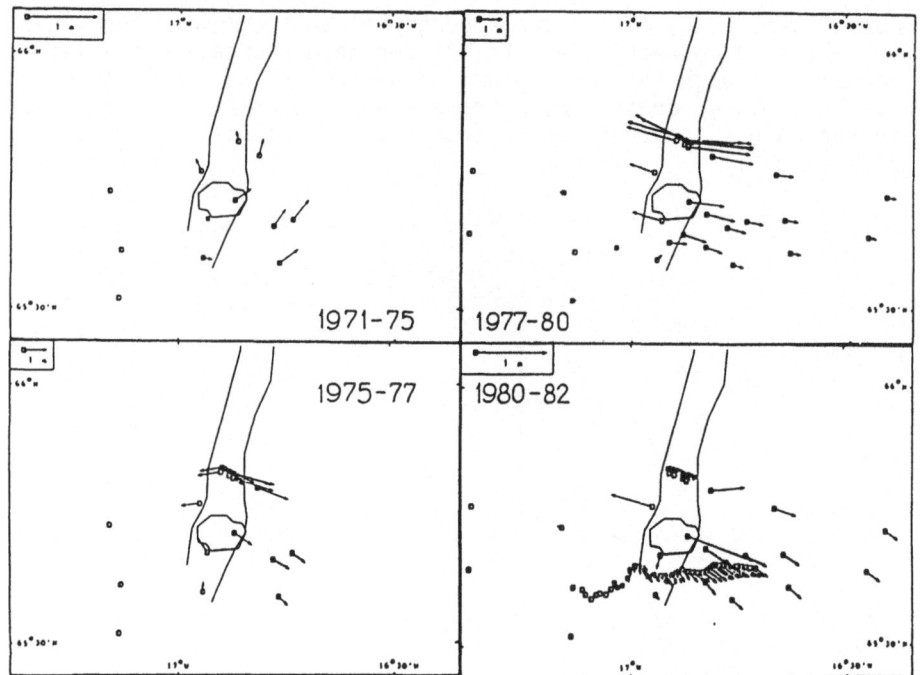

Fig. 6 The horizontal displacements observed for the time intervals 1971-1975-1977-1980-1982. Note that the scale of the displacement vectors differs from interval to interval. (after Möller & Ritter, 1980; Ritter, 1982; Wendt et al., 1985)

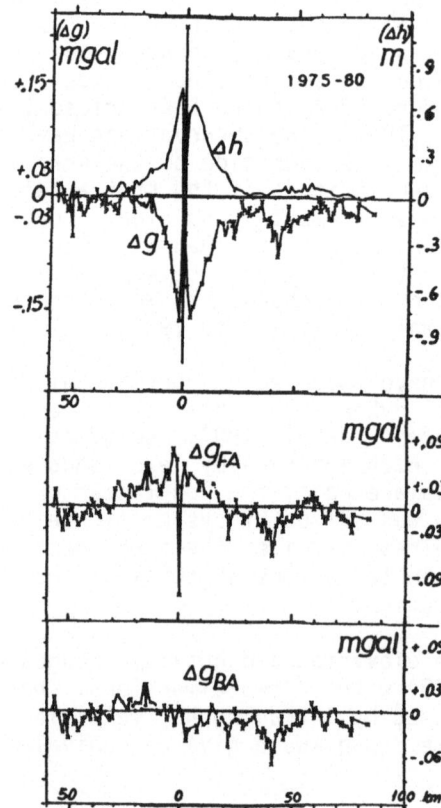

Fig. 7 The 1975 to 1980 change in observed gravity g and elevation h and the change of the Free Air anomaly g_{FA} and the Bouguer anomaly g_{BA} along a profile from Akureyri to Vopnafjördur (after Torge & Kanngieser, 1980; Kanngieser, 1982).

422

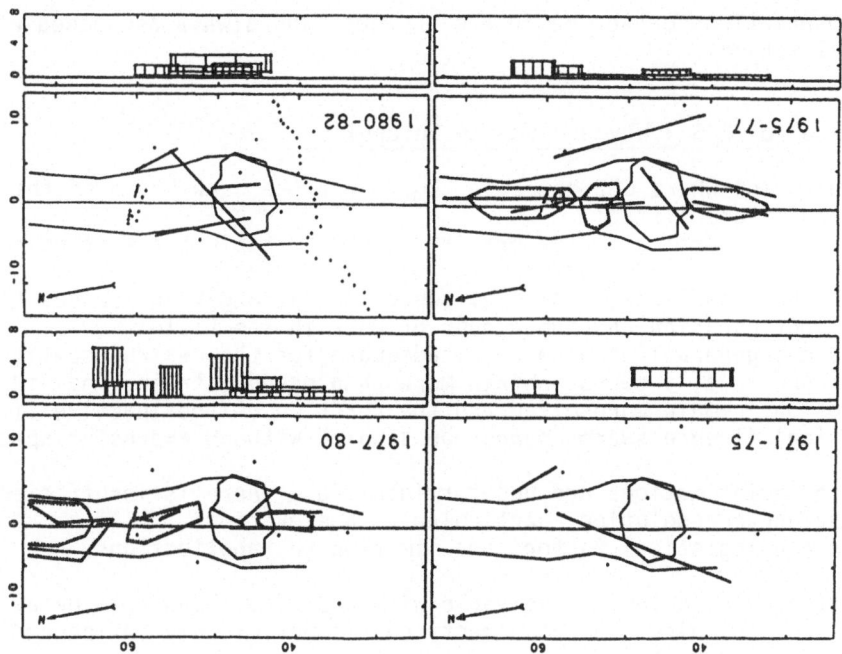

Fig. 8   Modelling results for the horizontal displacements. The best-fitting dykes (for the time intervals shown in map view and in side view (seen from the east); vertical (horizontal) lines denote dyke expansion (contraction); line density denotes relative source strength.

fit to filtered Bouguer anomaly change the depth of the source was found to be between 20 and 25 km, i.e. in the crust-mantle transition zone of a high concentration of partial melt, about 20% according to Schmeling (1985); the result is rather insensitive to the filtering of the observed data. Thus the geodetic-geophysical models are supported by the additional information on crust-mantle structure. However, there is a problem with this model: the amplitudes of the gravity and elevation change cannot be fitted at the same time; for the same model parameters either the gravity effect becomes too large or the elevation effect becomes too small. The discrepancy can be resolved by assuming either that in addition to a change of volume (expansion) there is a true mass change (e.g. removal by volcanism and dyke injection in the Krafla system) or that the crust and mantle are a layered structure which, of course, they are instead of an elastic half space; a "soft" layer with reduced E modulus and increased Poisson's ratio is needed in the source depth of crust mantle transition. Finite Element models (Altmann, 1987; Jacoby et al., 1988) have demonstrated that indeed the data can be fitted then satisfactorily. Thus we have learnt something not only about magma movements, but also about the medium in which the dynamic processes take place.

However, the presently available data are very limited in areal distribution, so that we cannot be absolutely certain about the interpretation. There is but one profile of gravity and elevation observations south of the center of activity and the two-dimensional model is hypothetical. For logistic reasons observations cannot be done everywhere and always as needed with terrestrial methods.

3. The case for GPS - Why and how to be applied

This is the point where GPS comes in, but first the draw-backs of the terrestrial methods are listed:
* horizontal and vertical (plus gravity) control stations are often not the same;
* line of sight considerations, i.e. topography strongly influences the network configuration and the amount of work in the field;
* networks are generally coarse and inadequate for the desired spatial resolution; one usually cannot obtain more than pseudo-strain; individual block motions remain unresolved and the reaction of the inactive (i.e. not rifting) fissure swarms cannot be studied without expensive special efforts;
* "absolute" point motions are undeterminable and there is insufficient long-wavelength resolution, particularly in elevation;
* economic and logistical factors are limiting the applications.

Most of the above problem are solvable with GPS or at least can be eased. However, when designing GPS experiments one is unavoidably influenced by one's conscious or subconscious ideas. In fact, one should be aware of what questions one wants to be answered. Thus, if the whole "plumbing system" is to be investigated, one needs a rather uniform observational network in the area, i.e. between Vatnajökull and the north coast. For logistic reasons it may be reasonable to combine GPS with photogrammetry. If the detailed reaction of the neovolcanic zone with the many en-echelon fissure swarms is to be studied detailled networks or profiles are to be set up. If individual deflation-rifting events, lasting a day or so, are to be observed in real time, GPS should be applied in the kinematic mode at critical points. If after the termination of the rifting episode proper the dissipation of strain into the plate interiors is to be investigated, a repetition of observations every 5 years on a network as wide as possible is required; such observations will be most useful information on the rheology of the special Icelandic lithosphere-asthenosphere system. All the above problems require the highest possible accuracy, but if the immediate rifting and magma chamber behaviour is the objectif, local cm or even dm precision may suffice, as demonstrated by Figs. 4, 6, and 7. Finally, it is mentioned that a combination of GPS point displacements with gravity measurements enhances the value for geodynamic studies which cannot be overestimated, as the above discussion will probably have demonstrated.

Two GPS experiments have already been conducted in Iceland in 1986 and 1987 (Foulger et al., 1987) by international groups. While the 1986 survey had the deformation of the whole of Iceland in focus with special interest in that of the South Iceland Seismic Zone (where a major earthquake is believed to be pending), the 1987 survey concentrated on the North Iceland region of recent rifting which is discussed in this report (P. Einarsson, G. Foulger, and G. Seeber, pers. comm., 1986-1988). The surveys were truly interdisciplinary and ideas, as put foreward here, were considered in the planning. (The author was only marginally involved in the planning discussions).

4. Conclusions

There is a good chance that by doing a well designed repeated GPS experiment over the coming years we should be able to resolve the sub-Iceland "plumbing system" in considerable detail and to better understand the driving mechanism of rifting. I should have a component in the large-scale plate motion and one in the local melt accumulation, the mutual importance of which is not known, however.

Geodynamics, in the mind of a geophysicist, is gravity and heat. Gravity provides the forcing and heat the energy. The former tends to produce equlibria and the latter tends to disturb them. It will be exciting to get a closer look at how geodynamics works in Iceland with the aid of GPS, just as exciting as a look at an erupting volcano. Fig. 9 shows such a look the auther was happy enough to have had.

Fig. 9 Scoria cone being produced by the eruption of the Krafla fissure swarm at its northern end; sketched by author on 8 Sep. 1984.

References

Altmann, U. (1987): Modellierung von Massenverschiebungen vor und während der Riftphase des Krafla-Spaltenschwarms in Nordisland anhand von Schwere- und Höhenänderungen, Ph.D. thesis, Universität Frankfurt

Arnason, K. (1981): Magnetotellurische Messungen auf einem Profil über dem Zentralvulkan Krafla in Nord-Ost-Island, Diplomarbeit, Universität München

Beblo, M. and A. Björnsson (1980): A model of electrical resistivity beneath NE Iceland, Correlation with temperature, J. Geophys., 47, 184-190

Beblo, M., A. Björnsson, K. Arnason, B. Stein and P. Wolfgram (1983): Electrical conductivity beneath Iceland - Constraints imposed by magnetotelluric results on temperature, partial melt, crust and mantle structure, J. Geophys., 53, 16-23

Björnsson, A. (1985): Dynamics of crustal rifting in NE-Iceland, J. Geophys. Res., 90, 10151-10162

Björnsson, A., G. Björnsson, A. Gunnarsson and G. Thorbergsson (1985): Breytingar á landhaed vid Kröflu 1974-1984, Orkustofnun Report OS-85019/JHD-05, Reykjavik

Foulger, G., Bilham, R., Morgan, W.J. and P. Einarsson (1987): The Iceland GPS geodetic field campaign 1986, EOS Dec. 29, 1809-1818

Gebrande, H., Miller, H. and Einarsson (1980): Seismic structure of Iceland along RRISP profile I, J. Geophys., 47, 239-249

Jacoby, W.R.(1979): Iceland and the North Atlantic: A review, Geojournal, 3.3, 253-262

Jacoby, W.R., U. Altmann and G. Marquart (1983): Structure and evolution of Iceland and the Reykjanes Ridge, Rep. Abschlußkolloquium Meteor Expedition 45, Inst. Geophys., Univ. Hamburg, pp. 67-83

Jacoby, W.R., H. Zdarsky and U. Altmann (1988): Geodetic and geophysical evidence for magma movement and dyke injection during the Krafla rifting episode in North Iceland. (submitted to American Geophys. Union)

Kanngieser, E. (1982): Untersuchungen zur Bestimmung tektonisch bedingter zeitlicher Schwere- und Höhenänderungen in Nordisland. Wiss. Arb. Fachricht. Vermessungswesen, Dissertation, Nr. 114, Univ. Hannover

Möller, D. and B. Ritter (1980): Geodetic measurements and horizontal crustal movements in the rift zone in NE Iceland. J. Geophys., 47, 110-119

Ritter, B. (1982): Untersuchungen geodätischer Netze in Island zur Analyse von Deformationen von 1965 - 1977, Deutsche Geodät. Komm., Ser. C, 271

RRISP Working Group (1980):Reykjanes Iceland Seismic Experiment (RRISP77), J. Geophys., 47,228-238

Schmeling, H. (1985): Partial melt below Iceland: A combined interpretation of seismic and conductivity data, J. Geophys. Res., 90, 10105-10116

Torge, W. and E. Kanngieser (1980): Gravity and height variations during the present rifting episode in northern Iceland, J. Geophys., 47, 125-131

Tryggvason, E (1986): Multiple magma reservoirs in a rift zone volcano: Ground deformation and magma transport during the September 1984 eruption of Krafla, Iceland, J. Volcan. Geotherm. Res., 28, 1-44

Tryggvason, E. (1987): Myvatn lake level observations 1984-1986 and ground deformation during a Krafla eruption, J. Volcan. Geotherm. Res., 31, 131-138

Wendt, K., D. Möller and B. Ritter (1985): Geodetic measurements of surface deformations during the present rifting episode in NE-Iceland. J. Geophys. Res., 90, 10163-10172

Zdarsky, H. (1987): Analytische Modelle für die Horizontalverschiebungen während der Riftepisode des Krafla-Spaltenschwarms in Nordost-Island. Diplomarb., Univ. Frankfurt

Present State of the Central Andean GPS-Traverse ANSA

J. Klotz and D. Lelgemann

Summary
Within the framework of an interdisciplinary geoscientific research group a GPS-traverse was established covering an area from ANtofagasta (Chile) to SAlta (Argentina) and further east into the Chaco. The first observation campaign was successfully completed in the spring of 1988. A short description of the state of the project is given.

1   Introduction

In May 1984 a research group "Mobility of Active Continental Margins" was established at the Freie and Technische Universität Berlin (FUB/TUBerlin) with special support from the Deutsche Forschungsgemeinschaft (DFG). Members of this research group are carrying out geoscientific investigations in a region crossing the Central Andes between 21o and 25o southern latitude, the area of the Transsecta 6 of the International Lithosphere Programm. The group is an interdisciplinary research group; participating are scientists from the following disciplines: Geology, Paleontology, Petrology, Geochemistry, Geochronology, Geophysics, Geomorphology and Geodesy (Giese et. al. 1987.).

Despite the fact that the general behaviour of geodynamic processes at a subduction zone such as the Andes are fairly understood, detailed investigations reveal a lot of new information of the geotectonic evolution and the ongoing tectonic and magmatic processes. Interdisciplinary interpretations are required to get a more detailed conception of the geodynamic evolution in particular of the continental margin and the mountain building processes. For the determination and the analysis of the contemporary rate of crustal movement only modern geodesy may provide quantitative information. The Central Andes have been recommended as a region to study in detail deformation at a convergent plate margin and regional tectonic movements accompanying earthquakes in (L.S. Walter 1983).

Based on interdisciplinery discussions to identify major geotectonic questions a geodetic network was designed, spanning the Andes from the Pacific to the Gran Chaco with an east-west extension of about 600 km. The network crosses the Andes from Antofagasta (Chile) to Salta (Argentina) and beyond into the Chaco, following the area of the Transsecta 6 of the International Lithosphere Program. Using repeated determinations of three-dimensional relative coordinates obtained from GPS-observations the mutual horizontal as well as vertical movements of the reference points will be derived after further observation campaigns.

2   Setup of the GPS-Traverse

2.2.   Cooperation

Geodetic contributions to geodynamic projects require extensive
cooperation on different levels and for different purposes

- cooperation with other geoscientific disciplines and institutes
  for the network design and interpretation of the results
- cooperation with institutes in those countries where the
  network
  is established
- cooperation with other geodetic institutes involved in
  GPS-observations
- cooperation with other geodetic institutes involved in
  processing the GPS-data
- cooperation with institutes gathering data relevant to our
  project.

Close cooperation with other geoscientific disciplines has been
ensured from the beginning, since the GPS-traverse has been
established within an interdisciplinary research group. This was
not only extremely important for the design of the network and
the selection of the reference points, but will be even more
important in the future for the interpretation of the geodetic
results in combination with the detailed information collected by
the other disciplines.

Members of the research group in Berlin had already developed
close contacts to southamerican universities formalized by
cooperation agreements with the Universidad de Chile and the
Universidad Nacional Salta (UNSa). In an annex to the cooperation
agreement between the FUB/TUB and UNSa the cooperation was
expanded to the GPS-network.

There was no cooperation with southamerican geodetic institutes
and the Berlin research group at the beginning. This situation
now has changed. Our first contacts to a group of geodesists at
the Universidad de Santiago de Chile (USACH), the only civil
organisation in Chile involved in geodetic research, led to a
cooperation agreement betwen USACH and the TUB. It is planned to
cooperate also with a group of the Universidad de La Plata in
Argentina.

The cooperation with scientific institutions in Argentina and
Chile are considered as very important for the project. Since the
southamerican scientists are most familiar with the local
situation substantial contributions to the design of the network,
site selection and monumentation are supplied by our
southamerican colleagues.

Usually, larger geodetic observation campaigns are carried out by
several geodetic institutes working together. In the moment the
geodetic institutes of the Universität Stuttgart, the Universität
der Bundeswehr in München and the Institut für Weltraumforschung
der Österreichischen Akademie der Wissenschaften (Graz) are
participating in the project.

The Institut für Physikalische Geodäsie der TU Darmstadt had
previously carried out high presision gravity observations
including reference points of the GPS-traverse. Repeated
observations in the future may result in an additional control of
vertical movement determinations.

2.2. Network Design and Site Selection

The design of the network as well as the individual site
selection required consideration of
- geoscientific aspects
- logistical aspects
- observational (geodetic) aspects.
As usual, the final result is a compromise.

The major aspects for the design, of course, arised from the
neotectonic formation in this region as identified by discussions
with geologists and geophysicists from Berlin, Argentina and
Chile. Based on the tectonic structures of the region a network
configuration was designed covering in particular regions in
which recent crustal movements are expected. Those regions are
connected in such a way that
- the GPS-traverse crosses completely the mountain range of the
  Andes
- not more than about 30 reference points are established
- the distance between the reference points does not exceed 50
  km.
The final configuration is given in fig 1. The numbers 9 and 14
are omitted; these numbers were used only in a preliminary
configuration.

Points 1-5 have been chosen to identify a (present?) strong
uplift of the coastal region as expected from geological evidence
as well as possible crustal movements due to the Atacama Fault
System.

Geological data indicates subsidence/extension in the area of the
Salar de Atacama as well as uplift/compression in the region west
of the Salar in the Cordillera del la Sal; therefore, the network
(points 11-16) was enlarged in this region.

The series of volcanoes in the West-Cordillera may be accompanied
by an uplift of this region (points 17-20).

Points 19-26 were chosen to observe movements connected to the
lineament of the Quebrada el Toro. This lineament describes the
site of an old megafault system, but there are different opinions
whether this megafault system is active today.

During the Pliocene there was certainly extremely strong
compressional tectonic in the area of the Subandin; obduction
from the east over younger rocks in the west may be an ongoing
process. Movements of points 27-32 may clarify the present
tectonic activities.
The logistic situation in this area is extremely difficult and
was an important consideration for the network design. All sites

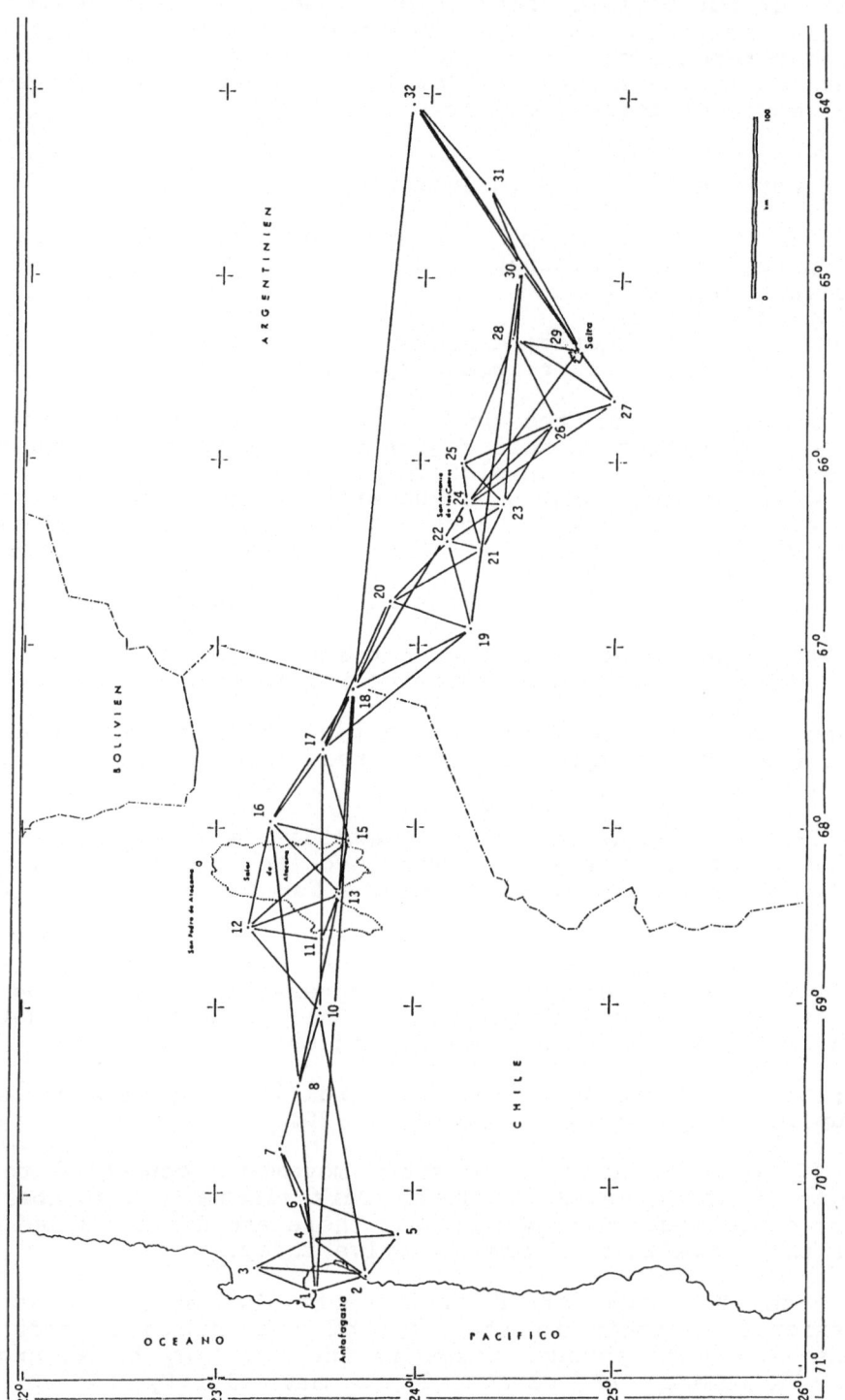

Fig. 1: The Central Andean GPS-Traverse ANSA

have been selected in such a way that they can be approached by cars able to traverse rough countries. As a general geodetic aspect it was required that the side length of the network should be uniform and less than 50 km. It is left to the future whether additional points may be required due to a refined analysis of error propagation or refined geologic information.

There are several requirements regarding the local surrounding of the sites. If possible, the marker should be placed into outcrop showing no young local deformation. Therefore, it was important that a geologist participated in the site selection. It was also required that the satellites are visible over 15° elevation. Finally, points should be chosen so that multi-path effects as well as abnormal tropospheric refraction effects are avoided, but this is difficult to judge.

2.3. Monumentation

Since GPS-geodynamic networks are considered as very long term projects (re-observation may be of interest even after decades), the monumentation is of special importance. The technical realisation has been discussed thoroughly with investigators of NASA's Crustal Dynamics Project from the University of Delft and with soil mechanic scientists from the TU Berlin. Several requirements for the monumentation have been identified
- additional stress in the surrounding rocks due to the monumentation process should remain minimal
- the excentricity vector between the phase-center of the GPS-antenna and the marker has to be determined with extremely high accuracy (+ 1mm)
- considering multi-path effects the antenna should be near the ground
- considering the danger of wanton damage the monumentation should be inconspicious.

Considering those requirements a marker is used as described in fig. 2. The top of the marker is designed in the form of a screw-thread (M 30x1.5 thread) protected by a removable cap. In case of observation, the cap is removed and a plate (see also fig.2) as a connecting link is placed on the marker and the antenna is placed on the connecting plate. In this way reobservations are always performed at the monumentation centre and therefore (in view of deformation analyses very dangerous) excentricity errors cannot occurre.

In general, it is only necessary to put a small hole (depth about 8 cm, diameter about 3 cm) into the rock. The marker is attached to the rock using a special concrete-metal adhesive. The requirement to place the marker into outcrop could not always be fulfilled. In such cases concret blocks had to be built first.

With underground security markers in the immediate vicinity of the main marker as well as security markers at adequate distances every reference point will be secured so that a reconstruction

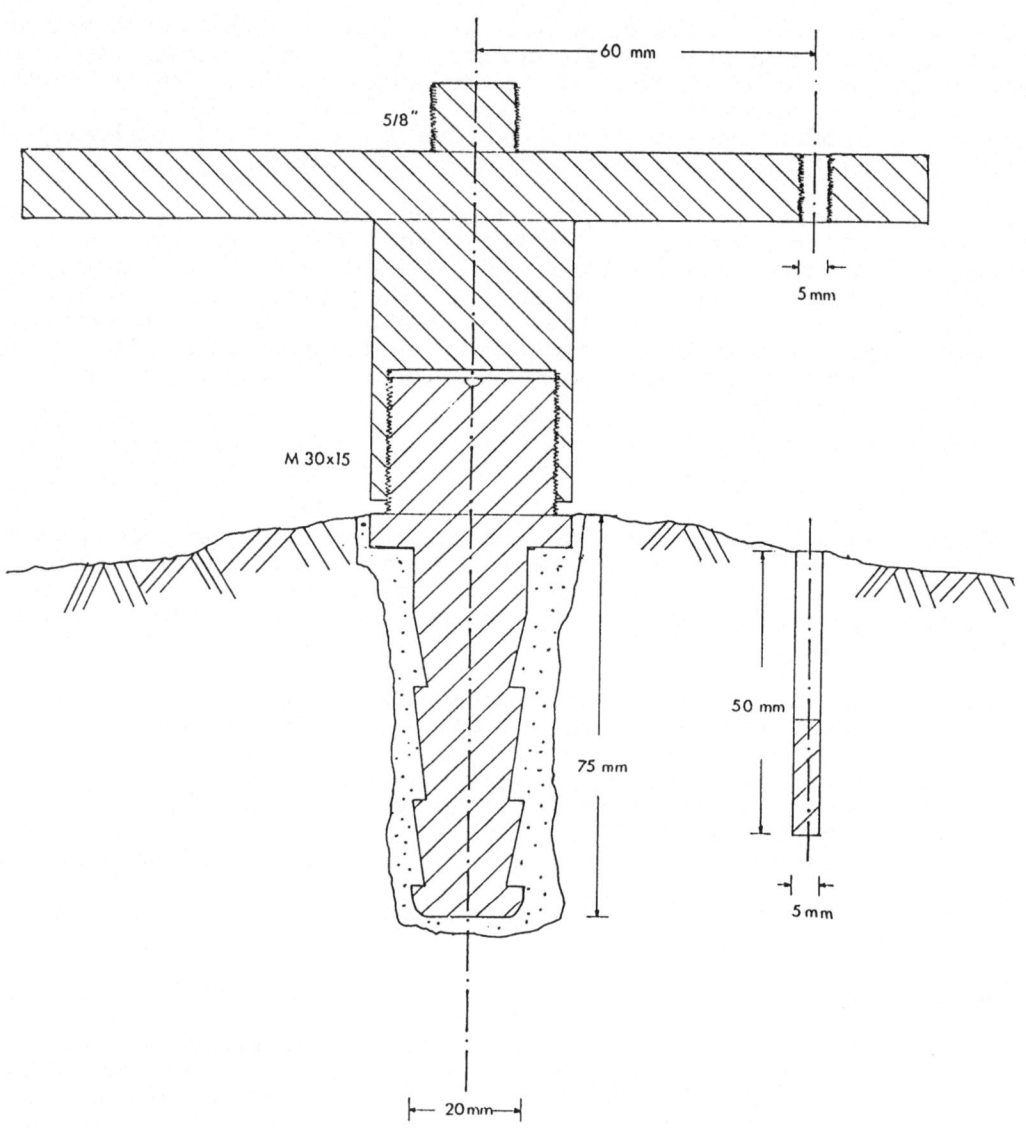

Fig. 2: Design of a marker

with an accuracy of at least 5mm will be possible in case of damage.

The monumentation of the reference points in the Chilenean part of the traverse was performed by the TU Berlin; security markers have been established and security observations already performed by the staff of the Universidad de Santiago de Chile.

According to the cooperation agreement the monumentation in the Argentinean part of the traverse was completed by the Geological Institute of UNSa.

The monumentation of all 30 points as described in fig. 1 was finished just in time for the first observation campaign.

3. The First Observation Campaign

The first observation campaign was conducted from March 17th to May 3th, 1988. Participating institutes were

- Universität Stuttgart (2 WM 101 single-frequency C/A Code receiver)

- Universität der Bw München (TI 4100 two-frequency P-Code receiver)

- Institut für Weltraumforschung Graz (T 4100 two frequency P-Code receiver)

- Technische Universität Berlin (2 WM 101 single frequency C/A Code receiver)

Since WM 101 two-frequency receivers are not available yet it was decided to perform simultaneous two-frequency observations in the vicinity of the WM 101 receivers with a TI 4100. As shown in (Georgiadou 1987) the single frequency data can be corrected for ionospheric refraction using the data from a two-frequency receiver. However, since such a technique cannot be considered as optimal further campaigns will be carried out using only two-frequency receivers.

The four WM 101 receivers have been at our disposal during the total observation campaign. One TI 4100 could be made available only from March 17th to April 20th and the other one only from March 27th to May 3th, 1988. Due to time restrictions the observation campaign (7 weeks) was structured according to the follwing schedule:
1.-2. week: Observations at the Chilenean reference points using four WM 101 and one TI 4100
3.-5. week: Observations of overlapping baselines using both TI 4100; parallel observations using the four WM 101; in the 4. week transition from Chile to Argentina
6.-7. week: Observations at the Argentinean reference points using four WM 101 and one TI 4100
Additional problems resulted from the malfunctioning of one the WM 101 receivers for about 5 weeks. Nevertheless, all necessary observations were conducted.

The satellite constellation during the observation period was remarkably good in the region of the geotraverse. We observed signals from satellite 3, 6, 8, 9, 11, 12 and 13. At least four of them have been visible during daily observation periods of 5 hours. Moreover, for about 3 hours even five satellites could be observed simultaneously.

The daily observation period started between 3.oo and 6.oo in the night; observations always took place during a 5-6 hours period each day, without changing the observation site.

34 days of observations were conducted during the total 46 days of the campaign. All the baselines as shown in fig. 1 have been observed at least once using WM 101 receivers.

All overlapping baselines are observed with TI 4100 receivers, e.g. 1-8, 8-13, 13-18, 18-24, 24-29, 29-32. Besides controlling the network these observations will deliver the scale of the network from two-frequency observations.

4. Data-Evaluation

For routine purposes the POPs-software from Wild-Magnavox as well as the Phaser-Software-System developed at the National Geodetic Survey of the USA are used. With Phaser we are able to process data from different receivers, also data from an WM 101. Comparisons of results have shown differences greater then acceptable for geodynamic investigations. Nevertheless, for the required highest accuracy a thorough investigation of GPS-data evaluation techniques must be undertaken by a user group in any case.

For several reasons a data evaluation technique based on single observations are considered by us as preferable despite the fact that for routine purposes the use of double differences may be prefered. Most important then is the relationship between the measurement, that is already a phase difference (between the phase of the oszillator of the receiver i and the received phase transmitted from satellite j), and a geometric quantity. In the case of GPS-phase observations the range between the position of the phase center of the antenna of satellite j at the transmitter epoch τ and the position of the phase centre of the antenna of the receiver i at the receiving epoch t should be considered

$$\tau_i^k = (t_i^k - \rho_{ij}/c)$$

In (Gehlich et.al.1988) the following (non-linear) observation equations are derived ($\lambda = c/f_s$ =wavelength)

a) Null (zero-epoch) observations at start epoch O

$$L_{ij}(t_i^0) = Q_{ij}^0 - \rho_{ij}(t_i^0)/\lambda$$

b) Further observations at epoch k

$$L_{ij}(t_i^k) = Q_{ij}^0 - \rho_{ij}(t_i^k)/\lambda + [S_i^k - R_i^k + f_{SE}(t_i^k - t_i^0)]$$

c) Last observations at epoch k=K

$$\Sigma S_i^K = 0 \qquad \text{(condition equation)}$$

where we have

Q_{ij}^0 =auxiliary unknowns; the sum of Q_{ij}^0 and $L_{ij}(t_i^0)$

corresponds to the range $\rho_{ij}(t_i^0)$

S_i^k, R_i^k = integrated frequency instabilities of the satellite or the receiver oscillator from epoch t_i^0 to epoch t_i^k

f_s =mean value of all satellite frequencies during the oberservation period. f_s cannot be estimated from the data; usually, the nominal frequency $L_1 \doteq f_s$ is used for data evaluation

$f_{SE} = f_S - f_E$; f_E = mean value of participating receiver frequencies.

Following the brilliant suggestion of (Goad 1985) the auxiliary unknowns can be substituted to some extent by integer valued unknowns ΔN_{ij}^0 (the so-called ambiguities) using

$$Q_{ij}^0 = \Delta N_{ij}^0 + Q_{i1}^0 + Q_{1j}^0 - Q_{11}^0$$

$$\Delta N_{ij}^0 = \begin{bmatrix} 0 & \text{for i=1 or j=1} \\ N_{ij}^0 + N_{11}^0 - N_{i1}^0 - N_{1j}^0 & \text{for i=1 and j=1} \end{bmatrix}$$

Number of Q_{ij}^0: (I*J)

Number of Q_{i1}^0 and Q_{1j}^0: (J+I-1)

Number of ambiguity unknowns ΔN_{ij}^0: (I*J+1-I-J)

It is well-known and often used that some of the unknowns may be eliminated by forming differences:

$$_i^{i+1}(L_{ij}^k) = {_i^{i+1}}(Q_{ij}^0) - {_i^{i+1}}(R_i^k) - {_i^{i+1}}(\rho_{ij}^k/\lambda) \quad \text{(Single diff)}$$

$$_j^{j+1}\,{_i^{i+1}}(L_{ij}^k) = {_j^{j+1}}\,{_i^{i+1}}(Q_{ij}^0) - {_j^{j+1}}\,{_i^{i+1}}(\rho_{ij}^k/\lambda) \quad \text{(Double diff)}$$

Recognize that the first term on the right-hand side of the last equation must be an integer value. Double-differences are often preferred as (correlated) pseudo-observations. Depending on the evaluation of the data we may distinguish between
    a) "pseudo-doppler data" estimation process.

435

b) "pseudo-range data"   estimation process.

If we solve for all auxiliary unknowns Q the estimation process corresponds to the traditional use of (integrated) Doppler-counts. This technique of data analysis may be suggested for orbit computations if data from a few globally distributed stations (but each data set covering a relatively long observation period) are at hand.

A more stable solution, however, may be obtained if it is possible to identify the expectation of the ambiguity unknowns N by statistical methods e.g. using a Student test. Consider the case that the ambiguities are already identified. Then we may interpret the phase-observations as (pseudo-) range observations with the requirement that the few auxiliary unknowns left have to be estimated in addition. This technique is certainly preferable for station position computations.

Based on this analysis of the GPS-phase observation process we are working at a critical examination of GPS-phase data processing algorithmns for the purpose of geotectonic investigations.

5 Future Activities

A main concern at the moment seems to be the data evaluation process which may be developed further in the next years to obtain optimal results. It is assumed that small, but possibly substantial progress will be attained in the future. Therefore, final evaluation of the data already observed will be accomplished in several years.

The next observation campaign is already financed and will be undertaken in spring 1990 when more satellites will provide better observation conditions.

References
- Giese, P., Haak, V., Jacobshagen, V., Reuther, K.J. (1987). Mobilisierung eines Kontinentalrandes, ein subduktionsindu- zierter Prozeß. DFG-Mitteilung XVI.
- Georgiadou,Y. and Kleusberg,A. (1987). Ionospheric refraction, Multipath and Phase center variation effects in GPS carrier phase observations. Gen. Assembly IUGG, Vancouver
- Gehlich, U., Angermann. D., Lelgemann, D., (1988). Zur Modellierung des Phasenmeßprozesses geodätischer GPS-Empfänger. Festschrift Rudolf Sigl, DGK Reihe B, Nr. 287, München
- Goad, C.C., (1985): Precise relative position determination using Global Positioning System carrier phase measurements in a non difference mode. Proceedings of the First Int.-Symp. on Precise Positioning with GPS, Vol. 1, p. 347-361, Rockville,
- Walter, L.S. (1983). Geodymics. NASA Conference Publication 2325

REMARKS TO THE ESTABLISHMENT OF A REGIONAL GPS-TRACKING NETWORK

by

Bernd Breuer and Wolfgang Schlüter

1. Introduction

The results in point positioning using GPS are influenced by

- environmental effects such as troposphere and ionosphere refraction
- orbital errors biased by insufficient force models and errors in station coordinates.

To minimize the orbital and ionospheric effects in geodetic applications, emphasis has to be placed in the establishment of traking networks for the GPS-satellites. Beside the global networks which are and will be set up by e.g. the

- Defense Mapping Agency
- National Geodetic Survey

- both national U.S. agencies- regional networks in continental or national extension are required independently. The regional networks will be set up for

- national survey applications,
- scientific applications

to derive coordinates or baselines with the highest achievable accuracy within a clearly defined geodetic reference frame under clear responsibilities. A regional geodetic reference frame represented by a regional tracking network and corresponding orbit information for the GPS satellites has to be realized in accordance with the global networks to provide a uniform world wide Datum. The realization and the follow on maintenance has to be carried out by regional or national authorities independently but in cooperation with the responsible agencies for the global networks to meet the regional or national requirements.

The Network will have to provide data for the determination of precise orbits and for relative positioning. Moreover the data also can be used to derive a regional ionospheric model to support survey with one-frequency receivers to obtain accurate ionospheric corrections.

IfAG, AIB and IWF have agreed to develop the necessary components of a regional tracking network and to demonstrade its feasibility

- to compute precise orbits and station coordinates,
- to set up and maintain a uniform geodetic reference frame
- to derive a model for the determination of ionospheric refraction.

Three main items have to be worked out and clearly defined:

1. the tracking facilities for a reference station
2. the communication between the tracking stations, the computing center and the users

3. the data analysis center with software for continous orbit determination and for the determination of an ionospheric model.

The work has started end of 1987 and should be finalized before the GPS will have its final configuration.

2. Tracking facilities for a reference station

The GPS will be set up with a constellation of 24 satellites. At a tracking station a changing scenario of up to 8 satellites will be observable at the same time. The observations will have to be carried out continously 24 hours a day all over the year, which requires unmanned, automatic tracking. The data provided by the tracking stations will have to include for each of the visible satellite

- epoch (receiver time) of the data point
- Satellite number (SV)
- Phase L1
- Phase L2
- Code L1
- Code L2 if available

as well as

- Status information
- Broadcast ephemeris
- Almanach
- meteorological information
- local site information

The data output format should be receiver independent. The tracking software of the receiver has to be split into a receiver dependent part and into a receiver independent part. The receiver dependant part will handle the receiver functions, the set up of the scenario and sampling rates, the epoch timing and the receiver dependand collection of data. That part of software also will convert receiver dependant data into a receiver independent format. The receiver independent software part should control the quality of the data, condense the data from the receiver data rate to a data rate requested by the analysis center and provide datafiles for various users which individually will have to be created on request. Handshake software for the communication will have to be implemented which will allow the data transfer via a communication link (e.g. X.25).

3. **Communication**

Communication links making use of electronic mail systems will have to be set up between the tracking stations and the computing centers, as well as between the tracking stations and some dedicated users. The provision of the orbit information and of the ionosphere-model should be spread by different communication techniques to a broad spectrum of users.

4. **Analysis Center**

The analysis center will need a software package for the determination of precise ephemeris (including earth rotation parameters) and ionospheric models operational on a continous basis by automatic collection of the data from the tracking stations. The center will have to provide the scenarios and the priorities of satellites to be tracked for the stations and will have to distribute the results to all users.

5. **Final remarks**

To establish a regional tracking network including an analysis center is a demand for the future work with GPS. The authors are members of agencies which have experiences in GPS-tracking and data reduction. Starting from the basis of the available software and tracking facilities a solution for a regional network has to be developed which meets all the regional requirements.

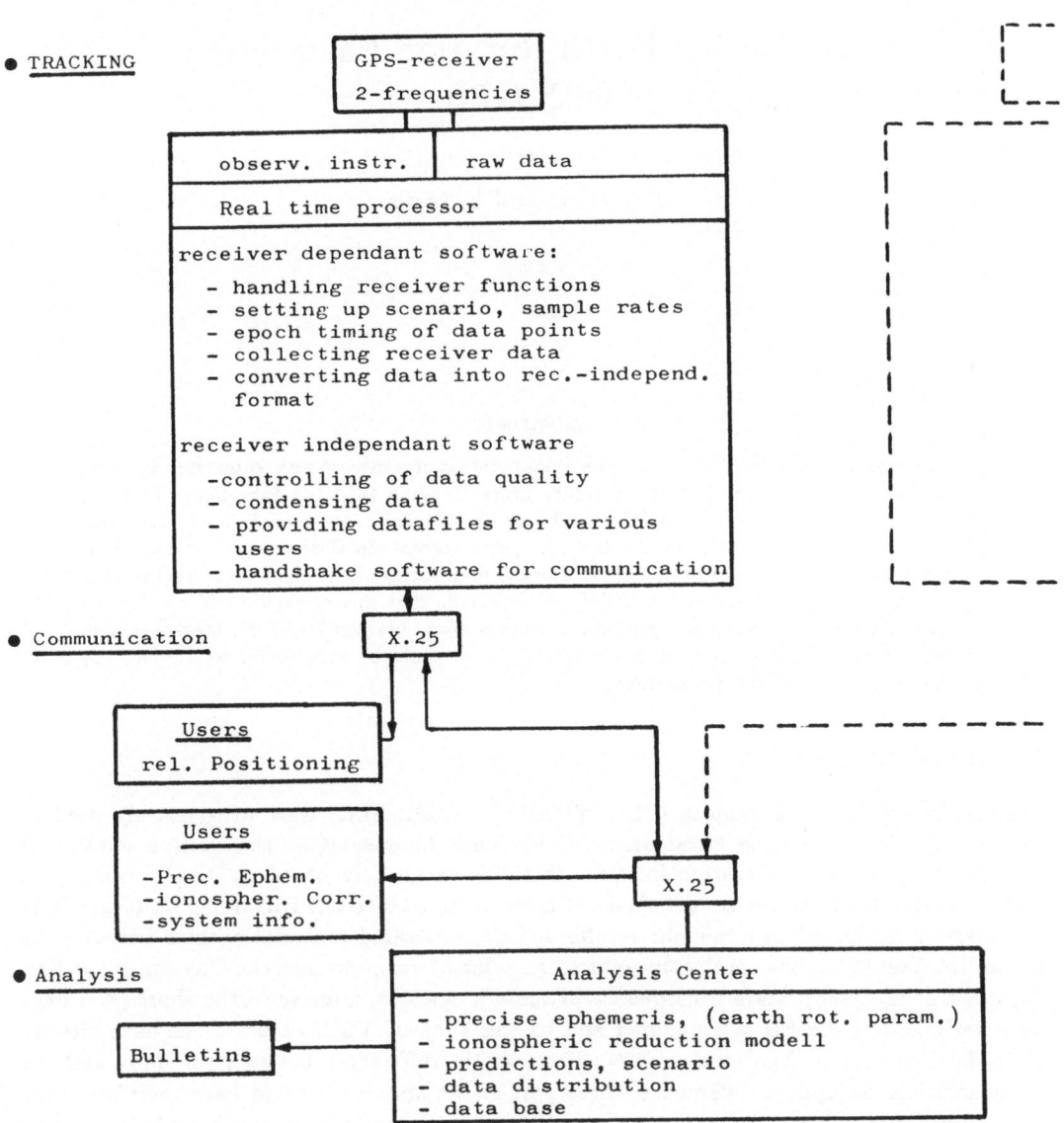

● TRACKING

GPS-receiver
2-frequencies

| observ. instr. | raw data |

Real time processor

receiver dependant software:

- handling receiver functions
- setting up scenario, sample rates
- epoch timing of data points
- collecting receiver data
- converting data into rec.-independ.
  format

receiver independant software

- controlling of data quality
- condensing data
- providing datafiles for various
  users
- handshake software for communication

● Communication

X.25

Users
rel. Positioning

Users

- Prec. Ephem.
- ionospher. Corr.
- system info.

X.25

● Analysis

Bulletins

Analysis Center

- precise ephemeris, (earth rot. param.)
- ionospheric reduction modell
- predictions, scenario
- data distribution
- data base

Recovering Earth Rotation Parameters with GPS.

P. Pâquet and L. Louis

Abstract

The monitoring of the Earth Rotation Parameters is presently mainly supported by Very Long Base Interferometry (VLBI), Satellite Laser Ranging (SLR), Lunar Laser Ranging (LLR). Among those techniques VLBI only seems able to access to the short variations of the Earth Rotation (ER). In the last few years several simulation were conducted to estimate the potential contribution of GPS to the monitoring of the short term fluctuations of the Earth Rotation Parameters (ERP). All conclude that this method could conduct to results whose the precisions are comparable to that ones obtained by VLBI. Moreover, due to the easiness of data collection and analysis, GPS would be very useful to support the services in charge of ERP predictions.

1 Introduction

Since 1969 new space techniques (LLR, TRANET, VLBI, SLR) were progressively used to monitor the Earth Rotation Parameters (ERP); since January 1988, the routine solution of the International Earth Rotation Service (IERS) is exclusively based on VLBI, LLR, SLR observations. Thanks to the contributions of these techniques to the BIH 5-day solution whose the analysis conducted to numerous results and demonstrated the relationships between the Earth Rotation (ER), the total Atmospheric Angular Momentum and the Physics of the Sun (Djurovic et al., 1988). As a consequence of these progresses, interest to the short ($\leq$ 1 day) term variations of the ER is expanding and on this subject, VLBI observations have already given detailed results (Luo et al., 1987). However the differences between the daily and the 5-day solutions let appear differences whose part of the amplitude could have their processes in unmodelled geophysical phenomema. Additional accurate techniques would be helpful to confirm the short term fluctuations as observed by VLBI techniques.

On the other hand the predictions of ERP are provided on a regular basis and the present predictions demonstrate that the Universal Time (UT) component is degrading three times faster than the predictions of the Polar Motion (PM) (Feissel et al., 1986; Mc Carthy, 1986). For this specific application and due to the easiness to deduce the ERP from satellite observations, the contribution of GPS would be also very valuable.

Recently several authors (Anderle et al., 1982; Zelensky et al., 1987; Pâquet et al., 1987) have investigated this aspect and they reached positive conclusions on the possibility to use GPS for ERP monitoring. The three approaches and methods are briefly reviewed.

2 Anderle's approach.

Anderle et al. (1982) analysed real data acquired by four stations on four GPS satellites and they provided one ERP solution every 7 days. The observables are pseudo-range on two frequencies and Doppler measurements for one frequency. The main parameters included in the solution are the six orbit constants, two scaling factors to scale the atmospheric drag and solar pressure, the three ERP (ΔUT_1, X_p, Y_p). With respect to the BIH solution, the observed PM exhibits random error of the order of 1.5 meter; this has been reduced to 30 cm by introducing bias parameters for each station and satellites clocks. The accuracy of ΔUT_1, 0.3 msec/day, was very promising. Although the results are not better than those obtained by other techniques currently used in 1982, this first experiment to recover ERP from real GPS data is encouraging and can certainly be improved by using more accurate observations obtained from more stations tracking more satellites.

3 Zelensky's approach.

Zelensky et al. (1987) developed a detailed simulation to estimate the error on ERP expected to be recovered from GPS data. With different type of observables (pseudo-range, differenced carrier phase, group delay), measured by stations distributed in the VLBI POLARIS sites and observing the 18 GPS satellites, different solutions which include orbit parameters, ground clock errors, station positions, tropospheric scaling and the ERP, conduct the authors to conclude that the *POLARIS like network should be able to determine ERP about as well as the current VLBI programme using single differenced carrier phase observations.*

The two days solution conducted to the precision of 0.5, 1.0, 0.8 mas for the pole coordinates X_p, Y_p and ΔUT_1 respectively.

4 Authors's approach

The simulation conducted by Pâquet et al. (1988) is based on the data expected to be collected by a world network of 10 stations (Fig.-1), tracking 18 satellites of the GPS operational constellation. The observables are the dual integrated Doppler shift, free of ionospheric refraction. The solution does not include all the usual parameters as in the Zelensky's solution but it contains only the ERP unknowns. The errors on the orbits, the network and the refraction and their contribution to the accuracy of the ERP parameters are estimated as follow.

- *Estimation of the mean error.*
  To compute from the variance-covariance matrix the error of each unknown, the mean error m_0 of the observations must be estimated. It is deduced from simulated "observed quantities" computed by pertubing the reference orbit and / or the consistency of the tracking network and / or the tropospheric refraction correction.

- *Perturbations.*

  - *(a)Perturbations of the ephemerides.* For the participating stations all different arcs observed are shifted in along track and range by constant quantities extracted from two series of random numbers whose the variance is given a-priori. The simulations were conducted with two a-priori variances fixed to 0.5 and 1.0 meters.

- *(b)Perturbation of the station coordinates.*

  Each station is supposed to observe all available passes for a maximum of 3 or 6 hours, symetrically distributed from rise time to set time. The minimum satellite elevation above the horizon is 10 degrees and the minimum elevation at closest approach is 15 degrees.

  To generate the simulated observed quantities, the adopted network coordinates are modified by adding to each component a number selected in a random series of 30 numbers whose the variance is given a-priori. For the simulations, the a-priori variance has been fixed to 10 cm which seems realistic according to the network errors estimated by C. Boucher and al (1987).

- *(c)Perturbation of the tropospheric refraction.*

  The meteorological parameters were kept the same for the whole network (Temp = 10^0, pressure = 1013 mb, humidity = 60%) and the adopted perturbation has been fixed in the range of 1% to 3%. At the zenith 2% corresponds to a range error of 3 cm while at 10^0 elevation it can reach 25 cm. The tropospheric refraction is estimated with the Hopfield's model (1969).

- *Precision of ERP*

  Table-1 gives the expected precision of ERP for different scenarios related to perturbations and length of the arc observed by each station. From Table-1 it can be seen that, *after 24 hours of observations, the pole position could be given with a sub-decimeter precision while after 6 hrs it is limited to 10 to 15 cm. In the same conditions, ΔUT_1 would be determined with a precision of .15ms to .30 ms.*

  Interesting results are given in Figure-2 which represents the expected precision on each ERP versus the number of tracking stations. Starting from 2 to 10 stations, the site location has always been designed to keep an homogeneous world distribution. It can be seen that a number of at least 10 stations is required to reach the precision of VLBI over time intervals of few hours. As the results are based on simulations, generaly considered as optimistic, for practical reasons and based on the experience of existing tracking network performances we believe that a realistic tracking network would have to be extended to 12 to 14 stations.

5 Conclusion

Two simulations, performed by two different groups using different methods, conclude that GPS could contribute to recover the short term component of the Earth Rotation fluctuations with a precision equivalent to that one obtained from the VLBI observations. As the SLR and LLR could not access such a high resolution the contribution of GPS to this window would be very useful to identify or confirm which part of the fluctuations could have geophysical origins or which part of the fluctuations has to be attributed to other sources.

Moreover, due to the easiness to collect and analyse data from Earth satellites, GPS would give a major support for the ERP predictions.

6 References

-Anderle R.J., Beuglass L.K.,Carr J.T., 1982. Earth's Rotation and Polar Motion based on Global Positioning System Satellite data. Proc. High Precision Earth Rotation and Earth-Moon Dynamics, Ed. O. Calame, Reidel Publ.

-Boucher C., 1987. Definition and realization of Terrestrial Reference Systems for monitoring Earth Rotation. Proc. of the IUGG Interdisciplinary Symposium on "Variations in Earth Rotation". Ed. W.E. Carter, D. McCarthy, P. Pâquet, AGU Publ. Series, in press.

-Djurovic D., Pâquet P., 1988. On the Solar Origin of the 50-day fluctuation of the Earth Rotation and Atmospheric Circulation. Accepted for publication in Astronomy and Astrophysics (in press).

-Feissel M., Gambis D., Vesperini T., 1986. Predicting weeks to months variations of the Earth Rotation. Proc. IAU Int. Symp. 128, The Earth Rotation and Reference Frame for Geodesy and Geodynamics. Ed. A. Babcock and G. Wilkins, Reidel Publ. in press.

-Hopfield S., 1969. Two-quartic Tropospheric refractivity Profile for correcting Satellite data. JGR 74 (18).

-Luo S., Zheng D., Robertson D.S., Carter W.E., 1987. Short-Period Variations in the Length of Day: Atmospheric Angular Momentum and Tidal Components. JGR, Vol 92, $N^0 B11$.

-Mc Carthy D., 1986. Predicting Earth Orientation. Proc. IAU Int. Symp. 128, The Earth Rotation and Reference Frame for Geodesy and Geodynamics. Ed. A. Babcock and G. Wilkins, Reidel Publ. in press.

-Pâquet P., Louis L., 1987. Simulations to Recover Earth Rotation Parameters with GPS System. Proc. of the IUGG Interdisciplinary Symposium on "Variations in Earth Rotation". Ed. W.E. Carter, D. McCarthy, P. Pâquet, AGU Publ. Series, in press.

-Zelensky N., Ray J., Liebrecht P.,1987. Error analysis for Earth Orientation recovery from GPS data. Proc. of the IUGG Interdisciplinary Symposium on "Variations in Earth Rotation". Ed. W.E. Carter, D. McCarthy, P. Pâquet, AGU Publ. Series, in press.

Table-1

Sensitivity of the ERP with respect to :
- length of the arc observed by each station
- orbit errors
- refraction error 2%
- network of 10 stations, consistency 10 cm

Observations : - 18 GPS satellites
                - Doppler count of 60 seconds

Solution :   - after 6, 12, 24 Hrs of observations

	Al. Tr., Range → .50 m				Al. Tr., Range → 1.0 m				
SOL. (hours)	ERP St. Dev. X_p (m)	Y_p	ΔUT_1 (ms)	m_0	ERP St. Dev. X_p (m)	Y_p	ΔUT_1 (ms)	m_0	Nber Equ.
(a) Each arc observed a maximum of 3 hours :									
06	.141	.122	.42	.055	.195	.170	.59	.077	10222
12	.100	.087	.30	.055	.139	.121	.42	.077	20444
24	.071	.061	.21	.055	.098	.085	.30	.077	40889
(a) Each arc observed a maximum of 6 hours :									
06	.100	.090	.29	.051	.140	.125	.40	.071	16147
12	.071	.064	.20	.051	.099	.089	.28	.071	32294
24	.050	.045	.15	.051	.070	.063	.20	.071	64589

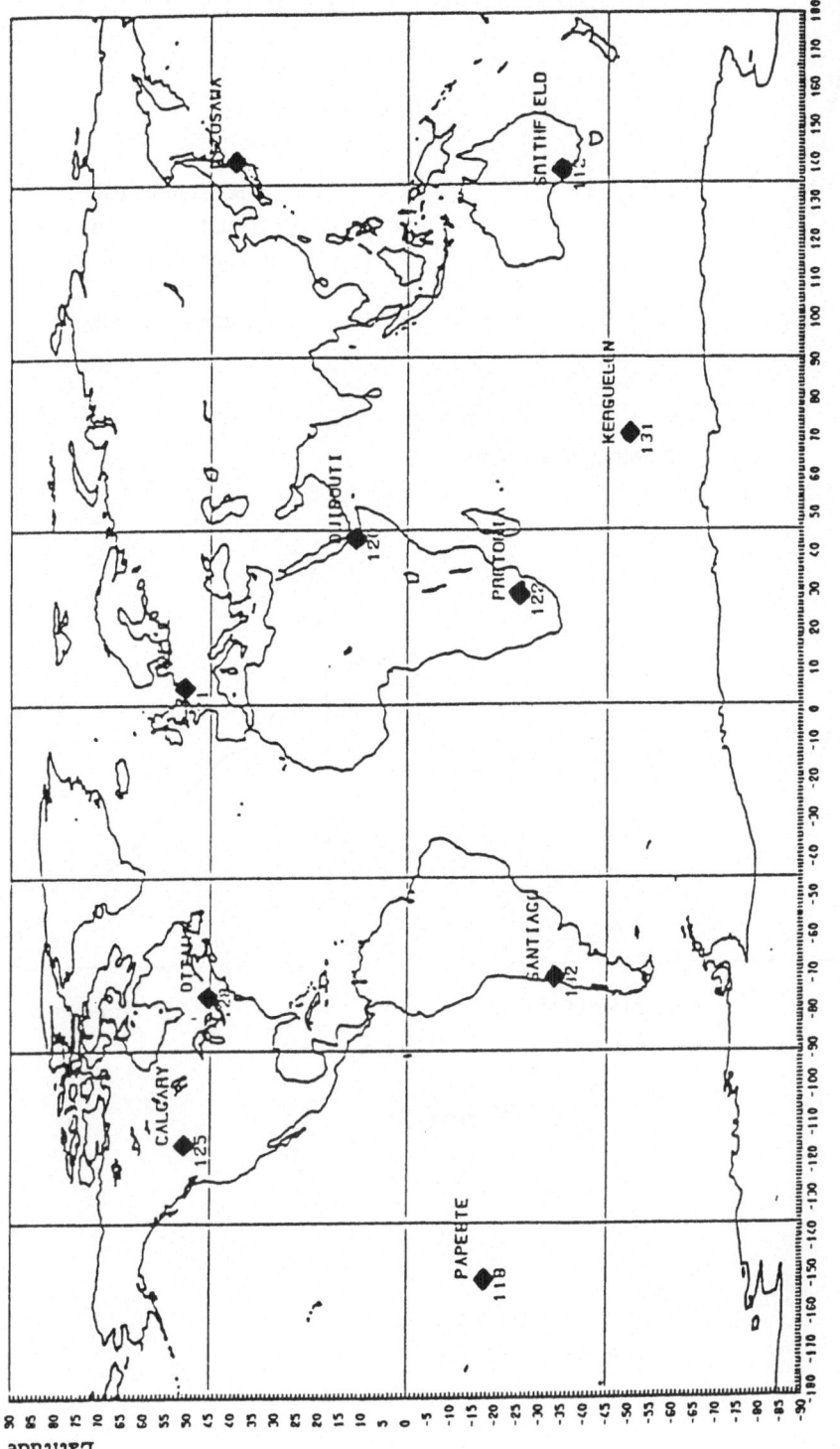

Figure-1: Station Network for GPS Simulation (10 Stations and 18 GPS satellites)

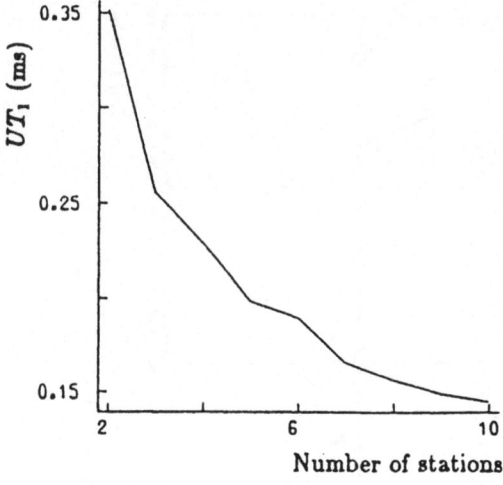

Precision of ERP
according to the number
of stations involved
in the GPS
tracking network

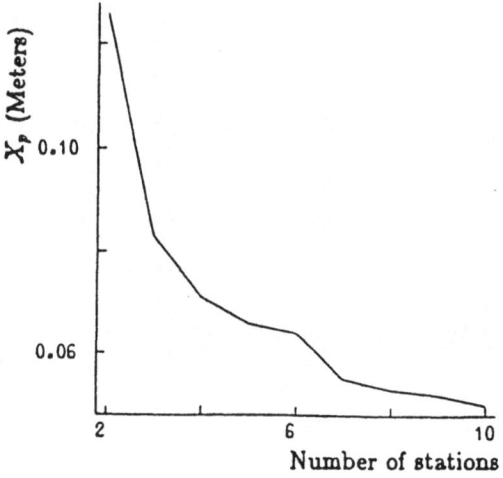

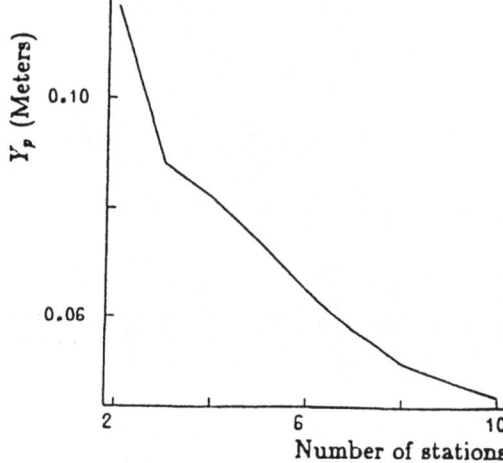

Figure-2

448

A NOTE ON
MONITORING TECTONIC PLATE MOTION USING GPS AND CLASSICAL TECHNIQUES

by

Erwin Groten

Abstract

Based on the assumption that tectonic plate motion is, in general, not
associated with substantial variations in the very high harmonics of
the gravity field the fundamental differences between coordinate
variation in classical geodetic data and GPS-results are pointed out.
The pecularities along plate margins and active seismic and volcanic
belts are discussed. Combination of different types of data with GPS-
measurements are outlined in order to obtain the complete information
on the geodynamic processes.

1. Introduction

There appears to exist a certain confusion in the literature as far as the detection of station shift and dis- or translocation due to tectonic motion is concerned. The reason for this seems to lie in the fact that purely geometric shift parameters as deduced from GPS or similar satellite and VLBI results are in some cases combined with triangulation data which are, of course, basically related to the plumb line and the gravity vector $\vec{g}$, respectively. Consequently, in view of increasing accuracy, also with <u>absolute</u> satellite positioning, a better understanding of the time-dependent phenomena inherent in the transition from classical triangulation to the modern satellite systems appears necessary.

2. Some geotectonic considerations

If local or regional intraplate motion is omitted here, which is usually also taken out in large-scale VLBI or SLR-considerations, we may assume that large-scale tectonic plate movement is detected by VLBI or SLR or similar large-distance methods; presently, it is attempted to improve GPS-orbits to such an extent that cm-accuracy will be available in the future for relative positioning also over distances of about 1000 km. If baseline techniques are directly applied to base lines between different tectonic plates such approaches give indeed the relative position changes of one plate with respect to the other.

This technique is, of course, also useful in studying intraplate or regional distortion or station movements. Typical examples are found in the Western part of the US where local triangulation measurements were repeatedly and successfully carried out over several decades, as in California, in order to monitor local plate motion. Also temporal variations in astronomical positioning data have been interpreted accordingly. However, great care and caution is necessary in interpreting quantities related to the plumb line such as deflections of the vertical $\vec{\vartheta}(\xi,\eta)$ or astronomical latitude φ and longitude λ with purely geometrical quantities such as geodetic coordinates (B,L) or satellite and VLBI coordinates (x_i).

If we assume that the low harmonics of the gravity field are <u>mainly</u> produced in the deep mantle whereas the high harmonic part is generated in the lithosphere and if, moreover, the tectonic plate motion is basically a lithosphere motion then, of course, any horizontal motion of the plate also involves a motion of the high-harmonics part of the gravity field; the latter moves along with the plate. This is not so at the plate margins where vertical motion is usually significant, contrary to the inner part of the large tectonic plates. In order to illustrate this we show in Fig. 1 the vertical motion in the european Alps of about 1 mm/year as well as the aforementioned local vertical motion in the southern part of West Germany (see Fig. 2) which have to be taken into account or eliminated if large-scale tectonic movements are investigated.

As the wave length of low order harmonics is of the order of several hundred kilometers they do not affect station shifts with speed of 8 cm/year or so even over a century. Therefore, by omitting plate

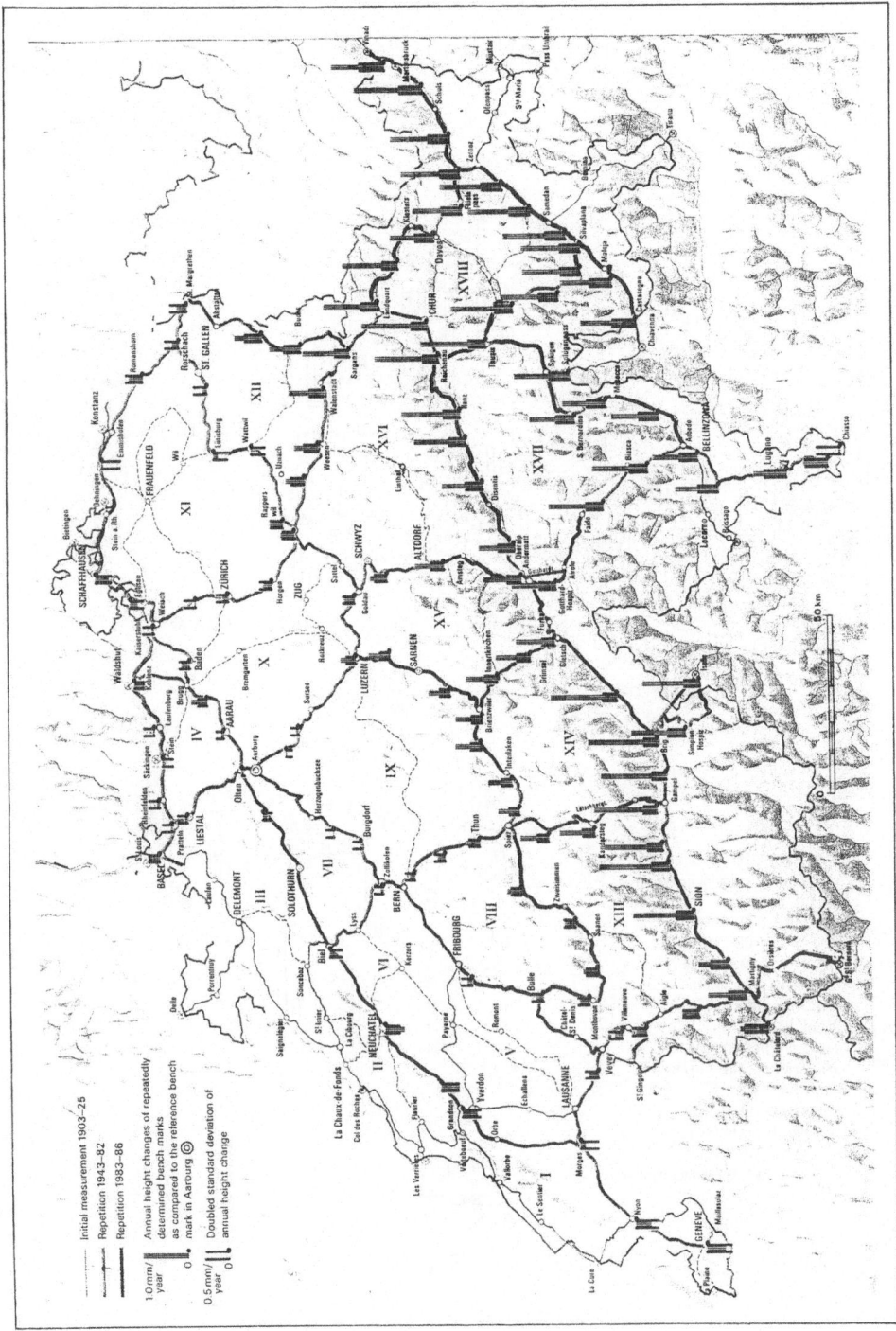

Fig. 1   First Order Levelling Net and Recent Crustal Movements

Reproduziert mit Bewilligung des Bundesamtes für Landestopographie vom 21.7.1988

FIRST (AND SECOND) ORDER LEVELLING NETWORK OF HESSEN (FRG)

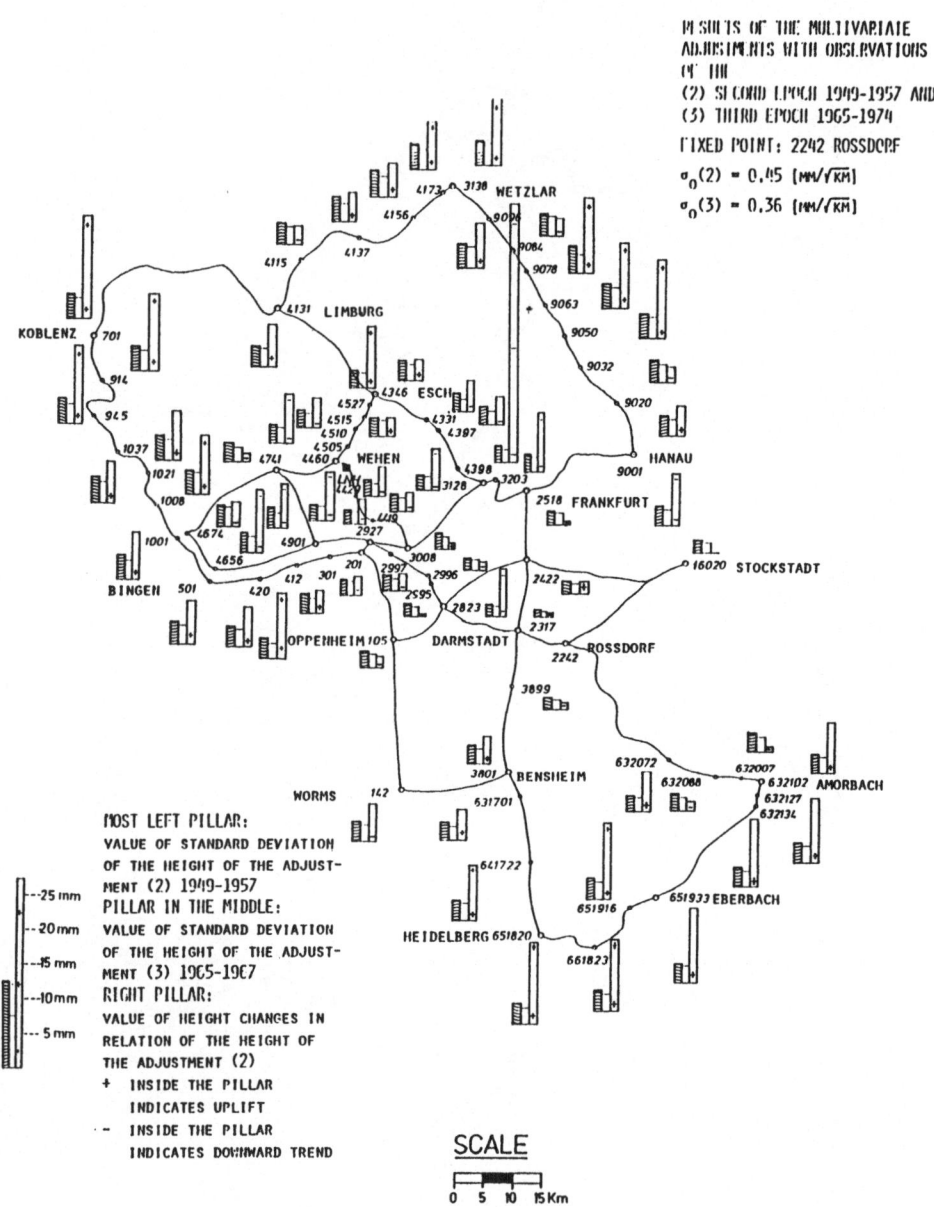

RESULTS OF THE MULTIVARIATE
ADJUSTMENTS WITH OBSERVATIONS
OF THE
(2) SECOND EPOCH 1949-1957 AND
(3) THIRD EPOCH 1965-1974
FIXED POINT: 2242 ROSSDORF
$\sigma_0(2) = 0.45$ [MM/$\sqrt{\text{KM}}$]
$\sigma_0(3) = 0.36$ [MM/$\sqrt{\text{KM}}$]

MOST LEFT PILLAR:
VALUE OF STANDARD DEVIATION
OF THE HEIGHT OF THE ADJUST-
MENT (2) 1949-1957
PILLAR IN THE MIDDLE:
VALUE OF STANDARD DEVIATION
OF THE HEIGHT OF THE ADJUST-
MENT (3) 1965-1974
RIGHT PILLAR:
VALUE OF HEIGHT CHANGES IN
RELATION OF THE HEIGHT OF
THE ADJUSTMENT (2)
+ INSIDE THE PILLAR
   INDICATES UPLIFT
- INSIDE THE PILLAR
   INDICATES DOWNWARD TREND

--25 mm
--20 mm
--15 mm
--10 mm
-- 5 mm

SCALE

0 5 10 15 Km

Figure 2

452

boundaries we may assume that the gravity vector of a station does not change significantly (neither in magnitude nor in direction) as a consequence of horizontal plate motion. Therefore, even if the station moves with the plate the astronomical coordinates (φ, λ) will not change significantly. Astronomical positioning is, consequently, no tool for detecting plate motion ! However, deflections $\vec{\vartheta}(\xi, \eta)$ of the plumb line do change because the geodetic coordinates vary as a consequence of the changed position with respect to the reference ellipsoid which is assumed to be held fixed.

In active tectonic areas such as plate margins or in volcanic belts the situation is rather different. Vertical motion implies high-harmonics variations in the topography and, consequently, in the gravity field. These variations in the immediate neighborhood of the stations may strongly affect the plumb line and lead to changes of $\bar{g}$ even if the station itself does not move at all. As VLBI and SLR stations are today often equipped with borehole tiltmeters which measure the variations of the plumb line it is noteworthy that tiltmeters measure a combined effect of the tilt of the terrain plus the plumb line and the separation of both is not straightforward. In the future kinematic GPS-techniques might help to separate both.

However, basically geometrical techniques such as VLBI, SLR etc. measure the pure geometrical relative shift whereas changes of (φ, λ) or (ξ, η) reveal quantities which are primarily of different nature.

In this connection also the fact needs to be considered that GPS does not give relative orthometric but ellipsoidal heights. As the associated change of the geoid is usually of the order of 10 percent of the surface deformation it can be neglected today, to some extent, as GPS-vertical data are seldom better now than ± 1 cm. In the future this aspect needs more detailed consideration.

3. <u>The transition from triangulation to GPS data</u>

With the accuracy available in the past the foregoing considerations did not deserve much interest but they will now and in the future be significant. The accuracy of astronomical positioning in first order triangulation in Germany is of the order of $\pm 0\overset{"}{.}5$ or worse which corresponds to ± 15 m; this is insignificant in view of plate motions in Europe of less than ± 10 cm/year.

Moreover, first order triangulation data had accuracy of relative positioning over large distances of not better than ± 20 or 30 cm.

Consequently, in geodetic datum transformations it was justified to neglect time and plate motion. However, if in the future different data sets, assuming higher precision, should be integrated in the sense of integrated geodesy then time has to be taken into account. In order to illustrate the associated details a few remarks are in order: Classical triangulation was established by setting geodetic coordinates (B,L) equal to the astronomical position (φ, λ) at the initial point of the triangulation by shifting the reference ellipsoid by an unknown (pure translation) vector $\vec{r}_0$ and keeping the parallelism of the axes in view of the Laplace condition. Moreover, the unknown geoid height N was put equal to zero at the initial station so that there we had

$$\xi - \varphi - B - \eta - (\lambda - L) \cos \varphi - H - h \tag{1}$$

where H - ellipsoidal and h - orthometric height, respectively. Half a century (or so) later we determined the geocentric position of the inertial station via satellite tracking and found $\vec{r}_o$; see Fig. 3. The time difference $(T_1 - T)$ between the installation of the triangulation where $(\xi - \eta - 0)$ was adopted and the time T_1 when the geometric satellite location for the initial station was determined was ignored because meanwhile the station had moved by $\Delta\vec{r}$ as seen in Fig. 4. If we presume that the station is not situated in an active zone we may suppose that $\vec{g}$ did not change in the interval $(T_1 - T)$. And assuming half a century for $(T_1 - T)$ and an accuracy of the astronomical measurements (φ, λ) of the order of $\pm0\overset{..}{.}3$ makes a station movement of 10 cm/year irrelevant as it leads to 5 m in 50 years which has to be compared with an accuracy of $\pm0\overset{..}{.}3 \overset{\triangle}{=} \pm10$ m.

As the direction of the ellipsoid normal $\vec{n}$ at epoch T is different from $\vec{n}(T_1)$ the deflection at the initial point is no longer $\vartheta - 0$ as illustrated in Fig. 4 even if $\vec{g}$ did not change at P during $(T_1 - T)$.

In the future we shall need very precise transition from GPS- to classical triangulation coordinates where much higher accuracies are necessary. We can do this by threedimensional Helmert-transformations in order to avoid the aforementioned difficulties. But if those results do correspond with datum transitions of the type

$$\vec{r}(P,T_1) - \vec{r}'(P,T_1) \tag{2}$$

it should be recognized that different results have to be expected, because the relation (2) is actually replaced by the incorrect relation

$$\vec{r}(P,T_1) - \vec{r}'(P, T) \tag{3}$$

the difference, i.e. the error in $\vec{r}_o$, being (see Fig. 4)

$$\Delta\vec{r}. \tag{4}$$

4. A remark on polar motion determination

Polar motion determination can, in principle, be done by using satellites, such as GPS, as well as classical techniques; in the first case it is a purely geometrical approach using coordinates whereas in the second case the determination is done with respect to the plumb line. If we adopt again the assumption that the high harmonic gravity field moves with the tectonic plate and low harmonics do not affect a dislocation of about ten meters then again, in the classical case, tectonic motion is not reflected in the plumb line directions; consequently, only in active plate marginal zones, volcanic belts etc., a variation of the plumb line might effect significantly the polar motion determination whereas the plate motion is fully inherent in the tracking station coordinate variations and, consequently, affects polar motion determination more strongly.

However, during a recent series of earthquakes (around the end of 1984) in the area of Santiago de Chile (Noel, priv. comm., 1987) a

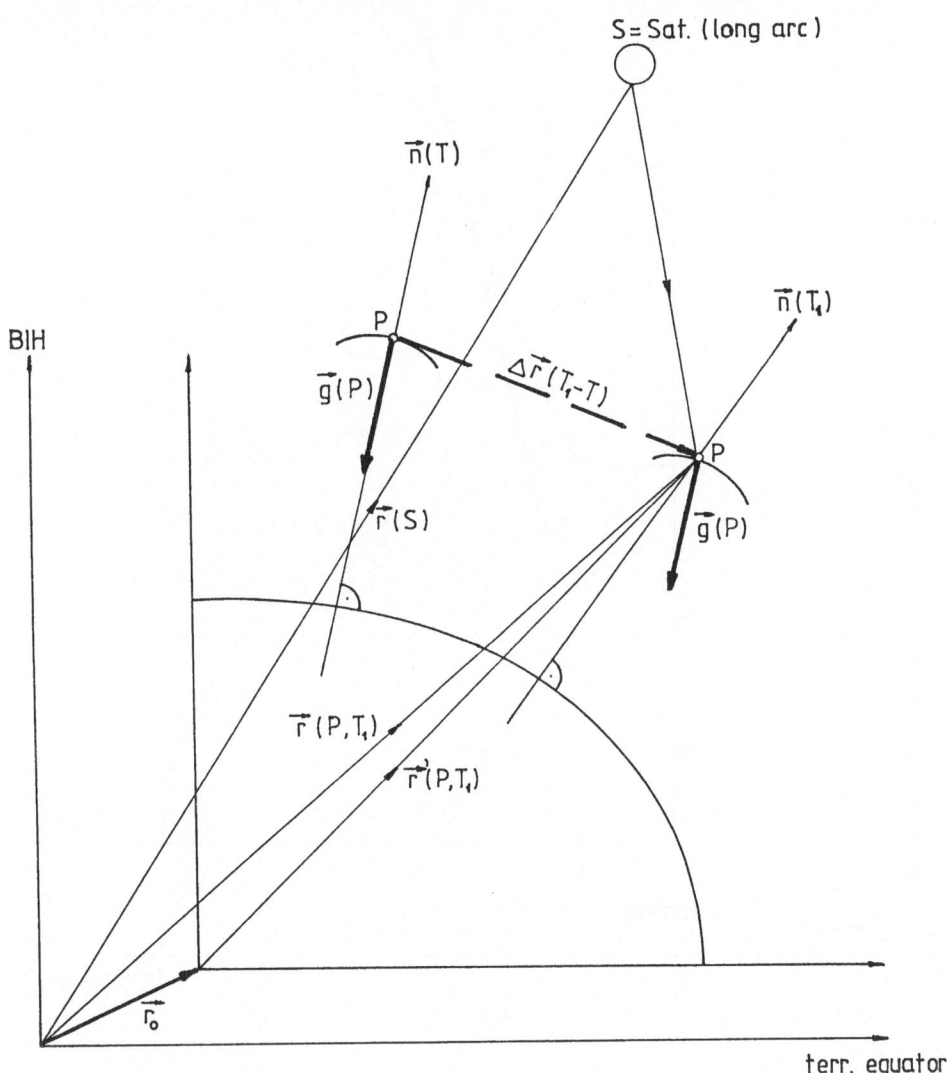

Figure 3

continuous series of astronomical polar motion determination at the
observatory of Santiago did not reveal any significant change after
the earthquakes which were quite remarkable; Santiago is not far from
the Pacific plate boundary. It seems that variations and possible
consequences of plumb lines with time are often overestimated.

5. Conclusions

As $\vec{r}(T_1,P)$ is independent of the gravity vector $\vec{g}(T_1,P)$ any change
of the plumb line during the interval $(T_1 - T)$ does not primarily
affect the transformation (2). However, as $\vec{n}(P,T_1) \neq \vec{n}(P,T)$ and
$\vec{g}(P,T_1) \neq \vec{g}(P,T)$ there is a double effect on $\vec{\vartheta}$ in any
transformations concerning $\vec{\vartheta}(T_1,P)$. Moreover, any transformation
involving the other triangulation stations belonging to the same
system will be affected.

455

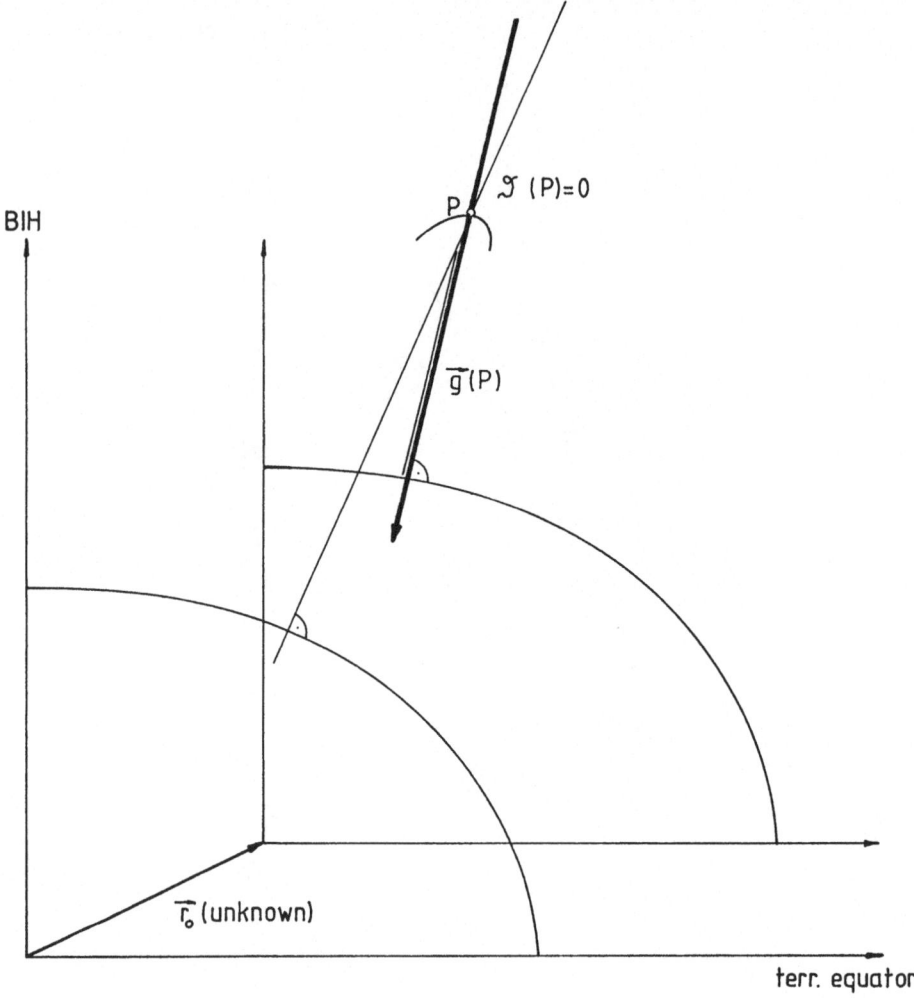

Figure 4

High-precision astronomical as well as satellite and VLBI or LLR
(lunar laser ranging) data are useful in monitoring _various_ and
different tectonic phenomena; also tiltmeter and extensometer data
(for purely local purposes) are, in addition, of interest. The
transition from one data set to the other is not unproblematic, in
general, and has to be carried out taking temporal changes into
account. The interpretation of variations in (φ, λ) in terms of static
shift is consequently misleading.

References

Groten, E., Schwarz, E. and Prinz, H.: Zur Problematik der Berechnung
    und Interpretation von Höhenänderungen, Zeitschrift für
    Vermessungswesen, Heft 9, September 1978, 103. Jahrgang
    (Fig. 2)
Report on the Geodetic Activities in the years 1983 to 1987, presented
    to the XIX General Assembly of the Intern. Union of Geodesy
    and Geophysics in Vancouver, August 1987, 1987 (Fig. 1).

Ninth session: Special Applications and Orbits

Chairman: Prof. Grafarend, Stuttgart

Determination of Azimuths from GPS

Measurements and Comparison with

Common Methods

by

Angela Beckmann

Hans-Jürgen Larisch

Otmar Schuster

Abstract

Since 1983, the use of GPS methods - especially together with the Macrometer system - has become a standard practice in the improvement of the trigonometric network in Nordrhein-Westfalen. To provide these advantages also to the lower order surveys, which are based on angular and distance measurements, an interface between GPS and conventional methods has to be set up. The determination of azimuths at the GPS stations gives a tool to achieve this connection.

Several methods to derive azimuths in a GPS network are discussed. Results of measurements in the network Elmpt and a practical way to handle the problems are shown.

1) Integration of GPS measurements in existing trigonometric networks

Although GPS baselines have been measured for quite a few years, the efficiency and the advantages of this method is not accepted by all possible customers (mostly surveying departments). This is especially true when you consider that measurements in higher order trigonometric networks were in the past the domain of state surveying departments.

GEOsat as a private company is now penetrating this domain, and is able to compete successfully with the conventional surveying methods used to establish and improve geodetic networks.

To be successful with GPS applications means therefore

- to develop methods which are superior to conventional surveying practices without changing the "geodetic philosophy" too much

- to convince the surveying departments that GPS measurements are more efficient than old methods when a thorough and true cost-benefit-calculation is carried out.

An example for the application of this theory is the combination of GPS observations with azimuths of other sources.

In this country, GPS application in a geodetic network means more often an improvement of existing networks than an establishment of new networks. Only in special cases and in lower order surveying it might be possible that new networks are established.

So the location of a trigonometric point is more or less static, and due to the history mostly a highly visible point like a church-tower etc.... – It was at least a highly visible point 100 years ago! Each following measurement must relate to already existing points. In the case of the church-tower the centric point can only be used as a remote target.

The renewal of an existing network with GPS methods results in new eccentric points as antenna stations. Generally the frequency of eccentric points increases in lower order networks. (When we do a GPS measurement in 4th order networks we can more or less use only eccentric points). Therefore the advantages of GPS methods can not be used thoroughly

- speed: The antenna stations can be connected very quickly with each other. The reduction to the centre is as unefficient as in the past.

- efficiency: The centring observations change the expected cost-benefit-ratio.

2) Arguments to combine GPS measurements with azimuths

There are two possibilities to solve the problems arising from
the integration of GPS measurements in existing networks (see
chapter 1).

- simplification of centring measurements

- establishment of antenna stations as new centres

In both cases the use of azimuths for the orientation of lower
order measurements would improve the efficiency of the method, but
the following questions have to be answered first.

- Can the orientations be derived directly from
  GPS measurements?

- Which other possibilities exist to get the orientations?

- Do the results of different methods correspond to each
  other?

To answer these (and further) questions the test-network "Elmpt"
has been set up.

Besides that, we combined GPS measurements with other survey
methods in the past ("Unterlüß") - without scientific interests
but because of no other possibilty to do the job.

3) Comparison of GPS azimuths with alternative methods

GPS baseline observations result in three-dimensional coordinate
difference defined in WGS72 or WGS84 from which azimuths are
derived. To make these results comparable with other methods, the
influence of the local gravity field has to be calculated. The
coordinates also have to be transformed to that coordinate system
that is used in common surveying (in Germany: Bessel-System
(DHDN)). The best value to be compared is an astronomical
azimuth. So all measured or calculated orientations are corrected
for the influence of the deflection of the vertical. This
procedure is shown in Fig.1 for different sources of azimuths.

An other source of azimuths - not mentioned in Fig. 1 - are
observations to the pole star (or the sun) which have not yet been
carried out (due to wheather conditions!). These measurements are
planned for the near future.

GPS	conventional	gyro theodolite
PHI, LAMBDA, RAD (WGS 72, 84) | B, L, h (DHDN) | ellips. azimuth | deflection of the vertical XI, ETA | astron. azimuth	B, L, h (DHDN) | ellips. azimuth | deflection of the vertical XI, ETA | astron. azimuth	astron. azimuth

Fig. 1: Determination of astronomical azimuths

The following tables show the results which could be obtained up to now:

1. test-line "Mintard" (comparison of GPS and gyro azimuths)

    GPS : Macrometer V-1000 (Aero Service)
    gyro: Gyromat (WBK)

GPS [gon]	gyro [gon]	type	diff. [mgon]	distance [m]	lateral error [m]
167.08874	167.0885	G	0.24	425	0.002
368.77262	368.7723	G	0.32	426	0.002

2. test-network "Elmpt" (comparison of GPS and gyro azimuths)

    GPS : Macrometer V-1000 (Aero Service)
    gyro: MK 15 (BGT)
          MK 12 (BGT)
          Gyromat (WBK)

GPS [gon]	gyro [gon]	type	diff. [mgon]	distance [m]	lateral error [m]
120.3317	120.3321	G	0.4	96	0.001
	120.3201	MK12	11.6	96	0.017
	120.3387	MK15	7.0	96	0.011
320.3331	320.3334	G	0.3	96	0.001
	320.3258	MK12	7.3	96	0.011
	320.3373	MK15	4.2	96	0.006
312.0457	312.0428	G	2.9	1032	0.047
	312.0477	MK12	2.0	1032	0.032
	312.0437	MK15	2.0	1032	0.032

3. network "Unterlüß" (comparison of GPS and gyro azimuths; determination of an excentric point)

   GPS : Macrometer V-1000 (Aero Service)
   gyro: MK 12 (BGT)

The use of a gyro theodolite in a GPS network can be shown with the network "Unterlüß" (UNLUMAC). The baseline measurements were carried out with Macrometer V-1000 and a gyro theodolite MK 12 was used to provide azimuths. With the help of the gyro the Macrometer network was connected to a known (starting) point. Also one eccentric antenna station was connected to the center using a gyro azimuth.

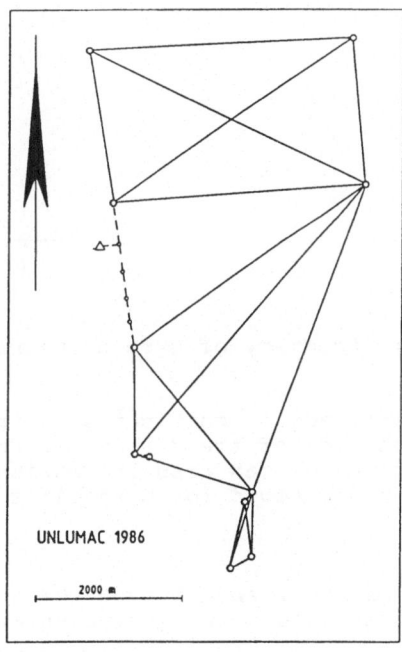

Fig. 2: Macrometer campaign "Unterlüß" (UNLUMAC)

463

Theoretical studies [SCHUCHARD 1986] show that azimuths derived
from GPS observations have a precision in the order of

      0.6 mgon < σ < 1.5 mgon   with distances of 100 m,
      0.1 mgon < σ < 0.2 mgon   with distances of 1000 m.

Our measurements with the gyro theodolites resulted in

      σ = ± 1.1 mgon (Gyromat),
      σ = ± 5.5 mgon (MK 12) ,
      σ = ± 2.9 mgon (MK 15) .

These results could be obtained within 75 min with Gyromat and MK
15, and it took 30 min to measure an azimuth with MK 12.

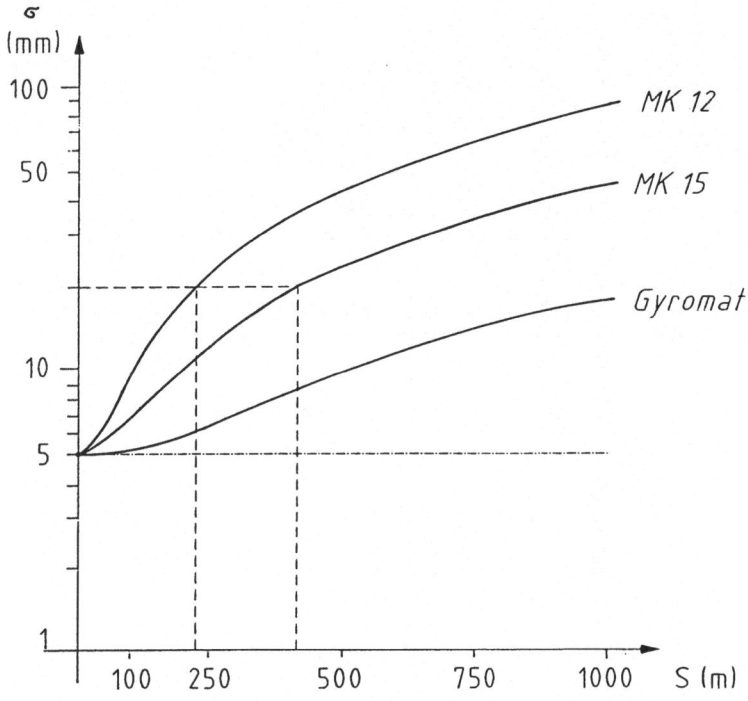

Fig. 3: Accuracy of gyro azimuths

Fig. 3 shows the gyro accuracy in combination with centring and
pointing errors as a function of the distances. For a given
standard error of position you can read the maximum length of
sight which should not be exceeded for a specific gyro theodolite.

If the length of sight is short enough even the gyro theodolite
MK 12 can be used which is able to determine an azimuth within
half an hour.

4) Consequences of these methods for future GPS applications

The main results of this investigation can be shown in the following statements:

- network coordinates determined from GPS baseline observations can be combined successfully with orientation measurement from other sources.
  The accuracy is not reduced!

- the planning of GPS networks should provide the possibility to include azimuth observations to facilitate the centring observations and the connection to lower order measurements!

- the determination of twin points in areas with no orientation can be done with GPS measurements. But it is more efficient to use a modern, automatic and user-friendly gyro-theodolite.

5) References

Beyer, W.; H.-J. Larisch; O. Schuster: Richtungsübertragungsgenauigkeit bei Zwillingspunkten, Interner Bericht (GEOsat), Mülheim 1986

Großmann, W.: Geodätische Rechnungen und Abbildungen in der Landesvermessung, Stuttgart 1976

Lindstrot, W.: Zur Orientierung örtlicher Messungen beim Einsatz von GPS-Empfängern im TP-Feld, in Vorbereitung

Schmidt, R.; D. Ehlert: Die Diagnoseausgleichung des Deutschen Hauptdreiecksnetzes, DGK Reihe B Nr. 262, Frankfurt 1982

Schuchardt, A: Hochgenaue Richtungsbestimmung mit Hilfe von GPS-Empfangsanlagen, Forschungsbericht am Institut für Erdmessung der Universität Hannover, 1986

THE EUROPEAN TRACKING NETWORK

by

Georg Groven
Gert Riemersma

INTRODUCTION

GPS Services A/S, a specialist GPS survey company in Norway, and
Aero Service, Houston, are setting up an European GPS Satellite
Tracking Network to supply post-processed Ephemerides for precise
GPS measurements. After extensive investigation we concluded that
a need for such a service existed as it would provide European
GPS users with Precise Ephemerides based on data collected in the
European Area.

The post-processed Ephemerides are needed for the codeless GPS
receivers, or when the survey requires a high degree of precision, for example in subsidence monitoring and control network
improvement.

In the future it might also be possible to provide Broadcast
Ephemerides for real-time positioning with an accuracy greater
than the degraded Broadcast Ephemeris which will be available
from the satellites.

POST-PROCESSED EPHEMERIDES

The post-processed Ephemeris of a satellite is determined by
computation of tracking observations. Tracking data is required
over a three day period from a number of sites of well known
position. These are then processed to obtain a best fit orbit.
Limitations on such Ephemerides are due to such factors as the
quality of the tracking data, the ability to model the ionesphere
and troposphere accurately, the location and number of tracking
sites, and the geometric and dynamic earth models used in the
computational procedures.

Precise Ephemerides can currently be obtained from NGS and from
Aero Service in Houston with an accuracy of 1-2 ppm of baseline
length.

BROADCAST EPHEMERIDES

The Broadcast Ephemerides are currently estimated to be accurate
at the 20 to 50 meter level, enabling baseline differential phase
observations accurate to a few ppm. However the Broadcast
Ephemerides can only be used by GPS receivers with a built-in
signal decoding facility.
As the Broadcast Ephemeris is intended to provide real-time
information for defence navigation purposes the quality of the
Broadcast Ephemeris available to the civilian user will be
degraded to the 100m level once GPS becomes fully operational in
1991.
Consequently, the 1 ppm precision now required of surveys will
only be available from Post-Processed Ephemerides by the end of
this decade.

THE ADVANTAGES OF AN EUROPEAN TRACKING NETWORK

The following points we consider to be a few of the benefits to
be gained by establishing a European GPS Satellite Tracking
Network.

a) Minimize dependance upon the US controlled provision of
    Ephemerides

b) Provide Post-processed Precise Ephemerides for European needs
    with data collected from the European Area, as opposed to
    data collected from America, as is the case at the moment.

    The accuracy of the Ephemerides will be around the 1 ppm
    level with futher accuracy improvements expected in years
    to come.

c)   Protection against the degradation of the Broadcast Ephemeris
     in 1991 and provide the ability to create an alternative more
     accurate Broadcast Ephemeris.

d)   Independent monitoring and research into ionospheric effects

e)   Independent monitoring of clock drift rates of the satellites

TRACKING NETWORK

The Tracking Network will consist of five Tracking Stations (see
figure 2). Their location needs to be sufficiently distributed in
the North-South, East-West directions to enable a reliable
solution for the Ephemerides to be computed for the whole
European area.
The Tracking Stations should ideally be located at laser stations, as the most important requirement is that the position of
the stations are known to a high degree of accuracy and on a
datum common to all stations, as it is the case with the laser
stations. Another requirement is the availability of an atomic
clock, to provide a stable frequency source for the receiver.

The following laser stations have been proposed as suitable
locations:

          Metsahovi   -   Finland
          Wettzell   -   Germany
          Dionysos   -   Greece
          San Fernando   -   Spain
          Madrid   -   Spain (Mark III VLBI to become opera-
                             tional in April 1988 (?) and will
                             provide a better location than
                             San Fernando)
          Tromsø   -   Norway (However this is not yet a mobile
                             laser station)

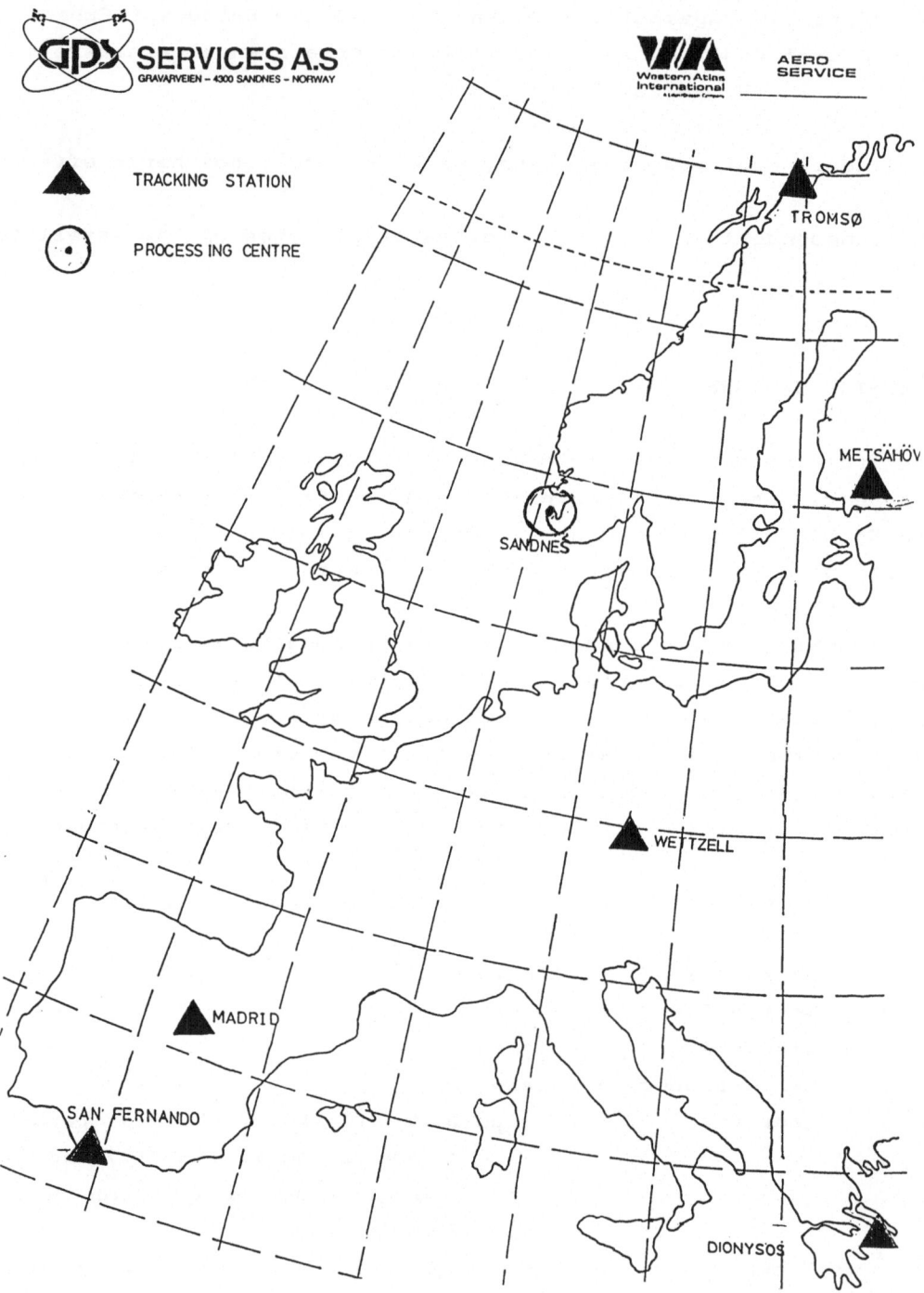

Figure 2: PROPOSED LOCATIONS OF TRACKING STATIONS IN THE
EUROPEAN TRACKING NETWORK

EQUIPMENT INSTALLED AT TRACKING STATIONS

The equipment installed at the Tracking stations will consist of
a dual frequency fully automated Mini-Mac 2816 AT GPS receiver, a
modem, a meteorological control unit and an atomic clock (see
figure 3).
The Mini-Mac receiver tracks up to eight satellites simultaneously and will record both the C/A-code pseudo-ranges and phase
data for processing. The 20 M-byte storage unit inside the
receiver will store the raw GPS data until it is accessed and can
be transferred to the Processing Centre (see figure 4).

The receiver is interfaced to the processing centre in Sandnes,
Norway, via a modem and telecommunication link. It will allow the
Processing Centre to control, monitor and transmit operational
commands to the receiver. It will also download the raw GPS data
once a day for post-processing.
The meterorological control unit is linked to an analog barometer, temperature and relative humidity sensors. This information is required for the tropospheric corrections and will be
transmitted along with the raw GPS data.

PROCESSING AND CONTROL CENTRE

The processing and Control Centre can be divided into three parts
(see figure 5):

1) The Data Management System
    It's Sole function is to operate and monitor the Tracking
    Station and to collects and store the raw GPS data for
    processing. The data will be collected once a day and the
    data transfer will last about an hour for each station.
    Several times during the day the Management System will also
    check on the operational status of the Tracking stations, to

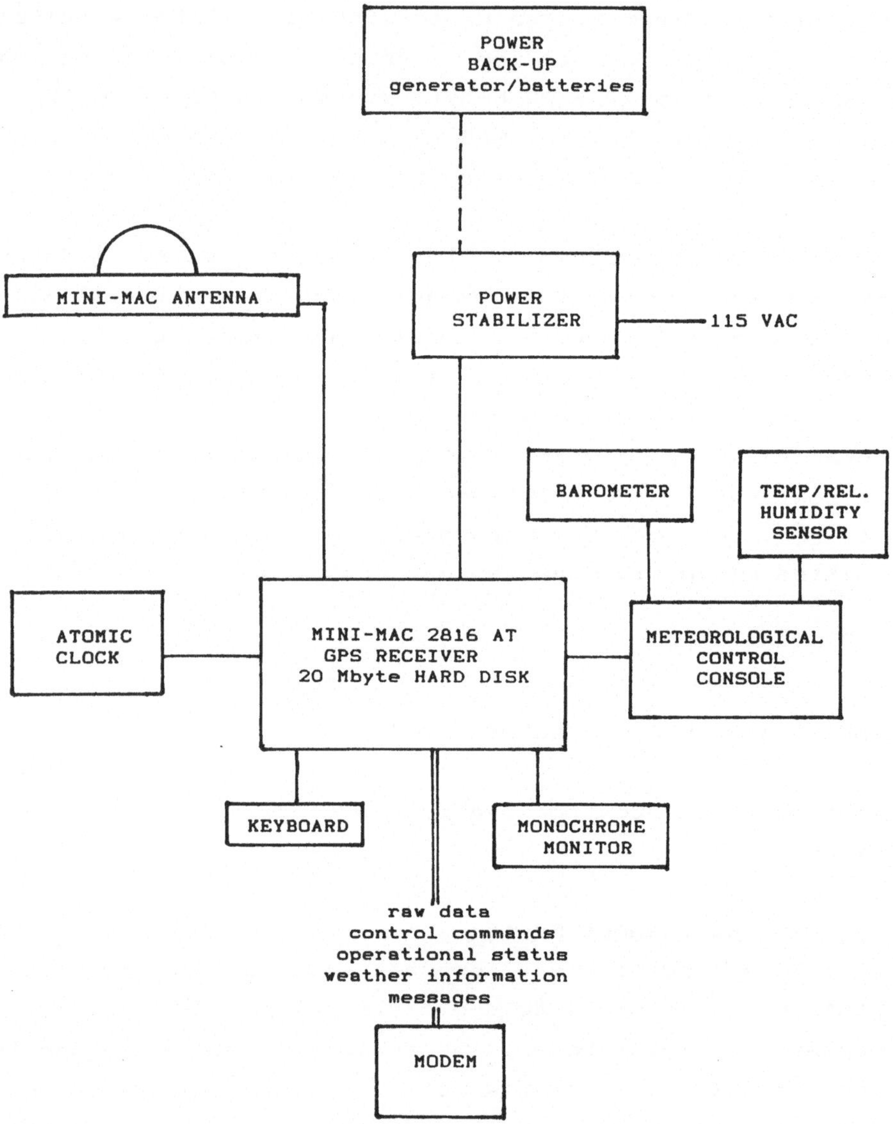

GPS SERVICES A.S
GRAVARVEIEN – 4300 SANDNES – NORWAY

Western Atlas
International
A Litton/Dresser Company

AERO
SERVIC

```
                              ┌──────────────────┐
                              │      POWER       │
                              │    BACK-UP       │
                              │ generator/batteries │
                              └──────────────────┘
                                       ┊
                                       ┊
  ┌──────────────────┐        ┌──────────────────┐
  │ MINI-MAC ANTENNA │        │      POWER       │────── 115 VAC
  └──────────────────┘        │   STABILIZER     │
                              └──────────────────┘
```

POWER
BACK-UP
generator/batteries

MINI-MAC ANTENNA

POWER
STABILIZER   ──► 115 VAC

BAROMETER

TEMP/REL.
HUMIDITY
SENSOR

ATOMIC
CLOCK

MINI-MAC 2816 AT
GPS RECEIVER
20 Mbyte HARD DISK

METEOROLOGICAL
CONTROL
CONSOLE

KEYBOARD

MONOCHROME
MONITOR

raw data
control commands
operational status
weather information
messages

MODEM

Figure 3:   EQUIPMENT INSTALLATION AT TRACKING STATIONS

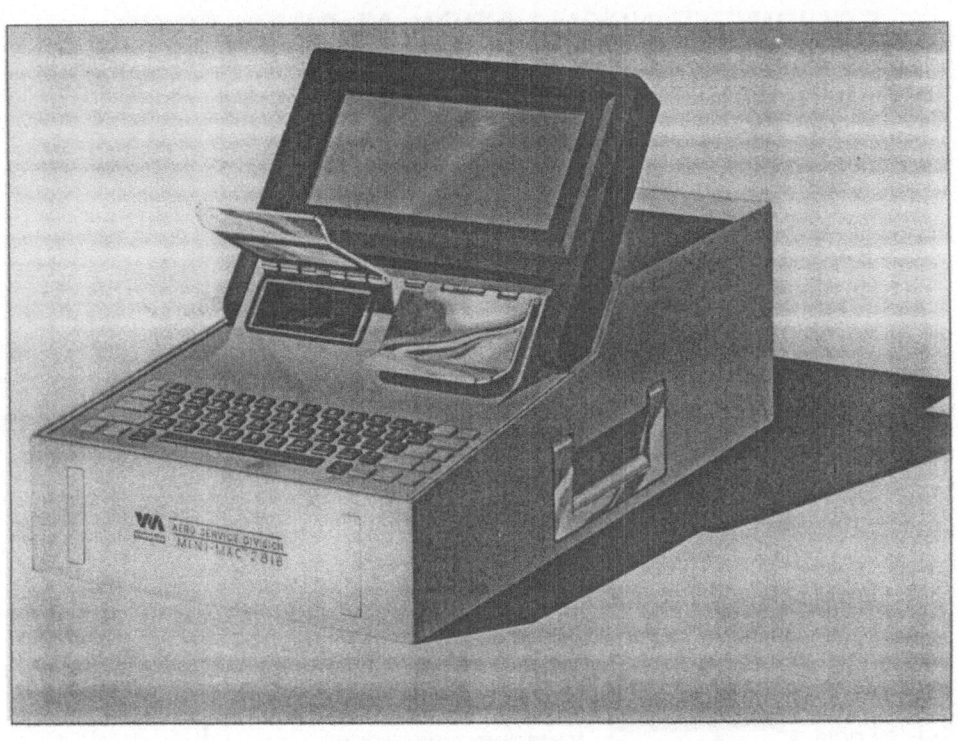

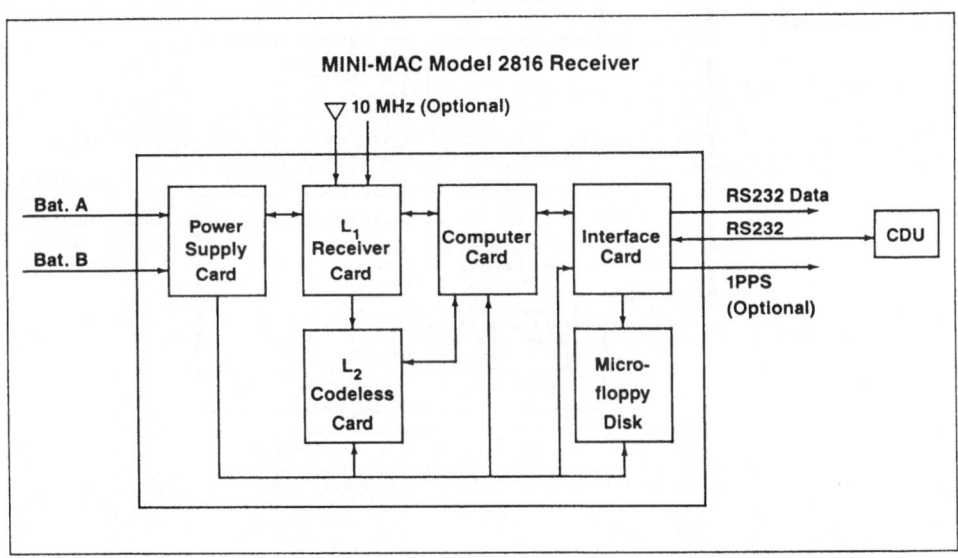

MINI-MAC Model 2816 Receiver

▽ 10 MHz (Optional)

Bat. A

Bat. B

Power Supply Card

L_1 Receiver Card

Computer Card

Interface Card

RS232 Data

RS232

CDU

1PPS (Optional)

L_2 Codeless Card

Micro-floppy Disk

Figure 4: MINI-MAC 2816AT GPS RECEIVER

EQUIPMENT INSTALLATION AT PROCESSING
AND CONTROL CENTRE

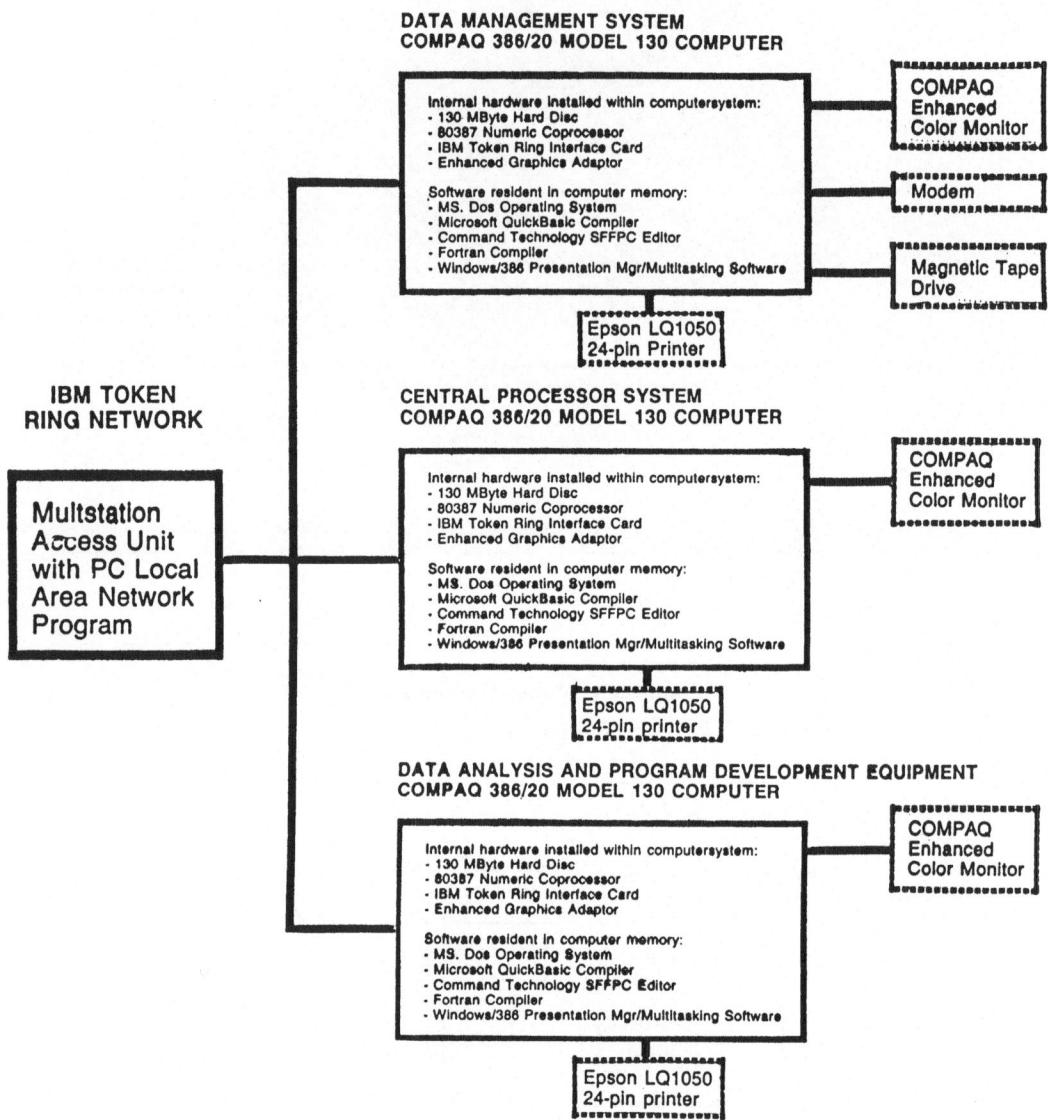

DATA MANAGEMENT SYSTEM
COMPAQ 386/20 MODEL 130 COMPUTER

Internal hardware installed within computer system:
- 130 MByte Hard Disc
- 80387 Numeric Coprocessor
- IBM Token Ring Interface Card
- Enhanced Graphics Adaptor

Software resident in computer memory:
- MS. Dos Operating System
- Microsoft QuickBasic Compiler
- Command Technology SFFPC Editor
- Fortran Compiler
- Windows/386 Presentation Mgr/Multitasking Software

COMPAQ Enhanced Color Monitor

Modem

Magnetic Tape Drive

Epson LQ1050 24-pin Printer

IBM TOKEN RING NETWORK

Multstation Access Unit with PC Local Area Network Program

CENTRAL PROCESSOR SYSTEM
COMPAQ 386/20 MODEL 130 COMPUTER

Internal hardware installed within computer system:
- 130 MByte Hard Disc
- 80387 Numeric Coprocessor
- IBM Token Ring Interface Card
- Enhanced Graphics Adaptor

Software resident in computer memory:
- MS. Dos Operating System
- Microsoft QuickBasic Compiler
- Command Technology SFFPC Editor
- Fortran Compiler
- Windows/386 Presentation Mgr/Multitasking Software

COMPAQ Enhanced Color Monitor

Epson LQ1050 24-pin printer

DATA ANALYSIS AND PROGRAM DEVELOPMENT EQUIPMENT
COMPAQ 386/20 MODEL 130 COMPUTER

Internal hardware installed within computer system:
- 130 MByte Hard Disc
- 80387 Numeric Coprocessor
- IBM Token Ring Interface Card
- Enhanced Graphics Adaptor

Software resident in computer memory:
- MS. Dos Operating System
- Microsoft QuickBasic Compiler
- Command Technology SFFPC Editor
- Fortran Compiler
- Windows/386 Presentation Mgr/Multitasking Software

COMPAQ Enhanced Color Monitor

Epson LQ1050 24-pin printer

Figure 5.   Processing and Control Centre.

ensure it is still functioning properly. The Data Management
System consist of Compaq 186/20 computer, magnetic tape drive
and modem.

2) The Central Processor System

It is linked to the Data Management System via a IBM token
ring Network, and consist of Compaq 186/20 computer, with
numeric coprocessor and MS.DOS operating system.
It's function is to process the raw GPS data and using the
Orbital Determination Software to compute the Precise
Ephemerides. It also produces an Broadcast Ephemerides for
use with codeless GPS receivers to provide a satellite alert
almanac for satellite tracking.

3) Data Analysis and Program Development Equipment

This provides a back-up computer in case of a failure to any
of the equipment above. It is also used for quality assurance
and analysis of the Ephemerides produced.

CONCLUDING REMARKS

The Ephemerides will be stored on data files and this can be
accessed by the User through a modem and personal password. This
facility to access the Ephemeris will be made available in
several different countries through a number of agents.
The Tracking Network is expected to become operational by the end
of this year, after we have performed an extensive testing and
quality control campaign to ensure the quality and accuracy of
the Ephemerides.

REFERENCES

- A European GPS Tracking Network: G. Groven / S. Matthes
- Norwegian Tracking System for GPS Satellites: S. Matthes
- Mini-Mac Model 2816 GPS Surveyor: J. Cain

WORLD GEODETIC SYSTEM 1984 – GEODETIC REFERENCE SYSTEM OF GPS ORBITS

by

Franz Josef Lohmar

Abstract

The World Geodetic System has been developed for the U.S. Department of Defense. Starting with WGS 60, the approximation of the figure of the earth and its gravity field has steadily converged to the most recent WGS 84 of the Defense Mapping Agency (DMA). WGS 84 is the reference system for both GPS broadcast and GPS precise ephemeris. Thus, WGS 84 will be of great importance, worldwide, for practical applications. The absolute accuracy of an actual WGS 84 position reference is in the order of 1 to 3 meters, depending on the method applied. If WGS 84 is intended to be used in horizontal control networks, the approach of a zero order network measured by satellite techniques is mandatory to fix the realization of the WGS 84 absolutely. All other GPS-based measurements are to be connected to these points to proceed with densification using the excellent relative accuracies of GPS. For the period of measuring the zero order network maximum quality of the orbital elements is required.

1. Introduction

The U.S. Defense Mapping Agency (DMA) produces numerous mapping, charting, digital and geodetic products in support of the Department of Defense in various parts of the world. The advantage of using one single global reference system for all products has resulted as early as in the late 1950s in the generation of a World Geodetic System, i.e. the World Geodetic System 1960 (WGS 60).

New measuring techniques, especially newly developed space techniques and greater capabilities of hardware and software, have since enabled a stepwise refinement of the earth's model. Due to the geodetic applications of the Navy Navigation Satellite System (NNSS), the WGS 72 reference system became well known. Anybody using the Doppler point positioning and translocation techniques has to be familiar with the global reference system and its relation to the various local reference systems.

For the purpose of Doppler point positioning and precise satellite orbit determination DMA has primarily used the geodetic reference system NWL9D and the NSWC 9Z-2 respectively, as the accuracy and consistency of WGS 72 were insufficient for this type of application. Thus, Doppler solutions based on DMA precise ephemeris have been given in a slightly different and more accurately defined reference system than those based on the NNSS broadcast ephemeris (Jenkins and Leroy 1979).

Today, DMA has released the "Department of Defense World Geodetic System 1984 (WGS 84)". Its defining parameters and relationships with several classical local geodetic systems are extensively documented in a DMA Technical Report (DMA 1987) and a supplement consisting of 3 volumes.

WGS 84 comprises the realization of a geocentric coordinate system, a mean earth ellipsoid, an earth gravitational model and transformation parameters to other geodetic datums. For WGS 84 the para - meters of the international-sanctioned Geodetic Reference System 1980 (GRS 80) have been adopted (Moritz 1980).

A huge amount of data has been incorporated in estimating the defining parameters, especially data from the following modern techniques (Decker 1986):
- Satellite Laser Ranging
- Lunar Laser Ranging
- Very Long Baseline Interferometry
- Satellite Doppler
- Satellite Radar Altimeter
- Surface Gravity

WGS 84 has in the meantime become the reference system for both GPS broadcast and GPS precise ephemeris. Thus, WGS 84 will be of great importance, worldwide, for practical applications.

2. WGS 84 reference frame and coordinate system

The WGS 84 coordinate system is an earth-centered, earth-fixed coordinate system. Its origin is the center of mass of the earth. The orientation of the system is chosen to be identical with the BIH definition realized by the adopted coordinates of the BIH stations. Thus, the Z-axis is parallel to the direction of the Conventional Terrestrial Pole (CTP) and the Zero Meridian is parallel to the BIH-defined Zero Meridian. The X-axis is the intersection of the WGS reference meridian with a plane parallel to the CTP equator incorporating the WGS 84 defined earth center of mass (figure 1).

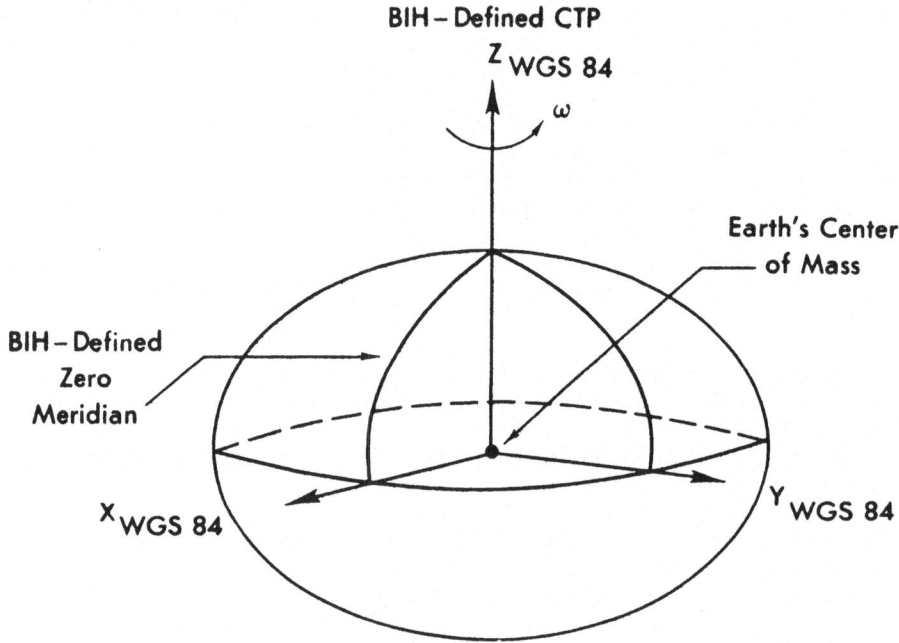

Figure 1: Definition of the WGS 84 Coordinate System (DMA 1987)

Apart from this theoretical definition of a coordinate system putting into practice by allocating coordinates to a set of earth-fixed stations is most important.

Thus, to find the WGS 84 coordinate system, DMA has modified the given NSWC 9Z-2 reference system realized and used as the basis for NNSS precise ephemeris (PE) and for PE-based Doppler point positioning with an absolute accuracy of 1-2m worldwide. The modifications have been derived from extensive comparison of Doppler point positioning results with other space techniques performed by DMA and several other groups around the globe (e.g., Hothem 1979).

The following parameters allow the conversion from the NSWC 9Z-2 system to WGS 84:

$$\begin{bmatrix} X_{WGS\ 84} \\ Y_{WGS\ 84} \\ Z_{WGS\ 84} \end{bmatrix} = \begin{bmatrix} 0 \\ 0 \\ 4.5 \end{bmatrix} + \begin{bmatrix} -0.6*10^{-6} & -0.814/RO & 0.0 \\ +0.814/RO & -0.6*10^{-6} & 0.0 \\ 0.0 & 0.0 & -0.6*10^{-6} \end{bmatrix} \cdot \begin{bmatrix} X_{XNSWC\ 9Z-2} \\ Y_{YNSWC\ 9Z-2} \\ Z_{ZNSWC\ 9Z-2} \end{bmatrix}$$

with   RO = 1 / sin 1"

In clear, this means (figure 2 and 3):

- lowering the NSWC 9Z-2 origin by 4.5 meters
- rotating the NSWC 9Z-2 Reference Meridian (Zero Meridian = X-axis) westward by 0.814"
- changing the NSWC 9Z-2 scale by -0.6 ppm.

With the aid of these conversion parameters all given Doppler point positioning results based on the DMA precise ephemeris referenced to NSWC 9Z-2 can be converted to WGS 84. This is a very important fact, as huge databases contain such coordinates!

Of course, the definition of an earth-centered, earth-fixed reference system cannot be performed separately from its relationship to inertial reference systems, because space techniques are inherently involved.

Without going into detail, it should be noted at this point, that the FK5 system referenced to epoch J2000.0 is the related Conventional Inertial System (CIS). The conversion from this earth-centered CIS to WGS 84 is determined by rotations due to polar motion, siderial time, nutation and precession.

The definition of a mean earth rotation rate is playing two different roles in the WGS 84: For the definition of the WGS 84 equipotential ellipsoid of revolution the GRS 80 angular velocity is adopted (see chapter 3):

ω = 7292115   radians / second ;

whereas for precise satellite applications the value is to be applied

$\omega*$ = 7292115.8553 $* 10^{-11}$ + 4.3 $* 10^{-15} *$ TU   radians / second

with TU = Julian Centuries from Epoch J2000.0.

This value is sanctioned by the International Astronomical Union (IAU) and includes the precession rate.

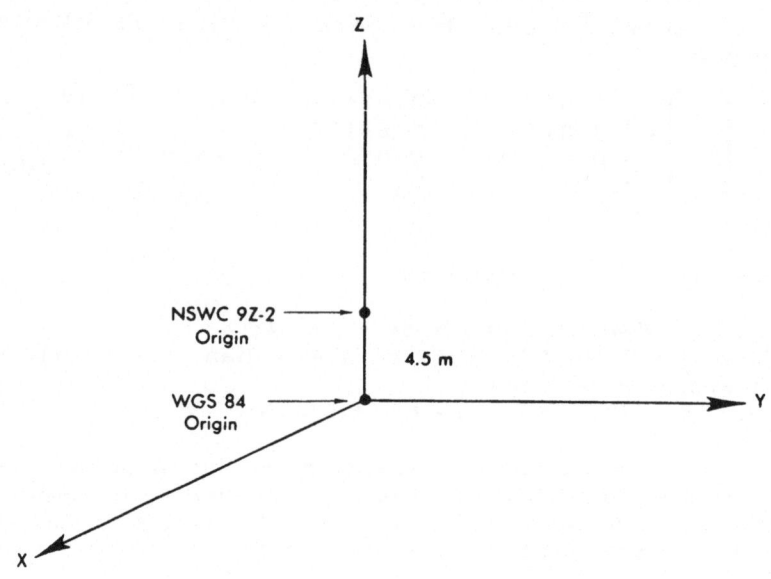

Figure 2: Difference between NSWC 9Z-2 and WGS 84 Reference Frame Origin (DMA 1987)

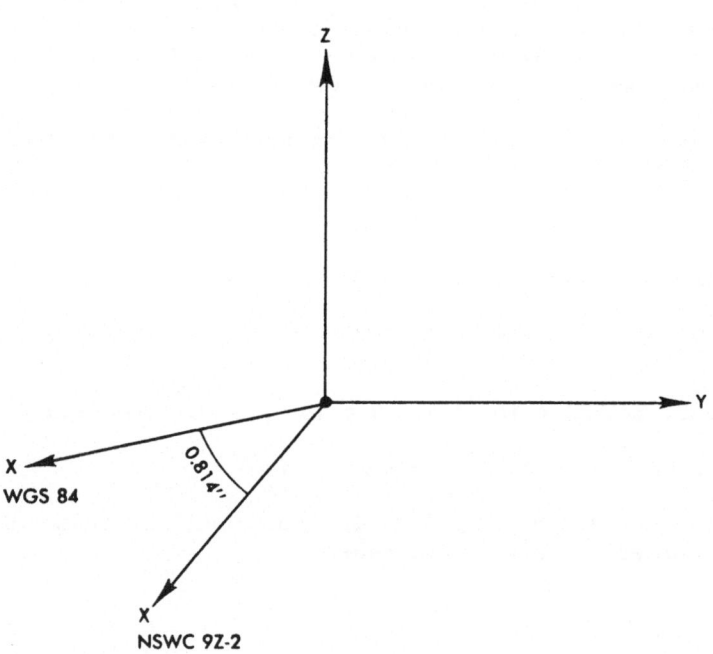

Figure 3: Differences between NSWC 9Z-2 and WGS 84 Longitude References (X-Axes) (DMA 1987)

3. WGS 84 Ellipsoid

The purpose of the WGS 84 Ellipsoid is not only to provide a geometric surface for horizontal positioning referencing but also and primarily to define a geocentric equipotential ellipsoid of revolution directly related to the WGS 84 earth gravity model (chapter 4). Of course, the origin and axis orientation of the WGS 84 Ellipsoid is identical with the WGS 84 coordinate system.

The parameters of the WGS 84 Ellipsoid are practically identical with the GRS 80 (Moritz 1980). The only exception is that the normalized second degree zonal gravitational coefficient $\overline{C}_{2,0}$ has been taken from the WGS 84 earth gravity model and, thus, slightly differs from the GRS 80 value. This discrepency ends up in a really insignificant difference in the semiminor axis of the corresponding ellipsoids:

$$b_{WGS\ 84} - b_{GRS\ 80} = +\ 0.1\ mm$$

The defining parameters of the WGS 84 Ellipsoid are listed in table 1.

4. WGS 84 Earth Gravitational Model (EGM)

The EGM and the directly corresponding WGS 84 geoid is perhaps the major improvement of WGS 84 vis-a-vis WGS 72. The WGS 84 EGM is a spherical harmonic expansion of the potential complete through degree and order 180. The coefficients through degree and order 18 are published, the others are CLASSIFIED.

As we have learned from Beutler et al. (1985), Landau and Hein (1986) for GPS satellite orbit calculations the EGM through degree and order 8 is sufficient even for highest demands.

5. WGS 84 relationships with other geodetic systems

For practical applications of WGS 84 with geodetic and geographic products the existence of conversion formulas and parameters from WGS 84 to conventional local geodetic datums e.g. North American Datum 1927 (NAD 27), European Datum 1950 (ED 50) is most important.

DMA has determined and published such parameters for 83 datums. They are derived from NNSS Doppler stations with given coordinates in WGS 84 (or predecessor NSWC 9Z-2) as well as in the respective local datums.

For NAD 27 and ED 50 such a great number of pass points have been available that a more sophisticated transformation formula than a 7-parameter transformation was considered to be appropriate: A Multiple Regression Equation has been chosen to perform the datum conversion in terms of geographic coordinates. In this manner, it was possible to account for some non-linear distortion in the local geodetic networks as well (e.g. figure 4 and 5).

Table 1: WGS.84 Ellipsoid, Defining Parameters (DMA 1987):

WGS 84 Ellipsoid
Four Defining Parameters

Parameters	Notation	Magnitude	Accuracy (1σ)
Semimajor Axis	a	6378137 m	±2 m
Normalized Second Degree Zonal Harmonic Coefficient of the Gravitational Potential	$\bar{C}_{2,0}$	$-484.16685 \times 10^{-6}$	$±1.30 \times 10^{-9}$
Angular Velocity of the Earth	ω	7292115×10^{-11} rad s^{-1}	$±0.1500 \times 10^{-11}$ rad s^{-1}
The Earth's Gravitational Constant (Mass of Earth's Atmosphere Included)	GM	3986005×10^{8} m^3 s^{-2}	$±0.6 \times 10^{8}$ m^3 s^{-2}
Parameter Values for Special Applications			
The Earth's Gravitational Constant (Mass of Earth's Atmosphere Not Included)	GM'	3986001.5×10^{8} m^3 s^{-2}	$±0.6 \times 10^{8}$ m^3 s^{-2}
Angular Velocity of the Earth (In a Precessing Reference Frame)	ω^*	$(7292115.8553 \times 10^{-11}$ $+ 4.3 \times 10^{-15}\ T_U)$ rad s^{-1}	$±0.1500 \times 10^{-11}$ rad s^{-1}

T_U = Julian Centuries From Epoch J2000.0

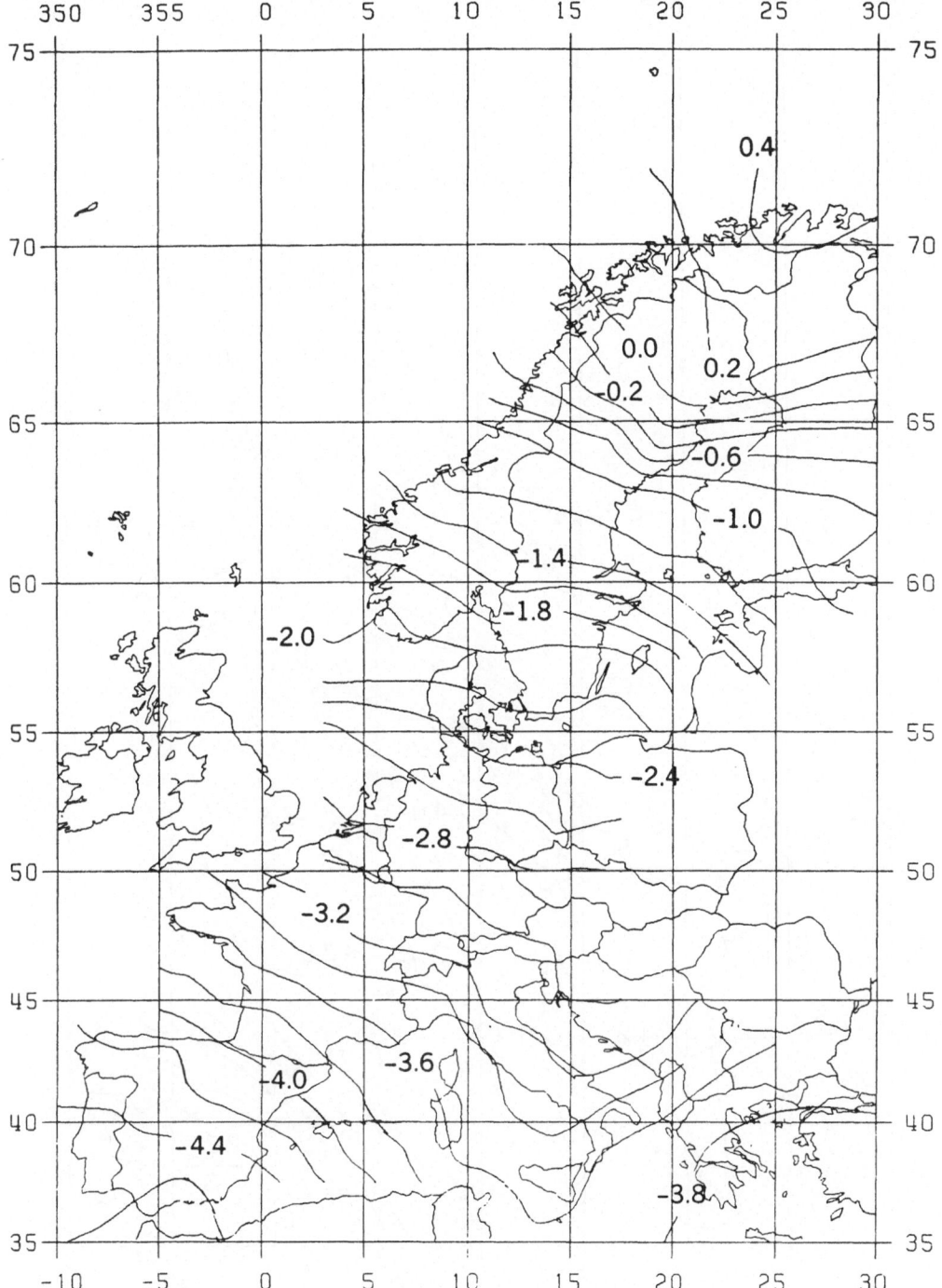

Figure 4: Latitude Differences WGS 84 minus ED 50,
Contour Interval = 0.2" (DMA 1987)

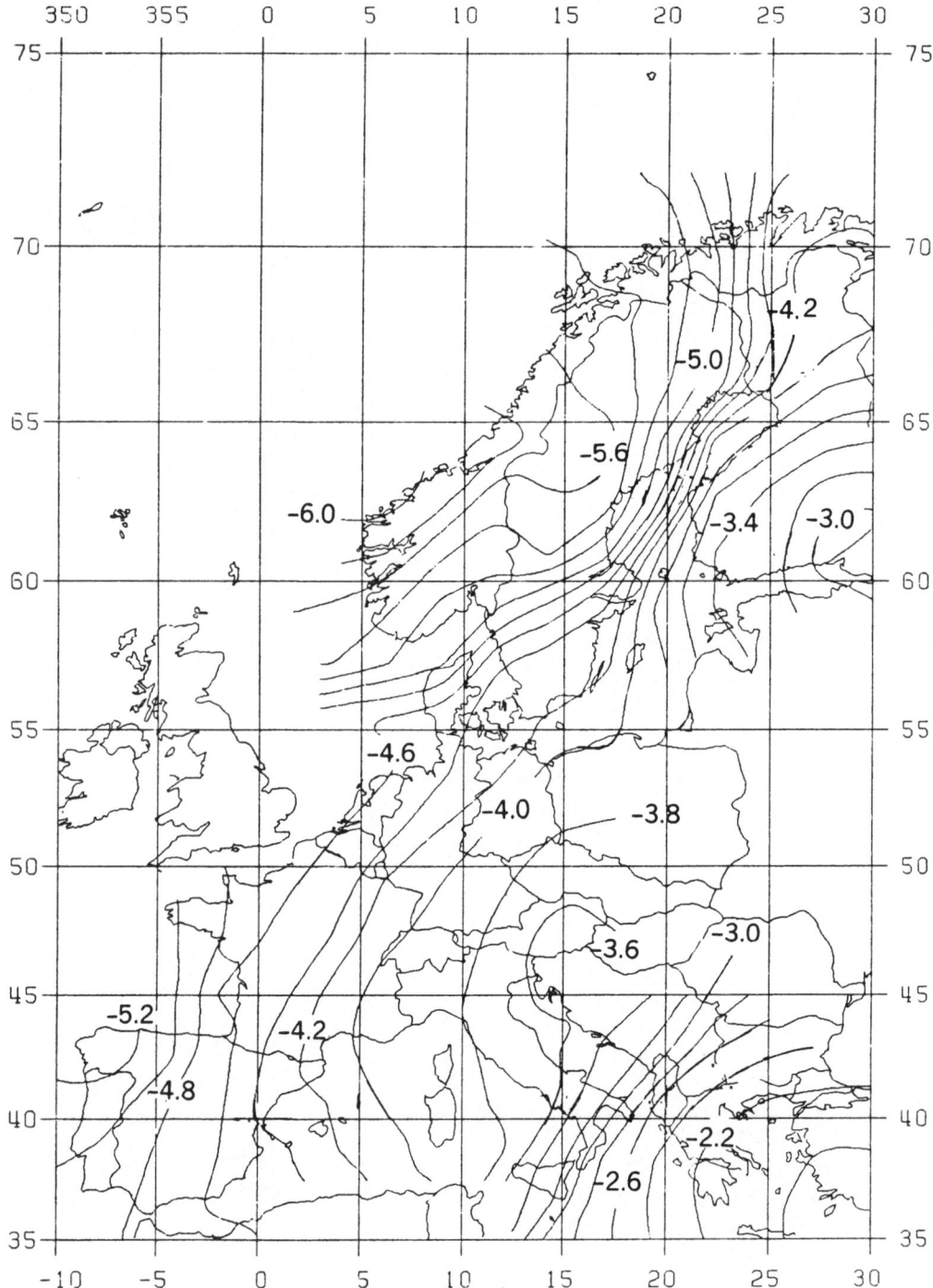

Figure 5: Longitude Differences WGS 84 minus ED 50,
Contour Interval = 0.2" (DMA 1987)

Of course, the accuracy of the resultant WGS 84 coordinates cannot be higher than the underlying Doppler point positioning results. Depending on the additional error sources, i.e. the residual distortions of the local networks, the absolute accuracy of WGS 84 transformation results are expected to be in the range of 2 to 5 meters.

6. WGS 84 – candidate as reference system for GPS applications in national geodetic control

It would be ideal to handle GPS measurements in national geodetic control networks in the same reference system as used for the satellite's coordinates, thus avoiding permanent datum transformations. In addition, this would be an opportunity to overcome the insufficiencies of the classical national reference systems (distortions, local scale changes). Furthermore, all data reduction could be performed in a three-dimensional cartesian system free of all hypothesis.

A precondition for this would be to start with the allocation of WGS 84 coordinates to some points of the net in an absolute sense. Thus, all consecutive GPS measurements should be connected relatively to this base-network.

The German Working Group of the Land Survey Administrations (AdV) has chosen this way (Strauß 1988) for their national geodetic control (DHDN).

The objectives of DMA are to ensure an absolute point positioning accuracy of 1m within WGS 84. DMA states that today the most accurate method of allocating WGS 84 coordinates is still an NNSS Doppler solution using NNSS precise ephemeris from DMA. When DMA's GPS precise ephemeris determination service is fully operational, this can also be achieved with GPS – in a shorter observation time, but not more accurately in the absolute sense.

Thus, especially for the determination of the base-network with GPS, high quality orbital elements in an absolute sense are required. The base-network will thus lay down the 1m uncertainty by definition for future applications. This first fundamental step can be called "realization of WGS 84 for the German DHDN".

For homogenity reasons, it would be highly desirable to perform this fundamental step not only in national areas but throughout the European continent.

7. References

Beutler, G., W. Gurtner, I. Bauersima and R. Langley (1985): Modelling and Estimating the Orbits of GPS Satellites. Proceedings First International Symposium on Precise Positioning with the Global Positioning System POSITIONING WITH GPS - 1985, Rockville, Maryland, pp. 99-111

Decker, L. (1986): World Geodetic System 1984, Proceedings of the Fourth International Geodetic Symposium on Satellite Positioning, Austin, Texas, pp. 69-92

DMA (1987): Department of Defense World Geodetic System 1984 - its Definition and Relationship with Local Geodetic Systems, DM Technical Report DMA TR 8350.2, Washington, DC

Hothem L. (1979): Determination of Accuracy, Orientation and Scale of Satellite Doppler Point Positioning Coordinates, Proceedings of the Second International Geodetic Symposium on Satellite Doppler Positioning, Austin, Texas, pp. 609-630

Jenkins, R.E. and C.F. Leroy (1979): "Broadcast" versus "Precise" Ephemeris - Apples and Oranges, Proceedings of the Second International Geodetic Symposium on Satellite Doppler Positioning, Austin, Texas, pp. 39-62

Landau, H. and G.W. Hein (1986): Preliminary Results of a Feasibility Study for a European GPS-Tracking Network. Proceedings of the Fourth International Geodetic Symposium on Satellite Positioning, Austin, Texas, pp. 337-353

Moritz, H. (1980): Geodetic Reference System 1980, Bulletin Geodesique Vol 54, pp. 395-405

Strauß, R (1988): Anwendung des Global Positioning System (GPS) in der Landesvermessung, Zeitschrift für Vermessungswesen 113, pp. 111-113

SEVERAL ASPECTS OF SOLAR RADIATION PRESSURE

by

Joachim Feltens

ABSTRACT

The disturbing influence of solar radiation pressure must be taken into account in modeling GPS orbits. The mathematical description of this disturbing force is complicated. There exist several models with distinct complexity. In the special case of GPS satellites there arise difficulties because of the irregular shape and surface of the satellites and by accounting for the alignment of the solar panels. These problems will be considered here. In connection with the alignment of the solar panels the orientation of the spacecraft coordinate system during the satellite's revolution will be examined under consideration of different positions of the Sun. In addition, possibilities to represent the irregular shape of the GPS satellites by parameters, which can be determined via orbit improvement, will be discussed.

1. Introduction

The magnitude of the acceleration caused by solar radiation pressure onto a GPS spacecraft is about $1 \cdot 10^{-7}$ m/s^2. Neglection of this force can cause errors of 1 km after 1-2 weeks of integration of the equations of motion.

This paper presents a simple model for the calculation of the solar radiation pressure effects onto a GPS satellite. Since construction- and reflection parameters of GPS satellites are only hardly obtainable, a model was designed, which allows to estimate some parameters, representing the spacecraft's feature, by an orbit improvement process without precise knowledge of the construction- and reflection parameters. Because of its use for an adjustment process, the model had to be kept as simple as possible.

The model is based on the purely geometric condition that the solar panel axis is permanently aligned perpendicular to the plane which is built by spacecraft, Sun and Earth (Fliegel et. al. 1985). Another aim of the conception of the model was a minimum need of external information (i.e. reflection coefficients etc.).

The model was numerically tested using some simple values, which might not be absolutely correct for a real GPS spacecraft. So the numerical results presented in this paper are only the results obtained by the calculations with these simple test values.

2. The spacecraft coordinate system

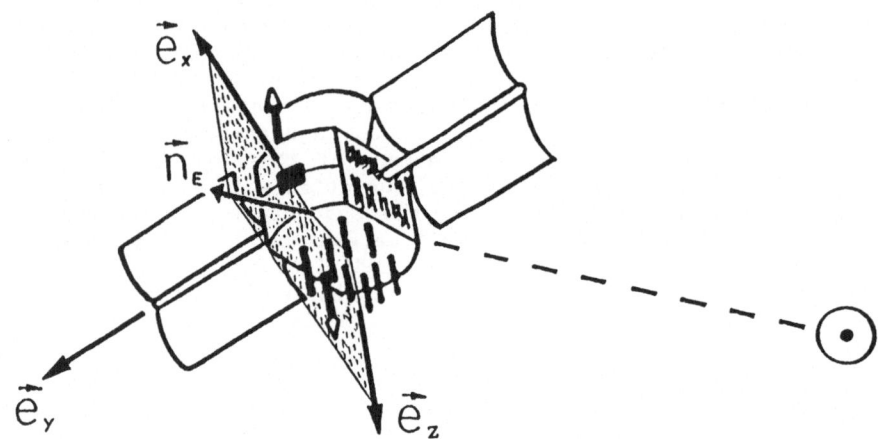

Figure 1. The spacecraft coordinate system

Before modeling the solar radiation pressure acceleration for a GPS satellite, the so called spacecraft coordinate system has to be considered (S.M.Lichten et. al. 1987, Fliegel et. al. 1985). The spacecraft coordinate system is satellite body-fixed and defined as follows (figure 1):

$\vec{e}_z$ - unit vector positive along the antenna directed towards the center of the Earth

$\vec{e}_y$ - unit vector along the solar panel support beam, normal to the spacecraft-Sun direction

$\vec{e}_x$ - unit vector perpendicular onto the other two axes in the sense of a right handed system

$\vec{n}_E$ - unit vector pointing from the Sun towards the spacecraft

During the satellite's revolution around the Earth the $\vec{e}_y$-axis is continuously aligned normal to the plane containing spacecraft, Sun and Earth (Fliegel et. al. 1985). The solar panels are then rotated around the $\vec{e}_y$-axis, always presenting their maximum surface area to the Sun. The vectors $\vec{e}_z$, $\vec{e}_x$ and $\vec{n}_E$ always lying in the plane satellite-Sun-Earth.

3. The spacecraft's changing orientation during its revolution around the Earth

The changing orientation of the spacecraft coordinate system with respect to the inertial reference frame is shown in figures 2a-2d. With respect to the Earth-Sun direction the orbit plane is subdivided into four quadrants. Point 1 is the Sun-nearest and point 3 is the Sun-farest orbit point. In these two points the plane satellite-Sun-Earth stands perpendicularly to the orbit plane. Because of the Sun's apparent revolution around the Earth, the quadrants are also being slowly moving. The angle γ, enclosed by the unit vectors $\vec{e}_z$ and $\vec{n}_E$, reaches its minimum value γ_{min} at point 1 and its maximum value γ_{max} at point 3. Assuming the vector $\vec{n}_E$ having the same direction in all orbit points, γ_{min} can easily be determined by:

$$\sin(\gamma_{min}) = |\vec{n}_E \cdot \vec{n}_o|$$

where $\vec{n}_o$ is the unit vector standing normal onto the orbit plane

$$\gamma_{max} = 180^0 - \gamma_{min}$$

Explanations to figures 2a-2d:

Figure 2a. The Sun stands below the orbit plane

- $\gamma_{min} \leq \gamma \leq \gamma_{max}$, $0^0 < \gamma_{min} < 90^0$, $90^0 < \gamma_{max} < 180^0$
- $\vec{e}_z$ always lying in the orbit plane pointing towards the Earth.
- $\vec{e}_y$ pointing above the orbit plane in the quadrants I and IV and below the orbit plane in the quadrants II and III. The projection of $\vec{e}_y$ onto the orbit plane permanentely pointing into the moving direction of the satellite.
- $\vec{e}_x$ always having a direction pointing above the orbit plane. In the orbit points 2 and 4 is $\vec{e}_x = \vec{n}_E$.

Figure 2b. The Sun stands above the orbit plane

- $\gamma_{min} \leq \gamma \leq \gamma_{max}$
- $\vec{e}_z$ always lying in the orbit plane pointing towards the Earth.
- $\vec{e}_y$ still pointing above the orbit plane in the quadrants I and IV and below the orbit plane in the quadrants II and III. But the projection of $\vec{e}_y$ onto the orbit plane now having the opposite direction than the spacecraft moves.
- $\vec{e}_x$ now always having a direction pointing below the orbit plane.

Figure 2c. The Sun is in the orbit plane

This event happens twice a year, when the Sun crosses the orbit plane during its apparent revolution around the Earth.
- $\gamma_{min} = 0^0$, $\gamma_{max} = 180^0$, $0^0 \leq \gamma \leq 180^0$
- $\vec{e}_z$ always pointing towards the Earth.
- $\vec{e}_y$ always being perpendicular to the orbit plane. In the quadrants I and IV $\vec{e}_y$ pointing above the orbit plane and in the quadrants II and III $\vec{e}_y$ pointing below the orbit plane.
- $\vec{e}_x$ permanently lying in the orbit plane.

In the orbit points 1 and 3, where γ is 0^0 and 180^0, respectively, the vectors $\vec{e}_x$ and $\vec{e}_y$ are not defined. In these two points $\vec{e}_y$ can arbitrarily be rotated around $\vec{e}_z$ always remaining perpendicularly onto $\vec{n}_E$. (The singularity at orbit point 3 ($\gamma = 180^0$) is irrelevant, because the spacecraft in this position stands in the Earth's shadow.)

Figure 2d. The Sun stands perpendicularly onto the orbit plane

- $\gamma = \gamma_{min} = \gamma_{max} = const \approx 90^0$
- $\vec{e}_z$ always pointing towards the Earth.
- $\vec{e}_y$ permanently lying in the orbit plane pointing towards the satellite's moving direction and, if the Sun stands above the orbit plane, against the satellite's moving direction.
- $\vec{e}_x \approx \vec{n}_E$ permanently

4. Modeling the solar radiation pressure acceleration

The direct solar radiation pressure acceleration $\vec{a}$ can be described in a first approximation by the following expression:

$$\vec{a} = k \cdot c_R \cdot p \cdot \left(\frac{a_E}{r_E}\right)^2 \frac{A}{m} \cdot \vec{n}_E \tag{1}$$

where: k - shadow factor; $k = 1$ for direct sunlight,
$k = 0$ for umbra and
$0 < k < 1$ for penumbra

c_R - reflectivity constant; depending on the spacecraft's surface features

p - solar pressure; $p = \dfrac{I}{c} \left[\dfrac{N}{m^2}\right]$

I - Intensity of radiation; its magnitude depends on the Sun's distance with respect to the Earth and on solar activity. It can be approximated by the following formula (Wakker et. al. 1983):

$$I = \frac{1358}{1.0004 + 0.0334 \cdot \cos(D)} \left[\frac{W}{m^2}\right] \tag{1a}$$

D - is the phase of the year measured from the 4th of July, when the Earth passes its aphelion
c - velocity of light
a_E - semimajor axis of the Earth's orbit around the Sun ($a_E \approx 1$ AU)

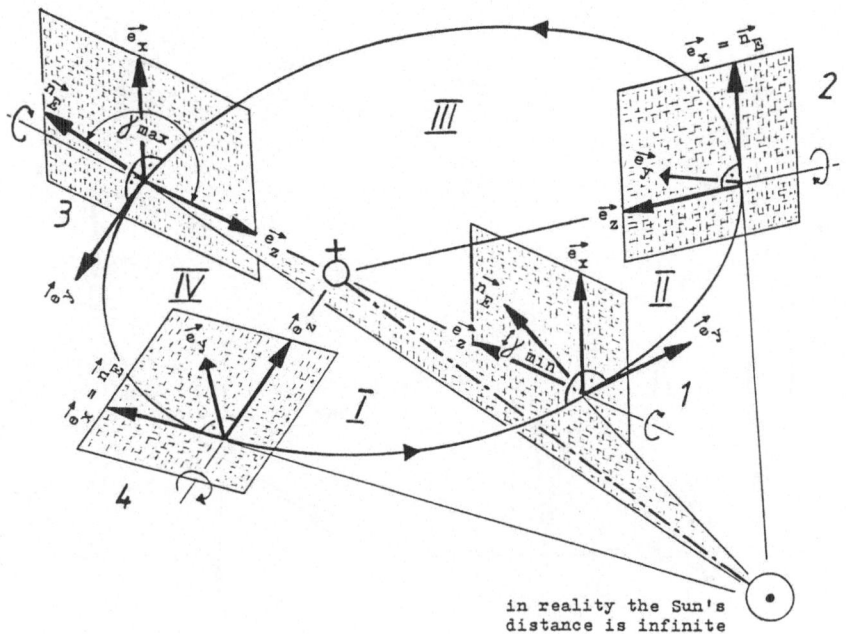

Figure 2a. The Sun stands below the orbit plane

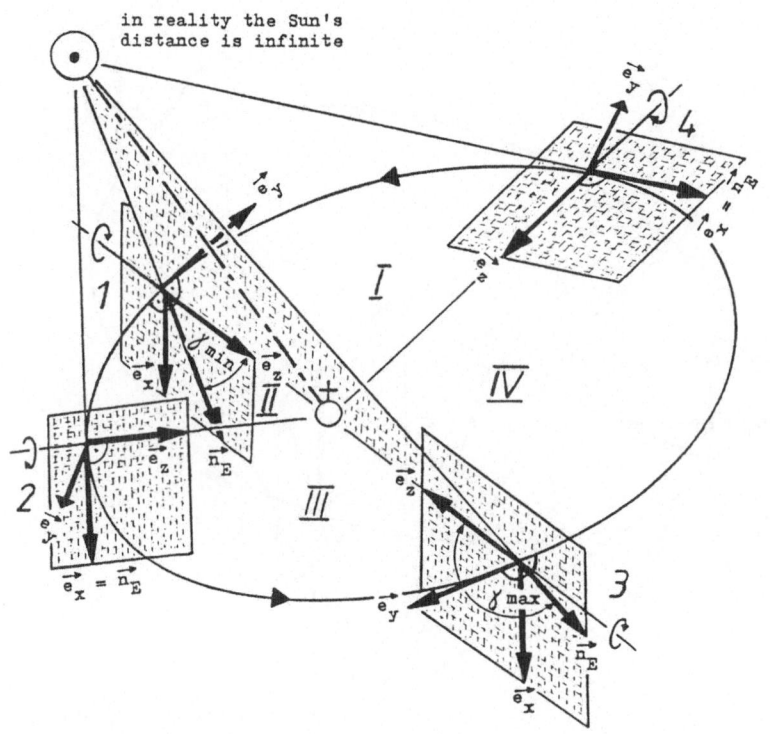

Figure 2b. The Sun stands above the orbit plane

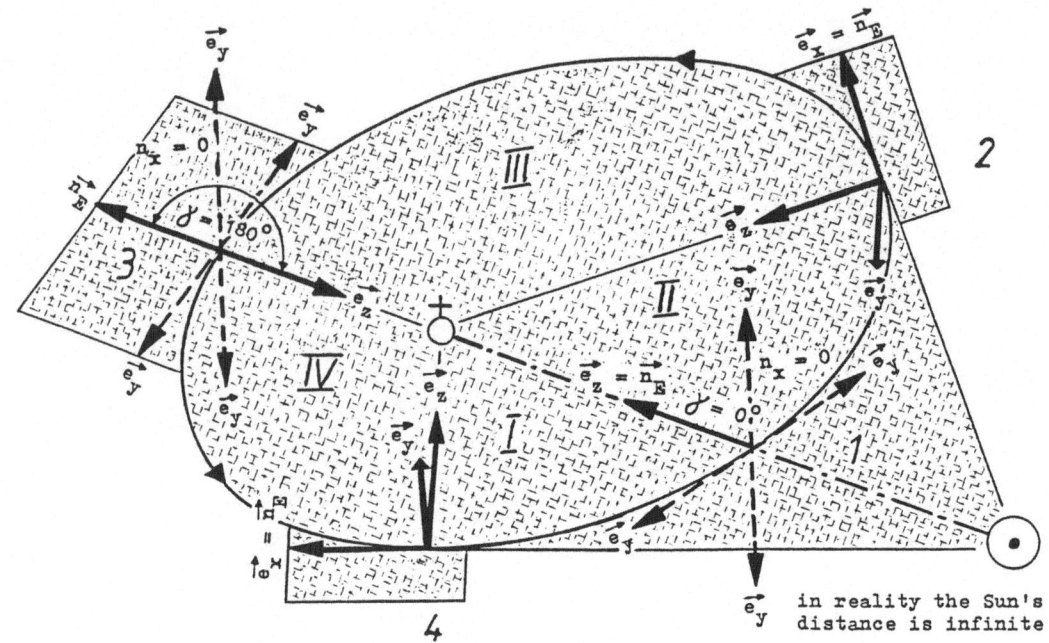

Figure 2c. The Sun is in the orbit plane

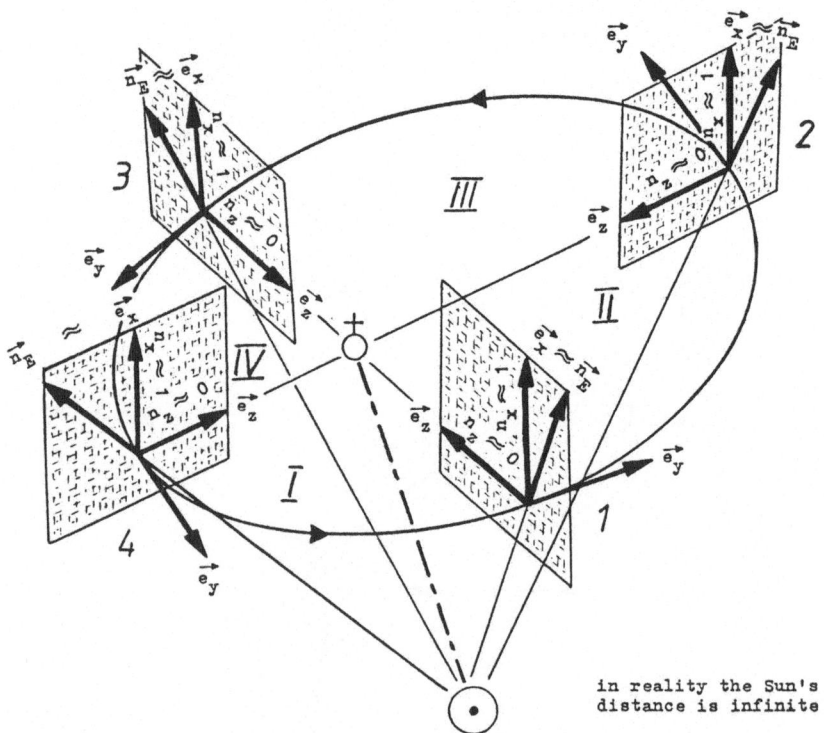

Figure 2d. The Sun stands perpendicular onto the orbit plane

r_E - distance between spacecraft and Sun
A - effective cross sectional area of the spacecraft
m - the spacecraft's mass
$\vec{n}_E$ - heliocentric unit vector (as explained in section 1)

The components of the disturbing acceleration $\vec{a}$ will be expressed now with respect to the spacecraft coordinate system. Since the projection of $\vec{n}_E$ onto the $\vec{e}_y$-direction is zero, $\vec{n}_E$ only having components in the $\vec{e}_z$- and the $\vec{e}_x$-direction:

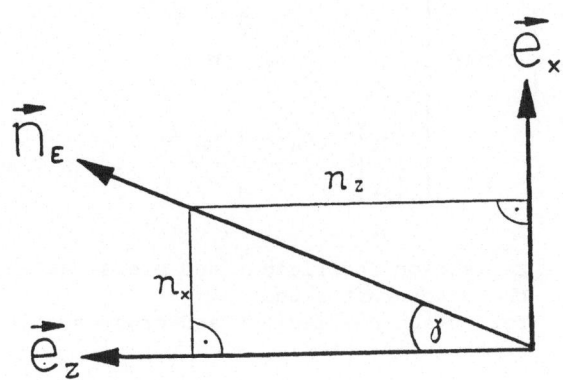

Figure 3. The components of $\vec{n}_E$ in the spacecraft coordinate system.

$$\vec{n}_E = \left\{ \begin{array}{l} n_x = \sin(\gamma) \\ n_y = 0 \; ; \; since \; \vec{e}_y \perp \vec{n}_E \\ n_z = \vec{e}_z \cdot \vec{n}_E = \cos(\gamma) \end{array} \right\} \tag{2}$$

Since $n_y = 0$, also the component a_y of the disturbing acceleration should be zero. But because of effects, such like misalignments of the solar panels etc., there also is a constant acceleration into the $\vec{e}_y$-direction, called the Y-bias (Fliegel et. al. 1985). The Y-bias acceleration can be described by the relation:

$$Y = 2 \cdot \overline{c}_R \cdot p \cdot \frac{\overline{A}}{m} \cdot Q \tag{3}$$

where: $\overline{c}_R$ - reflectivity constant of the solar panels
$\underline{p}$ - solar pressure
$\overline{A}$ - area of the solar panels
m - the spacecraft's mass
Q - sum of the angles d_1, d_2 and d_3:

$$Q = d_1 + \frac{d_2}{2} + d_3$$

d_1 - misalignment angle of the solar sensor
d_2 - angle of one solar panel with respect to the other
d_3 - yaw altitude control bias

Inserting (2) and (3) into (1) and assuming distinct values for the reflection coefficients for the solar panels and for the spacecraft's body yields in expression (4), which is related to the spacecraft coordinate system:

$$
\vec{a}(t) = \left\{ \begin{array}{c} a_x \\ a_y \\ a_z \end{array} \right\} = k \cdot p \cdot \left\{ \begin{array}{c} \left(\dfrac{a_E}{r_E}\right)^2 \cdot \dfrac{c_R \cdot A + \overline{c}_R \cdot \overline{A}}{m} \cdot \sin(\gamma(t)) \\[3mm] 2 \cdot \overline{c}_R \cdot \dfrac{\overline{A}}{m} \cdot Q \\[3mm] \left(\dfrac{a_E}{r_E}\right)^2 \cdot \dfrac{c_R \cdot A + \overline{c}_R \cdot \overline{A}}{m} \cdot \cos(\gamma(t)) \end{array} \right\} \qquad (4)
$$

where : c_R, A - reflection coefficient and cross sectional area of the spacecraft's body
$\overline{c}_R$, $\overline{A}$ - reflection coefficient and cross sectional area of the solar panels

For numerical integration purposes (4) has to be transformed into a corresponding expression related to the inertial reference frame. Starting from the simple vector relation

$$
\vec{a} = a_x \cdot \vec{e}_x + a_y \cdot \vec{e}_y + a_z \cdot \vec{e}_z \qquad (5)
$$

this can be achieved immediately by exchanging the components of the unit vectors $\vec{e}_x$, $\vec{e}_y$ and $\vec{e}_z$ related to the spacecraft coordinate system by their corresponding components connected to the inertial reference frame (table 1). Using the expressions in the left column of table 1 in (5) results in (4), inserting the relations given in the right column in (5) yields the expression referring to the inertial reference frame:

$$
\vec{a}(t) = \left\{ \begin{array}{ccc} e_{xx}(t) & e_{yx}(t) & e_{zx}(t) \\ e_{xy}(t) & e_{yy}(t) & e_{zy}(t) \\ e_{xz}(t) & e_{yz}(t) & e_{zz}(t) \end{array} \right\} \cdot \left\{ \begin{array}{c} a_x \\ a_y \\ a_z \end{array} \right\} \qquad (6)
$$

$$
\begin{array}{ccc} \updownarrow & \updownarrow & \updownarrow \\ \vec{e}_x(t) & \vec{e}_y(t) & \vec{e}_z(t) \end{array}
$$

Table 1. The unit vectors related to the spacecraft coordinate system and to the inertial reference frame ((t) stands for time dependence).

spacecraft coordinate system	inertial refence frame
$\vec{e}_z = (0,0,1)$	$\vec{e}_z(t) = - \dfrac{\vec{r}(t)}{|\vec{r}(t)|}$
$\vec{e}_y = (0,1,0)$	$\vec{e}_y(t) = \dfrac{\vec{e}_z(t) \times \vec{n}_E(t)}{|\vec{e}_z(t) \times \vec{n}_E(t)|}$
$\vec{e}_x = (1,0,0)$	$\vec{e}_x(t) = \dfrac{\vec{e}_y(t) \times \vec{e}_z(t)}{|\vec{e}_y(t) \times \vec{e}_z(t)|}$
	$= \dfrac{\vec{n}_E(t) - \vec{e}_z(t) \cdot \cos(\gamma(t))}{|\vec{n}_E(t) - \vec{e}_z(t) \cdot \cos(\gamma(t))|}$
$\vec{n}_E(t) = (\sin(\gamma(t)),0,\cos(\gamma(t)))$	$\vec{n}_E(t) = \dfrac{\vec{r}(t) - \vec{r}_\odot(t)}{|\vec{r}(t) - \vec{r}_\odot(t)|}$

$\vec{e}_x$ and $\vec{e}_y$ are not defined for $\gamma = 0^0$ and for $\gamma = 180^0$. $\vec{r}(t)$ and $\vec{r}_\odot(t)$ are the spacecraft's and the Sun's geocentric position-vector, respectively.

5. Determination of the effective cross sectional area - numerical calculations

5.1. The assumed shape and size for a GPS spacecraft

For testing the solar radiation pressure acceleration model, a reference orbit was created, assuming the strongly simplified model for the shape of a GPS satellite as shown in figure 4.

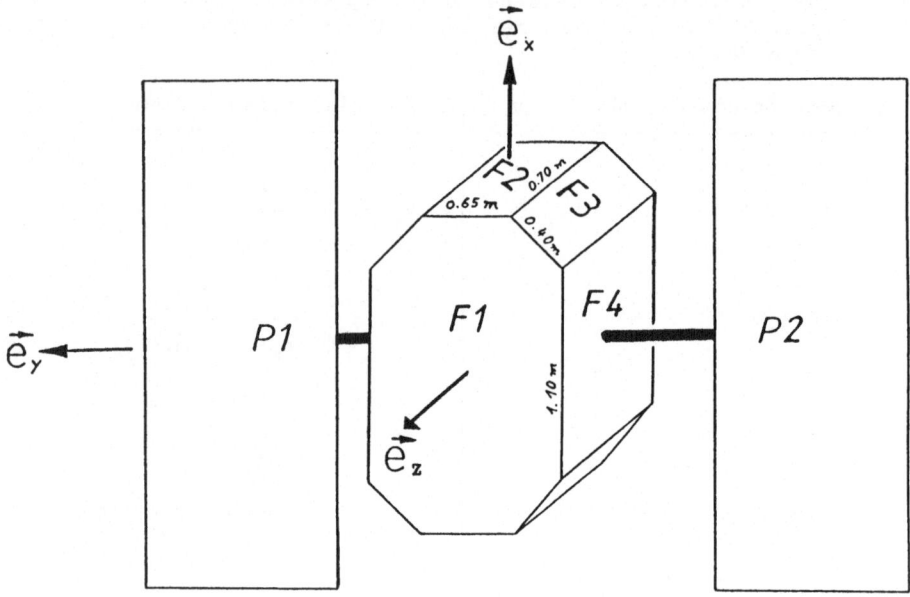

Figure 4. The assumed shape and dimensions of a GPS spacecraft.

Only flat surfaces were included and round shapes, i.e. the apogee
kick motor etc., were neglected. The numerical values being used,
might not be fully correct for a real GPS satellite, but for the
numerical tests of the solar radiation pressure acceleration model it
was supposed, that these values would be sufficient:

    - the areas forming the spacecraft's body: $F1 = 1.9$ m^2
                                                 $F2 = 0.5$ m^2
                                               $F3 = 0.3$ m^2
                                               $F4 = 0.8$ m^2
    - the area of the solar panels: $\bar{A} = P1 + P2 = 8.0$ m^2
    - the spacecraft's mass: $m = 815$ kg
    - reflectivity constant of the
      spacecraft's body: $c_R = 1.5$
    - reflectivity constant of the
      solar panels: $\bar{c}_R = 0.2$

5.2. Creating the reference orbit

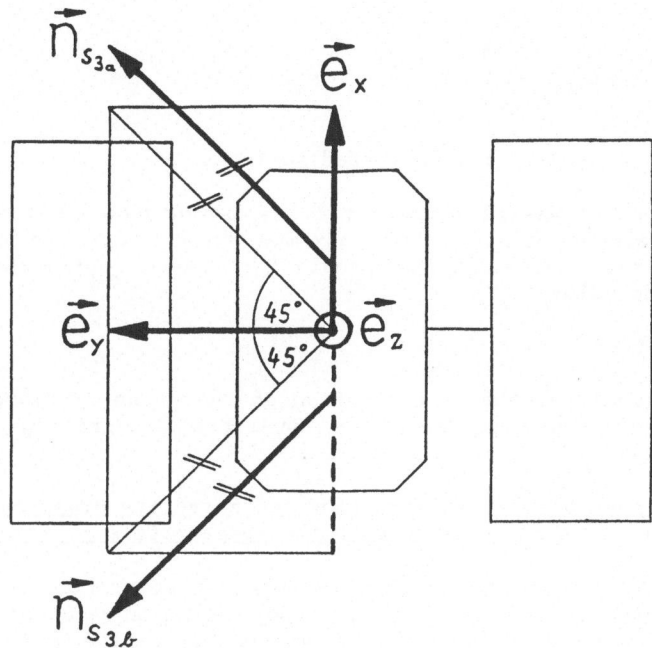

Figure 5. The surface normal vectors

For creating the reference orbit, the effective cross sectional area of the spacecraft's body was modeled in the following way: The effective area of a surface Fi of the satellite's body is given by the expression:

$$Fi_{eff} = Fi \cdot |\vec{n}_{Si} \cdot \vec{n}_E| \qquad ; \qquad i = 1, \ldots, 4 \qquad (7)$$

where: Fi_{eff} - effective area of the surface Fi
$\vec{n}_{Si}$ - unit vector standing normal onto the surface Fi

If the dot product is positive, the corresponding surface Fi lies on the shadow side of the spacecraft and might not be considered. For most of the surfaces Fi of the spacecraft's body the surface normal vectors $\vec{n}_{Si}$ are identical with the unit vectors $\vec{e}_x$, $\vec{e}_y$ and $\vec{e}_z$.
In the case of surface F3 the normal vectors can easily be built from these unit vectors. For simplicity it was assumed, that F3 encloses an angle of 45° with the surfaces F2 and F4. For the surface normal vectors the following expressions hold:

F1: $\quad \vec{n}_{S1} = \pm \vec{e}_z$ $\qquad\qquad$ F2: $\quad \vec{n}_{S2} = \pm \vec{e}_x$

F3: $\quad \vec{n}_{S3a} = \pm \dfrac{\vec{e}_x + \vec{e}_y}{|\vec{e}_x + \vec{e}_y|}$, $\qquad \vec{n}_{S3b} = \pm \dfrac{-\vec{e}_x + \vec{e}_y}{|-\vec{e}_x + \vec{e}_y|}$ $\qquad$ (7a)

F4: $\quad \vec{n}_{S4} = \pm \vec{e}_y$; since $\vec{e}_y \perp \vec{n}_E$ permanently, F4 will never be shone on by the Sun.

497

The effective cross sectional area of the whole spacecraft's body then is the sum of the effective areas of all surfaces Fi shone on by the Sun:

$$A = \sum_{i=1}^{n} Fi_{eff} = \sum_{i=1}^{n} Fi \cdot | \vec{n}_{Si} \cdot \vec{n}_E | \qquad (8)$$

where: n - number of illuminated surfaces

Since the solar panels permanently are orientated in such a manner, that they always present their maximum surface area to the Sun, the effective area $\overline{A}$ of the solar panels was assumed permanently having the constant value of

$$\overline{A} = 8.0 \text{ m}^2$$

With this model for the determination of the cross sectional area, the reference orbit was generated by numerical integration over a time span of 10 days.

5.3. Other possible ways of modeling the effective cross sectional area and their influence onto the spacecraft's position accuracy

The orbit was integrated again assuming a constant effective cross sectional area for the spacecraft's body having a value of $A = 1.5 \text{ m}^2$. This is a mean value of the time varying effective areas which were obtained from several steps of numerical integration of the reference orbit. Then the spacecraft's positions obtained from the reference orbit were compared with the corresponding positions which resulted from the integration making the assumption of a constant effective area. As figure 6 shows, the position deviation with respect to the reference orbit grows up to 50 m in the along track component. In the out of plane direction the deviation from the reference orbit is nearly zero, since in the used test dataset the Sun stood nearby the orbit plane.

Next the effective cross sectional area was approximated by a trigonometric expression. It was assumed that the spacecraft in certain periods would present the same side to the Sun. Thereby the time period would be dependent on the spacecraft's shape and on the revolution time. Since it was assumed (figure 4), that opposite sides of the spacecraft are identical, the satellite would present approximately the same size every half of its revolution time to the Sun. This would result in a time period of about 6 hours for a GPS spacecraft. For a real GPS satellite, having a totally irregular shaped form, the same size would be presented to the Sun approximately every full revolution period. Therefore, in this case, the time period would be about 12 hours.

The time dependent expression for modeling the effective area by a trigonometric expression was:

$$A(t) = A_0 + A_1 \cdot \cos\{\beta \cdot (t-t_0)\} + B_1 \cdot \sin\{\beta \cdot (t-t_0)\} \qquad (9)$$
$$+ A_2 \cdot \cos\{2 \cdot \beta \cdot (t-t_0)\} + B_2 \cdot \sin\{2 \cdot \beta \cdot (t-t_0)\}$$

$$\beta = \frac{2 \cdot \pi}{\Delta T}$$

constant effective area – position error

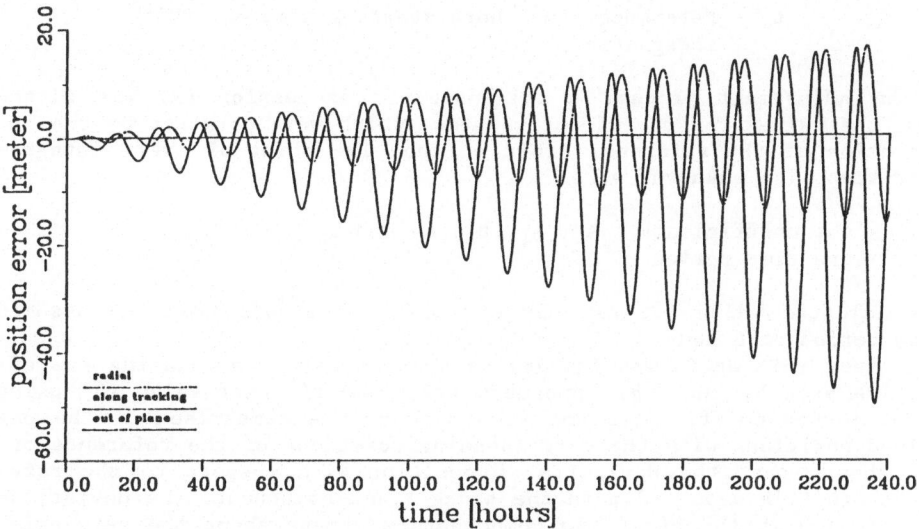

Figure 6. The position deviations between the reference orbit and the orbit integrated by assuming a constant effective area.

trigonometric expression – position error

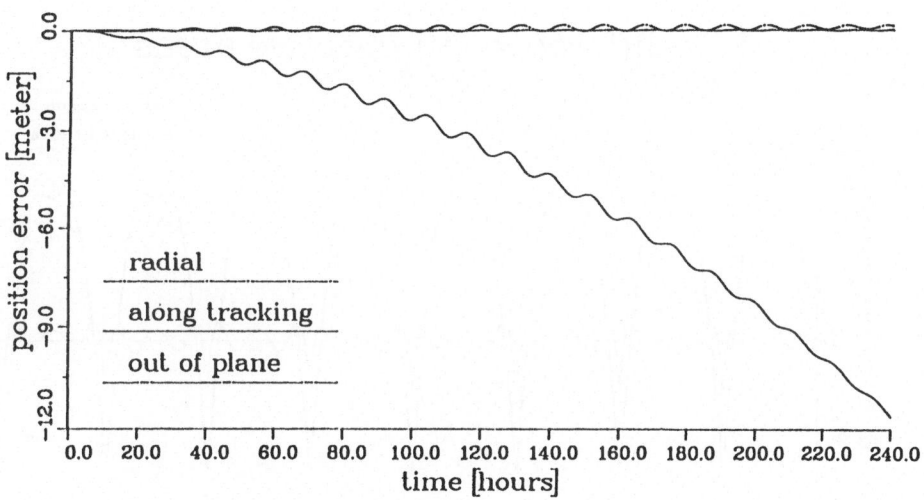

Figure 7. The position deviations between the reference orbit and the orbit integrated by approximating the effective area by a trigonometric expression.

ΔT - time period
t - actual time
t_0 - reference time (here starting time of orbit integration)

By an adjustment process a trigonometric expression (9) was fitted into the values for the effective area, which were obtained by the integration of the reference orbit at the time point of every integration step. The unknowns were:

- the coefficients A_0 , A_1 , B_1 , A_2 and B_2
- the time period   ΔT.

The adjusted value for ΔT only deviated 63 seconds from the assumed time period of 6 hours.
Then the orbit again was integrated over 10 days, calculating the effective area by the time dependent trigonometric expression (9) using the adjusted coefficients and time period. The comparison of the obtained positions with the corresponding positions of the reference orbit shows, that the deviation of position with respect to the reference orbit is about 12 m in the along track component. The deviations of position in the other two components stagnate about the zero-axis. So the deviation of position could be reduced to about 30 % of its former value, by approximating the effective area by a time dependent trigonometric expression.
In figure 8 the differences between the effective areas of the reference orbit and the approximations are shown. It can be seen, that the effective area obtained by a trigonometric expression only deviates about one third of the amount, the constant effective area deviates.

differences in effective areas

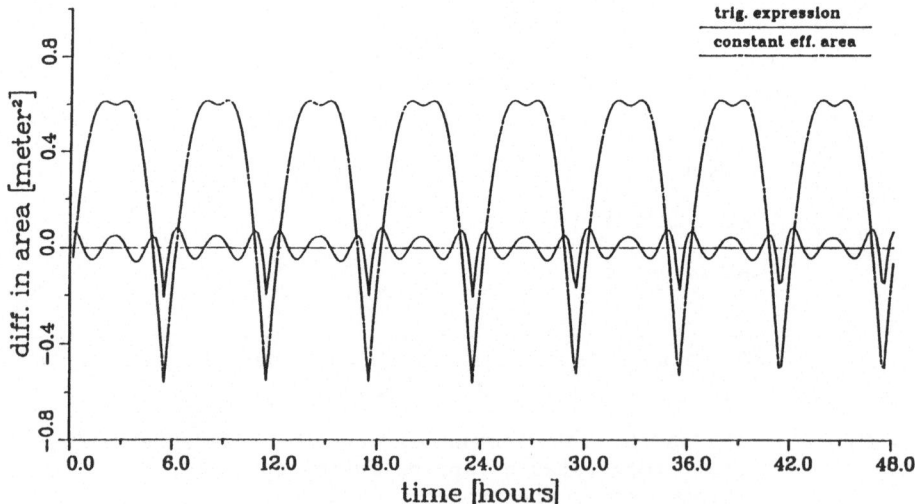

Figure 8. Differences between the correct effective area of the reference orbit and the approximation values.

6. Conclusions

- The changing orientation of the spacecraft coordinate system can easily be described by simple vector relations. The same vectors also can be used for modeling the effective area of an irregularly shaped satellite body.

- The orbit obtained by approximating the effective area by a trigonometric expression deviates in position from the reference orbit by an amount of 30 % of the values, for which the orbit, integrated with the assumption of a constant effective area, deviates.

Table 2. The reduction of position deviation in dependence of the degree of the used trigonometric expression in relation to the deviations of the orbit integrated by assuming a constant effective area.

degree of the trigonometric expression	deviation of position reduced to:
1	50 %
2	30 %

7. Further plans

In a next step it is intended to determine the coefficients of the trigonometric expression from real GPS phase data for real GPS satellites by an orbit improvement process. Similar to the Y-bias, every GPS spacecraft then could get its own set of coefficients, describing its effective cross sectional area. The values of the coefficients might be valid for about one week. Then a new set of coefficients might be determined.
But there still remain some difficulties when applying the model, as it was presented in this paper, onto real GPS satellites. The number of unknown parameters has to be kept as small as possible. In an orbit improvement process also other parameters have to be determined. If the number of unknowns is too large, the obtained values, especially for the parameters which come from small disturbing forces like solar radiation pressure, will not be significant. Therefore the number of unknowns in the trigonometric expression has to be reduced. So the trigonometric expression might only be expanded up to the first degree and the value of the time period could be built as the sum of the revolution time plus a time correction for the Sun's apparent movement during the spacecraft's revolution, and then held fixed.
Another problem is the consideration of the variation of the reflection coefficients. In the described model the reflection coefficient c_R of the whole spacecraft's body was kept constant having a value of 1.5 and only the effective area A was treated as time variable. But in reality the reflection coefficient also varies with time, since the radiated surfaces vary. Therefore, at an application to real GPS satellites, not only the effective area A but the product $c_R \cdot A$ should be approximated as a whole by the trigonometric expression, because the

reflectivity coefficient c_R changes with the same time period as the effective area A does.
So, im summary, it is obviously possible to improve the spacecraft's position accuracy by modeling the effective cross sectional area by using a trigonometric expression. But still tests and improvements of the current model are necessary to get a simple, but efficient model for the easy calculation of solar radiation pressure effects.

8. References

Fliegel, H.F., W.A. Feess, W.C. Layton and N.W. Rhodus (1985): The GPS radiation force model, Proceedings of the First International Symposium on Precise Positioning with the Global Positioning System, Vol. 1, pp. 113-119

Lichten, S.M. and J.S. Border (1987): Strategies for High-Precision Global Positioning System Orbit Determination, Journal of Geophysical Research, Vol. 92, No. B12, pp. 12751-12762

Wakker, K.F., B.A.C. Ambrosius and L. Aardoom (1983): Precise Orbit Determination for ERS-1, Delft University of Technology, Department of Aerospace Engineering, Section Orbital Mechanics, ESA Contract Report, pp. 84-85

TOPAS
A NEW GPS ADJUSTMENT SYSTEM
FOR MULTISTATION POSITIONING
AND ORBIT DETERMINATION

by

Herbert Landau

Abstract

A new software package is presented for positioning with the Global Positioning System in a multistation/multisession network approach. The program allows the determination of satellite orbits, clock and atmospheric parameters and terrestrial positions by using raw undifferenced carrier phase and code measurements in a sequential approach. All parameters are modelled and determined in an optimal filtering/smoothing technique based on U–D factorization methods resulting in precise terrestrial positions and satellite orbits.
The paper describes the software product and the method principally applied. Results of applications of TOPAS to different networks (DOENAV and GINFEST) are presented showing single–day repeatablities of about 5 cm in all components for baselines of 100 to 1600 km length. Comparisons between results obtained by using broadcast and precise ephemeris data and results derived from measurement improved orbital information are made. This shows the influence of the orbit improvement technique used by TOPAS on precise point positioning. Repeatability analysis of multi–day solutions results in accuracies of 0.01 to 0.03 ppm for baselines of 600 to 700 km length.

The presented material will be published in

Landau,H.(1988): *Zur Nutzung des Global Positioning Systems in Geodäsie und Geodynamik: Modellbildung, Softwareentwicklung und Analyse.* Ph.D. Thesis (in preparation). Institute of Astronomical and Physical Geodesy, University FAF, Munich

COVARIANCES IN 3D NETWORK RESULTING FROM
ORBITAL ERRORS

by

Janusz B. Zielinski

Abstract

The influence of orbital errors of GPS geodetic network may became significant if the measured distances are greater than 50 km. The proper input covariance matrix is then required for the adjustment. In the paper the numerical method for finding this covariance matrix is described, based on the principle of randomization of errors of the model parameters. Some examples of application of this method are given for estimation of correlations between the baselines and for construction of the variance-covariance matrix for the 5 stations test network.

1.Introduction.

The method often used in practice of the application of GPS to the network construction consists in the determination of the individual vectors and later combining them in a 3D network cf.*Bock et al.,(1985)*. In such solution the orbital elements are assumed as known (either from satellite message or from external sources). Errors of the assumend values of orbital elements not only produce errors of primary results (coordinates of vectors) as it was demonstrated in *Zielinski,(1987)*, but they also provoke correlations between differend parts of the network. Therefore it is necessary to find the proper covariance matrix which may be used in the adjustment process. Theoretical aspects of this problem were discussed by *Schaffrin,Bock,(1988)* and *Schaffrin,Zielinski,(1988)*, and they can be resumed as following:

Usual Gauss-Markov modeling of linearized observations is

$$Y = AX + V \qquad (1)$$

$$D(Y) = C_V \qquad (2)$$

where X - vector of coordinate increments
V - random errors vector
C_V- covariance matrix of measurements.

The solution is

$$X = (A^T C_V^{-1} A)^{-1} A^T C^{-1} Y \qquad (3)$$

Now we make an assumption that the measurement errors can be splitted in two parts

$$V = V_S + V_N \qquad (4)$$

where V_S - nonrandom part steming from errors in prior information (e.g. orbital elements)

$$V_S = f(S) \qquad (5)$$

V_N - random part steming from measured noises (normally distributed), eventually transformed by linear transformation (e.g.differencing)

$$V_N = g(N) \qquad (6)$$

Prior information errors S and random noises N have the dispersion respectively

$$D(S) = C_S \qquad (7)$$

$$D(N) = C_N = \sigma_o^2 I \qquad (8)$$

I - unit matrix.

If we denote the Jacobian of the prior information $\overset{o}{w}$ by

$$F = \frac{\partial Y}{\partial \overset{o}{w}} \qquad (9)$$

and of the original measurements z by

$$G = \frac{\partial Y}{\partial z} \qquad (10)$$

we find the dipersion of Y:

$$D(Y) = C_V = D(V_S) + D(V_N) = FC_S^{-1}F^T + GC_N^{-1}G^T \qquad (11)$$

which can be used in eq.(3). More details can be found in *Schaffrin,Zielinski,(1988)*.

In the present work the method is described of the determination of the matrix

$$D(V_S) = FC_S^{-1}F^T \qquad (12)$$

which reflects the <u>correlations</u> <u>originating</u> <u>from</u> <u>the</u> <u>prior</u> <u>information</u> <u>errors</u>. Numerical simulation approach was applied, nevertheless the method is of general character.

2. <u>Numerical</u> <u>simulation</u> <u>approach.</u>

As usually the simulation method is based on generating the simulated observations and processing them in a way analogous to the real data processing. However, as we are interested in the influence of the prior information errors we have to vary the model and to look at variations of results. To deal with the errors of the prior information vector $\overset{o}{w}$ the method of randomization of systematic errors is applied following *Kobayashi (1978),e.g.* It consists in the spreading of the synthetic value of the standard deviation on the number of samples. We generate K samples of <u>inceremental</u> <u>vectors</u> S_k, k being the sample index, to create realization of the random vector $\overset{o}{w}$, such that

$$\sum_{k=1}^{K} S_k = 0, \qquad (13a)$$

$$\frac{1}{K}\sum_{k=1}^{K} S_k S_k^T = D(\overset{o}{w}) =: C_S \qquad (31b)$$

and consequently for the random vector S itself

$$S \sim \mathbb{N}(0, C_S) \qquad (14)$$

holds true, where $\mathbb{N}$ denotes the <u>multivariate</u> <u>normal</u> <u>distribution.</u> Having now vectors $\overset{o}{w}_k$, $k \in (1,2....K)$, in hand we can calculate K times the matrix $F_k = F(\overset{o}{w}_k)$ and find as many solutions V_{S_k} . As an output we finally get a set of K baseline vector realizations Y_k.

Now, the vector Y was supposed to contain only those parameters which are of interest to the network adjustment, namely the components of the baseline vectors in the 3D coordinate system. We finally obtain, at least approximately,

$$\hat{D}(Y) \approx D\, (V_S) = \frac{1}{K}\sum_{k=1}^{K} V_{S_k} V_{S_k}^T \qquad (15)$$

taking the known expectation $E(V_S) = 0$ into account. .

However, keep in mind that for the development of the above formula (15) the assumption was made that the model is not changing and we have always the same expected prior information $E(\overset{o}{w})$. Actually it refers to orbital parameters belonging to one epoch only.

If we are dealing with two or more epochs we must also take the vector $\overset{o}{w}$ changing with correlations. Assume that we know $\mathrm{var}(\overset{o}{w}_{t_p})$ and $\mathrm{var}(\overset{o}{w}_{t_q})$ as well as $\mathrm{cov}(\overset{o}{w}_{t_p}, \overset{o}{w}_{t_q})$. Then the <u>correlation coefficient</u> may be calculated by

$$\rho(\overset{o}{w}_{t_p}, \overset{o}{w}_{t_q}) = \mathrm{cov}(\overset{o}{w}_{t_p}, \overset{o}{w}_{t_q})/(\mathrm{var}(\overset{o}{w}_{t_p})\,\mathrm{var}(\overset{o}{w}_{t_q}))^{\frac{1}{2}} \qquad (16a)$$

and the respective entries in the corrected covariance matrix $\hat{D}_c(Y)$ by

$$\mathrm{cov}(Y_1(t_p), Y_m(t_q)) = \mathrm{cov}(Y_1, Y_m)\cdot\rho \qquad (16b)$$

Here the indices 1 and m refer to the components numbering within the vector Y.

3. Numerical results.

The numerical simulation method enables us to analyse different aspects of the error propagation in GPS data processing. The first application of it was the estimation of the influence of the orbital parameter errors on the baseline lenght determination. It was published in *Zielinski,(1987)* where also some more details are presented about the algorithm of the calculation. The results are shown in the Tab.1 where the comparison also is made with the "rule of thumb"

$$\frac{\sigma_b}{\sigma_{sat}} \approx \frac{b}{r} \quad \begin{array}{l}\text{(baseline lenght)}\\[4pt]\text{(distance to satellite)}\end{array} \qquad (17)$$

We can see that computed variances are definitely smaller than estimated from "rule of thumb". More realistic estimation is

$$\frac{b}{10\,r} < \frac{\sigma_b}{\sigma_{sat}} < \frac{b}{4\,r} \qquad (18)$$

507

Tab 1. Variance σ_b of a baseline   resulting from the orbital
error $\sigma_{sat} \sim \pm 25.5$ m

Lat.	Azim.	Base line lenght		
		100 km	300 km	1000 km

Block 2 constellation, 6^h observation session

0°	0°	± 0.010 m	± 0.033 m	± 0.136 m
0	45	0.018	0.054	0.222
0	90	0.021	0.063	0.323
38	0	0.023	0.068	0.258
38	45	0.026	0.086	0.298
38	90	0.017	0.050	0.155
65	0	0.033	0.101	0.372
65	45	0.032	0.102	0.369
65	90	0.016	0.050	0.167

Block 1 constellation (7 satellites, 3^h observation)

65	0	0.023	0.069	0.243

"Rule of thumb"

		0.125	0.375	1.250

In this contribution we present two other examples of
application of our software: (1) calculation of correlations
resulting from orbital errors between the adjacent baselines;
(2) calculation of the covariance matrix reflecting the
influence of the orbital errors, for the fragment of DÖNAV
network.

The calculations have been done with the Fortran program
GPS2ST, implemented on the IBM 3081 computer, Rechenzentrum
Uni Stuttgart.

3.a. Baseline correlations.

The assumptions for this calculation are as following: The
Block   2 (18 satellites) constellation is assumed and the
6-hours observation session is supposed. During the session
the observation of all visible satellites (15° above the
horizon) from   all involved baselines is simulated,   so
sometimes 5 objects are tracked simultaneously. According to
a realistic estimation of the present day orbital accuracy,
the following   orbital errors has been   assumed (standard
deviation values):

$$\sigma_a = \pm 1.5 m \qquad \sigma_i = \pm 0.377 \times 10^{-6} rad \qquad \sigma_\Omega = \pm 0.377 \times 10^{-6} rad$$

$$\sigma_e = 0 \qquad \sigma_\omega = \pm 0.570 \times 10^{-6} rad \qquad \sigma_M = \pm 0.570 \times 10^{-6} rad.$$

These errors produce the displacement of the satellite in space (mean square root values averaged over the whole session and all observed satellites):

$$\sigma \quad \text{along track} \quad \pm 23. \text{ m}$$

$$\sigma \quad \text{across track} \quad \pm 10. \text{ m}$$

$$\sigma \quad \text{radial} \quad \pm 1.5 \text{ m}$$

$$\sigma \quad \text{total} \quad \pm 25.5 \text{ m}.$$

These values vary for each individual satellite and each individual observation from 0 to several tens of meters, plus or minus, for the reason of geometry and because of the randomization of errors. The randomization has been performed always with sample size 500. Baselines of lenght of 100 km and 300 km were taken, located in equatorial and in middle and high latitude regions ($\sim 37^o$ and 65°).

Using the above data and the equation (15) we calculate variances and covariance of the length of two baselines b_1 and b_m . Then we find the correlation:

$$cor(b_1,b_m) = cov(b_1,b_m)/(var(b_1) \cdot var(b_m))^{\frac{1}{2}} \qquad (17)$$

The results are presented in the Tables 2 – 4. The Tab.2 shows the correlation between the baseline b_o having south-north direction and the baseline b_i with the azimuth growing from 10° to 180°. The correlation is strongly changing in the range from 1.0 to -0.2, but the general pattern of changes is the same in all three cases (100 km and 65°,100 km and $37^o.5$,300 km and $37^o.5$). The opening angle giving the smallest correlation lies between 60° and 120°.

In the Tab.3 we have the correlation between two baselines forming the angle $\alpha = 30^o$ and these two arms are rotating by 10° increment in the 180° angle range. The same is in the Tab.4 for the angle $\alpha = 90^o$. In both cases two different lengths (100 km and 300 km) and two latitude positions were taken.

We can see that there is a substantial change in the correlation between the two arms of the angle, depending on the orientation of the whole figure. It can vary from 0.2 to 0.9 in case of $\alpha =+30^o$ or from 0.0 to -0.5 in the case of $\alpha = 90^o$. There is also a slight dependence from latitude, but on the other hand the correlations are practically independent from the lenght of the baseline.

3b. Network covariance matrix.

The second example of the application of the numerical simulation method is the construction of the covariance matrix for one fragment of the DONAV network consisting of 5 points, cf. *Seeber et al.(1987)*. The covariances also refer to the orbital errors and concern the baseline lenghts. The

Tab. 2. Correlation between baselines with growing
          azimuth

α	100 km $\phi = 65°$	100 km $\phi = 37°,5$	300 km $\phi = 37°,5$
10	0.917	0.922	0.913
20	0.691	0.711	0.686
30	0.491	0.486	0.433
40	0.313	0.286	0.233
50	0.173	0.124	0.081
60	+0.042	+0.101	-0.030
70	-0.081	-0.076	-0.109
80	-0.189	-0.130	-0.159
90	-0.220	-0.140	-0.156
100	-0.035	-0.073	-0.038
110	-0.015	+0.078	+0.065
120	+0.108	+0.256	+0.205
140	+0.367	+0.551	+0.476
150	+0.525	+0.655	+0.607
160	+0.696	+0.795	+0.735
170	+0.905	+0.913	+0.912
180	+0.990	+0.996	+0.992

Tab. 3. Correlation between two baselines with 30° opening angle

No	Baselines azimuths	100 km $\phi = 65°$	100 km $\phi = 37°,5$	300 km $\phi = 37°,5$
1	0° - 30°	+0.491	+0.486	+0.433
2	10 - 40	.678	.622	.600
3	20 - 50	.804	.745	.768
4	30 - 60	.839	.849	.851
5	40 - 70	.802	.897	.879
6	50 - 80	.693	.867	.885
7	60 - 90	.420	.834	.831
8	70 - 100	.202	.676	.667
9	80 - 110	.239	.463	.522
10	90 - 120	.490	.340	.416
11	100 - 130	.700	.413	.479
12	110 - 140	.810	.611	.608
13	120 - 150	.861	.747	.732
14	130 - 160	.841	.834	.875
15	140 - 170	.732	.815	.724
16	150 - 180	.895	.692	.653

distances in the network are in the range from 100 to 200
km. The approximate geographical coordinates of the points
are given in the Tab. 5. All points were observed
simultaneously on 7. Nov. 1986. 5 satellites of the Block 1
constellation were active: 6, 9, 11, 12, 13. The session

lasted about 3 hours. These actual conditions were imitated by the program. The accuracy of orbital elements was assumed as in chapter 3.a.

The Tab. 6 presents the mean square errors for individual baselines resulting from orbital errors and the comparison with the "rule of thumb".

The Tab. 7 presents the whole matrix of variances and cross-covariances scaled by the factor of the variance of the first baseline, being equal to

$$\sigma_1^2 = + 0.62002 \cdot 10^{-2} m = (0.079 \ m)^2$$

Tab. 4. Correlations between two baselines with $\alpha = 90°$ opening angle

No	Baselines azimuths	100 km $\phi = 65°$	100 km $\phi = 37°,5$	300 km $\phi = 37°,5$
1	0° - 90°	-0.220	-0.140	-0.156
2	10 - 100	-0.328	-0.005	+0.035
3	20 - 110	-0.499	-0.083	-0.039
4	30 - 120	-0.623	-0.274	-0.221
5	40 - 140	-0.676	-0.449	-0.372
6	50 - 140	-0.682	-0.525	-0.457
7	60 - 150	-0.635	-0.541	-0.457
8	70 - 160	-0.516	-0.498	-0.444
9	80 - 170	-0.344	-0.374	-0.388
10	90 - 180	-0.220	-0.182	-0.190

Tab. 5. Geographical coordinates of the 5 DONAV Network Stations

No	Name	ϕ	λ
1.	Hohenstein	49°35'11"	11° 25'20"
2.	Katzenbuckel	49 28 15	9 02 29
3.	Oberkochen	48 47 32	10 05 10
4.	Schweitenkirchen	48 30 25	11 36 24
5.	Reichberg	48 18 19	8 59 35

Tab.6.　σ_b　error of the baseline lenght　resulting from
　　　　orbital errors σ_{sat} = ± 25.5 m for test.network

Baseline	Distance km	σ b (m)	Rule of thumb (m)
1 - 2	173	+0.079	+0.220
1 - 3	131	.076	.167
2 - 3	107	.042	.136
2 - 4	121	.077	.154
3 - 4	116	.048	.148
4 - 5	195	.090	.248
3 - 5	97	.053	.124
2 - 5	130	.099	.166

Tab.7. The covariance matrix resulting from orbital errors
　　　　for baseline lenghts of the 5 stations network.

Base-lines	1-2	1-3	2-3	1-4	3-4	4-5	3-5	2-5
1-2	1.000	0.740	0.368	0.003	0.588	1.139	0.588	0.017
1-3		0.931	0.148	0.283	0.338	0.866	0.630	0.555
2-3			0.285	0.194	0.276	0.407	0.139	0.129
1-4				0.950	0.003	0.007	0.121	1.179
3-4					0.377	0.662	0.294	-0.051
4-5						1.299	0.682	0.039
3-5							0.445	0.262
2-5								1.588

$$\text{scaling factor } \sigma^2_{1-2} = (0.079m)^2$$

This matrix is the final result of the covariance
calculation procedure and can be directly used in the
equation (11) for the network adjustment.
　　The Tab. 8 contains the correlation matrix for the same
network. We can see that there exists very strong correlation
between the baselines with nearly parallel orientation. The
baselines may not be adjacent to be strongly correlated. On
the other hand there exist pairs which are practically
uncorrelated.

4. Conclusions.

　　The first group of conclusions concerns the method of
numerical simulation itself. In spite of seemingly great
computational effort (500 permutations of each solution),
the program is quite effective and needs no more than 0.5

Tab.8. The correlation matrix resulting from orbital errors
for baseline lenghts of the 5 stations network.

Base-lines	1-2	1-3	2-3	1-4	3-4	4-5	3-5	2-5
1-2	1.000	0.767	0.689	0.003	0.957	0.999	0.881	0.014
1-3		1.000	0.287	0.301	0.570	0.788	0.978	0.547
2-3			1.000	0.373	0.841	0.669	0.389	0.191
1-4				1.000	0.005	0.007	0.189	0.960
3-4					1.000	0.946	0.719	-0.066
4-5						1.000	0.897	0.027
3-5							1.000	0.311
2-5								1.000

min of computer time with the IBM 3081 mainframe for the one
baseline solution with 18 satellites and 6^h observation
time. It can be implemented also on the PC table computer.
   For practical applications it is necessary to have a good
and realistic estimation of the accuracy of the prior
information. Therefore this problem requires further
investigation concerning the orbital data errors as well as
ionospheric and tropospheric models. As it is usually the
case with simulation methods we must be careful in drawing
too far-reaching conclusions from the numerical results.
However some generalizations are possible even at this stage.
   The application of orbital error covariances is certainly
useful for larger networks with distances over 100 km with
the nowadays accuracy of orbital elements. It is meaningless
for short sides, say up to 30 km. Probably the 50 km baseline
length is the right turning point for application of the
orbital error covariances. With increasing accuracy of
orbital elements this value will rise.
   From the calculations described in chapter 3.a we can
conclude that the pattern of correlation between two
baselines is too complicated to be described by some simple
analytical function. Hence the advantage of the purely
numerical approach is obvious. In general, however, we can
accept the conclusion that the smaller the difference
between the orientation of two baselines, the greater is the
correlation. For opening angles between 60° and 120° the
correlation is approaching zero. This can give some
indication which baselines could be selected and which
eliminated from the network if we want to limit the number
of sides.
   From the network example in chapter 3.b we can see that
there exist very strong correlations because of the orbital
errors influence for baselines with parallel or nearly
parallel orientation. It proves that the error propagation of
this influence is not isotropic. We can expect directional
deformations of the network as a whole if we do not apply
the proper weighting within the variance-covariance matrix.
   The 5 points network example confirms the opinion presented
in Zielinski (1987) that the influence of the orbital errors
on the accuracy of the individual baselines is
overestimated by the so-called "rule of thumb". At the same

time it seems that the problem of correlations resulting from the same source was rather underestimated.

Acknowledgments.

This work has been done during the research visit of the author at the Geodetic Institute of Stuttgart University. This stay has been arranged upon the invitation of Prof. Erik Grafarend and supported by the A. v. Humboldt Stiftung. Theoretical foundations of the method were elaborated together with B. Schaffrin. Discussions with W. Lindlohr were helpful for clarifying the ideas. Support has also been provided by the DFG Sonderforschungsbereich 228 "Precise Navigation" and by the program CPBP 01.20 "Satellite Geodesy".

References.

1. Bock Y., R. I. Abbot, C. C. Counselman, S. A. Gourevitch, R. King: *Establishment of three dimensional geodetic control by interferometry with the Global Positioning System* J. Geophys. Res., 90, Nr. 8-9, pp. 7689-7703, 1985.
2. Kobayashi H.: *Modelling and Analysis.* Addison-Wesley Publ. 1978
3. Koch K. R.: *Parameterschätzung und Hypothesetests in linearen Modellen.* Duemmler, Bonn, 1980.
4. Lindlohr W., D. Wells: *GPS design using undifferenced carrier beat phase observations.* **Manuscripta Geodaetica, 10,** pp. 255-295, 1086.
5. Schaffrin B.: *A note on linear prediction within a Gauss Markov model linearized with respect to a random approximation.* Proc. of the 1st Tampere Seminar on Linear Models, Tampere, pp. 285-300, 1985.
6. Schaffrin B., E. W. Grafarend: *Generating classes of equivalent linear models by nuisance parameter elimination. Applications to GPS observations.* **Manuscripta Geodaetica, 11,** pp. 262-271, 1986.
7. Schaffrin B., Y. Bock: *A unified scheme for processing GPS dual band phase observation.* **Bull. Geodesique** (in press), 1988.
8. Schaffrin B., J. B. Zielinski: *Designing a covariance matrix for GPS baseline measurements.* Subm. to **Manuscripta Geodaetica,** 1988.
9. Seeber G., G. Wuebbena, A. Schuchardt: *Status report on DONAV.* Pres. at XIX Gen. Assembly of IUGG, Symposium 3, **Impact of GPS on Geophysics,** Vancouver, August, 1987.
10. Zielinski J. B.: *GPS baseline error caused by the orbit uncertainty.* Pres. at XIX Gen Assembly of IUGG, Symposium 3, **Impact of GPS on Geodynamics,** Vancouver, August, 1987.

Closing session

Chairmen: Prof. Goad, Columbus, and Prof. Hein, Munich

Under the chairmanship of C.C. Goad (Ohio State University, Columbus,
Ohio) and G.W. Hein (University FAF Munich) a panel discussion took
place attended by G. Blewitt (Jet Propulsion Laboratory, Pasadena, Ca.),
W. Frei (Wild Heerbrugg Co.), R. Hyatt (Trimble Navigation, Sunnyvale,
Ca.), B. Remondi (Ashtech Telesis, Sunnyvale, Ca.) and P. Vanicek (University of New Brunswick). Thereby the following questions were tried
to answer:

(1)   Are all problems of GPS solved in principle?

      - Hardware, Software, practical observation strategies.

(2)   How does geodesy look like in the future?

      - Is classical geodesy dead?

      - Can GPS replace Very Long Baseline Interferometry (VLBI) and
        Satellite Laser Ranging (SLR)?

(3)   Are new fields/applications open for surveyor and geodesists?

      - Navigation, geodynamics.

Without completeness some of the answers are summarized here briefly.

There is a clear trend to recognize that GPS receiver currently under
development will significantly improve with respect to number of channels, phase noise, kinematic use and also will show a dramatic reduction of price. An estimate given by JPL ranges in the order of US $
25000 for a digital dual-frequency instrument in 2 - 3 years. Software
development concentrates now more and more on the kinematic application
of GPS since only slight improvements can be expected with regard to
the static case. However, this does not hold for long baselines where
orbit improvement, atmospheric and clock error modelling still is under
investigation. Practical observation times can be reduced to a few minutes in static GPS.

There is no question that geodesy will be no more longer divided in
horizontal and vertical networks in the future. Since GPS provides the
opportunity to replace to a certain extent spirit levelling if precise
geoid heights are available, the gravity field and physical geodesy has
to be considered - even for practice - in a completely different manner. Classical surveying techniques will stay for a simple reason: GPS
cannot be used in urban areas where buildings, trees, etc. cause a shadowing and the satellite signals cannot be received. VLBI cannot be replaced by GPS since the high accuracy level of VLBI has not been
achieved by GPS for long baselines. In addition, visibility problems of
GPS between continents have still to be investigated and corresponding
data reduction software developed. However, with respect to medium
range baselines (500 - 1000 km) a serious competition may take place.

Laser ranging techniques are absolutely needed to provide precise geocentric station coordinates for a global frame in which the interpolation by GPS can be carried out. The problem of cycle slips in the GPS
phase data will lead to a combination of GPS with a inertial surveying
instrument, or a possible combination with traditional instrumentation
in urban areas. With the new GPS hardware and software generation a decrease from centimeter to millimeter accuracy can be expected.

With this revolution in instrumentation and achievable accuracy new
fields such as high-precision navigation and geodynamics are open for
surveyors and geodesists.

Günter W. Hein

Recommendations

RESOLUTION 1

RECOGNIZING the problems encountered by GPS users in employing receiver pre-processed data, the INTERNATIONAL GPS WORKSHOP, DARMSTADT 1988,

RECOMMENDS CSTG SUB-COMMISSION on GPS to prepare as quickly as possible standards for GPS data formatting to be communicated to receiver manufacturers.

RESOLUTION 2

RECOGNIZING that GPS observations and GPS derived coordinate differences have reached the level of accuracy where earth's temporal deformations have to be considered, the INTERNATIONAL GPS WORKSHOP, DARMSTADT 1988,

RECOMMENDS that IAG Section V (Geodynamics) take appropriate steps leading to a formulation of standard procedures for correcting the above quantities for the body tide and sea-tide loading effects. These standards should be published in the GEODESIST'S HANDBOOK as soon as possible.

RESOLUTION 3

RECOGNIZING the desirability of using common accuracy measures when comparing both results and potential results (pre-analyses), the INTERNATIONAL GPS WORK-SHOP, DARMSTADT 1988,

RECOMMENDS that CSTG SUB-COMMISSION on GPS prepares standards for quantifying accuracy and the contribution of geometry to the accuracy of GPS derived positions and position differences (covariance matrices? DoPs? error hyper-ellipsoids?).

RESOLUTION 4

RECOGNIZING the desirability of being able to compare the performance of different software packages for GPS data processing, the INTERNATIONAL GPS WORKSHOP, DARMSTADT 1988,

RECOMMENDS that the appropriate bodies of IAG prepare a set of standards for classifications and evaluations of GPS software.

RESOLUTION 5

RECOGNIZING the need for fast and efficent modern geodetic and surveying measurements, also in combination with inertial and similar methods, and

STATING the availability of new dual frequency GPS-receivers and similar high precision equipment, the INTERNATIONAL GPS WORKSHOP, DARMSTADT 1988,

RECOMMENDS to the NATIONAL COMMITTEES of IAG to study the combinations of high precision GPS-measurements with inertial, gravity, gravity gradiometer and similar observations as well as the joint applications of GPS with such high precision equipment in order to obtain precise informations of locations and the gravity field.

RESOLUTION 6

RECOGNIZING the effort of PROFESSOR DR. ING. E. GROTEN and his Organizing Committee, the participants of the INTERNATIONAL GPS WORKSHOP, DARMSTADT 1988,

RESOLVE   to thank and congratulate Professor Dr. Ing. E. GROTEN and his Organizing Committee on the success of the Workshop.

Appendix: Field Measurements

MINI-MAC 2816 DUAL FREQUENCY RECEIVER

by

Jim Cain

Abstract

MINI-MAC MODEL 2816 was demonstrated Tuesday, 12 April at Darmstadt 1580 station. Several survey and navigation observations were taken for demonstration purposes. In the afternoon it began to rain, thus offering a test of the environmental security of the MINI-MAC. It observed directly in the rain without any environmental protection being provided. No problems were encountered. Since only one instrument participated in the measurements, no baseline data were obtained. Aero Service can, however provide various test baseline data to interested parties. Feel free to contact Jim Cain of Aero Service if you wish to have such data.

KINEMATIC LAND SURVEY DEMONSTRATION
INTERNATIONAL GPS WORKSHOP - DARMSTADT, WEST GERMANY
APRIL 10-14, 1988

Ron C. Hyatt
Clyde C. Goad

INTRODUCTION

The Trimble 4000 series is an ideal product for dynamic and kinematic positioning and surveying. Since it continuously tracks all satellites in view, typical survey vehicle dynamics have little effect. Several tests have been done over the past year to demonstrate the ability to do rapid surveys in a kinematic mode.

One type of kinematic surveying that can substantially increase the number of points surveyed is "stop and go" kinematic surveying when only a few measurements need to be collected at each stop after an initial observation is completed to determined the starting baseline. The success of this type of surveying directly depends on the ability to maintain lock on at least three satellites (preferably four) at any time.

With the current constellation of satellites, there exists more than one hour of six satellites in view at one time in many locations throughout the world. This period is the optimum time for kinematic surveying in the stop and go mode. With six satellites in view, the likelihood of maintaining lock on three or four satellites is very high. There will always be one satellite low on the horizon, either rising or setting, and loss of lock is likely on such a satellite. Obstructions can also cause loss of lock on one or more satellites.

As the Block II satellites are launched, there will be more time each day when this type of surveying will be highly productive. Trimble has expanded its 4000 series receivers to operate with ten channels of L1 tracking for this application. With six or more satellites continuously tracked, kinematic surveying will become a practical surveying tool.

DEMONSTRATION

At the International Workshop on GPS in Darmstadt, Germany, a kinematic survey demonstration was done utilizing two ten-channel 4000SD Surveyors. This demonstration was done on the university campus with the usual trees and buildings as obstructions. The demonstration was done when five and six satellites were in view.

Figure 1 shows the stationary receiver at site 1600 and the rover starting at site 2120, moving to 1120, 1250, and returning to 2120. A second circuit included 1120, 1000, and returning to 2120. A repeat of the second circuit was also done. The antenna was moved from vehicle to tripod to vehicle at each stop, while connected to the receiver/data collector in the vehicle.

Table I shows the results of the tests. Repeatability better than 1 centimeter is evident at stations 2120, 1120, and 1000. Comparison with terrestial survey distances indicated 1 centimeter absolute accuracy also. Observation times at each of the stops were five minutes or less. The data collector was set to record measurements every 5 seconds to ensure sufficient epochs to verify the quality of the data. When the geometry is good, a single measurement can provide centimeter accuracy if multi-path effects are minimal.

Loss of lock occurred several times on either a rising or setting satellite, particularly when near an obstruction such as a tree or building in the direction of the low elevation satellite. There were ten stops during the test covering about 90 minutes including the initial 30 minute observation to determine the starting baseline and integer ambiguities.

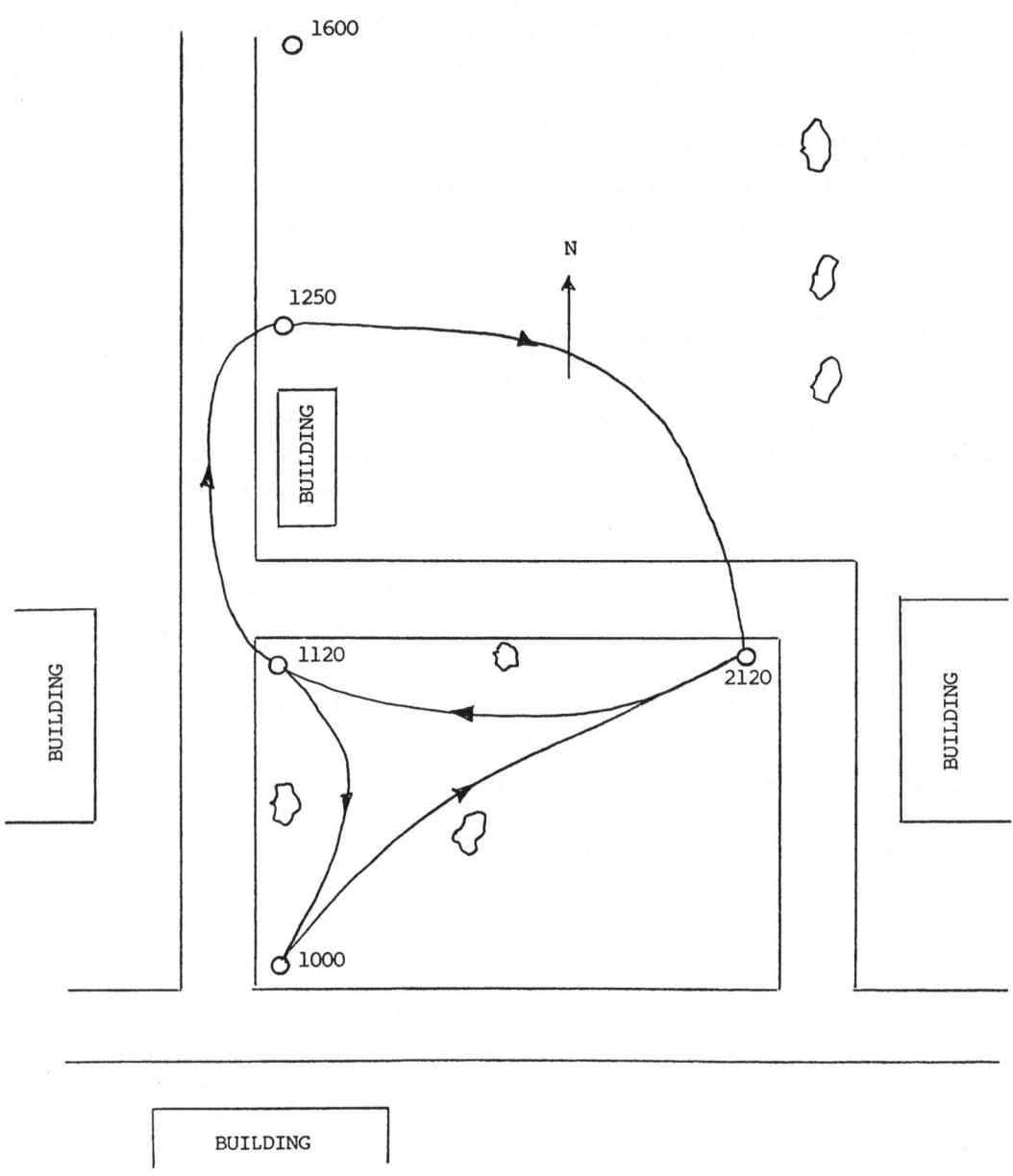

Figure 1

Start and stop times were logged each time the antenna was moved as well as the antenna height recorded. The post-processing is interactive and requires familiarity with the details of triple and double difference processing in TRIMVEC[tm]. The ten points were processed in about one hour by an experienced operator.

Station	GPS Delta X	GPS Delta Y	GPS Delta Z	GPS Distance	Terrestial Survey
2120-1	+355.079	+126.428	-305.280	485.037	485.031
2120-2	+355.074	+126.426	-305.286	485.037	
2120-3	+355.080	+126.431	-305.271	485.033	
2120-4	+355.071	+126.421	-305.284	485.032	
1120-1	+365.647	+56.560	-305.761	479.986	479.982
1120-1	+365.648	+56.562	-305.754	479.982	
1120-3	+365.647	+56.561	-305.756	479.983	
1000-1	+458.894	+70.984	-380.012	600.026	600.023
1000-2	+458.884	+70.979	-380.015	600.020	

The points were visited in their natural sequence, not in the order shown above.

Table I.

CONCLUSIONS

With the addition of ten-channel kinematic tracking, the 4000 series offers a complete range of capabilities for most all land surveying. Demonstration of kinematic land surveying shows how rapidly points can be surveyed to centimeter accuracy when tracking conditions are good and five or more satellites are in view.

Automatic data logging of start and stop times as well as antenna heights will be added to the next release of 4000 series software this summer. The post-processing will require more operator interaction than static long observations. Further improvements can be made in TRIMVEC[tm] to provide easier identification of cycle slips that will occur as satellites rise and set during the observation window. As Block II satellites are launched, kinematic land surveying will become a very practical land surveying technique.

The International GPS-Workshop Field Tests

compiled by

Hans-Jürgen Euler

On the afternoon of April 11 and April 12 GPS measurements were carried out at the university campus and two stations of the first order triangulation network nearby Darmstadt. On April 12 all participants had the possibility to see one instrument of each firm operating at the university campus.

Following receivers were presented at the workshop:

Mini-Mac 2816	(dual frequency)	Aero Service
Nr. 52	(single frequency)	Sercel
4000 SD	(dual frequency)	Trimble Navigation
WM 101	(single frequency)	Wild-Magnavox

The triangles sketched in figure 1 were measured at the first day using eight receivers, three WM 101, three Trimble 4000 SD and two Sercel Nr. 52. Unfortunately, Aero Service had problems with the power-supply of one of their receivers, so they did not have the chance to participate. During the second day only one long baseline was measured and the receivers were held fixed during the whole session. Table 1 shows the schedule for the measurements of both days.

The distance between the first order triangulation stations Melibokus and Feldberg was well determined in the past, using different terrestrial equipment for range measurements. Since the positions of the receivers at the university campus were interchanged during the first day, the distance of the high precision calibration line, which is known to the submillimeter, could be determined from baseline computations of consecutive measurements.

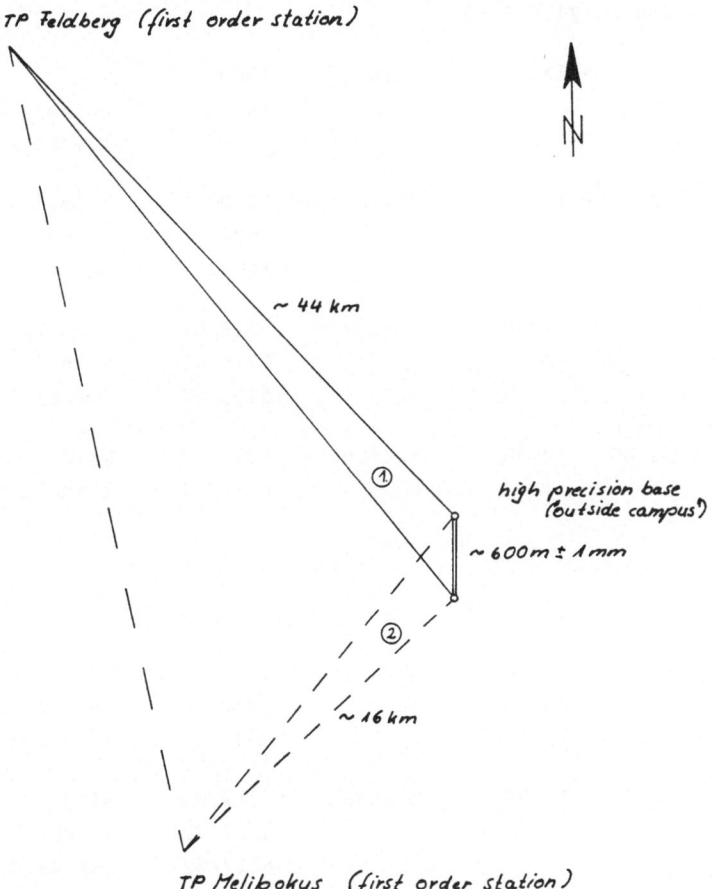

TP Feldberg (first order station)

N

~ 44 km

① high precision base
(outside campus)

~ 600 m ± 1 mm

②

~ 16 km

TP Melibokus (first order station)

Figure 1

From the manufacturers we got different types of computations. Wild-Magnavox sent us a complete network adjustment of all their measurements. The standard deviations of the distances are always few millimeters. Wild-Magnavox claims in the technical reference manual the post-processing relative position accuracy as 10 mm plus 2 ppm (126 mm for 58 km), which seems to be more suitable.

The high precision calibration line at the university campus was not observed directly by Wild-Magnavox, but the computed value is accurate within two millimeter.

Trimble carried out baseline solutions for the long distances. Since they did not compute all possible baselines, the independent check with the calibration line is not possible. On the second day the calibration line was measured directly during the kinematic test. The deviation from the accurate value is three millimeter (see also short note of Hyatt and Goad concerning the kinematic test).

Monday 11.4.1988

16.30 - 18.00	Darmstadt	1000	Wild
		1580	Sercel
		1600	Trimble
18.30 - 19.30	Darmstadt	1000	Trimble
		1020	Sercel
		1600	Wild
16.30 - 19.30	Feldberg	5716/02	Trimble
		5716/41	Wild
		5716/42	Sercel
16.30 - 19.30	Melibokus	6217/01	Wild
		6217/23	Trimble

Tuesday 12.4.1988

16.30 - 19.30	Darmstadt	1010	Wild
		1020	Sercel
		1600	Trimble
16.30 - 19.30	Melibokus	6217/01	Wild
		6217/21	Trimble
		6217/22	Sercel

Table 1   Time Schedule for Field Test

The slope distance between Melibokus - Feldberg computed by Trimble is 8.5 cm longer than the terrestrial measurements (Schmidt 1981, Ehlert 1984), the value of Wild-Magnavox is 21.9 cm longer. The Sercel distance was computed by combining measurements of two days and deviates 4.4 cm from the terrestrial value.

Wild stated in their covering letter, that the data of the first day are not as good as the data of the second day. The reason for bigger noise as usual was seen in the radio transmitters and the topography at the first order stations.

On the other hand we should mention that the distance Melibokus - Feldberg was measured using a Macrometer V-1000 in October 1983 (Hausch et. al. 1985, Schmidt 1986). The measurement of 1983 is between the results of Trimble and Wild-Magnavox.

However, the precision of the terrestrial measurement might be affected by an unknown scale factor and by transformations due to the eccentricities (at Feldberg, the center is about 100 m distant from the GPS observation site). Since the GPS measurements are close together, the comparison within GPS values seems to be justified (see table 2). All measurements deviate within the expected uncertainties from the mean value of all GPS measurements.

	distance (m)	diff. (mm)	R.M.S. (mm)
GPS mean value	58013.625	0	
Terrestrial reference	58013.508	-117	35
Macrometer 1983	58013.626	1	170
Sercel	58013.552	-73	7
Trimble	58013.593	-32	4
Wild-Magnavox	58013.727	102	1

Table 2   Centric Slope Distances Feldberg - Melibokus

(R.M.S. values as given in the computation outputs)

Due to the low signal to noise ratio in codeless L2 measurements and the fact that we have big radio frequency transmitters on Feldberg and Melibokus, the dual frequency measurements with the Trimble 4000 SD are strongly disturbed. The dual frequency measurements at university campus were more successful, but since good L2 data is only available at this station, the computations carried out by Trimble used only L1 measurements.

The GPS field test during the International GPS-Workshop Darmstadt gives not the possibility to judge the precision of currently available GPS receivers. Every receiver delivers the same result for the viewed slope distance within the expected uncertainties for the measurements of a single frequency receiver. Unfortunately, the comparison between dual frequency receivers and single frequency receivers was not possible, since the measurements of Trimble's 4000 SD on L2 were disturbed, and due to the problems with the power supply of the Mini-Mac.

However, an objective comparison between GPS receivers needs a more extended network and a homogeneous way for computations.

Ehlert, D. (1984) : Die Diagnoseausgleichung 1980 des Deutschen Haupt-
     dreiecksnetzes, Band IV Netzausgleichung, Deutsche Geodätische
     Kommission Reihe B

Hausch, W., Groten, E., Euler, H.-J., Strauß, R., Feltens, J. (1985) :
     Three-Dimensional Geodetic Control of a Regional Macrometer
     Network, Manuscripta Geodetica, pp 306-316

Schmidt, R. (1981) : Die Diagnoseausgleichung 1980 des Deutschen
     Hauptdreiecksnetzes, Band II Strecken, Deutsche Geodätische
     Kommission Reihe B

Schmidt, R. (1986) : Kontrolle des Deutschen Hauptdreiecksnetzes durch
     Macrometer-Messungen 1983-1985 - KONMAC -, Deutsche Geodätische
     Kommission Reihe B

Wild-Magnavox (1986) : WM 101 GPS Satellite Surveying Equipment, Tech-
     nical Reference Manual, WM Satellite Survey Company